AF781876

Jellyfish Blooms IV

Developments in Hydrobiology 220

Series editor

K. Martens

For other titles published in this series, go to
http://www.springer.com/series/5842

Jellyfish Blooms IV

Editors

Jennifer Purcell[1], Hermes Mianzan[2] & Jessica R. Frost[3]

[1]*Western Washington University, Shannon Point Marine Center, Shannon Point Road 1900 98221 Anacortes, Washington, USA*
[2]*Paseo V. Ocampo no. 1 7602 Mar del Plata, Argentina*
[3]*University of Florida School of Forest Resources and Conservat, Fisheries and Aquatic Science Program, NW 71st Street 7922 32653 Gainesville, Florida, USA*

Previously published in *Hydrobiologia,* Volume 690, 2012

Springer

Editors
Dr. Jennifer Purcell
Western Washington University
Shannon Point Marine Center
Shannon Point Road 1900
98221 Anacortes, Washington, USA

Jessica R. Frost
University of Florida
School of Forest Resources and Conservat
Fisheries and Aquatic Science Program
NW 71st Street 7922
32653 Gainesville, Florida, USA

Dr. Hermes Mianzan
Paseo V. Ocampo no. 1
7602 Mar del Plata, Argentina

ISBN 978-94-007-5315-0

Springer Dordrecht Heidelberg New York London

Library of Congress Control Number: 2012943979

© Springer Science+Business Media Dordrecht 2012
This work is subject to copyright. All rights are reserved by the Publisher, whether the whole or part of the material is concerned, specifically the rights of translation, reprinting, reuse of illustrations, recitation, broadcasting, reproduction on microfilms or in any other physical way, and transmission or information storage and retrieval, electronic adaptation, computer software, or by similar or dissimilar methodology now known or hereafter developed. Exempted from this legal reservation are brief excerpts in connection with reviews or scholarly analysis or material supplied specifically for the purpose of being entered and executed on a computer system, for exclusive use by the purchaser of the work. Duplication of this publication or parts thereof is permitted only under the provisions of the Copyright Law of the Publisher's location, in its current version, and permission for use must always be obtained from Springer. Permissions for use may be obtained through RightsLink at the Copyright Clearance Center. Violations are liable to prosecution under the respective Copyright Law.

The use of general descriptive names, registered names, trademarks, service marks, etc. in this publication does not imply, even in the absence of a specific statement, that such names are exempt from the relevant protective laws and regulations and therefore free for general use.

While the advice and information in this book are believed to be true and accurate at the date of publication, neither the authors nor the editors nor the publisher can accept any legal responsibility for any errors or omissions that may be made. The publisher makes no warranty, express or implied, with respect to the material contained herein.

Printed on acid-free paper.

Springer is part of Springer Science+Business Media (www.springer.com)

Contents

This Symposium was supported by the following grants: FONCyT RC2009-352 and PICT 01553; CONICET RC 2009-374; the Inter-American Institute for Global Change Research (IAI) CRN 2076, which is supported by the US National Science Foundation (grant GEO-0452325) and INIDEP.

Hydrobiologia (2012) 690:1–2
DOI 10.1007/s10750-012-1050-z

Preface: Jellyfish blooms: interactions with humans and fisheries

Hermes Mianzan · Jennifer E. Purcell · Jessica R. Frost

Published online: 15 March 2012
© Springer Science+Business Media B.V. 2012

There is a general impression that jellyfish and other gelatinous organisms are increasing in number. Media, TV, and newspapers contribute to this impression. So are increases in jelly populations real, or is this phenomenon just a biased perception? Answering this question is a difficult task because jelly populations normally fluctuate enormously, being everywhere some years, and impossible to find in others. It is also true that occasional swarms of great density have a notorious effect on many human economic activities such as tourism, fisheries, and various coastal industries.

Ten years ago the first Jellyfish bloom meeting was envisioned. Held at Gulf Shores, Alabama in January 2000, it was a response to the need to consider the ecological as well as societal aspects of jellyfish blooms. The main objective was to come together and find a unified voice that would direct new fields of research on the subject. The second International Jellyfish Blooms Symposium was held on the Gold Coast, Queensland, Australia in June 2007, and the general message was to examine the problem on a wider scale, encouraging people to consider the use of fisheries as well as molecular techniques for jellyfish research. And recently, special sessions on gelatinous plankton can be found in other general meetings like the Nice ASLO meeting in January 2009. In each of the three cases mentioned, about 60 talks and posters were presented and a special volume published.

What is our ultimate goal? To understand the dynamics and impacts of jellyfish blooms at a global scale. The Third Symposium, organized by Hermes Mianzan, Gabriel Genzano, Agustin Schiariti and Marcelo Acha, held 13–16 July 2010 in Mar del Plata, Argentina, reached us at the right time. We are facing clear examples that some jellyfish species are increasing

Guest editors: J. E. Purcell, H. Mianzan & J. R. Frost / Jellyfish Blooms: Interactions with Humans and Fisheries

Electronic supplementary material The online version of this article (doi:10.1007/s10750-012-1050-z) contains supplementary material, which is available to authorized users.

H. Mianzan
CONICET-Instituto Nacional de Investigación y, Desarrollo Pesquero (INIDEP), Paseo Victoria Ocampo, n_1, B7602HSA Mar del Plata, Argentina
e-mail: hermes@inidep.edu.ar

J. E. Purcell (✉)
Western Washington University, Shannon Point Marine Center, 1900 Shannon Point Road, Anacortes, WA 98221, USA
e-mail: purcelj3@wwu.edu

J. R. Frost
University of Hamburg, Institute for Hydrobiology and Fisheries Science, Olbersweg 24, 22767 Hamburg, Germany
e-mail: frost.jessica.r@gmail.com

Present Address:
J. R. Frost
Fisheries and Aquatic Science Program, School of Forest Resources and Conservation, University of Florida, 7922 NW 71st Street, Gainesville, FL 32653, USA

their frequency of occurrence, expanding their geographical distributional range, being introduced, with sometimes devastating consequences on human enterprises. *Mnemiopsis leidyi*, the ctenophore that gained a "bad reputation" by invading the Black Sea in the 1980s is continuously expanding its distributional range and every year it is found in new localities. *Nemopilema nomurai*, a giant Asian scyphozoan, undertakes interannual population explosions generating severe damage to the Japanese fishing industry in the last 10 years. Even nuclear power stations occasionally need to stop their activities due to jellies that have clogged their refrigerating water intakes. And more and more examples are continuously being reported.

Within this third meeting, attention was placed on fish–jellyfish interactions, and fisheries. More than a 100 talks and posters have been presented by the 98 delegates from 31 countries. Dr. Daniel Pauly offered the Keynote Address on "Changes of Jellyfish Abundance: Testing Hypotheses at the Large Marine Ecosystem Scale." Dr. Jenny Purcell run a special session on "Causes and consequences of Jellyfish Outbreaks and Aggregations" with invited speaker, Dr. Shin-ichi Uye, Dr. Rick Brodeur spoke in the second special session, "Interactions Between Jellyfish and Marine Fish and Fisheries: Insights into Ecosystem Functioning". A variety of topics were included in the General Session.

The papers presented in this volume are a subset of the 100+ talks and posters presented at the meeting (Supplement 1). They provide a good representation of the diversity of issues discussed and hopefully will stimulate researchers worldwide to continue research into how and why jellyfish blooms occur.

H. Mianzan (conference organizer and guest editor)
J. E. Purcell (guest editor)
J. R. Frost (guest editor)

Acknowledgments The guest editors and conference organisers wish to thank the sponsors of the symposium, specifically INIDEP, CONICET, Agencia, IAI, Mar del Plata National University, and AACIMA. The guest editors are also indebted to the numerous referees for their detailed and insightful reviews of manuscripts and, of course, to the contributing authors who stoically endured the numerous queries and requests of the guest editors.

Hydrobiologia (2012) 690:3–20
DOI 10.1007/s10750-012-1039-7

Increasing jellyfish populations: trends in Large Marine Ecosystems

Lucas Brotz · William W. L. Cheung · Kristin Kleisner · Evgeny Pakhomov · Daniel Pauly

Published online: 3 April 2012
© The Author(s) 2012. This article is published with open access at Springerlink.com

Abstract Although there are various indications and claims that jellyfish (i.e., scyphozoans, cubozoans, most hydrozoans, ctenophores, and salps) have been increasing at a global scale in recent decades, a rigorous demonstration of this has never been presented. Because this is mainly due to scarcity of quantitative time series of jellyfish abundance from scientific surveys, we attempt to complement such data with non-conventional information from other sources. This was accomplished using the analytical framework of fuzzy logic, which allows the combination of information with variable degrees of cardinality, reliability, and temporal and spatial coverage. Data were aggregated and analyzed at the scale of Large Marine Ecosystem (LME). Of the 66 LMEs defined thus far that cover the world's coastal waters and seas, trends of jellyfish abundance after 1950 (increasing, decreasing, or stable/variable) were identified for 45, with variable degrees of confidence. Of those 45 LMEs, the majority (28 or 62%) showed increasing trends. These changes are discussed in the context of possible sources of bias and uncertainty, along with previously proposed hypotheses to explain increases in jellyfish.

Keywords Jellyfishes · Gelatinous zooplankton · Blooms · Pelagic cnidarians · Ctenophores · Fuzzy logic

Guest editors: J. E. Purcell, H. Mianzan & J. R. Frost / Jellyfish Blooms: Interactions with Humans and Fisheries

L. Brotz (✉) · K. Kleisner · D. Pauly
Sea Around Us Project, Fisheries Centre, University of British Columbia, 2202 Main Mall, Vancouver, BC V6T 1Z4, Canada
e-mail: lucasbrotz@gmail.com

L. Brotz · E. Pakhomov
Department of Earth and Ocean Sciences, University of British Columbia, 6339 Stores Road, Vancouver, BC V6T 1Z4, Canada

W. W. L. Cheung
School of Environmental Sciences, University of East Anglia, Norwich NR4 7TJ, UK

W. W. L. Cheung
Fisheries Centre, University of British Columbia, 2202 Main Mall, Vancouver, BC V6T 1Z4, Canada

Introduction

Jellyfish are a conspicuous, but relatively little studied component of marine ecosystems, whose populations fluctuate widely with ocean climate and also experience sudden outbursts known as "blooms," followed by population crashes (Purcell, 2005). There are also recent suggestions that jellyfish may be synanthropic, specifically, benefiting from human interactions with the oceans, and thus may be increasing globally (Mills, 2001; Purcell et al., 2007; Pauly et al., 2009a; Richardson et al., 2009). Previous global reviews of jellyfish populations (e.g., Mills, 2001; Purcell et al., 2007; Chudnow, 2008) show evidence of numerous localized increases; however, for most ecosystems, long time series of abundance measures for jellyfish are lacking, and the perceived widespread or global increase in jellyfish still lacks a rigorous foundation.

Establishing abundance trends for jellyfish is difficult due to a number of factors. There is a dearth of historical information on jellyfish, because they were usually damaged or not recorded when caught in routine bottom trawl or zooplankton surveys (Pugh, 1989; Hay, 2006). In fact, the latter often used gear designed to exclude jellyfish from plankton samples (e.g., Heinle, 1965) or were based on methodologies that explicitly recommended their removal before analysis (e.g., Dovel, 1964). For example, a classic manual on zooplankton sampling published by UNESCO (1968) mentions jellyfish only once, i.e., "Gelatinous organisms and other animals [...] will occur in the catches and these must be considered separately from the main sample."

Moreover, jellyfish are difficult to sample even when targeted (Omori & Hamner, 1982; Pierce, 2009). As a result of their neglect in routine surveys and marine samples, jellyfish were generally perceived as a bothersome and unimportant component of marine ecosystems (Pauly et al., 2009a), which then justified their further neglect. Furthermore, despite recent advances in research and understanding of jellyfish ecology at local scales, such knowledge is rarely used to evaluate possible causes or consequences of jellyfish blooms at larger scales, or to make predictions (Purcell, 2009).

Their peculiar life cycles, which can result in extremely high temporal and spatial variability in abundance, peaking in the form of "blooms" (Mills, 2001; Purcell et al., 2007; Boero et al., 2008; Dawson & Hamner, 2009; Hamner & Dawson, 2009), also contribute to why jellyfish tend to be understudied. All cubozoans, as well as many hydrozoans and scyphozoans have a life history consisting of a sessile polyp phase and a planktonic medusa phase. Many polyps reproduce asexually through the process of strobilation, producing multiple ephyrae which join the zooplankton community (Arai, 1997) and rapidly grow to become medusae (Palomares & Pauly, 2009). For some species, the polyps may asexually bud more polyps or form dormant cysts capable of surviving harsh environmental conditions (Arai, 2009). These characteristic life history traits make jellyfish suited to highly variable environments, because they can survive when conditions are unfavorable and rapidly reproduce when conditions are favorable (Boero et al., 2008; Richardson et al., 2009). Siphonophores, ctenophores, and salps lack a polyp phase, but can also reproduce rapidly under favorable conditions (Alldredge & Madin, 1982; Purcell et al., 2007). Such varied reproductive strategies make it extremely difficult to assess jellyfish populations. Indeed, if few surveys have been conducted to quantify medusa abundance, even less is known about their polyps (Mills, 2001).

More attention has been paid to jellyfish in recent years because of their interference in human enterprises, their ecological importance, and their benefits to humans. Jellyfish directly interfere with many human activities (reviewed by Purcell et al., 2007; Richardson et al., 2009), specifically, through stings (beach closures, tourism impacts, injuries, deaths), clogging intakes (coastal power and desalination plants, mining and military operations, shipping, aquaria), interference with fishing (clogged and split nets, spoiled catch, stung fishers, damaged gear, capsized boats), aquaculture (fish deaths, pens fouled by polyps), and marine biological surveys (interference with trawls and acoustic surveys). Jellyfish also have ecosystem impacts with indirect effects on fisheries resources that are difficult to quantify, such as their roles as predators of zooplankton, fish eggs and ichthyoplankton, as vectors for parasites, as food for fish, and as refugia and food for some species of juvenile fish (interactions reviewed by Purcell & Arai, 2001).

Some jellyfish also benefit humans (reviewed in Purcell et al., 2007), notably as food (Hsieh et al., 2001), and potentially for use in drugs (e.g., Sugahara et al., 2006; Ohta et al., 2009). The discovery, isolation, and development of a fluorescent protein from jellyfish led to a revolution in biotechnology (Zimmer, 2005) and a Nobel Prize (Coleman, 2010); however, the proteins now are synthesized in the laboratory. Unfortunately, such benefits may be outweighed by the direct and indirect negative impacts of jellyfish blooms.

The lack of jellyfish population datasets covering large temporal and spatial scales limits the scope of inferences that can be drawn about jellyfish on a global basis. To compensate for this, we used analytic methods designed to allow for the inclusion of a wide variety of information, including "anecdotal data," whose value is often underestimated (Pauly, 1995). Because the majority of recently reported changes in jellyfish populations around the globe occur in coastal waters or semi-enclosed seas (Mills, 2001; Purcell et al., 2007), the Large Marine Ecosystem (LME) framework provided a suitable stratification scheme to investigate these trends. We used the established system of fuzzy logic to examine the evidence for changes in jellyfish populations over recent decades.

Materials and methods

Definition of "jellyfish"

Because the term "jellyfish" lacks a formal definition, we present an operational definition used in this analysis, which will be used to refer to both single and multiple species. Here, the word "jellyfish" refers to gelatinous zooplankton that include medusae of the phylum Cnidaria (scyphomedusae, hydromedusae, cubomedusae, and siphonophores) and planktonic members of the phylum Ctenophora. We also included the pelagic tunicates known as salps due to their gelatinous nature, pulsed life cycles, and apparent response to changing oceanic conditions (Loeb et al., 1997; Atkinson et al., 2004; Lee et al., 2010). Especially sparse time series data on pyrosomes and doliolids prevented their inclusion in the analysis.

Other types of gelatinous zooplankton, such as appendicularians, mollusks, and chaetognaths, were not included in our analysis for various reasons (their small size, life history, ecological roles, high carbon-to-weight ratio), and the fact that they are generally not considered jellyfish (see Mianzan & Guerrero, 2000; Graham & Bayha, 2007; Richardson et al., 2009). Pleustonic jellyfish, such those belonging to the genera *Physalia*, *Porpita*, and *Velella*, also were excluded because their local distribution is heavily influenced by wind patterns (Mackie, 1974). As such, locations reporting these species are frequently implicated in claims of "unprecedented" blooms and mass beach strandings lacking a historical context.

LME approach and the 1950 baseline

In order to examine and compare changes in jellyfish populations, data were stratified by LME. The LME framework defines boundaries based on ecological criteria rather than economic or political criteria (Sherman & Hempel, 2009). LMEs may extend from nearshore areas, including river basins and estuaries, out to the seaward boundaries of continental shelves or coastal currents (Sherman & Tang, 1999). Four sets of factors were considered when defining the physical extent of the LME boundaries, i.e., bathymetry, hydrography, productivity, and trophic relationships. LMEs range from 150,000 km^2 to more than 5 million km^2. To date, 66 LMEs have been described in terms of these parameters (Sherman & Hempel, 2009), with emphasis on fisheries (Pauly et al., 2009b, see also www.seaaroundus.org).

In order to examine changes in jellyfish populations, a baseline must be selected. For our analysis, changes were only considered if they occurred after 1950, notably because this was the first year for which the Food and Agriculture Organization of the United Nations (FAO) published its annual compendia of global fisheries catches (which now include jellyfish), part of an effort by the United Nations to "quantify the world" (Ward et al., 2004). Also, most of the reported changes in jellyfish populations stem from recent decades (Mills, 2001; Purcell et al., 2007); thus, a 1950 baseline provides the contrast required for comparison and testing of such reports. Finally, many of the anthropogenic factors that have been suggested as causes of recent increases in jellyfish populations have been quantified only since the mid-twentieth century, notably because they are derived from FAO data (e.g., Watson et al., 2004) and recently have been re-expressed at the LME scale (e.g., Maranger et al., 2008; Pauly et al., 2009b).

The jellyfish chronicles

The data used in this analysis were aggregated into "chronicles." Each chronicle consists of one or more pieces of evidence and has an associated "Abundance Trend" and "Confidence Index," calculated from scores for spatial and temporal extent, as well as reliability. The reliability score allowed us to consider and combine information from the scientific, peer-reviewed literature, as well as information gleaned from other sources (e.g., "anecdotes"). These chronicles were aggregated by LME and then analyzed using a fuzzy logic expert system (Zadeh, 1965) to generate a "Jellyfish Index" for each LME.

Multiple pieces of evidence covering similar temporal and spatial scales were included as one chronicle, and only data that referred to changes (or lack thereof) over several years or greater were included. Therefore, isolated references to "lots of jellyfish" or "more jellyfish than last year" would not qualify for inclusion due to low temporal coverage. The same rationale was applied to populations with decreasing or relatively stable trends, or those showing high variability.

Increasing or decreasing trends were reported to occur only if they were sustained. Thus, a population of jellyfish showing a prolonged increase followed by a similar decrease was classified as "variable." Because chronicles were scored on several components, those with no recent data (post-2000) received a lower temporal score in order to reflect the uncertainty of whether the identified trend has continued or not. Data for the North Sea LME chronicles were included here as an example. Details of all chronicles are in Brotz (2011).

Data selection

All direct comments or measurements indicating changes (or lack thereof) in jellyfish populations over several years were included in the analysis; however, indirect evidence was not included. Such indirect evidence consists of impacts of jellyfish on human activities such as sting events, clogging of intake pipes for power generation, shipping, or mining operations, as well as interference with aquaculture operations. Although changes in the frequency of these events may indicate changes in jellyfish populations (Purcell et al., 2007), there can also be a consequence of changes in sampling effort. For example, a jellyfish bloom that interferes with an industrial operation may actually represent a stable jellyfish population if the industrial operation is new to the region, rather than an actual increase in jellyfish. Therefore, isolated interference events with industrial operations were excluded from the analysis.

Individual events related to direct interference with fishing activities also were excluded. However, we included information that referred to the changing frequency of such events because we believed this to be a strong indication of a change in jellyfish abundance. For example, fishers in some locations reported increasing jellyfish by-catch over years or decades (e.g., Uye & Ueta, 2004). Because fishers generally have a keen understanding of the marine environment, such statements were assumed to be reliable. In addition, for most locations with extant fisheries, it is expected that fishers have improved their ability to avoid catching jellyfish over time (e.g., Kendall, 1990; Matsushita & Honda, 2006; Nagata et al., 2009). Thus, we believed that increases in jellyfish by-catch observed by fishers were likely to reflect increased jellyfish populations.

Sting data generally were not included in our analysis, because they are problematic due to a number of factors. First, an increase in the number of people participating in marine activities would increase encounter rates (Macrokanis et al., 2004). In addition, data showing an increase in sting events may simply be a reflection of increased reporting (Gershwin et al., 2010). As such, an increase in sting events may not necessarily represent an increase in the amount of jellyfish present. Conversely, awareness and educational campaigns, as well as the use of jellyfish deterrents or countermeasures, can result in a decrease in sting events without a concomitant reduction of the jellyfish population (Gershwin et al., 2010). Therefore, sting data were excluded from the analysis, except where they revealed temporal changes (e.g., increase in the stinger season) or spatial changes (e.g., increased distribution of jellyfish).

Abundance Trend

Each chronicle was assigned an "Abundance Trend" of increasing (+1), decreasing (−1), or stable/variable (0). The trend was identified by considering changes in integrated biomass (i.e., abundance and presence). Therefore, increases (or decreases) in any metric (overall biomass, frequency of occurrence, or duration of occurrence) were considered to be indications of an increase (or decrease). As such, more frequent blooms, larger blooms, longer-lasting blooms, and range expansions (and their converses) all were included. When knowledge was available on multiple species over similar scales, the overall biomass of jellyfish within the ecosystem was considered. In addition, small, non-abundant hydromedusae received lower scores due to the fact that they are less likely to affect the overall biomass of jellyfish in the ecosystem.

Supporting evidence for each chronicle consisted of either qualitative or quantitative information.

Chronicles with qualitative data as their primary source were classified based on the description of the jellyfish population in question (Table 1). For chronicles with quantitative records, such as multi-year datasets with values for relative abundance or biomass, a general linear regression analysis was performed. If the slope of the linear regression (abundance against time) was positive and significantly different from zero ($P < 0.05$), the dataset was considered to represent an increase. Conversely, a significant negative slope constituted a decrease. If the slope of the linear regression was not statistically significant, the dataset was classified as stable/variable.

Scoring the chronicles

Each chronicle was scored according to a set of rules (Table 1) based on temporal coverage ("Time score"), spatial coverage ("Space score"), and reliability ("Reliability score") with the reliability for invasive species scored differently. These scores were used to calculate the overall "Confidence Index," a measure of the level of certainty for each chronicle.

Invasive species

Here, we consider invasive species to represent those that have been declared as non-indigenous by experts. The presence of invasive species of jellyfish was assumed to represent an increase in jellyfish biomass (Abundance Trend = 1). With this assumption, it is clearly important to understand if an invasive species is truly established because some invaders can appear briefly in a particular area, but not be detected thereafter. Knowledge of such events was assumed to represent no change in a jellyfish population (Abundance Trend = 0), rather than an increase, as the excess biomass due to the invader presumably vanishes if the species is no longer detected. However, the possibility of repeated detection persists due to potential establishment by cryptic polyps or successive invasions, as is likely with *Phyllorhiza* sp. in the South Brazil Shelf LME (Haddad & Nogueira, 2006). The possibility also exists that invasive species of jellyfish could cause a reduction in native jellyfish biomass. However, no evidence of such an event was found (but see Brotz, 2011 for discussion).

Chronicles that pertained to invasive species were scored similarly to other chronicles on the basis of time and space, but differently for reliability. The contribution to an increase in jellyfish biomass due to an invader was weighted by the "Invasive reliability score" (Table 1) to provide a more accurate estimate of the total change in jellyfish biomass. The assumptions and weighting factors were designed to avoid an overemphasis on invasive species. However, the invasive jellyfish accounted for in this analysis represent a conservative estimate, because it is likely that far more invasions have occurred than have been documented due to incomplete treatment, unusual life histories, and species crypsis (Holland et al., 2004; Dawson et al., 2005; Graham & Bayha, 2007). Invasive species were treated separately during analyses, allowing assessment of their contribution to the results. Consistent with the baseline selected for the analysis, species that invaded regions prior to 1950 were excluded.

Fuzzy logic expert system

Scores and chronicles were combined using a series of rule sets and fuzzy logic. The steps are outlined below, and the methodology diagramed using the North Sea LME as an example (Fig. 1). Fuzzy set theory, originally developed by Zadeh (1965), is now firmly established in engineering and science (e.g., Lee, 1990; van der Werf & Zimmer, 1998; Cheung et al., 2007), and fuzzy models are increasingly being used for ecological applications (Jørgensen, 2008). Fuzzy set theory allows the representation of variables according to a gradation or degree of membership, rather than the classic "true" or "false" membership of conventional Boolean sets. In addition, fuzzy logic allows a conclusion to be reached with an associated gradation or degree of belief. As such, fuzzy set theory and logic provide an ideal system for combining information of variable cardinality and confidence. Adriaenssens et al. (2004) reviewed fuzzy set theory used in ecosystem studies and Brotz (2011) details the specific methodology used in our study.

Variables with differing degrees of confidence were combined using the "MYCIN" method, an asymptotic accumulation of the degree of belief, after Buchanan & Shortliffe (1984). This knowledge accumulation method is not affected by the order in which evidence is combined, and can be defined as:

Table 1 Rule sets defined for analysis of jellyfish population trends in LME

	Definition
Abundance Trend rule set	
Abundance Trend	
−1 (decrease)	Decrease in overall biomass, relative abundance, frequency of occurrence or duration of occurrence
0 (stable/ variable)	Stable or no obvious trend
+1 (increase)	Increase in overall biomass, relative abundance, frequency of occurrence or duration of occurrence
Time score rule set	
Time score	
Low	Multiyear trend < 5 years; recent and unrepeated bloom that has not occurred previously; unclear timeframe; no recent data (post-2000)
Medium	Short term (5–9 years)
High	Medium term (10–14 years)
Very high	Long term ($\geq$15 years)
Space score rule set	
Space score	
Low	Singular location or small region within LME (<200 km wide)
Medium	Large region or two disparate locations within LME (>200 km apart)
High	Three or more disparate locations within LME; wide-scale sampling in at least half of LME
Very high	Wide-scale sampling of LME
Reliability score rule set	
Reliability score	
Low	Lifeguard or NGO commentary; species unlikely to contribute significantly to biomass; high uncertainty; documented anthropogenic polyp habitat
Medium	Marine professional commentary (e.g., fisher)
High	Marine scientist commentary; synthesized knowledge; "bookend" (i.e., non-continuous) scientific data
Very high	Scientific data of numerous or dominant species; well-documented frequency of blooms
Invasive reliability score rule set	
Reliability score	
Low	Uncertainty of invasiveness or species is unlikely to contribute significantly to biomass (e.g., small hydromedusae)
Medium	Documented invasive species or newly-blooming species (without knowledge of other species in ecosystem) or unsuccessful establishment[a]
High	Thriving invasive species
Very high	Known dominant species

Rule sets here include: Abundance Trend, Time score, Space score, and Reliability scores for native and invasive species. Additional parameters are in Tables 2 and 3

[a] Abundance Trend = 1 in all invasive cases except for unsuccessful establishment (where Abundance Trend = 0 and Invasive reliability score = medium)

$$\text{Degree of belief}_{n+1} = \text{Evidence}_n + [(1 - \text{Evidence}_n) \times \text{Evidence}_{n+1}]$$

where Degree of belief_{n+1} is the membership in the conclusion after combining the membership from Evidence_n and Evidence_{n+1}. The membership for any number of pieces of evidence can thus be combined to yield a final membership (i.e., degree of belief) in the conclusion.

The three scores for each jellyfish chronicle (time, space, and reliability) were combined using a fuzzy rule set, or combination matrix, to yield a "Confidence Index" (Table 2). The combination matrix used treats all three scores equally, and therefore represents all

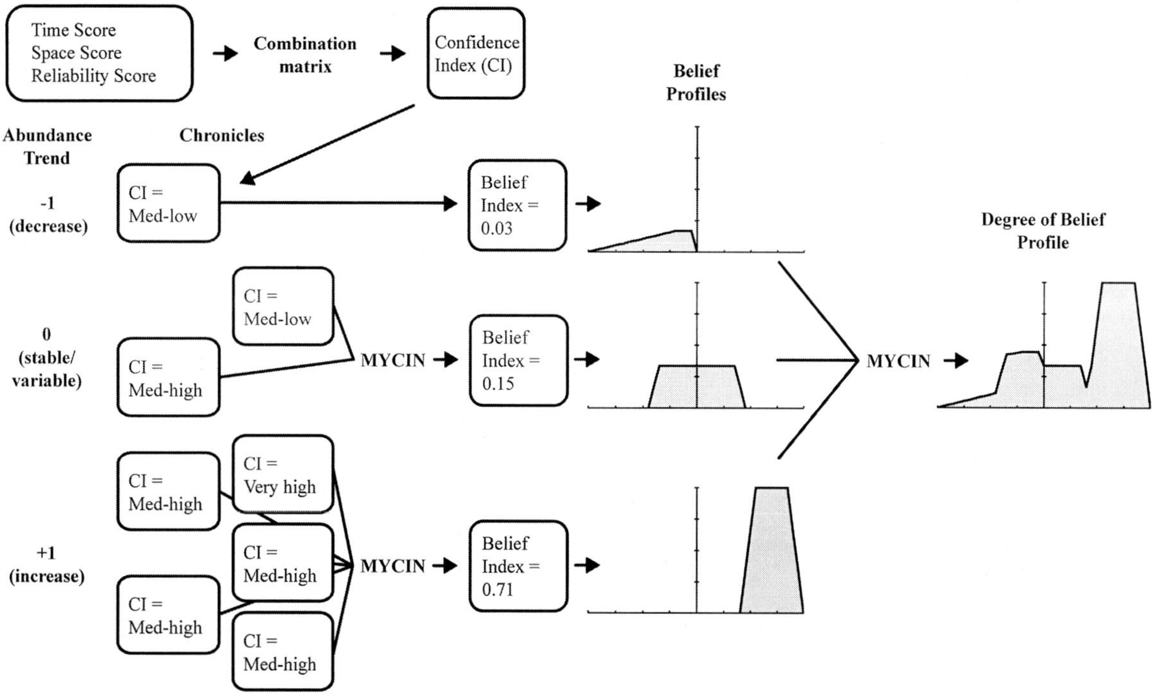

Fig. 1 Schematic diagram of the fuzzy expert system used in the analysis of jellyfish population trends by LME, with the North Sea LME represented as an example

possible combinations of scores. Thus, each chronicle has an associated Abundance Trend representing the direction of change for the jellyfish population in question and a Confidence Index representing the Degree of Belief. Chronicles included in the North Sea LME (Table 3) are depicted in the fuzzy expert system diagram (Fig. 1). Details of all chronicles used in this analysis are in Brotz (2011).

Within each LME, chronicles that had the same Abundance Trend were combined to yield a Belief Index, which was derived by converting the Confidence Index value for each chronicle into a membership (Degree of Belief; Table 4) and subsequently combining these memberships using MYCIN. The resulting Belief Indexes for each Abundance Trend were used to select an appropriate Belief Profile (Table 4). The Belief Profiles used in the fuzzy expert system were membership functions designed to represent the Degree of Belief over a continuous scale of -100 to $+100$, with negative scores representing declining jellyfish populations and positive scores representing increasing populations. These asymmetrical Belief Profiles therefore provide a representation of the accumulated evidence for each particular trend, including both the quantity and the relative certainty of the evidence. Within each LME, one profile was selected for each Abundance Trend, as long as there was supporting evidence (i.e., Belief Index >0). Thus, an LME could have 1, 2, or 3 profiles as inputs for the fuzzy expert system, depending on whether or not there were chronicles supporting each Abundance Trend. The Belief Profiles were combined using the MYCIN method to yield a final Degree of Belief profile for each LME. This profile contained information about the evidence within each LME over all Abundance Trends. To calculate a final Jellyfish Index, the centroid-weighted method (Cox, 1999) was used to "defuzzify" the final profile.

Uncertainty

The confidence in the Jellyfish Index was quantified by the Degree of Belief at the centroid value (the "Confidence Factor") and the associated values at Degree of Belief $= 0.25$ (the confidence limits). The difference between the confidence limits is defined as the "Confidence Interval." If a particular profile did not reach a Degree of Belief $= 0.25$ due to lack of

Table 2 Combination matrix used to combine scores, yielding a single Confidence Index

Score A	Score B	Score C	Confidence Index
Low	Low	Low	Low
Low	Low	Medium	Low
Low	Low	High	Medium-low
Low	Low	Very high	Medium-low
Low	Medium	Medium	Medium-low
Low	Medium	High	Medium
Low	Medium	Very high	Medium
Low	High	High	Medium
Low	High	Very high	Medium-high
Low	Very high	Very high	Medium-high
Medium	Medium	Medium	Medium
Medium	Medium	High	Medium-high
Medium	Medium	Very high	Medium-high
Medium	High	High	Medium-high
Medium	High	Very high	High
Medium	Very high	Very high	High
High	High	High	High
High	High	Very high	High
High	Very high	Very high	Very high
Very high	Very high	Very high	Very high

Scores for time, space, and reliability were treated equally, and therefore the matrix represents all possible combinations of scores

evidence (e.g., Gulf of California LME), the upper and lower confidence limits were selected where the Degree of Belief falls to zero. Use of these two measures of uncertainty (the Confidence Factor and the Confidence Interval) provided information about both the strength of the data within an LME and how consistent was the observed trend (if any). These would be similar to measures of "accuracy" and "precision," i.e., a high Confidence Factor represented a robust conclusion and could be interpreted as accurate. Similarly, a small Confidence Interval would indicate that the chronicles included in a particular LME have comparable trends and were therefore precise. The combination of these two measures ultimately defined the overall confidence in the Jellyfish Index for each LME; thus, we defined a "Confidence Quotient" as equal to the Confidence Factor divided by the Confidence Interval. Conclusions with a Confidence Quotient >1 were classified as "high certainty," while those with a Confidence Quotient <1 were classified as "low certainty."

Based on the Belief Profiles used in the analysis, Jellyfish Indexes could range from a minimum of −70 to a maximum of +70. LMEs with a Jellyfish Index of greater than +10 were classified as increases, while those with a Jellyfish Index less than −10 were classified as decreases. LMEs with a Jellyfish Index between −10 and +10 were classified as stable/variable, indicating they did not show an increasing or decreasing trend. These thresholds were chosen in order to ensure there was sufficient evidence to suggest a trend.

Results

A total of 138 jellyfish chronicles were included in the analysis, distributed unevenly over 45 LMEs. Results including both native and invasive species are presented in Table 5. Of the 45 LMEs, 28 (62%) showed increasing trends, while only 3 (7%) showed decreasing trends. The remaining 14 LMEs (31%) were classified as stable/variable, showing neither increasing nor decreasing trends (Fig. 2).

Out of the 28 LMEs exhibiting increases, 10 were classified as high certainty (Confidence Quotient > 1) and 18 as low certainty. Of the 14 LMEs with stable/variable trends, 4 were of high certainty and 10 were of low certainty. The Humboldt Current LME was the only system to exhibit a decrease associated with a high certainty.

The results are similar when normalized by area of the LMEs; 21% of the total area included represented regions with increases of high certainty, while increases of low certainty represented 45%. Stable/variable regions represented 28% of the total area included, while the remaining 6% was associated with decreases.

Effects of invasive species

Invasive species were separated from the analysis to examine their effects on the results (Tables 5, 6). Invasive species of jellyfish were reported in 21 LMEs. In eight of those, the inclusion of invasive species had a negligible contribution to the results and did not affect the Jellyfish Index. By contrast, the inclusion of invasive species was responsible for the conclusion of low certainty increases in four LMEs (Gulf of Mexico, Southeast U.S. Continental Shelf,

Table 3 Example of information constituting chronicles used in the analysis

Confidence Index	Abundance Trend	Time score	Space score	Reliability score	Country	Location	Dates	Species	Main source	Additional sources
Medium-low	−1	Low	Low	High	United Kingdom	Thames Estuary	Since 1985 (data 1977–1992)	*Aurelia* sp. (*Pleurobrachia pileus* is variable)	Attrill & Thomas (1996)	
Medium-high	0	Low	High	Very high	Numerous	Widescale sampling in half of LME	1971–1986	Numerous	Hay et al. (1990)	Lynam et al. (2004, 2005)
Medium-low	0	Very high	Low	Low[a]	Germany	Helgoland, German Bight	1975–1993; 1975–2002	Numerous	Greve et al. (1996, 2004)	Greve (1994) and Schlüter et al. (2010)
Very high	+1	Very high	Very high	High	Numerous	Entire LME	Since 1980s	Likely *Aglantha digitale*	Attrill et al. (2007) and Licandro et al. (2010)	Attrill & Edwards (2008) and Haddock (2008)
Medium-high	+1	Very high	Low	Very high	Netherlands	Texel Island	Recent decades	Numerous	van Walraven et al. (2010)	
Medium-high	+1	Very high	Low	High	Norway	Lurefjorden	Since 1970s	*Periphylla periphylla*	Fosså (1992)	Youngbluth & Båmstedt (2001) and Sørnes et al. (2007)
Medium-high	+1	Very high	Low	High	Denmark	Limfjorden	Since 1980s	*Aurelia* spp. (& others)	Riisgård et al. (2012)	Hoffmann (2005) and Møller & Riisgård (2007a, b)
Medium-high	+1	Medium	High	Medium	Numerous	Numerous	Since at least 2005	*Mnemiopsis leidyi* (invasive)	Oliveira (2007) and Tendal et al. (2007)	Faasse & Bayha (2006), Hansson (2006), Boersma et al. (2007), Hosia (2007), and Riisgård et al. (2007)

Details are presented here for chronicles used for the North Sea LME only

[a] See (Brotz, 2011) for discussion

Table 4 Rule sets used in the fuzzification process of the fuzzy expert system

Confidence Index	Degree of Belief (per chronicle)
Low	0.0156
Medium-low	0.0313
Medium	0.0625
Medium-high	0.125
High	0.25
Very high	0.5
Belief Index	**Belief Profile**
0	None
0.01–0.09	Low
0.1–0.19	Medium-low
0.2–0.34	Medium
0.35–0.49	Medium-high
0.50–0.59	High
0.60–1	Very high

Rule sets here include the Degree of Belief membership according to the Confidence Index for each chronicle and the Belief Profile selection according to the Belief Index. See text and Fig. 1 for additional information

Caribbean Sea, and Baltic Sea), because the exclusion of invaders changed the classification of these LMEs from increasing to stable/variable. Similarly, invaders were responsible for the low certainty increase in the East Brazil Shelf LME because there were no data for native species. The Insular Pacific-Hawaiian LME exhibited an increase due to native species; however, the inclusion of invasive species increased the certainty of the conclusion to high. In the remaining LMEs, the inclusion of invasive species increased the Jellyfish Index by variable amounts, but did not alter the conclusions.

Considering the effects of jellyfish overexploitation

Interestingly, several of the chronicles that were classified as decreases in the analysis (Abundance Trend $= -1$) concerned jellyfish species that have been harvested for food, science, or unique proteins, and have subsequently declined, possibly as a result of overfishing. Only four chronicles had a primary source of evidence that directly attributed a decrease to overexploitation; therefore, these chronicles were treated separately in the analysis. In the Arabian Sea LME, the inclusion of overfishing of jellyfish reduced the Jellyfish Index sufficiently to alter the trend conclusion from increasing to stable/variable (both conclusions of low certainty). Inclusion of overfishing of jellyfish for the Bay of Bengal LME resulted in no change to the Jellyfish Index. The South China Sea and East Central Australian Shelf LMEs showed a reduced Jellyfish Index when overfishing of jellyfish was included; however, this reduction was not sufficient to classify these LMEs as decreases and they remained classified as stable/variable. Thus, in the majority of locations where overfishing of jellyfish could be identified, it did not alter the conclusions of the analysis.

Discussion

This study represents the first rigorous demonstration that jellyfish populations appear to be increasing in coastal ecosystems worldwide, as previously suggested (Mills, 2001; Purcell et al., 2007; Pauly et al., 2009a; Richardson et al., 2009). Of the 45 LMEs included in our analysis, 28 (62%) showed increasing trends, while only 3 (7%) showed decreasing trends. The remaining 14 LMEs (31%) were classified as stable/variable, with no obvious trend. These results suggest that while increases of jellyfish populations are not universal, they are both numerous and widespread. Of the 21 LMEs that were not included in our analysis, most were from the Arctic (11), Australia (4), and the South Pacific (3). Therefore, our results represent extensive spatial coverage of the world's coastal ecosystems. While only 33% of the conclusions are of high certainty, the majority of those (10 of 15) were in LMEs that showed increasing trends. In addition to demonstrating that jellyfish populations have increased in numerous ecosystems around the world, our analysis also underscored the fact that information on jellyfish abundance is poor over much of the globe. Thus, we must strive to learn more about these important creatures, especially given the fact that they seem to be one of the few groups of organisms that may benefit from the continued anthropogenic impacts on the world's biosphere.

Defining an "increase"

Information used in the analysis was weighted by time, space, and reliability to reflect the relative contribution

Table 5 Results of analysis of jellyfish population trends by LME including both native and invasive species

LME ID	LME name	Trend conclusion	Conclusion certainty	Jellyfish Index	Confidence Quotient	Confidence Factor	Lower limit	Upper limit	Interval
1	East Bering Sea	Increase	High	61.84	1.47	0.83	34.50	91.00	56.50
2	Gulf of Alaska	Stable/ variable	Low	7.06	0.80	0.58	−35.00	37.24	72.24
3	California Current	Increase	Low	25.55	0.63	0.73	−31.25	85.00	116.25
4	Gulf of California	Increase	Low	35.87	0.13	0.13	0.00	100.00	100.00
5	Gulf of Mexico	Increase	Low	14.13	0.75	0.65	−35.00	51.25	86.25
6	Southeast US Continental Shelf	Increase	Low	14.13	0.75	0.65	−35.00	51.25	86.25
7	Northeast US Continental Shelf	Increase	High	52.52	1.58	0.83	43.75	96.25	52.50
8	Scotian Shelf	Stable/ variable	High	0.00	1.07	0.67	−31.25	31.25	62.50
9	Newfoundland-Labrador Shelf	Stable/ variable	High	0.00	1.54	0.83	−27.00	27.00	54.00
10	Insular Pacific-Hawaiian	Increase	High	54.84	1.13	0.67	25.63	85.00	59.37
11	Pacific Central-American Coastal	Increase	Low	41.74	0.77	0.30	12.50	51.25	38.75
12	Caribbean Sea	Increase	Low	13.60	0.81	0.31	3.00	41.26	38.26
13	Humboldt Current	Decrease	High	−42.80	1.26	0.71	−91.00	−34.50	56.50
14	Patagonian Shelf	Increase	Low	47.90	0.87	0.50	17.50	75.00	57.50
15	South Brazil Shelf	Stable/ variable	Low	7.06	0.80	0.58	−35.00	37.24	72.24
16	East Brazil Shelf	Increase	Low	35.87	0.13	0.13	0.00	100.00	100.00
18	West Greenland Shelf	Decrease	Low	−35.87	0.13	0.13	−100.00	0.00	100.00
21	Norwegian Sea	Increase	Low	41.74	0.70	0.27	12.50	51.25	38.75
22	North Sea	Increase	Low	35.89	0.22	0.30	−40.67	96.25	136.92
23	Baltic Sea	Increase	Low	14.13	0.75	0.65	−35.00	51.25	86.25
24	Celtic-Biscay Shelf	Increase	Low	36.94	0.44	0.56	−37.50	91.00	128.50
25	Iberian Coastal	Stable/ variable	Low	7.06	0.80	0.58	−35.00	37.24	72.24
26	Mediterranean Sea	Increase	Low	43.95	0.22	0.30	−37.50	96.25	133.75
28	Guinea Current	Increase	Low	35.87	0.13	0.13	0.00	100.00	100.00
29	Benguela Current	Increase	High	54.84	1.15	0.67	26.63	85.00	58.37
30	Agulhas Current	Stable/ variable	Low	0.00	0.71	0.50	−35.00	35.00	70.00
31	Somali Coastal Current	Stable/ variable	Low	0.00	0.44	0.33	−37.50	37.50	75.00
32	Arabian Sea	Increase	Low	14.13	0.75	0.65	−35.00	51.25	86.25
34	Bay of Bengal	Increase	Low	14.57	0.52	0.58	−37.24	75.00	112.24
35	Gulf of Thailand	Increase	Low	35.87	0.13	0.13	0.00	100.00	100.00
36	South China Sea	Stable/ variable	Low	8.86	0.56	0.44	−37.50	40.67	78.17
40	Northeast Australian Shelf	Increase	Low	35.87	0.13	0.13	0.00	100.00	100.00

Table 5 continued

LME ID	LME name	Trend conclusion	Conclusion certainty	Jellyfish Index	Confidence Quotient	Confidence Factor	Lower limit	Upper limit	Interval
41	East Central Australian Shelf	Stable/ variable	Low	0.00	0.71	0.50	−35.00	35.00	70.00
42	Southeast Australian Shelf	Stable/ variable	Low	8.86	0.56	0.44	−37.50	40.67	78.17
47	East China Sea	Increase	High	70.00	1.90	1.00	43.75	96.25	52.50
48	Yellow Sea	Increase	High	61.84	1.47	0.83	34.50	91.00	56.50
49	Kuroshio Current	Increase	High	35.34	1.13	0.67	25.63	85.00	59.37
50	Sea of Japan	Increase	High	61.84	1.47	0.83	34.50	91.00	56.50
51	Oyashio Current	Decrease	Low	−14.13	0.75	0.65	−51.25	35.00	86.25
52	Sea of Okhotsk	Stable/ variable	High	6.25	1.55	0.86	−27.00	28.56	55.56
53	West Bering Sea	Stable/ variable	Low	−7.49	0.40	0.50	−75.00	51.25	125.25
60	Faroe Plateau	Stable/ variable	High	0.00	1.54	0.83	−27.00	27.00	54.00
61	Antarctic	Increase	High	61.84	1.47	0.83	34.50	91.00	56.50
62	Black Sea	Increase	High	70.00	1.90	1.00	43.75	96.25	52.50
63	Hudson Bay	Stable/ variable	Low	0.00	0.44	0.33	−37.50	37.50	75.00

to a change in jellyfish populations within each LME. As a consequence of the methods used and the inclusion of anecdotal data, the results reflect the degree of belief that any particular jellyfish population has changed or not, rather than the magnitude of those changes. Therefore, observations of "more" jellyfish

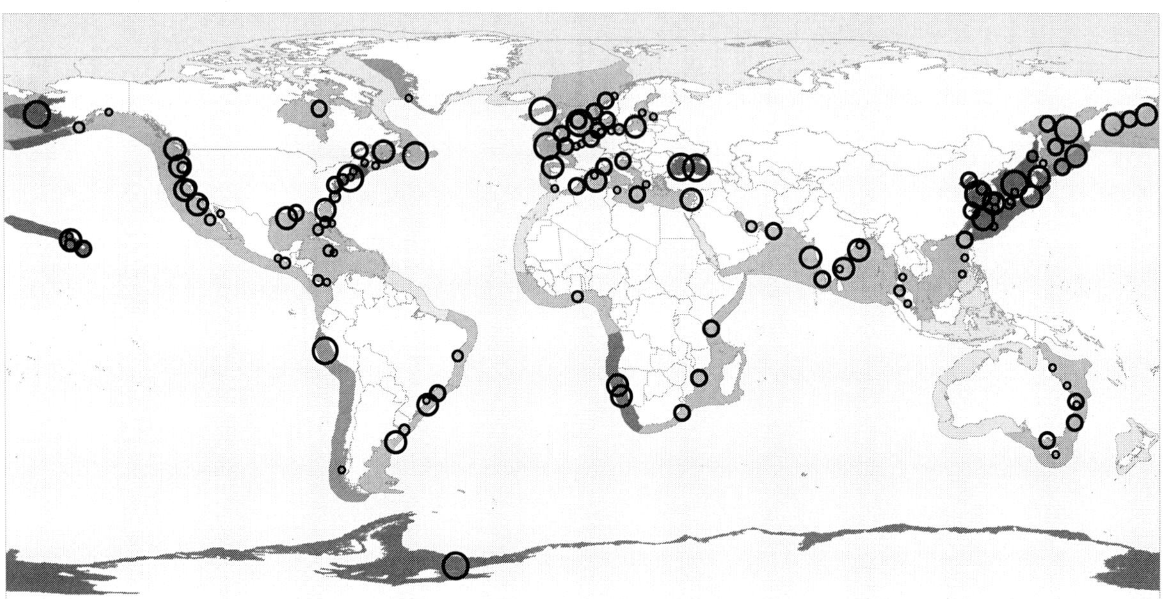

Fig. 2 Map of population trends of native and invasive species of jellyfish by LME. *Red* increase (high certainty), *orange* increase (low certainty), *green* stable/variable, *blue* decrease, *grey* no data. Circles represent discrete chronicles with relative sizes reflecting the Confidence Index. Circle locations are approximate, as some were shifted to avoid overlap; the circle for the Antarctic LME summarizes circumpolar observations

Table 6 Results of analysis of jellyfish population trends by LME including native species only (effects of invasive species excluded; only those LMEs that had invasive species are shown)

LME ID	LME name	Trend conclusion	Conclusion certainty	Jellyfish Index	Confidence Quotient	Confidence Factor	Lower limit	Upper limit	Interval
3	California Current	Increase	Low	19.82	0.73	0.78	−31.25	75.00	106.25
5	Gulf of Mexico	Stable/ variable	Low	7.06	0.80	0.58	−35.00	37.24	72.24
6	Southeast US Continental Shelf	Stable/ variable	Low	7.06	0.80	0.58	−35.00	37.24	72.24
7	Northeast US Continental Shelf	Increase	High	52.52	1.58	0.83	43.75	96.25	52.50
10	Insular Pacific-Hawaiian	Increase	Low	47.90	0.87	0.50	17.50	75.00	57.50
11	Pacific Central-American Coastal	Increase	Low	35.87	0.09	0.09	0.00	100.00	100.00
12	Caribbean Sea	Stable/ variable	Low	0.00	0.17	0.17	−50.00	50.00	100.00
13	Humboldt Current	Decrease	High	−61.84	1.47	0.83	−91.00	−34.50	56.50
14	Patagonian Shelf	Increase	Low	47.90	0.87	0.50	17.50	75.00	57.50
15	South Brazil Shelf	Stable/ variable	Low	0.00	0.71	0.50	−35.00	35.00	70.00
16	East Brazil Shelf	No data							
21	Norwegian Sea	Increase	Low	41.74	0.70	0.27	12.50	51.25	38.75
22	North Sea	Increase	Low	35.89	0.22	0.30	−40.67	96.25	136.92
23	Baltic Sea	Stable/ variable	Low	0.00	0.71	0.50	−35.00	35.00	70.00
25	Iberian Coastal	Stable/ variable	Low	0.00	0.71	0.50	−35.00	35.00	70.00
26	Mediterranean Sea	Increase	Low	31.02	0.54	0.66	−37.50	85.00	122.50
42	Southeast Australian Shelf	Stable/ variable	Low	8.86	0.56	0.44	−37.50	40.67	78.17
47	East China Sea	Increase	High	70.00	1.90	1.00	43.75	96.25	52.50
48	Yellow Sea	Increase	High	61.84	1.47	0.83	34.50	91.00	56.50
49	Kuroshio Current	Increase	High	35.34	1.13	0.67	25.63	85.00	59.37
62	Black Sea	Increase	High	61.84	1.47	0.83	34.50	91.00	56.50

may not necessarily mean there were really "more jellyfish" if the observations were not normalized by effort. Nonetheless, we expected that these factors were correlated, as changes of larger magnitude were assumed to be more noticeable and thus have more supporting evidence. Only after accepting this assumption should this analysis be considered to reflect real "increases" and "decreases."

Jellyfish populations are extremely variable on both temporal and spatial scales, due to their peculiar ecology. Thus, even LMEs showing pronounced increases in jellyfish populations with "high certainty" may also experience dramatic declines over short timescales. For example, the trend in the East Bering Sea LME was classified as an increase based on a regression analysis, but jellyfish in the Bering Sea declined dramatically after 2000 (Brodeur et al., 2008). Despite this decline, jellyfish abundance in this LME appears sustained above the levels observed in the 1980s and the increase remains significant. Other long-term studies show similar variability, such as the 37-year dataset from Peru (Quiñones et al., 2010). Jellyfish populations in that system appear tightly correlated with El Niño events, but the data exhibited a decline (see Brotz, 2011). Even the well-documented increase in blooms of the giant jellyfish

(*Nemopilema nomurai*) in East Asia has not been persistent, because blooms have not occurred every year (Uye et al., 2010). Clearly then, increases or decreases may actually represent a trend during only part of a cycle, and may reverse over longer timeframes (Purcell, 2012).

With such high population variability, poor sampling frequency in either the past or present could dramatically affect the detection of true trends. To account for these concerns, attempts were made to ensure chronicles used in the analysis were of sufficient duration and up to date wherever possible. Therefore, chronicles covering longer timescales and those with up-to-date information had more influence on the results. Nonetheless, few datasets of jellyfish abundance span multiple decades; therefore, our results represent only a rough estimate of true jellyfish population dynamics. Moreover, the possibility of a reporting bias, whereby newsworthy blooms or increases of jellyfish were reported, but absences and stable or declining populations were not, could tend to overestimate increases. However, the methods used in our analysis were designed to minimize this effect. For instance, episodic blooms were not included unless a temporal component of at least several years was identified. In addition, as mentioned above, these temporal components were scored based on whether they represent recent trends and are of significant duration. Interference events with human activities, which are typically newsworthy, also were not included unless the information was in a clear historical context. Finally, much of the anecdotal information used in the analysis was gleaned from targeted interviews (e.g., Uye & Ueta, 2004; Nagata et al., 2009; Pramod, 2010). Because numerous responses in those interviews indicated stable populations, they were assumed to represent a relatively unbiased source of information where scientific data were lacking.

The fact that jellyfish are typically part of the zooplankton makes them vulnerable to changes in oceanic current patterns. The presence or the absence of a bloom may simply be due to relocation; thus, an increase observed in one location may be concomitant with a decrease in another. If an increase is observed but a decrease is not, one may come to a false conclusion that jellyfish have increased. Whenever there was evidence of such an explanation, the chronicle was not included. An example is a recent quote from of a fisher in Florida who said he was seeing more sea nettles (*Chrysaora* sp.) now than in the preceding decades. However, this could be due to the relocation of the population normally observed elsewhere in the Gulf of Mexico (Spinner, 2010). Even without knowledge of such events, the analysis was not overly sensitive to that pitfall, because only multi-year data from the same location were used. As chronicles were either up-to-date or scored with low reliability, increases due to spatial redistributions would have to be sustained. In addition, chronicles based on information over short time periods or from single locations were also scored lower, thereby minimizing the effect on the results.

Possible causes of increasing jellyfish populations

Jellyfish have bloomed for hundreds of millions of years (Hagadorn et al., 2002; Young & Hagadorn, 2010) and are a natural presence in healthy ecosystems. Many jellyfish populations are known to fluctuate with oceanic climate (reviews in Purcell, 2005, 2012). There are also suggestions that jellyfish may benefit from anthropogenic pressures on the marine environment (Mills, 1995; 2001; Purcell et al., 2007; Pauly et al., 2009a; Richardson et al., 2009, Purcell, 2012). Suggested causes include eutrophication, overfishing, global warming, habitat modification, aquaculture, salinity changes, ocean acidification, and of course, translocation.

Invasive species of jellyfish were reported in 21 of 45 LMEs in this analysis (47% of the systems included). For the most part, invasive species were not responsible for the observed increases reflected in the results; however, the widespread detections demonstrate that jellyfish are truly global invaders of significant concern. Thriving populations of invasive jellyfish in systems like the Mediterranean and Black Seas should serve as warnings for other ecosystems around the globe, and it is likely that far more invasions have occurred than are reported (Holland et al., 2004; Dawson et al., 2005; Graham & Bayha, 2007).

There is clearly no single cause of increasing jellyfish blooms. For example, populations of *Aurelia* sp. appear tightly correlated with aquaculture operations in Tapong Bay, Taiwan (Lo et al., 2008), whereas recent increased blooms of *Aurelia* sp. in Tokyo Bay, Japan are more likely due to the effects of eutrophication (Nomura & Ishimaru, 1998; Ishii et al., 2008). In addition, possible causes may work in concert, synergistically creating conditions that benefit jellyfish (Purcell et al., 2007; Richardson et al., 2009; Pauly et al., 2009a; Purcell, 2012). Limited knowledge of jellyfish ecology, especially of benthic or sessile stages, inhibits our ability to draw conclusions regarding possible anthropogenic causes of jellyfish blooms. Nonetheless, the results of this analysis present a unique opportunity to examine commonalities using the LME framework. An important task will be to investigate possible linkages between anthropogenic stresses and increasing jellyfish populations as identified in this study.

Taxonomic concerns

The term "jellyfish," according the definition used here, refers to specimens from several phyla (Cnidaria, Ctenophora, and Chordata). Such organisms are obviously extremely distant phylogenetic relatives; therefore, grouping them under the umbrella term "jellyfish" is problematic. First, the use of such a term ignores taxonomy. The changes evident in the results of this analysis should not only be viewed in their entirety but also in the contexts of ecology and evolution. Without proper taxonomic resolution, a deeper and more meaningful understanding of the mechanisms and consequences involved may be unattainable (Haddock, 2004). Second, using a broad category also runs the risk of inferring attributes of a larger group of organisms based only on a handful of species. Such "errors of commission" (Dawson, 2010) could preclude robust conclusions if they are not made in the light of evolution. Generalizations concerning such a broad group of organisms will certainly have exceptions (Bayha & Dawson, 2010), and we must be careful not to ignore these differences by focusing only on commonalities.

Despite these concerns, there is also value in generalized results. Notwithstanding their phylogenetic diversity, jellyfish share many similarities. If the increasing trends identified in this analysis are indeed caused by anthropogenic factors, raising awareness of the issues and developing a deeper understanding of the mechanisms involved should be priorities.

Conclusions

Jellyfish populations appear to be increasing in the majority of the world's coastal ecosystems and seas. While these increases are conspicuous in several locations, even basic knowledge of jellyfish populations in most regions is poor. Many of the observed increases appear linked to human activities, but the mechanisms involved remain poorly understood. Because jellyfish populations can have important impacts on human activities and marine ecosystems, it is of paramount importance that we rapidly increase our understanding of these creatures.

Acknowledgments We thank the numerous scientists and marine professionals around the globe who kindly donated their knowledge and opinions of jellyfish populations. We also thank two anonymous reviewers and the guest editors of this volume whose comments improved the manuscript. This is a contribution from the *Sea Around Us* Project, a collaboration between the University of British Columbia's Fisheries Centre and the Pew Environment Group.

Open Access This article is distributed under the terms of the Creative Commons Attribution License which permits any use, distribution, and reproduction in any medium, provided the original author(s) and the source are credited.

References

Adriaenssens, V., B. D. Baets, P. L. M. Goethals & N. D. Pauw, 2004. Fuzzy rule-based models for decision support in ecosystem management. Science of the Total Environment 319: 1–12.

Alldredge, A. L. & L. P. Madin, 1982. Pelagic tunicates – unique herbivores in the marine plankton. BioScience 32: 655–663.

Arai, M. N., 1997. A Functional Biology of Scyphozoa. Chapman and Hall, London: 300 pp.

Arai, M. N., 2009. The potential importance of podocysts to the formation of scyphozoan blooms: a review. Hydrobiologia 616: 241–246.

Atkinson, A., V. Siegel, E. Pakhomov & P. Rothery, 2004. Long-term decline in krill stock and increase in salps within the Southern Ocean. Nature 432: 100–103.

Attrill, M. J. & M. Edwards, 2008. Reply to Haddock, S. H. D. Reconsidering evidence for potential climate-related increases in jellyfish. Limnology and Oceanography 53: 2763–2766.

Attrill, M. J. & R. M. Thomas, 1996. Long-term distribution patterns of mobile estuarine invertebrates (Ctenophora, Cnidaria, Crustacea: Decapoda) in relation to hydrological parameters. Marine Ecology Progress Series 143: 25–36.

Attrill, M. J., J. Wright & M. Edwards, 2007. Climate-related increases in jellyfish frequency suggest a more gelatinous future for the North Sea. Limnology and Oceanography 52: 480–485.

Bayha, K. M. & M. D. Dawson, 2010. New family of allomorphic jellyfishes, Drymonematidae, emphasizes evolution in the

functional morphology and trophic ecology of gelatinous zooplankton. Biological Bulletin 219: 249–267.
Boero, F., J. Bouillon, C. Gravili, M. P. Miglietta, T. Parsons & S. Piraino, 2008. Gelatinous plankton: irregularities rule the world (sometimes). Marine Ecology Progress Series 356: 299–310.
Boersma, M., A. M. Malzahn, W. Greve & J. Javidpour, 2007. The first occurrence of the ctenophore *Mnemiopsis leidyi* in the North Sea. Helgoland Marine Research 61: 153–155.
Brodeur, R. D., M. B. Decker, L. Ciannelli, J. E. Purcell, N. A. Bond, P. J. Stabeno, E. Acuna & G. L. Hunt, 2008. Rise and fall of jellyfish in the eastern Bering Sea in relation to climate regime shifts. Progress in Oceanography 77: 103–111.
Brotz, L., 2011. Changing jellyfish populations: trends in Large Marine Ecosystems. Fisheries Centre Research Report 19(5), Fisheries Centre, University of British Columbia, Vancouver.
Buchanan, B. G. & E. H. Shortliffe (eds), 1984. Rule-Based Expert Systems – The MYCIN Experiments of the Stanford Heurisitic Programming Project. Addison-Wesley, Menlo Park.
Cheung, W. W. L., R. Watson, T. Morato, T. J. Pitcher & D. Pauly, 2007. Intrinsic vulnerability in the global fish catch. Marine Ecology Progress Series 333: 1–12.
Chudnow, R., 2008. Are jellyfish populations increasing worldwide (and why)? Honours Bachelor's Thesis, Dalhousie University, Halifax.
Coleman, R., 2010. Jellyfish, fluorescent proteins, Nobel Prizes and pioneers in histochemistry. Acta Histochemica 112: 113–117.
Cox, E., 1999. The Fuzzy Systems Handbook: A Practitioner's Guide to Building, Using, and Maintaining Fuzzy Systems. AP Professional, San Diego.
Dawson, M., 2010. Cryptic ecology. Presentation, Third International Jellyfish Blooms Symposium, Mar del Plata, July 16, 2010.
Dawson, M. N. & W. M. Hamner, 2009. A character-based analysis of the evolution of jellyfish blooms: adaptation and exaptation. Hydrobiologia 616: 193–215.
Dawson, M. N., A. Sen Gupta & M. H. England, 2005. Coupled biophysical global ocean model and molecular genetic analyses identify multiple introductions of cryptogenic species. Proceedings of the National Academy of Sciences of the United States of America 102: 11968–11973.
Dovel, W. L., 1964. An approach to sampling estuarine macroplankton. Chesapeake Science 5: 77–90.
Faasse, M. A. & K. M. Bayha, 2006. The ctenophore *Mnemiopsis leidyi* in coastal waters of the Netherlands: an unrecognized invasion? Aquatic Invasions 1: 270–277.
Fosså, J. H., 1992. Mass occurrence of *Periphylla periphylla* in a Norwegian fjord. Sarsia 77: 237–251.
Gershwin, L. A., M. De Nardi, K. D. Winkel & P. J. Fenner, 2010. Marine stingers: review of an under-recognized global coastal management issue. Coastal Management 38: 22–41.
Graham, W. M. & K. M. Bayha, 2007. Biological invasions by marine jellyfish. In Nentwig, W. (ed.), Biological Invasions. Ecological Studies, Vol. 193. Springer, Berlin: 239–255.
Greve, W., 1994. The 1989 German invasion of *Muggiaea atlantica*. ICES Journal of Marine Science 51: 355–358.
Greve, W., F. Reiners & J. Nast, 1996. Biocoenotic changes of the zooplankton in the German Bight: the possible effects of eutrophication and climate. ICES Journal of Marine Science 53: 951–956.
Greve, W., F. Reiners, J. Nast & S. Hoffmann, 2004. Helgoland Roads meso- and macrozooplankton time-series 1974 to 2004: lessons from 30 years of single spot, high frequency sampling at the only off-shore island of the North Sea. Helgoland Marine Research 58: 274–288.
Haddad, M. A. & M. Nogueira, 2006. Reappearance and seasonality of *Phyllorhiza punctata* medusae in southern Brazil. Revista Brasileira De Zoologia 23: 824–831.
Haddock, S. H. D., 2004. A golden age of gelata: past and future research on planktonic ctenophores and cnidarians. Hydrobiologia 530: 549–556.
Haddock, S. H. D., 2008. Reconsidering evidence for potential climate-related increases in jellyfish. Limnology and Oceanography 53: 2759–2762.
Hagadorn, J. W., R. H. Dott & D. Damrow, 2002. Stranded on a late Cambrian shoreline: medusae from central Wisconsin. Geology 30: 147–150.
Hamner, W. M. & M. N. Dawson, 2009. A review and synthesis on the systematics and evolution of jellyfish blooms: advantageous aggregations and adaptive assemblages. Hydrobiologia 616: 161–191.
Hansson, H. G., 2006. Ctenophores of the Baltic and adjacent Seas – the invader *Mnemiopsis* is here! Aquatic Invasions 1: 295–298.
Hay, S., 2006. Marine ecology: gelatinous bells may ring change in marine ecosystems. Current Biology 16: R679–R682.
Hay, S. J., J. R. G. Hislop & A. M. Shanks, 1990. North Sea scyphomedusae – summer distribution, estimated biomass and significance particularly for 0-group Gadoid fish. Netherlands Journal of Sea Research 25: 113–130.
Heinle, D. R., 1965. A screen for excluding jellyfish and ctenophores from Clarke-Bumpus plankton samples. Chesapeake Science 6: 231–232.
Hoffmann, E., 2005. Fisk, fiskeri og epifauna – Limfjorden 1984–2004. Report 147-05, Danish Institute for Fisheries Research, Charlottenlund: 19 pp (in Danish).
Holland, B. S., M. N. Dawson, G. L. Crow & D. K. Hofmann, 2004. Global phylogeography of *Cassiopea*: molecular evidence for cryptic species and multiple invasions of the Hawaiian Islands. Marine Biology 145: 1119–1128.
Hosia, A., 2007. Gelatinous zooplankton in western Norwegian fjords: ecology, systematics and comparisons with adjacent waters. PhD Thesis, University of Bergen, Bergen.
Hsieh, Y. H. P., F. M. Leong & J. Rudloe, 2001. Jellyfish as food. Hydrobiologia 451: 11–17.
Ishii, H., T. Ohba & T. Kobayasi, 2008. Effects of low dissolved oxygen on planula settlement, polyp growth and asexual reproduction of *Aurelia aurita*. Plankton & Benthos Research 3: 107–113.
Jørgensen, S. E., 2008. Overview of the model types available for development of ecological models. Ecological Modelling 215: 3–9.
Kendall, D., 1990. Shrimp retention characteristics of the Morrison soft TED – a selective webbing exclusion panel inserted in a shrimp trawl net. Fisheries Research 9: 13–21.
Lee, C. C., 1990. Fuzzy logic in control systems: fuzzy logic controller – Part I. IEEE Transactions on Systems, Man, and Cybernetics 20: 404–418.

Lee, C. I., E. Pakhomov, A. Atkinson & V. Siegel, 2010. Long-term relationships between the marine environment, krill and salps in the Southern Ocean. Journal of Marine Biology 2010, Article ID 410129: 18 pp.

Licandro, P., D. V. P. Conway, M. N. D. Yahia, M. L. F. de Puelles, S. Gasparini, J. H. Hecq, P. Tranter & R. R. Kirby, 2010. A blooming jellyfish in the northeast Atlantic and Mediterranean. Biology Letters 6: 688–691.

Lo, W. T., J. E. Purcell, J. J. Hung, H. M. Su & P. K. Hsu, 2008. Enhancement of jellyfish (*Aurelia aurita*) populations by extensive aquaculture rafts in a coastal lagoon in Taiwan. ICES Journal of Marine Science 65: 453–461.

Loeb, V., V. Siegel, O. Holm-Hansen, R. Hewitt, W. Fraser, W. Trivelpiece & S. Trivelpiece, 1997. Effects of sea-ice extent and krill or salp dominance on the Antarctic food web. Nature 387: 897–900.

Lynam, C. P., S. J. Hay & A. S. Brierley, 2004. Interannual variability in abundance of North Sea jellyfish and links to the North Atlantic Oscillation. Limnology and Oceanography 49: 637–643.

Lynam, C. P., S. J. Hay & A. S. Brierley, 2005. Jellyfish abundance and climatic variation: contrasting responses in oceanographically distinct regions of the North Sea, and possible implications for fisheries. Journal of the Marine Biological Association of the United Kingdom 85: 435–450.

Mackie, G. O., 1974. Locomotion, floatation, and dispersal. In Muscatine, L. & H. M. Lenhoff (eds), Coelenterate Biology: Reviews and New Perspectives. Academic Press, New York: 313–357.

Macrokanis, C. J., N. L. Hall & J. K. Mein, 2004. Irukandji syndrome in northern Western Australia: an emerging health problem. Medical Journal of Australia 181: 699–702.

Maranger, R., N. Caraco, J. Duhamel & M. Amyot, 2008. Nitrogen transfer from sea to land via commercial fisheries. Nature Geoscience 1: 111–113.

Matsushita, Y. & N. Honda, 2006. Method of designing and manufacturing JET (Jellyfish Excluder for Towed fishing gear) for various towed fishing gears. Bulletin of Fisheries Research Agency 16: 19–27 (in Japanese with English abstract).

Mianzan, H. W. & R. A. Guerrero, 2000. Environmental patterns and biomass distribution of gelatinous macrozooplankton. Three study cases in the South-western Atlantic Ocean. Scientia Marina 64: 215–224.

Mills, C. E., 1995. Medusae, siphonophores, and ctenophores as planktivorous predators in changing global ecosystems. ICES Journal of Marine Science 52: 575–581.

Mills, C. E., 2001. Jellyfish blooms: are populations increasing globally in response to changing ocean conditions? Hydrobiologia 451: 55–68.

Møller, L. F. & H. U. Riisgård, 2007a. Impact of jellyfish and mussels on algal blooms caused by seasonal oxygen depletion and nutrient release from the sediment in a Danish fjord. Journal of Experimental Marine Biology and Ecology 351: 92–105.

Møller, L. F. & H. U. Riisgård, 2007b. Population dynamics, growth and predation impact of the common jellyfish *Aurelia aurita* and two hydromedusae, *Sarsia tubulosa*, and *Aequorea vitrina* in Limfjorden (Denmark). Marine Ecology Progress Series 346: 153–165.

Nagata, R. M., M. A. Haddad & M. Nogueira, 2009. The nuisance of medusae to shrimp trawls in central part of southern Brazilian Bight, from the perspective of artisanal fishermen. Pan-American Journal of Aquatic Sciences 4: 312–325.

Nomura, H. & T. Ishimaru, 1998. Monitoring the occurrence of medusae and ctenophores in Tokyo Bay, central Japan, in recent 15 years. Oceanography in Japan 7: 99–104 (in Japanese with English abstract).

Ohta, N., M. Sato, K. Ushida, M. Kokubo, T. Baba, K. Taniguchi, M. Urai, K. Kihira & J. Mochida, 2009. Jellyfish mucin may have potential disease-modifying effects on osteoarthritis. BMC Biotechnology 9: 98.

Oliveira, O. M. P., 2007. The presence of the ctenophore *Mnemiopsis leidyi* in the Oslofjorden and considerations on the initial invasion pathways to the North and Baltic Seas. Aquatic Invasions 2: 185–189.

Omori, M. & W. M. Hamner, 1982. Patchy distribution of zooplankton: behavior, population assessment and sampling problems. Marine Biology 72: 193–200.

Palomares, M. L. D. & D. Pauly, 2009. The growth of jellyfishes. Hydrobiologia 616: 11–21.

Pauly, D., 1995. Anecdotes and the shifting base-line syndrome of fisheries. Trends in Ecology & Evolution 10: 430.

Pauly, D., W. M. Graham, S. Libralato, L. Morissette & M. L. D. Palomares, 2009a. Jellyfish in ecosystems, online databases, and ecosystem models. Hydrobiologia 616: 67–85.

Pauly, D., J. Alder, S. Booth, W. W. L. Cheung, V. Christensen, C. Close, U. R. Sumaila, W. Swartz, A. Tavakolie, R. Watson, L. Wood & D. Zeller, 2009b. Fisheries in Large Marine Ecosystems: descriptions and diagnoses. In Sherman, K. & G. Hempel (eds), The UNEP Large Marine Ecosystem Report: A Perspective on Changing Conditions in LMEs of the World's Regional Seas. United Nations Environment Programme Regional Seas Report and Studies No. 182, Nairobi: 23–40.

Pierce, J., 2009. Prediction, location, collection and transport of jellyfish (Cnidaria) and their polyps. Zoo Biology 28: 163–176.

Pramod, G., 2010. Illegal, unreported and unregulated marine fish catches in the Indian Exclusive Economic Zone. Field Report, Policy and Ecosystem Restoration in Fisheries, Fisheries Centre, University of British Columbia, Vancouver: 30 pp.

Pugh, P. R., 1989. Gelatinous zooplankton – the forgotten fauna. Progress in Underwater Science 14: 67–78.

Purcell, J. E., 2005. Climate effects on formation of jellyfish and ctenophore blooms: a review. Journal of the Marine Biological Association of the United Kingdom 85: 461–476.

Purcell, J. E., 2009. Extension of methods for jellyfish and ctenophore trophic ecology to large-scale research. Hydrobiologia 616: 23–50.

Purcell, J. E., 2012. Jellyfish and ctenophore blooms coincide with human proliferations and environmental perturbations. Annual Review of Marine Science 4: 209–235.

Purcell, J. E. & M. N. Arai, 2001. Interactions of pelagic cnidarians and ctenophores with fish: a review. Hydrobiologia 451: 27–44.

Purcell, J. E., S. Uye & W. T. Lo, 2007. Anthropogenic causes of jellyfish blooms and their direct consequences for humans: a review. Marine Ecology Progress Series 350: 153–174.

Quiñones, J., H. Mianzan, M. Acha, S. Purca & W. M. Graham, 2010. El Viejo, El Niño and jellyfishes: dramatic environmental-driven changes in jellyfish abundances off Peru. Presentation, Third International Jellyfish Blooms Symposium, Mar del Plata, July 15, 2010.

Richardson, A. J., A. Bakun, G. C. Hays & M. J. Gibbons, 2009. The jellyfish joyride: causes, consequences and management responses to a more gelatinous future. Trends in Ecology & Evolution 24: 312–322.

Riisgård, H. U., L. Bøttiger, C. V. Madsen & J. Purcell, 2007. Invasive ctenophore *Mnemiopsis leidyi* in Limfjorden (Denmark) in late summer 2007 – assessment of abundance and predation effects. Aquatic Invasions 2: 395–401.

Riisgård, H. U., P. Andersen & E. Hoffmann, 2012. From fish to jellyfish in the eutrophicated Limfjorden (Denmark). Estuaries and Coasts. doi:10.1007/s12237-012-9480-4.

Schlüter, M. H., A. Merico, M. Reginatto, M. Boersma, K. H. Wiltshire & W. Greve, 2010. Phenological shifts of three interacting zooplankton groups in relation to climate change. Global Change Biology 16: 3144–3153.

Sherman, K. & G. Hempel (eds), 2009. The UNEP Large Marine Ecosystem Report: A Perspective on Changing Conditions in LMEs of the World's Regional Seas. United Nations Environment Programme Regional Seas Report and Studies No. 182, Nairobi: 872 pp.

Sherman, K. & Q. Tang (eds), 1999. Large Marine Ecosystems of the Pacific Rim: Assessment, Sustainability, and Management. Blackwell Science, Cambridge: 465 pp.

Sørnes, T. A., D. L. Aksnes, U. Båmstedt & M. J. Youngbluth, 2007. Causes for mass occurrences of the jellyfish *Periphylla periphylla*: a hypothesis that involves optically conditioned retention. Journal of Plankton Research 29: 157–167.

Spinner, K., 2010. Currents steering sea nettles our way. The Herald Tribune, Sarasota County, October 20, 2010.

Sugahara, T., M. Ueno, Y. Goto, R. Shiraishi, M. Doi, K. Akiyama & S. Yamauchi, 2006. Immunostimulation effect of jellyfish collagen. Bioscience, Biotechnology, and Biochemistry 70: 2131–2137.

Tendal, O. S., K. R. Jensen & H. U. Riisgård, 2007. Invasive ctenophore *Mnemiopsis leidyi* widely distributed in Danish waters. Aquatic Invasions 2: 455–460.

UNESCO, 1968. Zooplankton sampling. United Nations Educational, Scientific and Cultural Organization, Paris: 174 pp.

Uye, S. & U. Ueta, 2004. Recent increase of jellyfish populations and their nuisance to fisheries in the Inland Sea of Japan. Bulletin of the Japanese Society of Fisheries Oceanography 68: 9–19. (in Japanese with English abstract).

Uye, S., M. Shimizu & T. Watanabe, 2010. Tackling the giant jellyfish (*Nemopilema nomurai*) plague: cause, forecast and countermeasure. Plenary presentation, Third International Jellyfish Blooms Symposium, Mar del Plata, July 14, 2010.

van der Werf, H. M. G. & C. Zimmer, 1998. An indicator of pesticide environmental impact based on a fuzzy expert system. Chemosphere 36: 2225–2249.

van Walraven, L., V. T. Langenberg & H. W. van der Veer, 2010. Major changes in occurrence, seasonal patterns and species composition of gelatinous zooplankton in the western Dutch Wadden Sea from 1960 until present. Presentation, Third International Jellyfish Blooms Symposium, Mar del Plata, July 14, 2010.

Ward, J. M., J. E. Kirkley, R. Metzner & S. Pascoe, 2004. Measuring and assessing capacity in fisheries. 1. Basic concepts and management options. FAO Fisheries Technical Paper No. 433/1. Food and Agriculture Organization, Rome: 40 pp.

Watson, R., A. Kitchingman, A. Gelchu & D. Pauly, 2004. Mapping global fisheries: sharpening our focus. Fish and Fisheries 5: 168–177.

Young, G. A. & J. W. Hagadorn, 2010. The fossil record of cnidarian medusae. Palaeoworld 19: 212–221.

Youngbluth, M. J. & U. Båmstedt, 2001. Distribution, abundance, behavior and metabolism of *Periphylla periphylla*, a mesopelagic coronate medusa in a Norwegian fjord. Hydrobiologia 451: 321–333.

Zadeh, L. A., 1965. Fuzzy sets. Information and Control 8: 338–353.

Zimmer, M., 2005. Glowing Genes: A Revolution in Biotechnology. Prometheus Books, Amherst.

Hydrobiologia (2012) 690:21–46
DOI 10.1007/s10750-012-1037-9

Review Paper

Transitions of *Mnemiopsis leidyi* (Ctenophora: Lobata) from a native to an exotic species: a review

J. H. Costello · K. M. Bayha · H. W. Mianzan · T. A. Shiganova · J. E. Purcell

Published online: 21 March 2012
© Springer Science+Business Media B.V. 2012

Abstract The genus *Mnemiopsis* is comprised of a single species, *Mnemiopsis leidyi* A. Agassiz, 1865, that has recently made the transition from a distribution limited to the Atlantic coasts of North and South America to an invasive range that includes the Black, Caspian, Mediterranean, North, and Baltic seas. We review the foundations of the ctenophore's invasive success, which include the source-sink dynamics that characterize *Mnemiopsis* populations in temperate coastal waters where the ctenophore achieves its highest biomass levels and ecosystem impacts. Within its native temperate range, *Mnemiopsis* is frequently a dominant, seasonal, colonizing species with limited dispersal capacities. Cross-oceanic transport within ballast waters of intercontinental shipping vessels has altered this dispersal limitation and initiated a rapid global spread of *Mnemiopsis*. Owing to continuing transport via transoceanic shipping, we anticipate continued range expansion and review the variables most likely to determine whether introduction of *Mnemiopsis* to a novel community results in an inconspicuous addition or a disruptive invasion.

Keywords Invasion · Source-sink · Ballast transport · Niche flexibility · Range expansion

Guest editors: J. E. Purcell, H. Mianzan & J. R. Frost / Jellyfish Blooms: Interactions with Humans and Fisheries

J. H. Costello (✉)
Biology Department, Providence College, Providence, RI 02918, USA
e-mail: costello@providence.edu

K. M. Bayha
Dauphin Island Sea Lab, 101 Bienville Blvd., Dauphin Island, AL 36528, USA

H. W. Mianzan
CONICET-INIDEP, P. Victoria Ocampo No1, 7600, Mar del Plata, Argentina

T. A. Shiganova
Shirshov Institute of Oceanology, Russian Academy of Sciences, Moscow, Russia 117997

J. E. Purcell
Shannon Point Marine Center, Western Washington University, 1900 Shannon Point Road, Anacortes, WA 98221, USA

Introduction

The lobate ctenophore, *Mnemiopsis leidyi* A. Agassiz, 1865, has an established record of ecological importance within its native range, but has most recently gained notoriety for its expansion into exotic habitats (reviewed in Purcell et al., 2001). Before the invasion of the Black Sea, there was little discussion of the invasive capabilities of *Mnemiopsis*. Yet this ctenophore has proven to be a highly successful invader and, consequently, the future of its expansion is an important issue for marine planktonic communities. Our goal here is to examine the factors promoting and limiting invasive success of *Mnemiopsis* in order to review its potential for continued ecological range expansion.

Time course of invasive introductions

Range expansion of *Mnemiopsis* came into focus after the ctenophore was introduced to the Black Sea and surrounding areas (Vinogradov et al., 1989; Studenikina et al., 1991; Shiganova, 1993; Shiganova et al., 2001b; Shiganova & Malej, 2009) where fisheries' collapses and ecosystem disruptions were reported to be related to the introduction (Kideys, 2002; Knowler, 2005; Oguz et al., 2008). *Mnemiopsis* apparently was first transported accidentally in ballast water to the Black Sea (Ghabooli et al., 2010; Reusch et al., 2010). The ctenophores were first found in Sudak Bay in November, 1982 (Pereladov, 1988). By summer–autumn 1988, it had spread throughout the Black Sea, with average biomasses of up to 1 kg WW m^{-2} (40 g WW m^{-3}) and average numbers of up to 310 ctenophores m^{-2} (12.4 m^{-3}) (Vinogradov et al., 1989). Subsequently, *Mnemiopsis* moved through straits to adjacent basins (Fig. 1). It was first observed in the Sea of Azov in August, 1988 (Studenikina et al., 1991). Because *Mnemiopsis* cannot survive the winter low temperatures in the Sea of Azov, it must be re-introduced annually through the Kerch Strait from the Black Sea. The *Mnemiopsis* population spread from the Black Sea in the upper Bosporus current into the sea of Marmara, where it occurs all year in the upper water layer. It proceeded from the Sea of Marmara to the Mediterranean Sea, where it was first recorded in 1990 in the Aegean sea (Shiganova et al., 2001b). In subsequent accidental introductions, the ctenophore was transported from the Black Sea to the Caspian Sea in 1999 (Ivanov et al., 2000; Shiganova et al., 2001b) and from the northwestern Atlantic to the North and Baltic seas, where they were first reported in 2006 (Faasse & Bayha, 2006; Javidpour et al., 2006; Ghabooli et al., 2010; Reusch et al., 2010). It subsequently became apparent that *Mnemiopsis* was widely distributed in those waters (Hansson, 2006; Tendal et al., 2007; Schaber et al., 2011a, b).

Within the Mediterranean Sea, *Mnemiopsis* rapidly spread from the Aegean Sea to adjacent waters of the eastern Mediterranean (Levantine Sea), where it was found in Mersin Bay in spring 1992 (Kideys & Niermann, 1994) and in Syrian coastal waters in October 1993 (Shiganova, 1997). Until recently, *Mnemiopsis* was not reported from new locations in the Mediterranean. It was reported in the Northern Adriatic Sea in 2005 and from coastal waters of France in 2006 (Shiganova & Malej, 2009). Siapatis et al. (2008) developed a predictive model based on environmental conditions and water depth to identify the potential habitats of *Mnemiopsis* in the Mediterranean basin. Their model showed that many regions within the Mediterranean were potentially viable habitats for *Mnemiopsis* invasion. In 2009, blooms of *Mnemiopsis* were reported in waters of Israel (Galil et al., 2009; Fuentes et al., 2010), Italy (Boero et al., 2009), and Spain (Fuentes et al., 2010). *Mnemiopsis* from these locations genotypically resembled those from the northern Gulf of Mexico and the Black Sea (Fuentes et al., 2010). Both currents and shipping are probable methods of transport of *Mnemiopsis* within the Mediterranean Sea (Fuentes et al., 2010).

An uncorroborated report of *Mnemiopsis* came from the Indian Ocean (Sai Sastry & Chandramohan, 1989), while a more recent report, corroborated with photographs, comes from the Australian coast (Bayha, pers. obs.; Fig. 1). These reports are early indicators of *Mnemiopsis* presence in those regions but do not include biomass distributions or ecological interactions.

Genetic and physiological identity of *Mnemiopsis leidyi* global distributions

Knowledge of the specific identity of *Mnemiopsis* is an important starting point in order to insure that comparisons from different locations involve the same species. However, the taxonomic history of the genus over the past two centuries has been complicated and requires clarification regarding the true species diversity in the genus. Ctenophores closely resembling *Mnemiopsis* along the eastern coastline of the Americas have been described as three different genera: *Mnemia* (Eschscholtz, 1825), *Alcinoe* (Rang, 1828; Mertens, 1833) and *Mnemiopsis* (Agassiz, 1860, 1865; Mayer, 1900). While three *Mnemiopsis* species (*M. gardeni* L. Agassiz, 1860, *M. leidyi*, and *M. mccradyi* Mayer, 1900) are currently taxonomically valid (Cairns et al., 2002), *M. leidyi* and *M. mccradyi* are the only two species recognized in the recent literature (Harbison & Volovik, 1994). The species descriptions indicate that *M. leidyi* occurs north of Charleston,

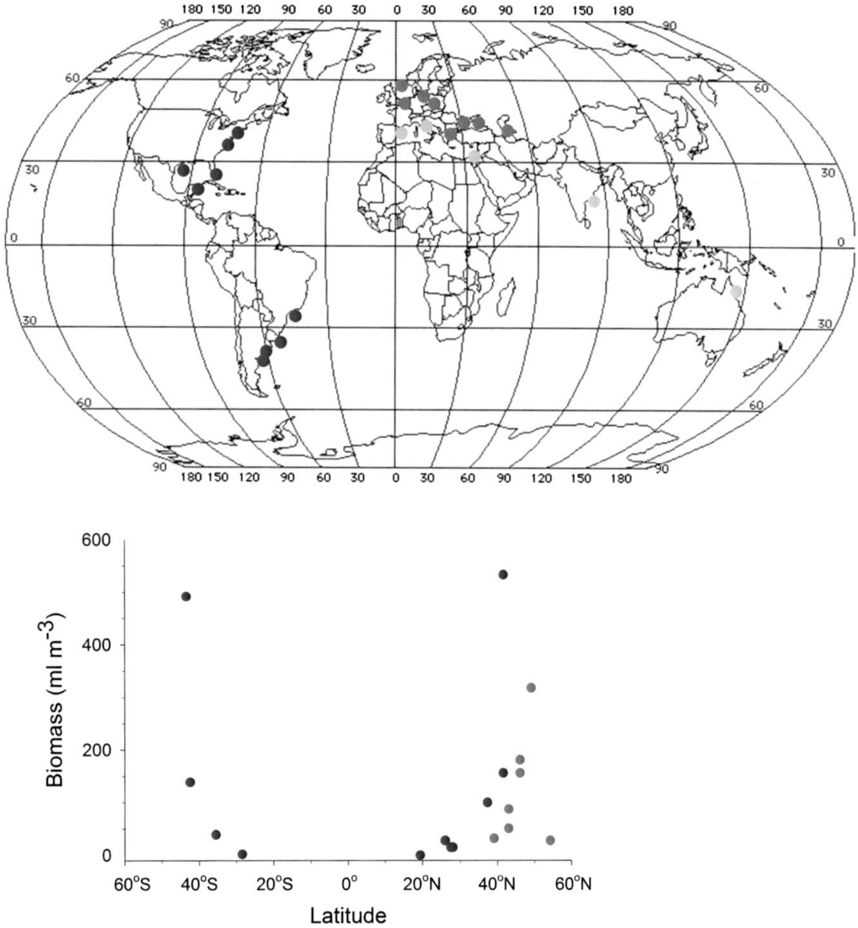

Fig. 1 Contemporary global distribution and average peak biomass of *Mnemiopsis leidyi*. *Top* global distribution with *blue circles* representing native range, *red circles* representing the invasive range. *Orange circles* represent invasive locations for which no ctenophore biomass estimate is available. *Bottom* Latitudinal range of average peak biomass levels. *Colors* represent the same data points as in the *top panel*. (Color figure online)

South Carolina, USA and *M. mccradyi* occurs from there south (Agassiz, 1865; Mayer, 1900); however, the main morphological character delineating the two species, the presence of papillate warts in *M. mccradyi*, was not in the original description (Mayer, 1900), but added later (Mayer, 1912). Harbison & Volovik (1994) and Seravin (1994a, b) effectively encapsulated doubts in the field regarding the establishment of two separate *Mnemiopsis* species, with Seravin (1994a, b) declaring *Mnemiopsis* to be monospecific, albeit-based solely on the examination of invasive animals in the Black Sea. Because the original invasive animals in the Black sea were initially identified alternatively as *M. mccradyi* (Zaika & Sergeeva, 1990) or *M. leidyi* (Vinogradov et al., 1989), the actual identity of ctenophores described in exotic regions has remained problematic. Taxonomic uncertainties in the Baltic sea (Gorokhova et al., 2009; Gorokhova & Lehtiniemi, 2010; Javidpour et al., 2010) have underscored the importance of resolving species identification.

Although taxonomic conclusions based on morphological studies have indicated multiple species of *Mnemiopsis*, none of the genetic studies performed on *Mnemiopsis* to date have revealed evidence of more than one *Mnemiopsis* species (Bayha, 2005; Ghabooli et al., 2010; Reusch et al., 2010). While there is no standard for what degree of genetic variation separates species, in the absence of morphological data or when an animal's morphology renders morphological species delineation questionable, a common practice is to compare the genetic divergence between two specimens with that between recognized species of similar taxa (Schroth et al., 2002; Dawson, 2004). This technique has been especially prevalent with gelatinous zooplankton (Dawson & Jacobs, 2001; Bayha et al., 2004; Miranda et al., 2010) and, for species other than *Mnemiopsis*, genetic studies have indicated significantly greater species diversity than was revealed based on morphology (Dawson & Jacobs, 2001; Bayha et al., 2004; Holland et al., 2004).

As of date, three studies have surveyed sequence divergence in the nuclear ribosomal internal transcribed spacer regions (ITS) to examine species diversity in *Mnemiopsis*. None of those studies found extensive sequence divergence among any of the worldwide populations that would be indicative of multiple species (Bayha, 2005; Ghabooli et al., 2010; Reusch et al., 2010). Given the extremely low divergence found among ctenophores for ribosomal genes (Podar et al., 2001), small divergence values would be expected, but Bayha (2005) showed that values were significantly lower than that found in other ctenophores. Additionally, sequence divergence values in *Mnemiopsis* cytochrome *b* (cyt*b*) also were lower than what is typically seen between invertebrate species, including other ctenophores (Bayha, 2005). In addition to indicating that *Mnemiopsis* is monospecific, all three genetic studies indicated that the invasive populations originated from the NW Atlantic, with the Black/Caspian population(s) from the vicinity of the Gulf of Mexico area (Bayha, 2005; Ghabooli et al., 2010; Reusch et al., 2010) and the northern European populations(s) from the northeastern coast of the USA (Reusch et al., 2010; Ghabooli et al., 2010; Fig. 2). These studies are consistent with the conclusion that only one species of *Mnemiopsis* occurs worldwide and that any morphological differences observed among native or invasive regions can be attributed to phenotypic plasticity.

Physiological evidence is also consistent with a monospecific identity of *Mnemiopsis* occupying diverse geographical regions. Comparisons between species previously described as *M. mccradyi* and *M. leidyi* revealed that physiological rates, including respiration, excretion, egg production, feeding, and growth, were indistinguishable at comparable conditions between the putative species (Kremer, 1994). Just as comparisons between populations within the endemic range of *Mnemiopsis* have yielded similar physiological patterns under comparable conditions, so also have comparisons between endemic and invasive populations indicated similar physiological patterns under comparable conditions. As a consequence, physiological traits of both endemic and invasive *Mnemiopsis* populations overlap to the extent that they are indistinguishable (Purcell et al., 2001). These results suggest that the phenological and ecological variations found between regions reflect flexible responses of one species to a range of environmental conditions. Importantly, a monospecific view of *Mnemiopsis* allows evaluation of data from variable locations to be used for examination of broad patterns within a flexible, but single, species.

Population dynamics: the source-sink perspective

Mnemiopsis reaches its maximal biomass and ecological impact in temperate latitudes. Within its native range along the North and South American Atlantic coasts, the average values of peak seasonal biomass increase with latitude until the middle 40° latitudes both north and south of the equator (Fig. 1). In these regions, *Mnemiopsis* seasonally dominates the planktonic biomass (N. America—Deason, 1982; Condon & Steinberg, 2008; S. America—Mianzan & Guerrero, 2000) and planktonic community structure (Deason & Smayda, 1982; Purcell & Decker, 2005; Sullivan et al., 2008). Likewise, *Mnemiopsis* can dominate temperate planktonic communities within its invasive range (Purcell et al., 2001; Finenko et al., 2006; Shiganova et al., 2001a, b).

Temperate coastal regions vary seasonally between warm spring–fall periods capable of supporting extensive *Mnemiopsis* biomass, and cold winter periods when *Mnemiopsis* is unable to reproduce. During the non-reproductive winter months, *Mnemiopsis* populations cannot replace losses to advective flows with the result that local circulation patterns can flush *Mnemiopsis* populations from large portions of its coastal habitat. Because retention times of coast water systems are often of much shorter duration than *Mnemiopsis* winter non-reproductive periods (Costello et al., 2006a), the winter months can cause local disappearance of *Mnemiopsis* over extensive areas of its temperate range. Seasonal elimination from areas has important implications for *Mnemiopsis* distributions because *Mnemiopsis* is a holoplanktonic species (Fig. 3) with no known benthic resting eggs, cysts, or specialized overwintering stages (Hyman, 1940; Brusca & Brusca, 2003). This contrasts with many coastal jellyfish and copepods (e.g., Sullivan & MacManus, 1986; Marcus & Boero, 1998), which possess either benthic resting eggs or life stages that allow species persistence during periods when the

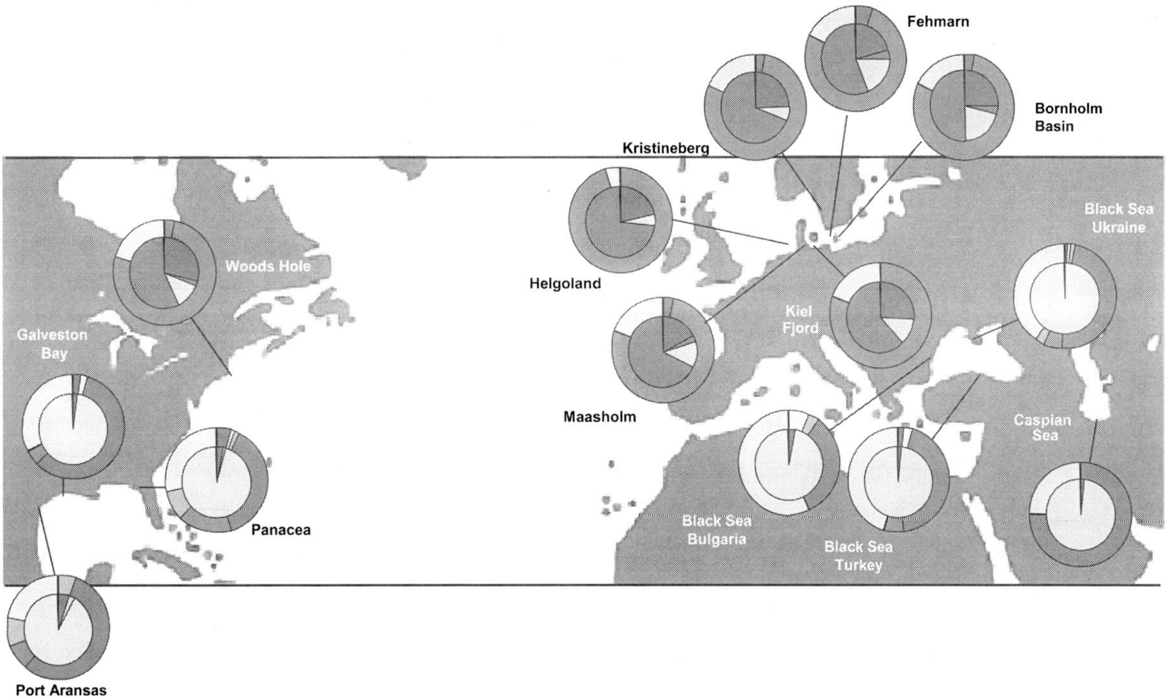

Fig. 2 Sampling locations of *Mnemiopsis leidyi* in their native distribution range along the North American Coast and in exotic locations within Eurasia. *Pie-diagrams* depict allele frequencies of two microsatellite loci that display five alleles (*inner circle*, MnleC1583) and seven alleles (MnleL13, *outer circle*), respectively. Note the overlap in the common alleles which suggests that both gene pools are not completely separated (from Reusch et al., 2010)

adult members of the species are not present in the water column. By contrast, overwintering *Mnemiopsis* populations persist in low advection regions, such as coastal embayments characterized by low water exchange rates with surrounding areas (Costello et al., 2006a). These regions maintain persistent *Mnemiopsis* populations and are termed source regions (see Hanski (1999) for a discussion of metapopulation dynamics). When favorable temperature and feeding conditions arise during the temperate-zone spring, these overwintering refugia serve as sources for *Mnemiopsis* inocula that seed population growth throughout non-overwintering areas. The latter non-overwintering areas are termed sinks because they do not harbor persistent, reproducing populations. Instead, sink areas are characterized by a regular, annual pattern of local *Mnemiopsis* elimination and require re-inoculation each year to initiate seasonal growth (Fig. 4). Local currents provide the transport mechanism from source to sink regions and the seasonal expansion from source regions, evident as both the rate of *Mnemiopsis* distribution changes and the location of primary reproduction, follows predominant local circulation patterns (Kremer & Nixon, 1976; Deason, 1982; Condon & Steinberg, 2008). The expansion from source regions during favorable environmental periods may encompass multiple sink

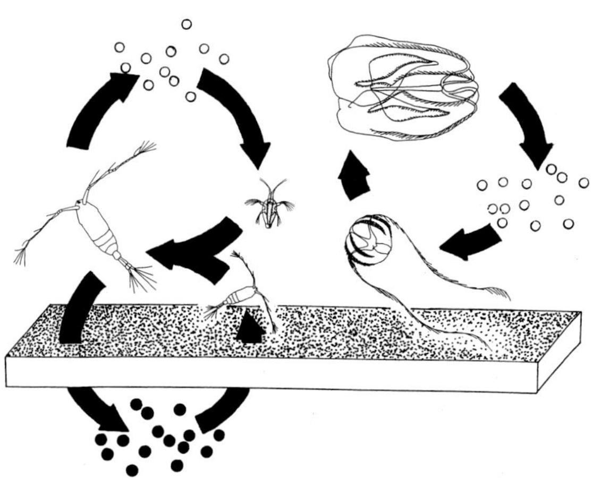

Fig. 3 Variations in life histories of holoplanktonic genera. *Left* the genus *Acartia* (*black circles* represent resting eggs, *clear circles* represent planktonic eggs). *Right* the genus *Mnemiopsis* (only planktonic eggs)

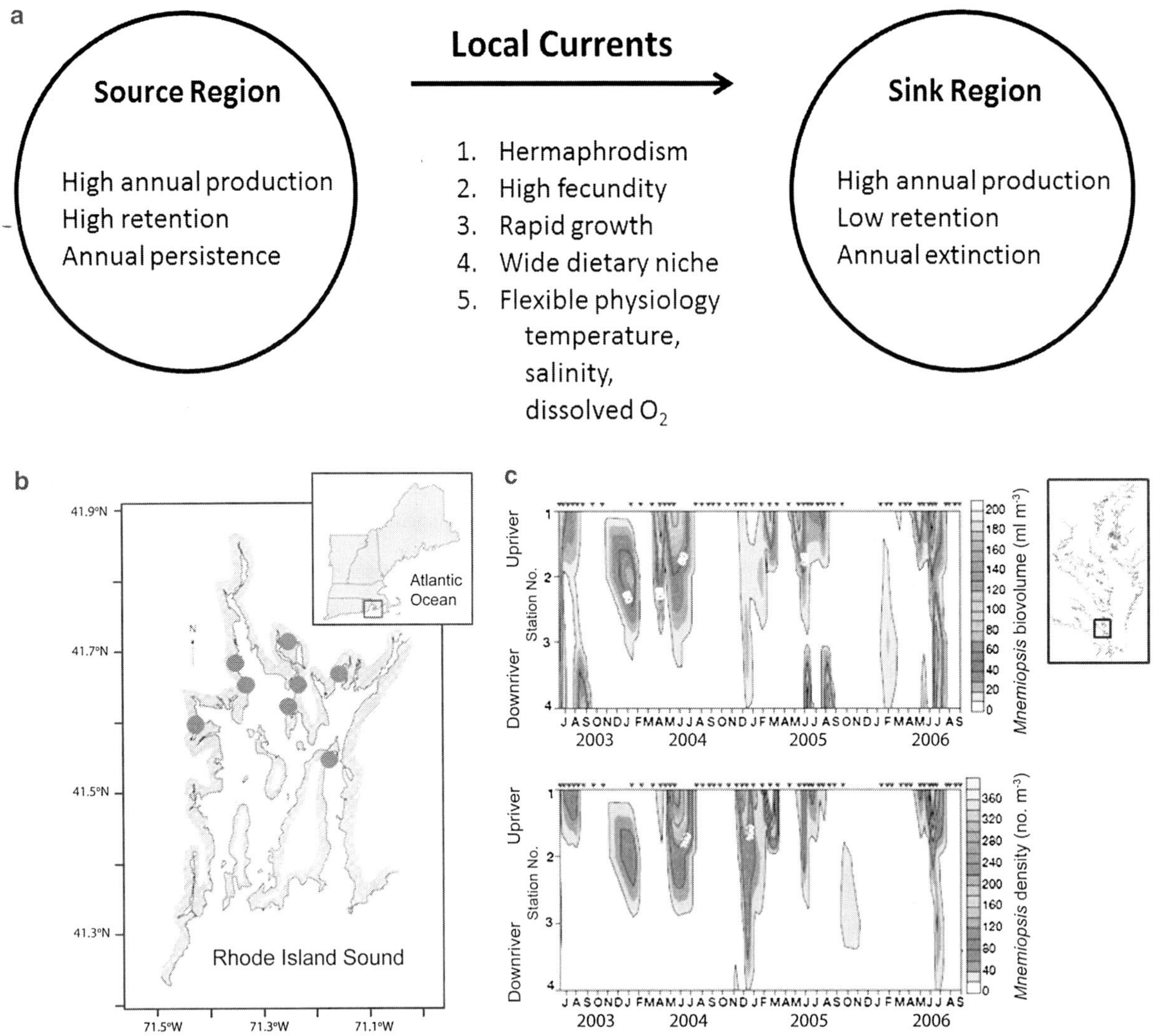

Fig. 4 *Mnemiopsis leidyi* metapopulation patterns. **a** Relationships between source regions and the factors enabling population growth in sink regions following dispersal via local currents. **b** Source regions in the native locale of Narragansett Bay, Rhode Island, USA. *Red dots* indicate embayments bordering the main Bay which can serve as source regions. *Mnemiopsis* populations persisted within these embayments throughout the winter of 2002 while no ctenophores were found within the central Bay during the same time period (Costello et al., unpublished data). **c** Seasonal source-sink distribution patterns within one embayment area, the York River estuary of Chesapeake Bay, USA (site location illustrated by inset panel). *Top* contour plots of average *Mnemiopsis* biomass (biovolume, ml ctenophore m^{-3}) and *bottom* density (no. ctenophores m^{-3}) for four stations plotted along an up-downstream gradient in the York River estuary. Note the predominance of *Mnemiopsis* presence upstream with periodic extensions of ctenophore biomass downstream, toward the estuary's convergence with the greater Chesapeake Bay (from Condon & Steinberg, 2008). (Color figure online)

regions and a large proportion of the total metapopulation range can exist in sink habitats if the source regions are sufficiently productive to subsidize the larger sink regions. The result of these interactions is a dynamic population distribution pattern involving seasonal refugia, current-driven dispersal, rapid population expansion and decline—all occurring within a mosaic of coastal source and sink areas. *Mnemiopsis* has achieved its greatest productivity levels (Fig. 1) with these source-sink population patterns. The dynamic nature of these population fluctuations favor life history traits that enable rapid conversion of plankton to ctenophore biomass under a range of local conditions occurring within temperate, coastal habitats. The traits evolved by *Mnemiopsis* that have allowed it to successfully navigate these population dynamics affect the ctenophore's success in both native and exotic regions. This leads us to a question—which traits promote, and which constrain, this dynamic life history pattern?

Characteristics promoting the success of *Mnemiopsis* as an invader

Population growth capacities

Mnemiopsis has evolved a suite of life history traits enabling rapid population growth. Simultaneous, self-compatible hermaphrodism (Pianka, 1974; Reeve & Walter, 1978) permits production of fertile larvae by all egg-producing members of the population. Fecundity can be high—frequently in excess of 2,000 eggs ctenophore day^{-1} (Costello et al., 2006a) and as high as 12,000 eggs ctenophore day^{-1} for *Mnemiopsis* taken directly from Narragansett Bay (Baker & Reeve, 1974; Kremer, 1976). Egg production rates of similar magnitude are reported from the Black (Purcell et al., 2001) and Caspian (Finenko et al., 2006) seas. When grown under favorable temperature (15–30°C) and food (>25 μg C l^{-1}) conditions, larvae are characterized by high ratios of growth to metabolism (>2) and high-gross growth efficiencies (>30%) that permit rapid development (Kremer & Reeve, 1989). Generation times can be short and, at favorable temperatures and food levels, eggs can hatch and develop into reproducing adults within 14 days (Reeve & Walter, 1978). High fecundity, rapid growth, and short-generation times are common for colonizing species (Funk & Vitousik, 2007) and important components of the metapopulation dynamics underlying high-biomass production of *Mnemiopsis* in temperate regions (Fig. 4).

Broad physiological tolerance levels

Mnemiopsis has broad physiological tolerances to temperature, salinity, and dissolved oxygen (DO) levels (Purcell et al., 2001; Table 1). Nevertheless, few organisms are unaffected by alterations in physical regime and each physical factor plays an important role modifying *Mnemiopsis* distribution and abundance patterns.

The capacity to tolerate temperatures between 0 and 32°C permits *Mnemiopsis* to occupy a diverse geographical range that includes temperate through tropical marine communities (Harbison et al., 1978; Mianzan, 1999); however, within these broad temperature limits, temperature thresholds affect *Mnemiopsis* population dynamics. Most generally, the upper and lower temperature tolerances determine survival of individuals within habitats. There has been little research on the upper temperature limits of *Mnemiopsis*; however, it occurs in native habitats and now in eastern Mediterranean waters where summer temperatures reach 32°C (Table 1). Importantly, the Q_{10} estimated when temperatures changed seasonally (1.3) were much lower than those determined when temperatures were changed in the laboratory (≥3.4) (Purcell, 2009), indicating considerable physiological flexibility within temperature variations characterizing field distributions of *Mnemiopsis*.

The lower temperature limit for *Mnemiopsis* persistence appears to be around freezing. The precise level of the survival temperature threshold varies by region and may depend upon salinity levels. In Narragansett Bay, USA salinities varied between 22 and 33 and *Mnemiopsis* was collected from waters as low as −1°C by breaking holes in surface ice (Costello et al., 2006a); however, at lower salinities in the shallow sea of Azov (surface salinity 0–14), *Mnemiopsis* may not survive below ~4°C (Purcell et al., 2001). Similarly, in the northern Caspian Sea, *Mnemiopsis* cannot survive when salinity is lower than 4 (Shiganova et al., 2004b). *Mnemiopsis* populations disappear in the Sea of Azov when water temperatures become colder than 3°C (Shiganova et al., 2001b, 2003). These reports indicate that low-salinity levels can adversely impact winter survival of *Mnemiopsis* populations.

A second temperature threshold directly affects *Mnemiopsis* population growth—the reproductive temperature threshold. Purcell et al. (2001) reported egg production of *Mnemiopsis* from Chesapeake Bay to occur between the temperatures of 12–29°C and results from Costello et al. (2006a) in Narragansett Bay broadly match those results (Fig. 5), with minor egg release at temperatures as low as 6°C. Conservatively, we expect that 10°C is an approximate lower temperature threshold for successful egg production by a developing *Mnemiopsis* population and egg production rates increase with higher temperatures,

Table 1 Comparison of systems over the native and invasive range of *Mnemiopsis leidyi*

Location	Native or invaded (year)	Temp. (°C)	Salinity	Predators	Zooplankton biomass or density (mg C or no. m^{-3})		Ctenophore biomass or density (mg C, ml WW, or no. m^{-3})			Reference
					Peak season	Range	Peak (present)	Range	Years (No.)	
Narragansett Bay, RI	Native	1–25	25–32	*Beroe*	June–July	30–110 C[a]	August–September (Spring–Fall)	6–100 C	>8	Hulsizer (1976), Kremer (1976), Kremer & Nixon (1976), Durbin & Durbin (1981), Deason (1982), Deason & Smayda (1982), Smayda (1988)
Mid Chesapeake Bay, MD	Native	2–30	5–16	Bay group	Summer	30–180 C[b]	June–September (all year)	10–100 C	16	Lonsdale (1981), Olson (1987), Purcell et al. (1994)
Biscayne Bay, FL	Native	18–32	14–45	*Beroe*	Fall to Winter	11 C[c]	Fall (all year)	ND	1+	Baker (1973)
Nueces Estuary, TX	Native	7–31	20–38		Variable	50 C	Summer (all year)	8–20 C	1	Buskey (1993)
Rio de la Plata estuary, ARG	Native	7.5–25	9–24	Several fish species	Spring	37 C	Spring	2–15 C	1	Mianzan et al. (1996), Sorarrain (1998)
Blanca bay, ARG	Native	5–24	24–38	*Beroe*	Spring–Summer	40,000 m^{-3}	Spring and Fall	ND	3	Mianzan & Sabatini (1985), Mianzan (1986)
Nord Patagonic Tidal front, ARG	Native	10–16	33	Several fish species			Summer	140 ml	1	Mianzan et al. (1996, 2010), Mianzan pers. Obs
Black Sea (before *Beroe* arrival)	Invaded 1982	0–27	12–22	*Beroe*	March–May; July–August	33,000 m^{-3} [d] 243–418 mg m^{-3}	All year	0.5–130 C	12	Purcell et al. (2001), Shiganova & Malej (2009)
Sea of Azov (before *Beroe* arrival)	Invaded 1988	−0.8 to 30	0–14	*Beroe*	May–June; July–August	350–390 mg m^{-3}	Spring–Fall	67–143 C	12	Shiganova et al. (2001b), Shiganova & Malej (2009)
Northern Caspian	Invaded 1999	0–28	0.1–11	No indigenous gelatinous predators	April–May; July–August	289 ± 296 mg WW m^{-3}	August–November	0.32–105 C	7–8	Shiganova et al. (2004b)
Middle		0–25	12.6–13			37.6 ± 58 mg WW m^{-3}	June–November			
Southern Caspian		10–30	12.6–13			66 ± 72 mg WW m^{-3}	All year			
Sea of Marmara	Invaded	8–29	18–29	*Beroe*	July; September–October		All year		11	Shiganova et al. (2001b)

Table 1 continued

Location	Native or invaded (year)	Temp. (°C)	Salinity	Predators	Zooplankton biomass or density (mg C or no. m^{-3})		Ctenophore biomass or density (mg C, ml WW, or no. m^{-3})			Reference
					Peak season	Range	Peak (present)	Range	Years (No.)	
Aegean Sea	Invaded 1990	13–29	33–40	Med group	Spring; Summer	0.8–6 mg DW m^{-3}	All year	0.1–20	11	Shiganova et al. (2004a)
Turkey (Mediterranean coast)	Invaded 1992	23–24	~38	Med group		0.02 ml m^{-3}	ND			Kideys & Niermann (1994)
Gulf of Trieste	Invaded 2005	10–26	32–38	Med group		3.6–9 C	Not established			Shiganova & Malej (2009)
France	Invaded 2006	?–31.5	39.5	Med group			ND		4	Shiganova & Malej (2009)
Catalan Sea	Invaded 2009	12–26	37–39	Med group	Spring; Fall	500–8000 m^{-3} [d]	Spring			Fuentes et al. (2010)
Italy	Invaded 2009	13–26	37.5–38	Med group		500–4000 m^{-3} [d]	ND			Boero et al. (2009)
Israel	Invaded 2009	16–32	39–40	Med group		0–2598 m^{-3}	ND			Fuentes et al. (2010)
North Sea Helgoland	Invaded 2006	7–14	~36	N Sea group	Spring–Summer	<40,000 m^{-3} [d]	ND	<0.3	3	Greve et al. (2004), Hamer et al. (2011)
Baltic Sea Limfjorden	Invaded 2006	5–15	32–19	N Sea group	Summer	<250 C	Summer	<80 C	3	Riisgård et al. (2012), Javidpour et al. (2009b), Schaber et al. (2011a)
Kiel Bight		5–17	11–22	N Sea group	Spring	<2 C	Spring–Fall (all year)	<75 C		

Bay Group = *Chrysaora quinquecirrha*, *Beroe ovata*, *Cyanea capillata* (Linnaeus)

Med. group = *Beroe cucumis* Fabricius, *Beroe forskalii* Milne Edwards, *Chrysaora hysoscella* Eschscholtz, *Pelagia noctiluca* (Forsskål) *Aequorea forskalea* Peron & Lesueur

N Sea group = *B. ovata*, *C. capillata*, *C. hysoscella*, *Aequorea vitrina* Gosse

C carbon, *ND* no data

[a] >153 μm fraction, assuming C = 35%DW, or 1 ml displacement volume = 60 mg C (Kremer 1994)

[b] Converted from counts assuming 3 μg C per copepodite or adult (Kremer 1994)

[c] >202 μm fraction, assuming C = 35%DW (Kremer 1994)

[d] Before *Mnemiopsis*

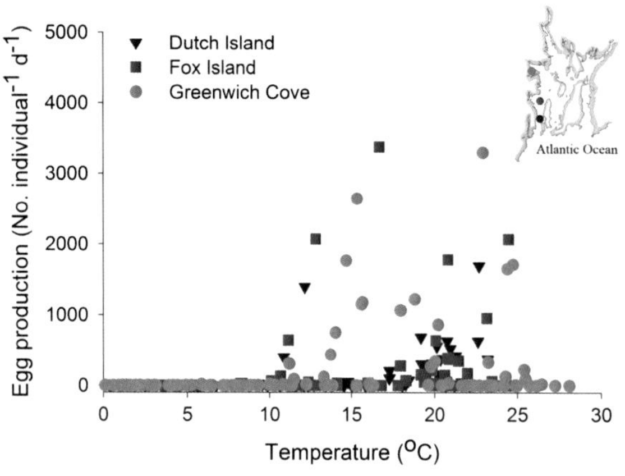

Fig. 5 Egg production by the ctenophore *Mnemiopsis leidyi* in Narragansett Bay during weekly sampling between the years 2001–2003. *Circles* of *different color* represent sites of *similar color* on the station map of Narragansett Bay, *upper right corner of figure* (from Costello et al., 2006a)

with maxima occurring between 15 and 30°C (Fig. 5; Purcell et al., 2001). These data also suggest several important relationships between temperature and *Mnemiopsis* population growth. First, optimal temperatures are a necessary but insufficient condition for *Mnemiopsis* population growth. Temperatures in the reproductive range of *Mnemiopsis* alone are not effective predictors of population growth; many sampling dates with adequate temperatures supported little or no egg production by *Mnemiopsis* field populations (Fig. 5; Purcell et al., 2001). Likewise, many regions with favorable temperature regimes in the subtropics and tropics do not generally support high *Mnemiopsis* biomass levels (Table 1; Fig. 1). Instead, favorable temperature levels may be viewed as a condition that permits high-population growth, but only when combined with sufficient prey concentrations and limited predation pressure (Kremer, 1994; Purcell et al., 2001). Second, during several months of the year, temperatures of temperate zone waters are below the reproductive threshold for *Mnemiopsis*. As noted previously, this affects annual distribution patterns and overwintering survival of *Mnemiopsis* populations. Third, climate change is altering the annual duration of this overwintering period in temperate waters. The number of days per year that are too cold for *Mnemiopsis* reproduction has decreased in recent years (Fig. 6a), and the coastal areas most affected by this climactic trend are inshore embayments that serve as *Mnemiopsis* source regions (Fig. 6b). One result of this trend is that *Mnemiopsis* population growth may now often begin earlier and persist longer on a seasonal basis in temperate coastal systems than during previously recorded periods (Costello et al., 2006b; Condon & Steinberg, 2008).

Mnemiopsis also has extremely wide salinity tolerances, from nearly freshwater to hypersaline lagoons (Table 1). A recent physiological study showed *Mnemiopsis* to be a hyper-osmoconformer (Yazdani Foshtomi et al., 2007). Its broad salinity tolerance has several important effects. First, it created confusion about identification of *M. leidyi*, which generally lacks warts in low-salinity environments (*M. leidyi*) but is firmer-bodied and generally has warts in high-salinity environments (mistakenly called *M. mccradyi*). Second, because dry weights (DWs) of *Mnemiopsis* reflect the salinity of its environment, physiological rates standardized by DW can appear to differ widely among habitats; thus, standardization by DW should be avoided and salinities should always be reported (Purcell, 2009). Third, its wide salinity tolerance allows the ctenophores to extend from offshore regions into embayments that experience wide fluctuations influenced by rain and runoff (Table 1; Kremer, 1994; Purcell et al., 2001). These low-salinity habitats serve as important refuges from less-euryhaline predators, such as *Chrysaora quinquecirrha* Desor, 1848 and *Beroe* spp. Gronov, 1760 (Purcell et al., 2001). This physiological flexibility has led to a perception that *Mnemiopsis* populations are not constrained by salinity variations (Reeve et al., 1989; Kremer, 1994; Purcell & Decker, 2005). However, although *Mnemiopsis* has wide salinity tolerances, low salinities can lead to reduced low-temperature survival, smaller maximum body size (Purcell et al., 2001), and decreased reproductive success (C. Jaspers, pers. comm.).

The capacity to function over a wide range of DO concentrations is an additional physiological trait with important adaptive advantages for *Mnemiopsis*. Low DO concentrations generally occur in shallow marine systems during summer months when water column stratification limits mixing and aeration of bottom waters. A variety of coastal mesoplankton are adversely affected by low O_2 levels (<3 mg O_2 l^{-1}), but *Mnemiopsis* is tolerant of low DO levels. *Mnemiopsis* feeding rates on copepods are undiminished at low DO levels and such large, lobate ctenophores actually experience elevated clearance rates in low DO

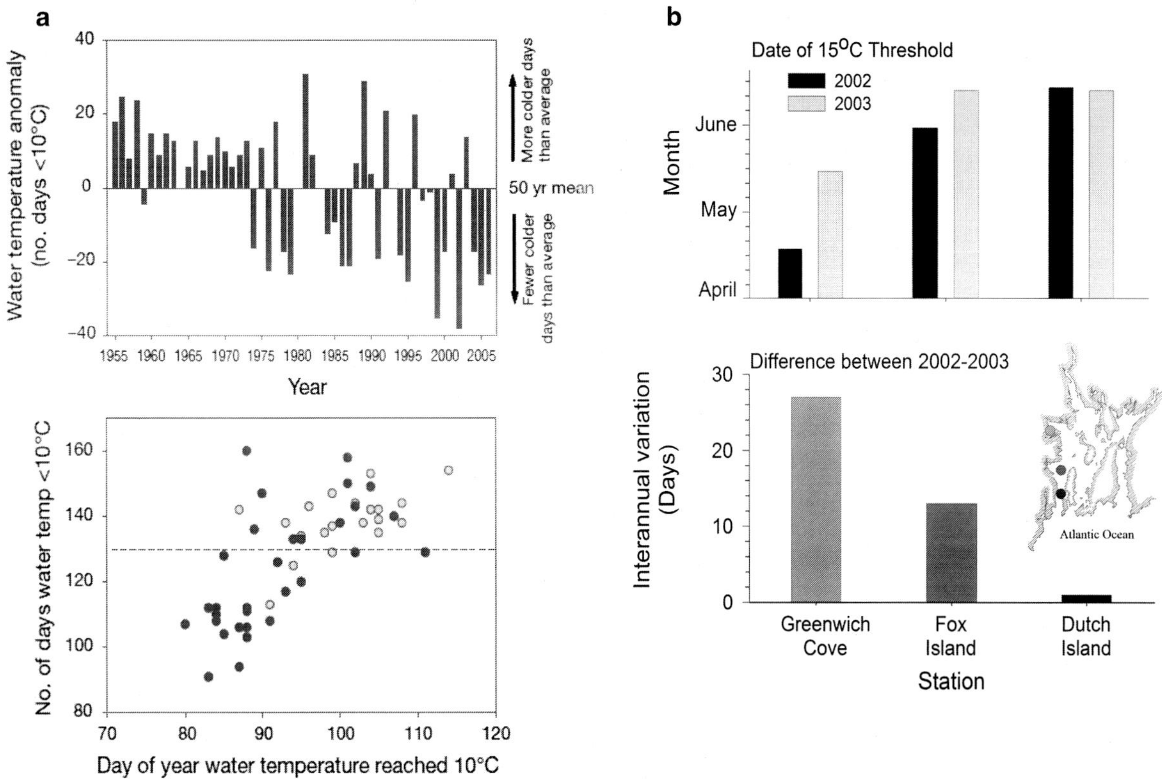

Fig. 6 The impact of climate change on threshold temperatures for *Mnemiopsis leidyi* population growth in North American, temperate habitats in the ctenophore's native range. **a** Warming temperatures in the York River estuary of Chesapeake Bay. *Upper left panel* comparison of water temperature anomaly from 1955 to 2006 against 50 year mean. York River water temperature anomaly was defined as the number of days per year winter–spring water temperatures were <10°C, minus the 50 year annual mean. Negative anomalies reflect increased water temperatures over the winter–spring period. *Lower left panel* Frequency of cold days (<10°C) in relation to when water temperature increased to and remained above the 10°C threshold (*x*-axis). Note the recent trend toward years with earlier warming and consequently fewer cold (<10°C), non-reproductive days. *Grey circle* 1955–1974; *black circle* 1975–2006; *dotted line* 50 year mean of the water temperature anomaly (from Condon & Steinberg, 2008). **b** Amplification of temperature warming within shallow embayments of Narragansett Bay, USA. *Upper right panel* date at which the 15°C threshold was reached during the spring months of 2002 and 2003 for three stations in Narragansett Bay (station locations illustrated by colored points of inset map). *Lower right panel* the advance, in days, of the 15°C threshold in the warm spring of 2002 relative to the colder spring of 2003. Note that whereas the timing of warming at the seaward-most station is relatively unaffected between years, the shallow embayment at Greenwich Cove is strongly affected. Greenwich Cove is a *Mnemiopsis* source population location (from Costello et al., 2006a)

conditions (Decker et al., 2004). Large (but not small) *Mnemiopsis* had lower growth and egg production in low DO concentrations (1.5 and 2.5 mg O_2 l^{-1}) than in saturated DO (Grove & Breitburg, 2005). Clearance rates of *Mnemiopsis* on fish eggs and larvae were the same at low- and high-DO concentrations (1.5 and 7.0 mg O_2 l^{-1}), and ctenophore densities were high in the bottom layer even in low DO (Kolesar et al., 2010). Tolerance to low DO levels provides *Mnemiopsis* a predatory advantage over prey experiencing impaired escape performance in low DO and a competitive advantage over zooplanktivorous fish with similar diets and higher sensitivity to hypoxia (Purcell et al., 2007). Thus, tolerance of hypoxia is a beneficial trait that enables *Mnemiopsis* to inhabit highly eutrophic coastal habitats.

Wide dietary niche

Dietary flexibility allows *Mnemiopsis* to exploit a variety of planktonic food sources, including microplankton, mesozooplankton, and ichthyoplankton, in environments characterized by diverse assemblages. The annual population growth cycle of *Mnemiopsis* in temperate waters of its native range involves transitions between regions characterized by different spectra of available prey. For example, in Narragansett Bay, USA, overwintering embayments are often highly productive environments (Fig. 7a), with more diverse metazoan planktonic assemblages than the central Bay regions. Whereas copepods typically dominate the mesozooplankton assemblages in the more central Bay waters (Fig. 7b), a variety of invertebrate larvae and other groups (e.g., molluscs, barnacles, polychaetes, ascidian larvae, rotifers) can numerically predominate in shallow embayments so that copepods may be a minority of prey encountered (Fig. 7c) by ctenophores in these embayments. The flexible feeding capacity of *Mnemiopsis* allows it to successfully exploit the variety of prey environments encountered during the regular annual population expansion cycle from embayments to the central Bay. A result of this dietary flexibility is that *Mnemiopsis* ingestion patterns vary widely depending upon the available prey and, consequently, these variations are reflected in the literature on *Mnemiopsis* (Table 2). Although characteristic of *Mnemiopsis* feeding patterns in its native environment, dietary flexibility is also an essential trait associated with invasive success by introduced species (Caut et al., 2008; Zhang et al., 2010).

The dietary breadth of *Mnemiopsis* is, however, life-stage dependent. Eggs are small, about 0.3 mm in diameter and the cydippid larval stage that hatches from an egg is of similar small dimensions and possesses delicate tentacles (Fig. 8) used for prey capture. The small size and low organic structure of newly hatched larvae render their tentacles vulnerable to physical damage during encounters with larger, more powerful metazoan prey (Greve, 1977; Stanlaw et al., 1981). Although all sizes of cydippid larvae are capable of capturing nauplii of the copepod *Acartia tonsa*, encounters of cydippid larvae less than 0.65 mm diameter with *A. tonsa* nauplii (NI–NII) often result in loss of the delicate cydippid tentacles. For small larvae (0.3–2.0 mm diameter), retention of nauplii was related to cydippid diameter (Waggett & Sullivan, 2006). Larger than 2.0 diameter, *Mnemiopsis* larvae retain copepod nauplii effectively (~90%) and larvae >2.5 mm retained >60% of *A. tonsa* copepodites. During the earliest cydippid stages, ingestion of a wide array of protists including both autotrophic and heterotrophic prey—diatoms,

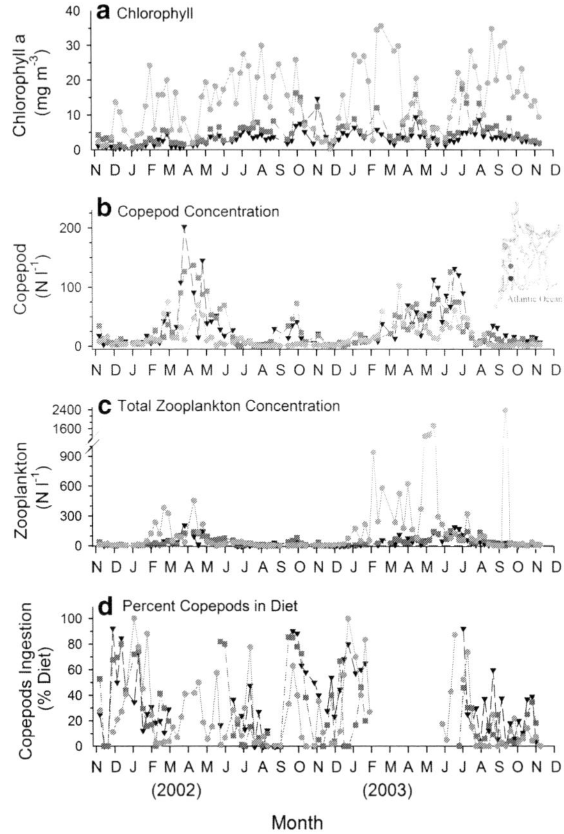

Fig. 7 Variation in plankton community composition and diet of *Mnemiopsis leidyi* at three locations (locations shown by colors within *inset* of panel **b**) during 2 years in Narragansett Bay, USA. The three sites possessed different **a** chlorophyll biomass, **b** different relative proportions of copepods and **c** other types of zooplankton. Note that diet composition, measured as the (**d**) proportion of the diet comprised of copepods, often varied between sites within Narragansett Bay on the same sample dates. Samples from different sites were taken within a 4 h period on each sample date (**a**–**c** from Costello et al., 2006a; **d** from Costello et al., unpublished)

dinoflagellates, euglenoids, aloricate, and tintinnid ciliates—are an important nutritional resource for *Mnemiopsis* (Sullivan & Gifford, 2004). These prey do not mechanically damage the cydippid larvae and provide a safe nutritional alternative to more powerful metazoan plankton. Immediately upon hatching, cydippid larvae begin consuming protistan microplankton. Protistan microplankton at densities representative of temperate coastal waters can provide sufficient nutrition for growth up to approximately 5 mm in diameter (Sullivan & Gifford, 2007), at which size *Mnemiopsis* begins the morphological transition to the lobate phase (Rapoza et al., 2005).

Table 2 *Mnemiopsis leidyi* prey ingestion based on in situ gut contents from various geographical locations

Site	Dominant prey	Reference
Indian River estuary, FL, USA	Copepod nauplii, barnacle nauplii, mollusc veligers, *Acartia* sp. adults & copepodites, *Oithona* sp.	Larson (1987)
Narragansett Bay, RI, USA	Copepod nauplii, *Acartia* sp. adults & copepodites, mollusc veligers, barnacle nauplii, rotifers	Newton et al., (2009)
Barnegat Bay, NJ, USA	Mollusc larvae, invertebrate larvae	Nelson (1925)
Woods Hole, MA, USA	*Acartia* sp. adults & copepodites, copepod nauplii, mollusc larvae, cladocera (*Penilia* sp.)	Rapoza et al. (2005)
Northern Black Sea, summer	Copepods (mainly *Acartia* sp. and *Calanus* sp.), *Penilia* sp., copepod nauplii, mollusc veligers, barnacle cyprids	Tzikhon-Lukanina et al. (1991)
Northern Black Sea, summer	Cladocerans, mollusc veligers, copepods, appendicularians, tintinnids, cyprid, gastropod and polychaete larvae, fish eggs	Zaika & Revkov (1998)
Southern Black Sea, summer	Copepods (mainly *Acartia* sp. and *Calanus* sp.), *Oithona* sp., *Pseudocalanus* sp., *Paracalanus* sp.	Mutlu (1999)
Southern Black Sea, winter	Copepods (*Acartia* sp., *Pseudocalanus* sp., *Calanus* sp., *Oithona* sp., mollusc larvae	Mutlu (1999)
Kiel Bight, Baltic Sea	Barnacle nauplii, copepods (*Acartia* sp., *Pseudocalanus* sp.), cladocera, scyphozoan planula larvae, ctenophore larvae	Javidpour et al. (2009a)
Gullmar Fjord, Baltic Sea	Tintinnids, appendicularians, *Penilia* sp., *Acartia* sp., copepodites, copepod nauplii, mollusc veligers, dinoflagellates, *Sagitta* sp.,	Granhag et al. (2011)

Capture and ingestion of protists continues throughout development of the lobate stage, but the transition from cydippid to lobate forms entails a dramatic broadening in diversity of metazoan prey consumed (Fig. 9). Small prey continue to be ingested by lobate *Mnemiopsis*, but probably are not a substantial nutritional source for larger lobate stages (Stanlaw et al., 1981). Although ingestion of metazoan prey has been the primary focus of *Mnemiopsis* trophic impacts (e.g., Table 2), microzooplanktonic feeding by *Mnemiopsis*, particularly larvae, may play an important role in microzooplankton dynamics when larvae are abundant (Stoecker et al., 1987; Sullivan & Gifford, 2004). Unfortunately, the short digestion times (2–6 min for copepod nauplii, <2 min for aloricate ciliates; Sullivan, 2010) make ctenophore feeding on microzooplankton difficult to quantify using conventional in situ gut content methods.

The wide dietary breadth of lobate stage *Mnemiopsis* is based on structurally simple but functionally complex feeding mechanisms. The simple component of the feeding system is the structural basis of encounter with prey. Prey entrained within a relatively uniform, laminar feeding current (Fig. 10) that provides transport to two major capture surfaces—the tentillae and the inside surfaces of the oral lobes. The functional elegance of the *Mnemiopsis* feeding system relies upon the matching of feeding current hydrodynamic traits with the sensory systems of zooplankton prey. Slowly swimming and low mobility prey (e.g., many larvae of crustaceans, molluscs, invertebrates, and fish eggs) are simply entrained in the feeding current and collected by sticky colloblasts on the tentillae (Waggett & Costello, 1999; Waggett & Buskey, 2006) and transported to the mouth via ciliary currents within a food groove (Main, 1928; Moss et al., 2004). Larger prey with sophisticated sensory and escape capabilities (e.g., copepods and fish larvae) are also entrained in the flows but are usually unresponsive to transport within these flows until the prey are surrounded by the ctenophore's lobes and escape probabilities are greatly reduced (Costello et al., 1999). The inability of prey to detect their transport prior to capture is due to the low-shear profiles of the feeding current flows generated by *Mnemiopsis* (Fig. 11). Prey, such as copepods, do eventually detect the presence of stronger shear gradients adjacent to the ctenophore's body, but attempted escape by these prey usually entails contact with the sticky inner oral lobe surfaces where the prey are captured by the ctenophore. The unique combination of morphological structure and hydrodynamic stealth allows *Mnemiopsis* a dietary

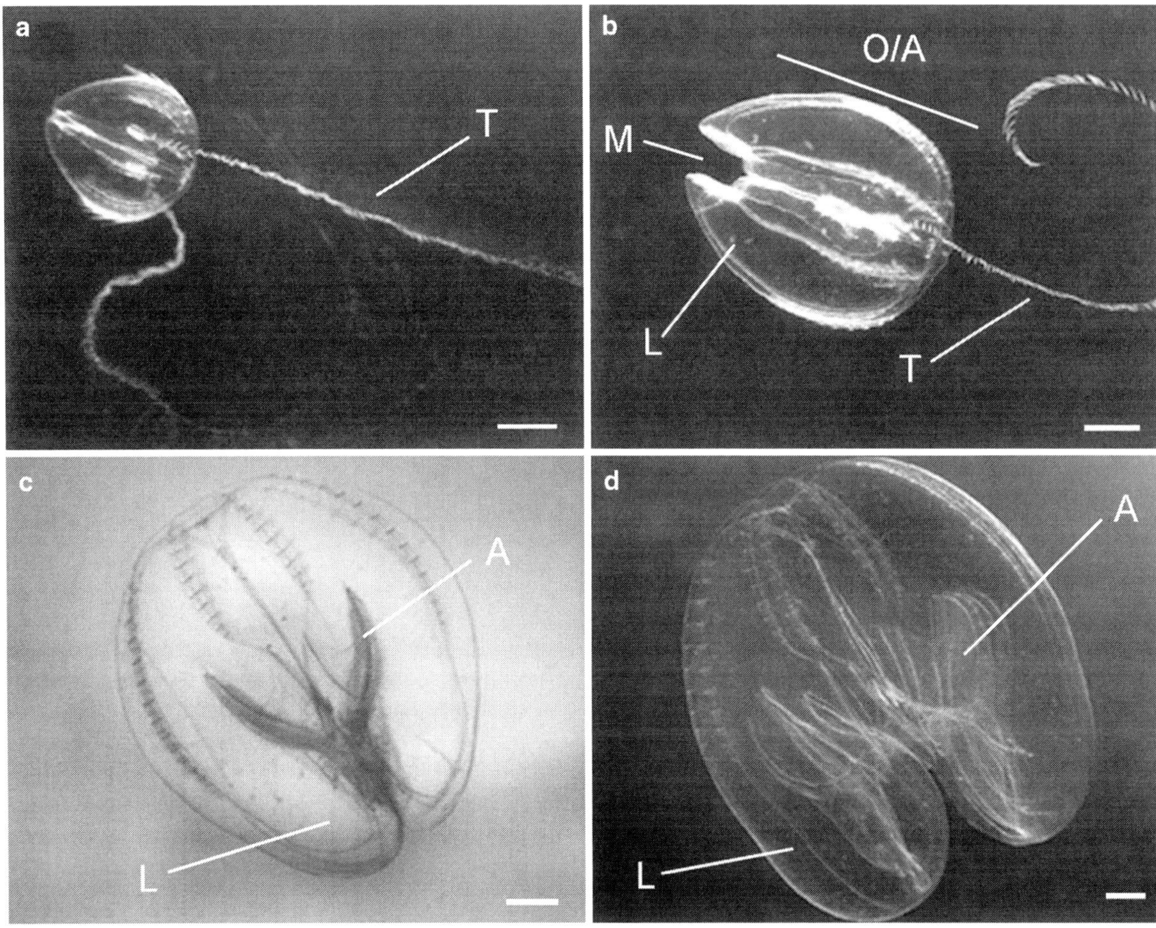

Fig. 8 *Mnemiopsis leidyi* life history stages. **a** Tentaculate-stage cydippid larva with trailing tentacles (T). **b** Transition-stage larva with tentacles and small oral lobes (L). Only one of the two tentacles is in focus. *O/A* oral–aboral axis; *M* mouth. **c** Lobate-stage larva, with developing auricles (**a**) and oral lobes. **d** Post-larval *Mnemiopsis* with completely developed auricles and oral lobes. *Scale bars* are 1.0 mm (from Sullivan & Gifford, 2004)

breadth that encompasses a wide portion of the diverse prey spectra it frequently encounters in its native and exotic ranges.

One consequence of the delicate hydrodynamic equilibrium involved in prey capture is that *Mnemiopsis* predation is likely to be highly sensitive to variations in ambient hydrodynamic conditions, such as turbulent mixing. The delicate morphology and very low shear levels observed in the feeding current of *Mnemiopsis* suggests that even low levels of ambient turbulence could potentially interfere with prey entrainment and encounter processes, reducing feeding proficiency. Field data (Fig. 12) suggest that *Mnemiopsis* avoids highly mixed regions when possible and can migrate vertically to minimize exposure to turbulent mixing (Miller, 1974; Costello & Mianzan, 2003; Mianzan et al., 2010). The interactions between mixing processes, ctenophore feeding currents, and prey escape behavior involves a variety of undocumented interactions that, when quantified, may provide insight in variations in predation patterns related to physical conditions.

The combination of effective feeding and rapid growth potential provide *Mnemiopsis* the ability to strongly impact planktonic communities. Early quantitative estimates suggested relatively low average (5–10% day^{-1}) capacities of *Mnemiopsis* to crop copepod standing stocks (Kremer, 1979). However, more recent estimates indicate substantially higher predatory potential (>100% of zooplankton standing stock day^{-1}; Table 3). These higher estimates of predatory potential are consistent with rapid declines

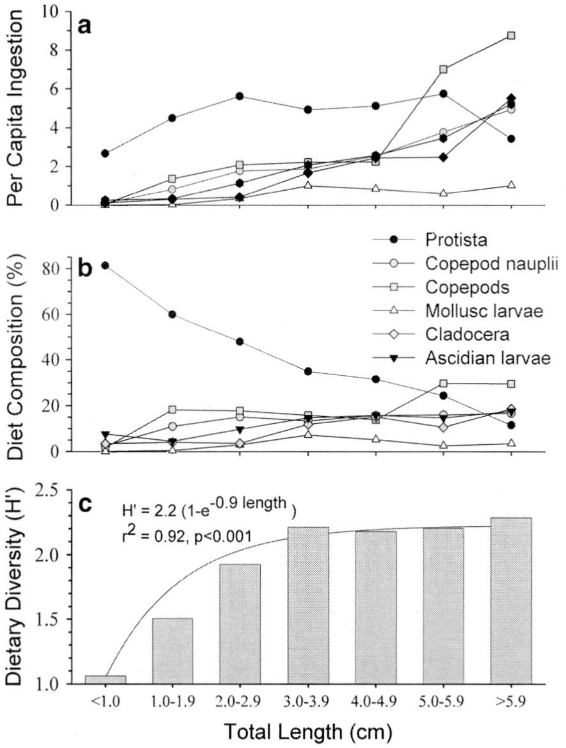

Fig. 9 Dietary patterns in relation to body size of *Mnemiopsis leidyi*. **a** Per capita consumption of different prey categories, **b** relative proportion of diet as reflected by in situ gut contents and **c** the diversity (H') of the diet (from Rapoza et al., 2005)

in zooplankton populations that often accompany increases in *Mnemiopsis* biomass (e.g., Fig. 13). The increasing appreciation of *Mnemiopsis* predatory impacts is related to methodological changes, particularly the use of larger volume experimental feeding containers and field ingestion estimates. For example, experiments with 10–50 mm sized ctenophores in containers ranging from 3.5 to 1,000 l demonstrated that ratios of container volume to ctenophore volume of <2,500:1 resulted in reduced ctenophore clearance rates. Clearance rates were greater in the larger containers and greatest in 1,000-l containers (Purcell, 2009). In addition, feeding rates determined from field gut contents and digestion rates were generally higher than from containers (reviewed in Purcell, 1997). For example, clearance rates of 40-mm-long (~15 g WW) *Mnemiopsis* on *Acartia* sp. copepods estimated from gut contents in the Baltic sea (8.3 l ind.$^{-1}$ h^{-1}) were 4-times those of similarly sized ctenophores measured in 1,000-l containers (Granhag et al., 2011). These high-predation rates make *Mnemiopsis* a competitive threat to fish larvae and zooplanktivorous (forage) fish species when their diets overlap (Darvishi et al., 2004). Indeed, competition for zooplankton prey has been assumed to be the main cause of inverse abundances of ctenophore and forage fish in the Black Sea region (e.g., Purcell et al., 2001; Oguz, 2005; Daskalov et al., 2007; Oguz & Gilbert, 2007; Oguz et al., 2008; Mutlu, 2009).

The role of *Mnemiopsis* as a direct fish predator currently presents a more complex picture. A variety of studies demonstrate direct predation on fish eggs and larvae (reviewed by Purcell & Arai, 2001; Purcell et al., 2001). However, recent studies in the Baltic region describe differing distributions of *Mnemiopsis* relative to fish eggs and larvae (Haslob et al., 2007; Schaber et al., 2011b) and a low clearance rate of the ctenophores on Baltic cod eggs (Jaspers et al., 2011). The variation in these results suggests that behavioral details of ctenophore–prey interactions need greater examination for more complete understanding. These details will help clarify the pathways through which ctenophore predatory impacts cascade through planktonic communities in native (Deason & Smayda, 1982; Purcell & Decker, 2005; Sullivan et al., 2008) and exotic (Shiganova et al., 2004a, b; Finenko et al., 2006; Kideys et al., 2008) habitats.

Constraints on the invasiveness of *Mnemiopsis*

Requirement for high-prey availability

Rapid population growth of *Mnemiopsis* requires high prey ingestion rates. Kremer & Reeve (1989) estimated that a minimum prey biomass of >24 μg C l^{-1} is required to support population growth of field populations. This is roughly an order of magnitude greater than average prey concentrations found in more oligotrophic oceanic waters (Kremer et al., 1986). *Mnemiopsis* is rarely found in waters less than 3 μg C l^{-1} (Kremer, 1994). Instead, a genus with many structural similarities but lower feeding effort at high-prey concentrations, *Bolinopsis* L. Agassiz, 1860, appears to replace *Mnemiopsis* in many less productive waters (Kremer et al., 1986). Reproduction by *Mnemiopsis* is sensitive to food supply and egg production declines within 24 h when feeding stops. No eggs are produced after 3–4 days of starvation (Reeve et al., 1989). The protein-dominated body

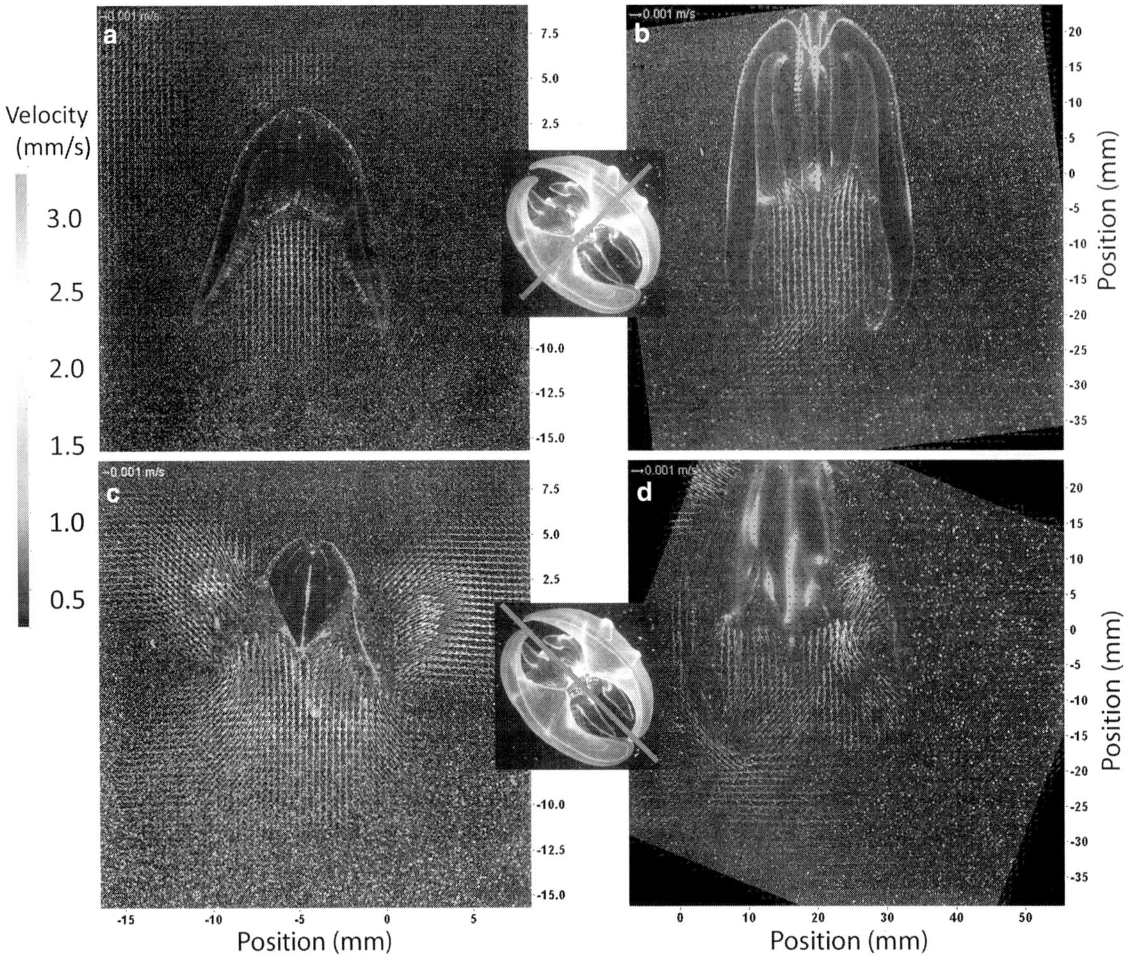

Fig. 10 Representative velocity vector fields around a small (1.3 cm long; **a**, **c**) and large (4.8 cm long; **b**, **d**) *Mnemiopsis leidyi*. Both ctenophores were stationary (i.e., swimming velocity of 0) and actively entrained fluid between their lobes. The laser sheet used for digital particle image velocimetry (DPIV) was directed through the center of the ctenophore at two perpendicular orientations (laser orientation illustrated by *red line*, *insets*). DPIV is shown with the laser directed through the lobes (**a**, **b**) and between the lobes (**c**, **d**). This view is through the transparent lobe to show particle velocities between the lobes. *White vectors* represent velocities greater than 3.5 mm s^{-1} (from Colin et al., 2010). (Color figure online)

tissues of *Mnemiopsis* (Kremer, 1976; Anninsky et al., 2005) are not well suited to low food supplies. *Mnemiopsis* possesses very limited lipid and carbohydrate body reserves, which are quickly utilized during non-feeding periods (Anninsky et al., 2005). The use of body proteins to satisfy metabolic demands leads to organic dilution of tissues (Reeve et al., 1989) and, after several days of starvation, body shrinkage (reviewed in Reeve & Walter, 1978). Loss of organic weight by *Mnemiopsis* during starvation averages 5.9% day^{-1} at 12°C (Anninsky et al., 2005), indicating that body shrinkage can be rapid at low prey concentrations. As a consequence of high-reproductive sensitivity to low-food levels and very limited starvation tolerance, *Mnemiopsis* appears incapable of extending population growth into regions of low-prey concentrations, such as oceanic waters.

Vulnerability to predation

Mnemiopsis biomass levels can be limited by influential predators. A wide array of predators consume *Mnemiopsis* (Table 1), including vertebrate (Mianzan et al., 1996) and gelatinous (reviewed in Purcell et al., 2001; Arai, 2005) predators. Gelatinous predators appear to be particularly influential (Purcell & Cowan,

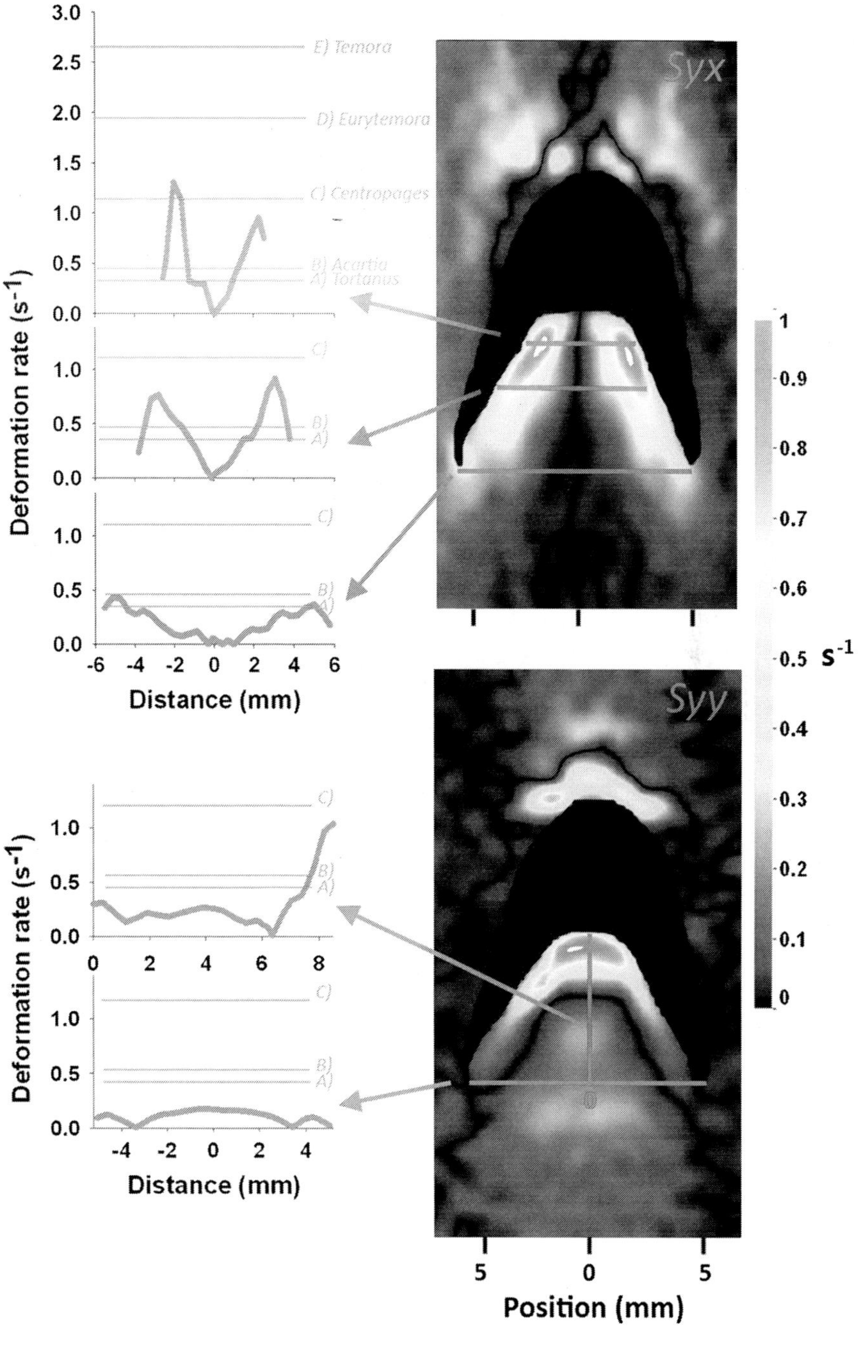

Fig. 11 Shear deformation rates of the two largest components of deformation in different regions of the feeding current of a small stationary *Mnemiopsis leidyi* (1.3 cm long). *Top* S_{yx} represents alterations in u_x (*x* component of fluid velocity) along the *y* axis. Three transects at outer, middle, and inner lobe positions (*top*, *right*) are compared with minimum threshold deformation rates that elicit escape responses of common coastal copepods (indicated by *green lines with letters* designating different copepod species). Deformation rate thresholds are from Kiørboe et al., 1999 (*Acartia*), Burdick et al., 2007 (*Centropages*, *Temora*, *Tortanus*), and Green et al., 2003 (*Eurytemora*). *Bottom* S_{yy} represents alterations in u_y (*y* component of fluid velocity) along the *y* axis. Two transects depict S_{yy} across the lobe opening and along a central axis from the lobe opening to the ctenophore's mouth (indicated by *red lines*, bottom). The observed deformation rates for this small ctenophore are large compared with those of larger ctenophores. Despite this, much of the feeding current is undetectable to prey. We would expect a greater portion of the feeding current of larger ctenophores to be below the threshold of prey detection (from Colin et al., 2010). (Color figure online)

1995; Shiganova et al., 2001a, 2004a; Finenko et al., 2003; Purcell & Decker, 2005; Condon & Steinberg, 2008), although little information exists on predation by fish. The regulatory effect of these gelatinous predators can dominate *Mnemiopsis* population biomass and have cascading effects through the plankton community (Fig. 14). The dominant influence of gelatinous predators on *Mnemiopsis* biomass is remarkable in light of the rapid population growth potential of *Mnemiopsis* and the high frequency (>90%) with which the ctenophores may evade predators (Kreps et al., 1997; Hosia & Titelman, 2011; Titelman et al., 2012). However, as evaluated by Condon and Steinberg (2008), the ability of predators

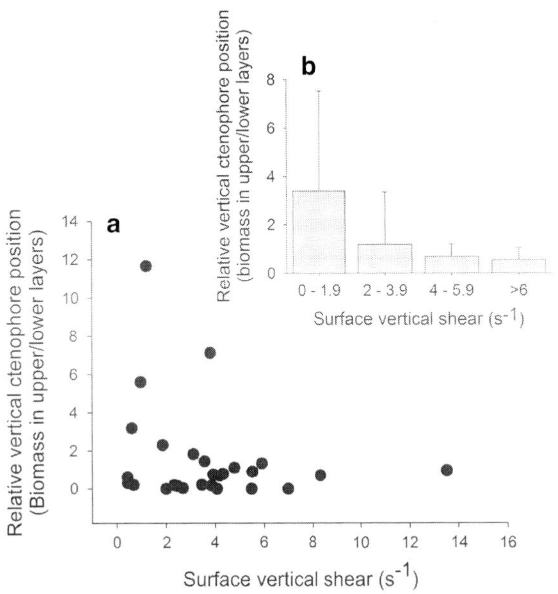

Fig. 12 Relative vertical position of *Mnemiopsis leidyi* biomass versus surface vertical shear in the Peninsula Valdes region during December, 1989. **a** Data for all individual stations, and **b** grouping of stations by shear levels allowing comparison relative positions of *Mnemiopsis* biomass in relation to surface vertical shear. Note that the proportion of ctenophores in the surface layers were always low when surface vertical shear was high (>4.0 s^{-1}); however, when surface vertical shear was low, the vertical positions of ctenophore biomass were more variable (from Mianzan et al., 2010)

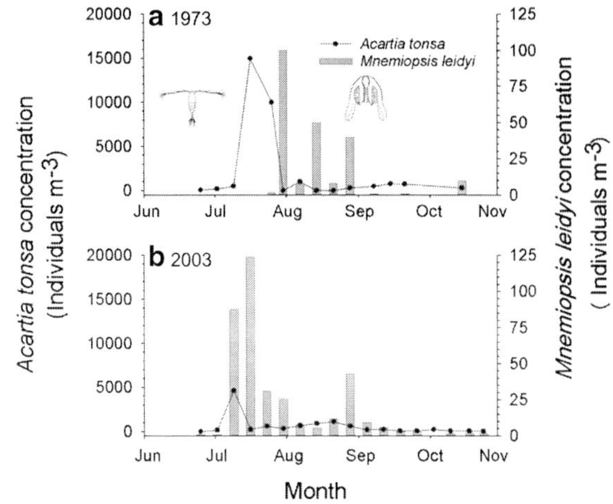

Fig. 13 Relationship between timing of population maxima for the copepod *Acartia tonsa* and the ctenophore *Mnemiopsis leidyi* in Narragansett Bay, RI, USA. *A. tonsa* has historically been the dominant summer copepod in Narragansett Bay and the copepod's population peaked before the onset of *Mnemiopsis* population growth (e.g., **a**; see also Deason, 1982). However, in recent years, *Mnemiopsis* populations have increased earlier (Costello et al., 2006b), and *A. tonsa* has become rare in Narragansett Bay during much of the summer (from Sullivan et al., 2007)

such as *C. quinquecirrha* to crop *Mnemiopsis* biomass exceeds the population growth potential of *Mnemiopsis* in even the most favorable food and temperature

Table 3 Predation rates of *Mnemiopsis leidyi* on copepods and *Chrysaora quinquecirrha* on *Mnemiopsis*

Month	Size (mm)	B_p	D_p	I	DPP	DC
Mnemiopsis consuming copepods						
April	43–43	0.7–10.7	119–1775	<0.1–0.5	<1–21	8–84
May	10–39	0.7–9.1	110–1520	0.7–2.5	27–127	144–697
June	12–13	0.0–6.4	81–1065	0.2–1.8	24–208	39–299
July	22–63	2.9–7.6	477–1271	0.0–1.2	0–43	0–206
August	0–34	3.6–113.6	594–18929	0.0–0.1	0–2	0–23
Chrysaora quinquecirrha consuming *Mnemiopsis*						
April	0	0.2–35.2	<1–38	0	0	0
May	0	19.9–32.8	32–425	0	0	0
June	23–133	5.3–50.2	82–170	0.1–13.5	2–27	0–3
July	76–152	0.0–12.5	0–86	0.6–72.7	37–242	<3–425
August	80–134	0.0–17.7	0–28	<0.1–28.1	<1–159	<1–107

Predicted monthly carbon (C) ingestion rates for populations at a upriver York River station. Ingestion rates based on mean-sized predator (mm), and predator and prey C standing stocks (mg C m^{-3}). Values are upper and lower monthly range estimates for April–August 2003–2006. Calculations were made using equations listed in Condon & Steinberg (2008). B_p = biomass of prey (mg C m^{-3}); D_p = density of prey (no. prey m^{-3}); I = population ingestion rates (mg C m^{-3} day^{-1}); DPP = daily population predation pressure rates (% prey C day^{-1}); DC = C-based daily prey consumption rates (no. prey m^{-3} day^{-1}) (From Condon & Steinberg, 2008) conditions for the ctenophore (Table 3). *Mnemiopsis* introductions have been comparatively devastating in regions without other gelatinous predators, such as the Caspian sea (Roohi et al., 2008, 2009), but they may be moderated in the presence of established indigenous gelatinous predators, such as in recently invaded European regions (Hosia & Titelman, 2011) or where other gelatinous predators have been introduced, such as *Beroe ovata* Bruguiève, 1789 in the Black Sea (Finenko et al., 2003; Shiganova et al., 2003). Consequently, the presence of predators, particularly gelatinous carnivores, can override other favorable conditions and diminish *Mnemiopsis* biomass both in its native (Kremer & Nixon, 1976; Purcell & Decker, 2005; Condon & Steinberg, 2008) and exotic ranges (Finenko et al., 2003; Shiganova et al., 2003).

Population dynamics and invasive patterns of *Mnemiopsis*

The same traits enabling high-biomass production and influential ecological impacts by *Mnemiopsis* in temperate regions of its native habitat are the basis for the ctenophore's success as an invader in exotic habitats. *Mnemiopsis* possesses many traits associated with "weed" species—wide physiological tolerances, wide dietary niche, rapid growth, short-generation times, and high fecundity (Sakai et al., 2001). These traits have been a consistent feature of *Mnemiopsis* population dynamics within its native range and are now shared by the ctenophore's populations in its exotic range. From this perspective, the high-invasive success of *Mnemiopsis* in exotic habitats can be viewed as an extension of the source-sink population dynamics enabling the ctenophore to successfully dominate temperate regions of its native range.

The transition from a historically stable to a contemporary invasive distribution was initiated by reducing limitations on dispersal of *Mnemiopsis*. Although dispersal between source and sink regions has historically been limited by local and regional circulation patterns, the contemporary marine environment also features trans-oceanic transport vectors in the form of ballast tanks within commercial sea vessels (Fig. 15). Molecular markers trace the pathways of these trans-oceanic *Mnemiopsis* introductions (Fig. 2) and similar patterns have been confirmed by multiple, independent studies (Ghabooli et al., 2010; Reusch et al., 2010). These data demonstrate an ongoing pattern that includes relatively recent introductions from North America to areas such as the North and Baltic Seas.

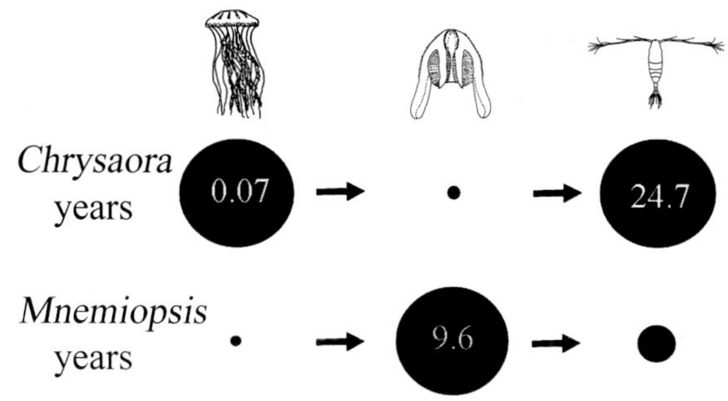

Fig. 14 The role of selective feeding by the scyphomedusa, *Chrysaora quinquecirrha*, on a planktonic community in mesohaline regions of Chesapeake Bay, USA, that contain the ctenophore *Mnemiopsis leidyi* and copepods. Data based on Purcell & Decker (2005; annual variations detailed therein). The circumference of the spheres under each organism represents the relative average proportions of those species in the plankton during years of high abundance in individuals of *C. quinquecirrha* (*Chrysaora* years: 1987–1990 and 1995) or *Mnemiopsis* (*Mnemiopsis* years: 1996–2000). The maximum concentrations of each organismal group are normalized to the same circumferences. Within each organismal group, the relative circumferences of the two time periods are proportionately dimensioned and the average abundances of each group (no. m^{-3} for *C. quinquecirrha* and *Mnemiopsis*, no. l^{-1} for copepods) are listed within the *circles*. Values for *smaller circles* (*C. quinquecirrha*: 0.007 m^{-3}, *Mnemiopsis*: 1.1 m^{-3}, copepods: 7.7 l^{-1}) were not listed in the figure. *Arrows* represent a simplification of trophic interactions because members of *C. quinquecirrha* prey upon both individuals of *Mnemiopsis* and copepods, but selectively prey upon ctenophores relative to copepods. Predation by individuals of *C. quinquecirrha* upon the ctenophore *Mnemiopsis* reduces the latter with a cascading effect on the ctenophore's principle prey items, the copepods. Consequently, the relative abundance of copepods in the plankton is dominated by trophic interactions that depend on the prey selection characteristics of the scyphomedusa *C. quinquecirrha*

It is likely that *Mnemiopsis* will continue to expand into exotic areas in the near future. The reasons for this projection are based on the processes driving invasive expansion by the ctenophore. Successful invasion of a novel area by *Mnemiopsis* is dependent both on the recipient environment (the area's "invasibility"—Leung & Mandrake, 2007) and on the ability to reach these new areas (the ctenophore's "propagule pressure"—Lockwood et al., 2005). Upon arrival, invaders must persist in the new habitat and persistence depends upon the match between the individual

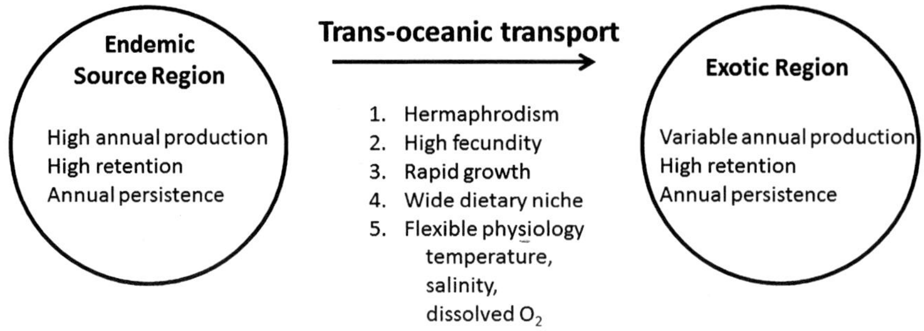

Fig. 15 The role of source-sink life history organization on invasive patterns of the ctenophore *Mnemiopsis leidyi*. Note that the critical variable distinguishing this scheme from that in Fig. 4 is the expansion of dispersal beyond the limitations of local currents by inclusion of long-range dispersal via human-related transport

species' traits and the new environment. The broad physiological tolerances of *Mnemiopsis* (Table 1) suggest that a wide array of productive coastal environments have high-invasibility levels for *Mnemiopsis* and could potentially be suitable habitats for the ctenophore; however, the ctenophore first has to reach those habitats. Historically, the expanse of low-productivity oceanic waters likely has prevented extension of *Mnemiopsis* beyond its native coastlines of the Atlantic North and South Americas (Harbison & Volovik, 1994). However, this historical limitation has been relaxed by ballast water transport via contemporary transoceanic shipping. Ballast water regulation is a developing field with limited prospects for reducing transfer of inocula in the near future (David & Gollasch, 2008). Hence, the key obstacle to *Mnemiopsis* invasion of new regions is relaxed during a period when the number of source regions for inocula has increased. Increasing the number of source regions can dramatically increase overall invasion rates—within 50 years of initial invasion, a new source region may supply inocula for invasion to an additional 300 ports (Kaluza et al., 2010). This combination of factors—a wide variety of high invasibility regions, reduction of dispersal limitation, and increasing propagule pressure—favors continued range expansion by *Mnemiopsis*.

We expect that the ecological role played by introduced *Mnemiopsis* populations will depend upon community structure in the novel environments. Within its native range, the ctenophore's ecological role is constrained by the variables previously considered (i.e., temperature and production regimes, predator dynamics). These same constraints will influence invasive populations of *Mnemiopsis* in exotic habitats. In a variety of native habitats, *Mnemiopsis* is a persistent but relatively inconspicuous community member (Kremer et al., 1986; Kremer, 1994). Even in areas that experience periodic, high *Mnemiopsis* biomass, fluctuations in ctenophore biomass depend upon predator population dynamics (Fig. 14). The dramatic effects following *Mnemiopsis* introductions documented in the Black (e.g., Kideys, 2002; Shiganova et al., 2004a) and Caspian (e.g., Shiganova et al., 2004b; Roohi et al., 2008, 2009) seas occurred in habitats that lacked gelatinous predators (Purcell et al., 2001). Recent introductions to the Baltic and North Seas occurred in habitats containing potentially influential gelatinous predators (Hosia & Titelman, 2011) that may impact the eventual role of *Mnemiopsis* in these communities. We expect that the variables favoring and constraining *Mnemiopsis* population dynamics in previously studied habitats will provide insight into the fate of introduced populations as the world community adjusts to the ctenophore's expanded the presence in coastal marine communities.

Acknowledgments We gratefully acknowledge support for this work from the US National Science Foundation (OCE-0350834, OCE-0623508) and the US Office of Naval Research (N000140810654) to J. H. C., project EC SESAME to T. A. S., ANPCyT-1553 to H. W. M., and NCEAS-12479, supported by NSF (Grant #DEB-94-21535), the University of California at Santa Barbara, and the State of California.

References

Agassiz, L., 1860. Contributions to the Natural History of the United States of America, Vol. 3. Little, Brown & Co., Boston.

Agassiz, A., 1865. Illustrated Catalogue of the Museum of Comparative Zoology. Vol. II. North American Acalephae. Welch, Bigelow & Co, Cambridge, MA.

Anninsky, B. E., G. A. Finenko, G. I. Abolmasova, E. S. Hubareva, L. S. Svetlichny, L. Bat & A. E. Kideys, 2005. Effect of starvation on the biochemical compositions and respiration rates of ctenophores *Mnemiopsis leidyi* and *Beroe ovata* in the Black Sea. Journal of the Marine Biological Association of the United Kingdom 85: 549–561.

Arai, M. N., 2005. Predation on pelagic coelenterates: a review. Journal of the Marine Biological Association of the United Kingdom 85: 523–536.

Baker, L. D., 1973. The ecology of the ctenophore *Mnemiospsis mccradyi* in Biscayne Bay, Florida. University of Miami Technical Report, University of Miami-Rosenstiel School of Marine and Atmospheric Science: 131 pp.

Baker, I. D. & M. R. Reeve, 1974. Laboratory culture of the lobate ctenophore *Mnemiopsis mccradyi* with notes on feeding and fecundity. Marine Biology 26: 57–62.

Bayha, K., 2005. The molecular systematics and population genetics of four coastal ctenophores and scyphozoan jellyfish of the US Atlantic and Gulf of Mexico. PhD Dissertation. University of Delaware, Newark.

Bayha, K. M., G. R. Harbison, J. H. McDonald & P. M. Gaffney, 2004. Preliminary investigation on the molecular systematic of the invasive ctenophore *Beroe ovata*. In Dumont, H., T. A. Shiganova & U. Niermann (eds), Aquatic Invasions in the Black. Caspian and Mediterranean Seas. Kluwer Academic Publishers, Dordrecht: 167–175.

Boero, F., M. Putti, E. Trainito, E. Prontera, S. Piraino & T. Shiganova, 2009. Recent changes in Western Mediterranean Sea biodiversity: the establishment of *Mnemiopsis leidyi* (Ctenophora) and the arrival of *Phyllorhiza punctata* (Cnidaria). Aquatic Invasions 4: 675–680.

Brusca, R. C. & G. J. Brusca, 2003. Invertebrates. Sinauer Associates, Sunderland, USA.

Burdick, D. S., D. K. Hartline & P. H. Lenz, 2007. Escape strategies in co-occurring calanoids copepods. Limnology and Oceanography 52: 2373–2385.

Buskey, E. J., 1993. Annual pattern of micro- and mesozooplankton abundance and biomass in a subtropical estuary. Journal of Plankton Research 15: 907–924.

Cairns, S. D., D. R. Calder, A. Brinckmann-Voss, C. B. Castro, D. G. Fautin, P. R. Pugh, C. E. Mills, W. C. Jaap, M. N. Arai, S. H. D. Haddock & D. M. Opresko, 2002. Common and Scientific Names of Aquatic Invertebrates from the United States and Canada: Cnidaria and Ctenophora, 2nd ed. American Fisheries Society Special Publication 28, Bethesda: 1–115.

Caut, S., E. Angulo & F. Courchamp, 2008. Dietary shift of an invasive predator: rats, seabirds and sea turtles. Journal of Applied Ecology 45: 428–437.

Colin, S. P., J. H. Costello, L. J. Hansson, J. Titelman & J. O. Dabiri, 2010. Stealth predation: the ecological success of the invasive ctenophore *Mnemiopsis leidyi*. Proceedings of the National Academy of Sciences of the United States of America 107: 17223–17227.

Condon, R. H. & D. K. Steinberg, 2008. Development, biological regulation, and fate of ctenophore blooms in the York River estuary, Chesapeake Bay. Marine Ecology Progress Series 369: 153–168.

Costello, J. H. & H. W. Mianzan, 2003. Sampling field distributions of *Mnemiopsis leidyi* (Ctenophora, Lobata): planktonic or benthic methods? Journal of Plankton Research 25: 455–459.

Costello, J. H., R. Loftus & R. Waggett, 1999. Influence of prey detection on capture success for the ctenophore *Mnemiopsis leidyi* feeding upon adult *Acartia tonsa* and *Oithona colcarva* copepods. Marine Ecology Progress Series 191: 207–216.

Costello, J. H., B. K. Sullivan, D. J. Gifford, D. Van Keuren & L. J. Sullivan, 2006a. Seasonal refugia, shoreward thermal amplification and metapopulation dynamics of the ctenophore *Mnemiopsis leidyi* in Narragansett Bay, Rhode Island. Limnology and Oceanography 51: 1819–1831.

Costello, J. H., B. K. Sullivan & D. J. Gifford, 2006b. A physical-biological interaction underlying variable phenological responses to climate change by coastal zooplankton. Journal of Plankton Research 28: 1099–1105.

Darvishi, F., A. Esmaili Sari & P. Farshchi, 2004. Dietary overlap between invasion ctenophora (*M. leidyi* and anchovy (*C. engrauliformis*) in the southern parts of Caspian Sea. International Journal of Environmental Science and Technology 1: 89–95.

Daskalov, G. M., A. N. Grishin, S. Rodionov & V. Mihneva, 2007. Trophic cascades triggered by overfishing reveal possible mechanisms of ecosystem regime shifts. Proceedings of the National Academy of Science of the United States of America 104: 10518–10523.

David, M. & S. Gollasch, 2008. EU shipping in the dawn of managing the ballast water issue. Marine Pollution Bulletin 56: 1966–1972.

Dawson, M. N., 2004. Some implications of molecular phylogenetics for understanding biodiversity in jellyfishes, with emphasis on Scyphozoa. Hydrobiologia 530/531: 249–260.

Dawson, M. N. & D. Jacobs, 2001. Molecular evidence for cryptic species of *Aurelia aurita* (Cnidaria, Scyphozoa). Biological Bulletin 200: 92–96.

Deason, E. E., 1982. Mnemiopsis leidyi (Ctenophora) in Narragansett Bay, 1975–79: abundance, size composition and estimation of grazing. Estuarine, Coastal and Shelf Science 15: 121–134.

Deason, E. E. & T. J. Smayda, 1982. Ctenophore-zooplankton-phytoplankton interactions in Narragansett Bay, Rhode Island, USA, during 1972–1977. Journal of Plankton Research 4: 203–217.

Decker, M. B., D. L. Breitburg & J. E. Purcell, 2004. Effects of low dissolved oxygen on zooplankton predation by the ctenophore *Mnemiopsis leidyi*. Marine Ecology Progress Series 280: 163–172.

Durbin, A. G. & E. G. Durbin, 1981. Standing stock and estimated production rates of phytoplankton and zooplankton in Narragansett Bay, Rhode Island. Estuaries 4: 24–41.

Eschsholtz, F., 1825. Berichtüber die zoologischeAusbeutewährend der Reise von Kronstadtbis St. Peter und Paul. Isis 1: 733–747.

Faasse, M. A. & K. M. Bayha, 2006. The ctenophore *Mnemiopsis leidyi* A. Agassiz 1865 in coastal waters of the Netherlands: an unrecognized invasion? Aquatic Invasions 1: 270–277.

Finenko, G. A., Z. A. Romanova, G. I. Abolmasova, B. E. Anninsky, L. S. Svetlichny, E. S. Hubareva, L. Bati & A. E. Kideys, 2003. Population dynamics, ingestion, growth and reproduction rates of the invader *Beroe ovata* and its impact on plankton community in Sevastopol Bay, the Black Sea. Journal of Plankton Research 25: 539–549.

Finenko, G. A., A. E. Kideys, B. E. Anninsky, T. A. Shiganova, A. Roohi, M. R. Tabari, H. Rostami & S. Bagheri, 2006. Invasive ctenophore *Mnemiopsis leidyi* in the Caspian Sea: feeding, respiration, reproduction and predatory impact on the zooplankton community. Marine Ecology Progress Series 314: 171–185.

Fuentes, V. L., D. L. Angel, K. M. Bayha, D. Atienza, D. Edelist, C. Bordehore, J.-M. Gili & J. E. Purcell, 2010. Blooms of the invasive ctenophore, *Mnemiopsis leidyi*, span the Mediterranean Sea in 2009. Hydrobiologia 645: 23–37.

Funk, J. L. & P. M. Vitousik, 2007. Resource-use efficiency and plant invasion in low-resource systems. Nature 446: 1079–1081.

Galil, B. S., N. Kress & T. A. Shiganova, 2009. First record of *Mnemiopsis leidyi* A. Agassiz, 1865 (Ctenophora Lobata Mnemiidae) off the Mediterranean coast of Israel. Aquatic Invasions 4: 356–362.

Ghabooli, S., T. A. Shiganova, A. Zhan, M. Cristescu, P. Eghtesadi-Araghi & H. MacIsaac, 2010. Multiple introductions and invasion pathways for the invasive ctenophore *Mnemiopsis leidyi* in Eurasia. Biological Invasions. doi: 10.1007/s10530-010-9859-8.

Gorokhova, E. & M. Lehtiniemi, 2010. Comment: Reconsidering evidence for *Mnemiopsis* invasion in European waters. Journal of Plankton Research 32: 93–95.

Gorokhova, E., M. Lehtiniemi, S. Viitasalo-Frosen & S. H. D. Haddock, 2009. Molecular evidence for the occurrence of ctenophore *Mertensia ovum* in the northern Baltic Sea and implications for the status of the *Mnemiopsis leidyi* invasion. Limnology and Oceanography 54: 2025–2033.

Granhag, L., L. F. Møller & L. J. Hansson, 2011. Size-specific clearance rates of the ctenophore *Mnemiopsis leidyi* based on in situ gut content analyses. Journal of Plankton Research doi. doi:10.1093/plankt/fbr010.

Green, S., A. W. Visser, J. Titelman & T. Kiørboe, 2003. Escape responses of copepod nauplii in the flow field of the blue mussel, *Mytilus edulis*. Marine Biology 142: 727–733.

Greve, W., 1977. Interspecific interaction: the analysis of complex structures in carnivorous zooplankton populations. Helgolander Wissenschaftliche Meeresunters 30: 83–91.

Greve, W., F. Reiners, J. Nast & S. Hoffmann, 2004. Helgoland Roads meso- and macrozooplankton time-series 1974 to 2004: lessons from 30 years of single spot, high frequency sampling at the only off-shore island of the North Sea. Helgoland Marine Research 58: 274–288.

Grove, M. & D. L. Breitburg, 2005. Growth and reproduction of gelatinous zooplankton exposed to low dissolved oxygen. Marine Ecology Progress Series 301: 185–198.

Hamer, H. H., A. M. Malzahn & M. Boersma, 2011. The invasive ctenophore *Mnemiopsis leidyi*: a threat to fish recruitment in the North Sea? Journal of Plankton Research 33: 137–144.

Hanski, I., 1999. Metapopulation Ecology. Oxford University Press, New York, USA.

Hansson, H. G., 2006. Ctenophores of the Baltic and adjacent Seas – the invader *Mnemiopsis* is here! Aquatic Invasions 1: 295–298.

Harbison, G. R. & S. P. Volovik, 1994. The ctenophore, *Mnemiopsis leidyi*, in the Black Sea: a holoplanktonic organism transported in the ballast water of ships. In Nonindigenous Estuarine and Marine Organisms (NEMO), Proceedings of the Conference and Workshop, Seattle, WA, April 1993. U. S. Department of Commerce, National Oceanic and Atmospheric Administration, Washington, DC: 25–36.

Harbison, G. R., L. P. Madin & N. R. Swanberg, 1978. On the natural history and distribution of oceanic ctenophores. Deep-Sea Research 25: 233–256.

Haslob, H., C. Clemmesen, M. Schaber, H. H. Hinrichsen, J. O. Schmidt, R. Voss, G. Kraus & F. W. Köster, 2007. Invading *Mnemiopsis leidyi* as a potential threat to Baltic fish. Marine Ecology Progress Series 349: 303–306.

Holland, B., M. N. Dawson, G. Crow & D. Hofmann, 2004. Global phylogeography of *Cassiopea* (Scyphozoa: Rhizostomeae): molecular evidence for cryptic species and multiple invasions of the Hawaiian Islands. Marine Biology 145: 1119–1128.

Hosia, A. & J. Titelman, 2011. Intraguild predation between the native North Sea jellyfish *Cyanea capillata* and the invasive ctenophore *Mnemiopsis leidyi*. Journal of Plankton Research 33: 535–540.

Hulsizer, E. E., 1976. Zooplankton of lower Narragansett Bay 1972–1973. Chesapeake Science 17: 260–270.

Hyman, L. H., 1940. The Invertebrates: Protozoa Through Ctenophora. McGraw-Hill Co., New York.

Ivanov, V. P., A. M. Kamakin, V. B. Ushivtsev, T. A. Shiganova, O. P. Zhukova, N. Aladin, S. I. Wilson, G. R. Harbison & H. J. Dumont, 2000. Invasion of the Caspian Sea by the comb jellyfish *Mnemiopsis leidyi* (Ctenophora). Biological Invasions 2: 255–258.

Jaspers, C., J. Titelman, L. J. Hansson, M. Haraldsson & C. R. Ditlefsen, 2011. The invasive ctenophore *Mnemiopsis leidyi* poses no direct threat to Baltic cod eggs and larvae. Limnology and Oceanography 56: 431–439.

Javidpour, J., U. Sommer & T. A. Shiganova, 2006. First record of *Mnemiopsis leidyi* in the Baltic Sea. Aquatic Invasions 1: 299–302.

Javidpour, J., J. C. Molinero, A. Lehmann, T. Hansen & U. Sommer, 2009a. Annual assessment of the predation of *Mnemiopsis leidyi* in a new invaded environment, the Kiel Fjord (Western Baltic Sea): a matter of concern? Journal of Plankton Research 31: 729–738.

Javidpour, J., J. C. Molinero, J. Peschutter & U. Sommer, 2009b. Seasonal changes and population dynamics of the ctenophore *Mnemiopsis leidyi* after its first year of invasion in the Kiel Fjord, Western Baltic Sea. Biological Invasions 11: 873–882.

Javidpour, J., J. C. Molinero & U. Sommer, 2010. Reconsidering evidence for *Mnemiopsis* invasion in European waters: reply. Journal of Plankton Research 32: 97–98.

Kaluza, P., A. Kölzsch, M. T. Gastner & B. Blasius, 2010. The complex network of global cargo ship movements. Journal of the Royal Society Interface. doi:10.1098/rsif.2009.0495.

Kideys, A. E., 2002. Fall and rise of the Black Sea ecosystem. Science 297: 1482–1484.

Kideys, A. E. & U. Niermann, 1994. Occurrence of *Mnemiopsis* along the Turkish coast (from northeastern Mediterranean to Istanbul). ICES Journal of Marine Science 51: 423–427.

Kideys, A. E., A. Roohi, E. Eker-Develi, F. Melin & D. Beare, 2008. Increased chlorophyll levels in the Southern Caspian Sea following an invasion of jellyfish. Research Letters in Ecology. doi:10.1155/2008/185642.

Kiørboe, T., E. Saiz & A. Visser, 1999. Hydrodynamic signal perception in the copepod *Acartia tonsa*. Marine Ecology Progress Series 179: 97–111.

Knowler, D., 2005. Reassessing the costs of biological invasion: *Mnemiopsis leidyi* in the Black sea. Ecological Economics 52: 187–199.

Kolesar, S. E., D. L. Breitburg, J. E. Purcell & M. B. Decker, 2010. Effects of hypoxia on *Mnemiopsis leidyi*, ichthyoplankton and copepods: clearance rates and vertical habitat overlap. Marine Ecology Progress Series 411: 173–188.

Kremer, P., 1976. Population dynamics and ecological energetics of a pulsed zooplankton predator, the ctenophore *Mnemiopsis leidyi*. In Wiley, M. L. (ed.), Estuarine Processes, Vol. 1. Academic Press, New York: 197–215.

Kremer, P., 1979. Predation by the ctenophore *Mnemiopsis leidyi* in Narragansett Bay, Rhode Island. Estuaries 2: 97–105.

Kremer, P., 1994. Patterns of abundance for *Mnemiopsis* in U.S. coastal waters: a comparative overview. ICES Journal of Marine Science 51: 347–354.

Kremer, P. & S. Nixon, 1976. Distribution and abundance of the ctenophore *Mnemiopsis leidyi* in Narragansett Bay. Estuarine and Coastal Marine Science 4: 627–639.

Kremer, P. & M. R. Reeve, 1989. Growth dynamics of a ctenophore (*Mnemiopsis*) in relation to variable food supply. II. Carbon budgets and growth model. Journal of Plankton Research 11: 553–574.

Kremer, P., M. R. Reeve & M. A. Syms, 1986. The nutritional ecology of the ctenophore *Bolinopsis vitrea*: comparison with *Mnemiopsis mccradyi* from the same region. Journal of Plankton Research 8: 1197–1208.

Kreps, T. A., J. E. Purcell & K. B. Heidelberg, 1997. Escape of the ctenophore, *Mnemiopsis leidyi* from the scyphomedusa predator, *Chrysaora quinquecirrha*. Marine Biology 128: 441–446.

Larson, R. J., 1987. In situ feeding rates of the ctenophore *Mnemiopsis mccradyi*. Estuaries 10: 87–91.

Leung, B. & N. E. Mandrake, 2007. The risk of establishment of aquatic invasive species: joining invasibility and propagule pressure. Proceedings of the Royal Society B 274: 2603–2609.

Lockwood, J. L., P. Cassey & T. Blackburn, 2005. The role of propagule pressure in explaining species invasions. Trends in Ecology & Evolution 20: 223–228.

Lonsdale, D. J., 1981. Regulatory role of physical factors and predation for two Chesapeake Bay copepod species. Marine Ecology Progress Series 5: 341–351.

Main, R. J., 1928. Observations of the feeding mechanism of a ctenophore, *Mnemiopsis leidyi*. Biological Bulletin 55: 69–78.

Marcus, N. H. & F. Boero, 1998. Minireview: the importance of benthic-pelagic coupling and the forgotten role of life cycles in coastal aquatic systems. Limnology and Oceanography 43: 763–768.

Mayer, A. G., 1900. Descriptions of new and little-known medusae from the Western Atlantic. Bulletin of the Museum of Comparative Zoology Harvard 37: 1–9.

Mayer, A. G., 1912. Ctenophores of the Atlantic Coast of North America. Carnegie Institute Publishing, Washington.

Mertens, H., 1833. Beobachtungen und Untersuchungenüber die beroeartigenAkalephen. Mémoires de L'AcademieImperiale des Sciences de St. Petersbourg (6th Sér.) 2: 479–543.

Mianzan, H. W., 1986. *Beroe ovata* en aguas de la bahía Blanca, Argentina (Ctenophora). Spheniscus 2: 29–32.

Mianzan, H., 1999. Ctenophora. In Boltovskoy, D. (ed.), South Atlantic Zooplankton. Backhuys Publishers, Leiden: 561–573.

Mianzan, H. W. & R. Guerrero, 2000. Environmental patterns and biomass distribution of gelatinous macrozooplankton. Three study cases in the Southwestern Atlantic. Scientia Marina 64(Supl. 1): 215–224.

Mianzan, H. W. & M. Sabatini, 1985. Estudio preliminar sobre distribución y abundancia de *Mnemiopsis maccradyi* en el estuario de bahía Blanca, Argentina (Ctenophora). Spheniscus 1: 53–68.

Mianzan, H. W., N. Mari, B. Prenski & F. Sanchez, 1996. Fish predation on neritic ctenophores from the Argentine continental shelf: a neglected food resource? Fisheries Research 27: 69–79.

Mianzan, H. W., P. Martos, J. H. Costello & R. Guerro, 2010. Avoidance of hydrodynamically mixed environments by *Mnemiopsis leidyi* (Ctenophora: Lobata) in open-sea populations from Patagonia, Argentina. Hydrobiologia 645: 113–124.

Miller, R. J., 1974. Distribution and biomass of an estuarine ctenophore population, *Mnemiopsis leidyi* (A. Agassiz). Chesapeake Science 15: 1–8.

Miranda, L., A. Collins, A. Marques & S. Vollmer, 2010. Molecules clarify a Cnidarian life cycle: the "Hydrozoan" *Microhydrulalimopsicola* is an early life stage of the Staurozoan *Haliclystu santarcticus*. PLoS ONE 5: e10182.

Moss, A. G., B. Wells & L. Muellner, 2004. A mechanosensory system that controls feeding in adult *Mnemiopsis*. Hydrobiologia 530(531): 145–153.

Mutlu, E., 1999. Distribution and abundance of ctenophores and their zooplankton food in the Black Sea. II. *Mnemiopsis leidyi*. Marine Biology 135: 603–614.

Mutlu, E., 2009. Recent distribution and size structure of gelatinous organisms in the southern Black Sea and their interactions with fish catches. Marine Biology 156: 935–957.

Nelson, T. C., 1925. On the occurrence and food habits of ctenophores in New Jersey inland coastal waters. Biological Bulletin 48: 92–111.

Newton, C., J. H. Costello, B. K. Sullivan & S. P. Colin, 2009. Feeding mechanisms and prey selection by the ctenophore *Mnemiopsis leidyi*. ASLO, abstract.

Oguz, T., 2005. Long-term impacts of anthropogenic forcing on the Black Sea ecosystem. Oceanography 18: 112–121.

Oguz, T. & D. Gilbert, 2007. Abrupt transitions of the top-down controlled Black Sea pelagic ecosystem during 1960–2000: evidence for regime-shifts under strong fishery exploitation and nutrient enrichment modulated by climate-induced variations. Deep-Sea Research Part I 54: 220–242.

Oguz, T., B. Fach & B. Salihoglu, 2008. Invasion dynamics of the alien ctenophore *Mnemiopsis leidyi* and its impact on anchovy collapse in the Black Sea. Journal of Plankton Research 30: 1385–1397.

Olson, M. M., 1987. Zooplankton. In Heck, K. L., Jr. (ed.), Ecological Studies in the Middle Reach of Chesapeake Bay. Springer-Verlag, New York: 38–81.

Pereladov, M. V., 1988. Some observations for biota of Sudak Bay of the Black Sea. III All-Russian conference of marine biology. Naukova Dumka, Kiev 1: 237–238. (in Russian).

Pianka, H. D., 1974. Ctenophora. In Giese, A. C. & J. S. Pearse (eds), Reproduction of Marine Invertebrates, Vol. I. Academic Press, New York: 201–265.

Podar, M., S. H. D. Haddock, M. L. Sogin & G. R. Harbison, 2001. A molecular genetic framework for the phylum Ctenophora using 18 s rRNA genes. Molecular Phylogenetics and Evolution 21: 218–230.

Purcell, J. E., 1997. Pelagic cnidarians and ctenophores as predators: selective predation, feeding rates and effects on prey populations. Annales de l'Institut océanographique, Paris 73: 125–137.

Purcell, J. E., 2009. Extension of methods for jellyfish and ctenophore trophic ecology to large-scale research. Hydrobiologia 616: 23–50.

Purcell, J. E. & M. N. Arai, 2001. Interactions of pelagic cnidarians and ctenophores with fishes: a review. Hydrobiologia 451 (Developments in Hydrobiology 155): 27–44.

Purcell, J. E. & J. H. Cowan, 1995. Predation by the scyphomedusan *Chrysaora quinquecirrha* on *Mnemiopsis leidyi* ctenophores. Marine Ecology Progress Series 129: 63–70.

Purcell, J. E. & M. B. Decker, 2005. Effects of climate on relative predation by scyphomedusae and ctenophores on copepods in Chesapeake Bay during 1987–2000. Limnology and Oceanography 50: 376–387.

Purcell, J. E., J. R. White & M. R. Roman, 1994. Predation by gelatinous zooplankton and resource limitation as potential controls of *Acartia tonsa* copepod populations in Chesapeake Bay. Limnology and Oceanography 39: 263–278.

Purcell, J. E., T. A. Shiganova, M. B. Decker & E. D. Houde, 2001. The ctenophore *Mnemiopsis* in native and exotic habitats: U.S. estuaries versus the Black Sea basin. Hydrobiologia 451: 145–176.

Purcell, J. E., S.-I. Uye & W.-T. Lo, 2007. Anthropogenic causes of jellyfish blooms and direct consequences for humans: a review. Marine Ecology Progress Series 350: 153–174.

Rang, P. C. A. L., 1828. Établissement de la famille des Béroïdesdansl'ordre des acalèphes libres, et description de deux genres nouveaux qui luiappartiennent Mémoires. Société d'Histoire Naturelle de Paris 4: 166–173.

Rapoza, R., D. Novak & J. H. Costello, 2005. Life-stage dependent, in situ dietary patterns of the lobate ctenophore *Mnemiopsis leidyi* Agassiz 1865. Journal of Plankton Research 27: 951–956.

Reeve, M. R. & M. A. Walter, 1978. Nutritional ecology of ctenophores – a review of recent research. Advances in Marine Biology 15: 249–287.

Reeve, M. R., M. A. Syms & P. Kremer, 1989. Growth dynamics of a ctenophore (*Mnemiopsis*) in relation to variable food supply. I. Carbon biomass, feeding, egg production, growth, and assimilation efficiency. Journal of Plankton Research 11: 535–552.

Reusch, T., S. Bolte, M. Sparwell, A. Moss & J. Javidpour, 2010. Microsatellites reveal origin and genetic diversity of Eurasian invasions by one of the world's most notorious marine invader, *Mnemiopsis leidyi* (Ctenophora). Molecular Ecology 19: 2690–2699.

Riisgård, H. U., C. V. Madsen, C. Barth-Jensen, & J. E. Purcell, 2012. Population dynamics and zooplankton-predation impact of the indigenous scyphozoan *Aurelia aurita* and the invasive ctenophore *Mnemiopsis leidyi* in Limfjorden (Denmark). Aquatic Invasions (in press).

Roohi, A., Z. Yasin, A. E. Kideys, A. T. S. Hwai, A. G. Khanari & E. Eker-Develi, 2008. Impact of a new invasive ctenophore (*Mnemiopsis leidyi*) on the zooplankton community of the Southern Caspian Sea. Marine Ecology 29: 421–434.

Roohi, A., A. E. Kideys, A. Sajjadi, A. Hashemian, R. Pourgholam, H. Fazli, A. G. Khanari & E. Eker-Develi, 2009. Changes in biodiversity of phytoplankton, zooplankton, fishes and macrobenthos in the southern Caspian Sea after the invasion of the ctenophore *Mnemiopsis leidyi*. Biological Invasions. doi:10.1007/s10530-009-9648-4.

Sai Sastry, A. G. R. & P. Chandramohan, 1989. Planktonic coelenterates of Vasishta Godavari estuary, east coast of India. Indian Journal of Marine Sciences 18: 160–164.

Sakai, A. K., F. W. Allendorf, J. S. Holt, D. M. Lodge, J. Molofsky, K. A. With, S. Baughman, R. J. Cabin, J. E. Cohen, N. C. Ellstrand, D. E. McCauley, P. O'Neil, I. M. Parker, J. N. Thompson & S. G. Weller, 2001. The population biology of invasive species. Annual Review of Ecology and Systematics 32: 305–332.

Schaber, M., H. Haslob, B. Huwer, A. Harjes, H.-H. Hinrichsen, F. W. Köster, M. Storr-Paulsen, J. O. Schmidt & R. Voss, 2011a. The invasive ctenophore *Mnemiopsis leidyi* in the central Baltic Sea: seasonal phenology and hydrographic influence on spatio-temporal distribution patterns. Journal of Plankton Research. doi:10.1093/plankt/fbq167.

Schaber, M., H. Haslob, B. Huwer, A. Harjes, H.-H. Hinrichsen, M. Storr-Paulsen, J. O. Schmidt, R. Voss, V. Neumann & F. W. Köster, 2011b. Spatio-temporal overlap of the alien invasive ctenophore *Mnemiopsis leidyi* and ichthyoplankton in the Bornholm Basin (Baltic Sea). Biological Invasions. doi:10.1007/s10530-011-9936-7.

Schroth, W., G. Jarms, B. Streit & B. Schierwater, 2002. Speciation and phylogeography in the cosmopolitan marine moon jelly, *Aurelia* sp. BMC Evolutionary Biology 2: 1.

Seravin, L. N., 1994a. The systematic revision of the genus *Mnemiopsis* (Ctenophora, Lobata). 1. The history of description and study of species belonging to the genus *Mnemiopsis*. Zoologichesky Zhurnal 73: 9–18.

Seravin, L. N., 1994b. The systematic revision of the genus *Mnemiopsis* (Ctenophora, Lobata). 2. Species attribution on *Mnemiopsis* from the Black Sea and the species composition of the genus *Mnemiopsis*. Zoologichesky Zhurnal 73: 19–34.

Shiganova, T. A., 1993. Ctenophore *Mnemiopsis leidyi* and ichthyoplankton in the Sea of Marmara in October of 1992. Oceanology 33: 900–903.

Shiganova, T. A., 1997. *Mnemiopsis leidyi* abundance in the Black Sea and its impact on the pelagic community. In Ozsoy, E. & A. Mikaelyan (eds), Sensitivity of North Sea, Baltic Sea and Black Sea to Anthropogenic and Climatic

Changes. Kluwer Academic Publishers, Dordrecht: 117–130.
Shiganova, T. A. & A. Malej, 2009. Native and non-native ctenophores in the Gulf of Trieste, Northern Adriatic Sea. Journal of Plankton Research 31: 61–71.
Shiganova, T. A., Y. V. Bulgakova, S. P. Volovik, Z. A. Mirzoyan & S. I. Dudkin, 2001a. The new invader *Beroe ovata* Brown 1756 and its effect on the ecosystem in the northeastern Black Sea in August-September 1999. Hydrobiologia 451: 187–197.
Shiganova, T. A., Z. A. Mirzoyan, E. A. Studenikina, S. P. Volovik, I. Siokou-Frangou, S. Zervoudaki, E. D. Christou, A. Y. Skirta & H. J. Dumont, 2001b. Population development of the invader ctenophore *Mnemiopsis leidyi* in the Black Sea and other seas of the Mediterranean basin. Marine Biology 139: 431–445.
Shiganova, T. A., E. I. Musaeva, Y. V. Bulgakova, Z. A. Mirzoyan & M. L. Martynuk, 2003. Ctenophores invaders *Mnemiopsis leidyi* (A. Agassiz) and *Beroe ovata* Mayer 1912, and their effect on the pelagic ecosystem of the north-eastern Black Sea. Biological Bulletin Russian Academy of Sciences, Moscow 2: 225–235.
Shiganova, T. A., H. J. Dumont, A. Mikaelyan, D. M. Glazov, Y. V. Bulgakova, E. I. Musaeva, P. Y. Sorokin, L. A. Pautova, Z. A. Mirzoyan & E. I. Studenikina, 2004a. Interaction between the invading ctenophores *Mnemiopsis leidyi* (A. Agassiz) and *Beroe ovata* Mayer 1912, and their Influence on the pelagic ecosystem of the northeastern Black Sea. In Dumont, H., T. A. Shiganova & U. Niermann (eds), The Ctenophore *Mnemiopsis leidyi* in the Black, Caspian and Mediterranean Seas and Other Aquatic Invasions. NATO Science Series: IV. Earth and Environmental Sciences, Vol. 35. Kluwer Academic Publishers, Dordrecht: 33–70.
Shiganova, T. A., H. J. Dumont, A. F. Sokolsky, A. M. Kamakin, D. Tinenkova & E. K. Durasheva, 2004b. Population dynamics of *Mnemiopsis leidyi* in the Caspian Sea, and effects on the Caspian ecosystem. In Dumont, H., T. A. Shiganova & U. Niermann (eds), The Ctenophore *Mnemiopsis leidyi* in the Black, Caspian and Mediterranean Seas and Other Aquatic Invasions. NATO Science Series: IV. Earth and Environmental Sciences, Vol. 35. Kluwer Academic Publishers, Dordrecht: 71–111.
Siapatis, A., M. Giannoulaki, V. D. Valavanis, A. Palialexis, E. Schismenou, A. Machias & S. Somarakis, 2008. Modeling potential habitat of the invasive ctenophore *Mnemiopsis leidyi* in Aegean Sea. Hydrobiologia 612: 281–295.
Smayda, T. J., 1988. Final report: environmental conditions and plankton dynamics in Narragansett Bay during an annual cycle characterized by a brown tide. Narragansett Bay Project.
Sorarrain, D., 1998. Cambios estacionales en la biomasa de organismos gelatinosos en relación con otros zooplácteres en la bahía Samborombón. MSc Thesis, Facultad de Ciencias Exactas y Naturales, Departamento de Ciencias Marinas, Universidad Nacional de Mar del Plata, Argentina, Mar del Plata: 35 pp.
Stanlaw, K. A., M. R. Reeve & M. A. Walter, 1981. Growth, food and vulnerability to damage of the ctenophore *Mnemiopsis mccradyi* in its early life history stages. Limnology and Oceanography 26: 224–234.
Stoecker, D. K., P. G. Verity, A. E. Michaels & L. H. Davis, 1987. Feeding by larval and post-larval ctenophores on microzooplankton. Journal of Plankton Research 9: 667–683.
Studenikina, E. I., S. P. Volovik, I. A. Miryozan & G. I. Luts, 1991. The ctenophore *Mnemiopsis leidyi* in the Sea of Azov. Oceanology 3: 722–725.
Sullivan, L. J., 2010. Gut evacuation of larval *Mnemiopsis leidyi* A. Agassiz (Ctenophora, Lobata). Journal of Plankton Research 32: 69–74.
Sullivan, L. J. & D. J. Gifford, 2004. Diet of the larval ctenophore *Mnemiopsis leidyi* A. Agassiz (Ctenophora, Lobata). Journal of Plankton Research 26: 417–431.
Sullivan, L. J. & D. J. Gifford, 2007. Growth and feeding rates of the newly hatched larval ctenophore *Mnemiopsis leidyi* A. Agassiz (Ctenophora, Lobata). Journal of Plankton Research 29: 949–965.
Sullivan, B. K. & L. T. MacManus, 1986. Factors controlling seasonal succession of the copepods *Acartia hudsonica* and *A. tonsa* in Narragansett Bay, RI: temperature and resting egg production. Marine Ecology Progress Series 28: 121–128.
Sullivan, B. K., J. H. Costello & D. VanKeuen, 2007. Seasonality of the copepods *Acartia hudsonica* and *Acartia tonsa* in Narragansett Bay, RI, USA during a period of climate change. Estuarine, Coastal and Shelf Science 73: 259–267.
Sullivan, B. K., D. J. Gifford, J. H. Costello & J. R. Graff, 2008. Narragansett Bay ctenophore-zooplankton-phytoplankton dynamics in a changing climate. In Desbonnet, A. & B. A. Costa-Pierce (eds), Science for Ecosystem-Based Management: Narragansett Bay in the 21st Century. Springer, New York: 485–498.
Tendal, O. S., K. R. Jensen & H. U. Riisgård, 2007. Invasive ctenophore *Mnemiopsis leidyi* widely distributed in Danish waters. Aquatic Invasions 2: 272–277.
Titelman, J., L. J. Hansson, T. Nilsen, S. P. Colin & J. H. Costello, 2012. Predator-induced vertical behavior of a ctenophore. Hydrobiologia. doi:10.1007/s10750-012-1056-6.
Tzikhon-Lukanina, E. A., O. G. Reznichenko & T. A. Lukasheva, 1991. Quantitative aspects of feeding in the Black Sea ctenophore *Mnemiopsis leidyi*. Oceanology 31: 272–276.
Vinogradov, M. E., E. A. Shushkina, E. I. Musaeva & P. Yu. Sorokin, 1989. Ctenophore *Mnemiopsis leidyi* (A. Agassiz) (Ctenophora: Lobata) – new settlers in the Black Sea. Oceanology 29: 293–298.
Waggett, R. J. & E. J. Buskey, 2006. Copepod sensitivity to flow fields: detection by copepods of predatory ctenophores. Marine Ecology Progress Series 323: 205–211.
Waggett, R. & J. H. Costello, 1999. Different mechanisms used by the lobate ctenophore, *Mnemiopsis leidyi*, to capture nauplii and adult life stages of the copepod *Acartia tonsa*. Journal of Plankton Research 21: 2037–2052.
Waggett, R. J. & L. J. Sullivan, 2006. Feeding efficiency of the larval ctenophore *Mnemiopsis leidyi* A. Agassiz (Ctenophora, Lobata). Journal of Plankton Research 28: 719–723.
Yazdani Foshtomi, M., B. Abtahi, A. Esmaili Sari & M. Taheri, 2007. Ion composition and osmolarity of Caspian Sea ctenophore, *Mnemiopsis leidyi*, in different salinities.

Journal of Experimental Marine Biology and Ecology 352: 28–34.

Zaika, V. E. & N. K. Revkov, 1998. Nutrition of the Black Sea ctenophore in dependence on zooplankton abundance. Gydrobiologichesky Zhurnal 34: 29–35. (in Russian).

Zaika, V. Ye. & N. G. Sergeeva, 1990. Morphology and development of *Mnemiopsis mccradyi* (Ctenophora, Lobata) in the Black Sea. Zoologicheskiy Zhurnal 69: 5–11. (in Russian).

Zhang, W., P. F. Hendrix, B. A. Snyder, M. Molina, J. Li, X. Rao, E. Siemann & S. Fu, 2010. Dietary flexibility aids Asian earthworm invasion in North American forests. Ecology 91: 2070–2079.

Hydrobiologia (2012) 690:47–55
DOI 10.1007/s10750-012-1052-x

JELLYFISH BLOOMS

Foods of *Velella velella* (Cnidaria: Hydrozoa) in algal rafts and its distribution in Irish seas

Jennifer E. Purcell · Emmett Clarkin · Thomas K. Doyle

Published online: 15 March 2012
© Springer Science+Business Media B.V. 2012

Abstract The pleustonic hydrozoan, *Velella velella*, occurs throughout tropical to cold-temperate oceans of the world and sometimes are stranded in masses along hundreds of kilometers of beaches. In June 2009, we encountered algal rafts in the Celtic Sea containing many *V. velella* that we immediately preserved for gut content analysis. Available prey were enumerated from raft-associated fauna and zooplankton sampled nearby. The identifiable prey (331) in *V. velella* comprised 78% raft-associated prey (primarily harpacticoid copepods, cumaceans, small fish) and 22% pelagic prey (calanoid copepods, barnacle nauplii, fish eggs). Comparison of ingested with available prey showed selection for fish eggs and small fish, among others; therefore, the null hypothesis, that *V. velella* consumed all available prey equally, was rejected. Transport by wind and water concentrate *Velella* spp. in convergences with algal rafts, which suggests that they are important predators of raft—as well as pelagic fauna. A visual survey in September 2004 across the Celtic Sea and beach-stranding data show that *V. velella* is very abundant in Irish waters at times. Its circumpolar abundance, consumption of pelagic prey and production from symbiotic zooxanthellae, and mass deposition on beaches suggest that *V. velella* is important in open-ocean carbon cycling and in transport of pelagic production to landmasses.

Keywords Jellyfish · Fish · Fish eggs · Zooplankton · Rafting · Convergence

Guest editors: J. E. Purcell, H. Mianzan & J. R. Frost / Jellyfish Blooms: Interactions with Humans and Fisheries

J. E. Purcell (✉)
Western Washington University, Shannon Point Marine Center, 1900 Shannon Point Road, Anacortes, WA 98221, USA
e-mail: purcelj3@wwu.edu

J. E. Purcell · T. K. Doyle
Coastal and Marine Research Centre, ERI, University College Cork, Naval Base, Haulbowline Island, Cobh, Co. Cork, Ireland

E. Clarkin
School of Biological Sciences, Queen's University, MBC, Lisburn Road, Belfast BT9 7BL, Northern Ireland, UK

Present Address:
E. Clarkin
Australian Institute of Marine Science PMB 3, Townsville MC, Townsville 4810, Queensland, Australia

Introduction

Velella velella (Linnaeus) is an unusual pleustonic cnidarian found globally in tropical to temperate open ocean waters (reviewed in Daniel, 1976; Bieri, 1977; McGrath, 1985; Evans, 1986; see also Mianzan & Girola, 1990; Flux, 2008). It is a holoplanktonic anthomedusan, previously placed in a separate order (Chondrophora) because of its chitinous sail. Its life cycle includes the colonial asexual stage that is visible at the ocean surface and a medusa stage that sinks to 600–1,000 m depths before reproducing sexually (Woltereck, 1904). The larvae then ascend to the surface and develop the sail (Bieri, 1977; Larson, 1980). Beneath the sail, the colonial stage has a large central stomach surrounded by numerous small gonozooids, all of which are involved in feeding and digestion of animal prey (Kirkpatrick & Pugh, 1984). Around the edge of the colony is a ring of sturdy tentacles used in prey capture. In addition to active predation, *V. velella* contains symbiotic zooxanthellae, analogous to the coral- and medusa-zooxanthella symbioses that provide the cnidarian hosts with nutrition.

Because their sail protrudes into the air, *V. velella* colonies are blown by winds and accumulate at the surface of wind-generated convergences of Langmuir cells, along with flotsam, weeds, and other animals, such as young fish and crabs (Shenker, 1988). Such convergences are known to be zones of intensified trophic activity beneath and at the water surface (Kingsford & Choat, 1985; Thiel & Gutow, 2005).

Ocean rafts, composed of coastally derived detached floating seaweed, are an important habitat for neustonic and intertidal marine organisms in terms of food, shelter, substrate attachment, and passive dispersal within a harsh ocean surface environment (Thiel & Gutow, 2005; Vandendriessche et al., 2006). After the initial detachment from the shore by storms, grazing, or seasonal release of thalli, macroalgal rafts form a unique habitat drifting passively, controlled by ocean currents, frontal systems, tidal regimes, and surface winds, and forming patches and lines of drift on the sea surface (Davenport & Rees, 1993; Ingólfsson, 1995, 1998; Vandendriessche et al., 2007).

The process of rafting is of great importance for several reasons. It has the potential to disperse marine species beyond their expected biogeographic ranges and also to connect populations at broad and local scales (Ó Foighil et al., 1999; Matthew & Orth, 2002; McCormick et al., 2008; Waters, 2008). Many of these studies investigated patterns of marine invertebrate dispersal with special interest in species lacking a pelagic larval phase in their life history. The importance of rafts to attractant juvenile fish has been increasingly highlighted with particular interest in juveniles of some commercial fish species (Komatsu et al., 2007). Juvenile fish are well known to inhabit ocean surface waters (Kingsford & Choat, 1985) and have a tendency to aggregate around drift material (Davenport & Rees, 1993; Vandendriessche et al., 2007). The diversity of fish was highest below drifting seaweed when compared to other floating objects at sea (Fedoryako, 1989; Kingsford, 1995; Vandendriessche et al., 2007). The presence of raft-associated macrofauna with the seaweed would contribute to the attraction of fish for foraging as well as for shelter from aerial- and aquatic-predators in the exposed neustonic zone.

Because of their open-ocean existence, little is known about the ecology of *V. velella*. Their occurrence at sea has been reported mostly as a curiosity or as parts of studies on other organisms (e.g., Daniel, 1976; Evans, 1986; García et al., 2003; Parker et al., 2005; Thiel & Gutow, 2005; Emmett & Krutzkowsky, 2008). Most knowledge of their spatial and temporal occurrences comes from strandings on beaches (e.g., Bieri, 1977; McGrath, 1985, 1994). They are renowned for massive beach strandings. The organic material deposited on an northeastern Pacific (Oregon, USA) beach by one mass stranding averaged 2,573 g ash-free dry weight per meter of shoreline, containing 1,223 g m^{-1} of carbon and 347 g m^{-1} of nitrogen (Kemp, 1986).

From a human perspective, *V. velella* is important as a predator or food of commercially important fish and crabs, as well as birds and sea turtles (Bieri, 1961; Wickham, 1979; Vermeer & Devito, 1988; Parker et al., 2005; Emmett & Krutzkowsky, 2008). The single previous study of their diet showed substantial predation on fish eggs off California, USA; 3.5 and 48% of the prey items in the colonies were anchovy (*Engraulis mordax* Girard) and jack mackerel (*Trachurus symmetricus* (Ayres)) eggs, respectively (Bieri, 1961). Regurgitations of 28% of Leach's storm-petrels (*Oceanodroma leucorhoa* (Vieillot)) contained *V. velella* during August 1983 in the NE Pacific (Vermeer & Devito, 1988). Stomachs of 25% of loggerhead sea turtles sampled in the central North Pacific contained *V. velella*, which composed 10.6% of the gut-content volume (Parker et al., 2005). Stomachs of 0.2% of the Pacific hake (*Merluccius productus* (Ayres)) in 2004 and of 0.4% of the jack mackerel contained *V. velella* only in 1999 out of the 7 years sampled off the NW US coast (Emmett & Krutzkowsky, 2008).

Here, we report the gut contents of *V. velella* colonies opportunistically collected in an algal raft in the Celtic Sea. We tested the null hypothesis that *V. velella* consumed all available prey equally by calculation of prey selection from comparisons of eaten prey with prey available in the water column and on the raft. We discuss the potential importance of *V. velella* as predators in algal raft communities. In order to illustrate the abundance of *V. velella*, we also present sightings from a ferry survey in 2004 and summarize available data on the dates and sizes of *V. velella* in beach strandings around Ireland.

Materials and methods

Samples for this study were collected in the Celtic Sea during a cruise of the Marine Institute's Celtic Voyager research vessel 13–22 June 2009 (Fig. 1). The purpose of the cruise was to investigate how fronts and coastal processes affect the dispersal and entrainment of jellyfish, zooplankton, and rafts.

Rafts

Raft samples were collected opportunistically during the 10-day cruise. During daylight hours, two observers stood on the bridge of the ship looking for floating seaweed patches ahead of the research vessel track; when a raft was located the ship slowed down and maneuvered carefully toward it. A large specially designed dip net (80 × 80 cm mouth with 0.25 × 0.25 cm mesh) was deployed over the starboard side of the ship and the seaweed rafts gently scooped up to minimize disturbance of the associated flora and fauna. When a seaweed raft was successfully brought aboard, it was placed in one or more 20-l bucket, preserved in 4% formaldehyde in seawater, labeled, and sealed with a lid until later analysis in the laboratory.

Each sample was processed within a large fume hood. First, the 4% formaldehyde solution was drained through a 500-μm-mesh sieve. Then all associated fauna were removed from the algae in a large white sorting tray filled with seawater. All organisms were sorted into major taxonomic groups and preserved in 74% industrial methylated spirits and seawater for later identification to species, if possible, with the aid of a dissecting microscope. All major seaweed species composing the raft were surface dried with absorbent paper towels for wet weight analysis to the nearest 0.1 g. All algae were identified, sorted, weighed, and returned to storage in 4% formaldehyde and seawater.

The large raft from which the *V. velella* colonies were counted, sail lengths measured, and gut contents analyzed was collected 17 June 2009 in the Celtic Gyre circulation (52.216°N, −5.798°W) (Fig. 1). Additional rafts were collected 14 June 2009 off Clear Island (51.440°N, −9.438°W) and all *V. velella* counted and sail lengths measured; gut contents were not analysed because the time between collection and preservation, which would affect recognition of consumed prey, was unknown for these small rafts.

V. velella

All visible *V. velella* were collected immediately from the large raft onboard the ship and then preserved in

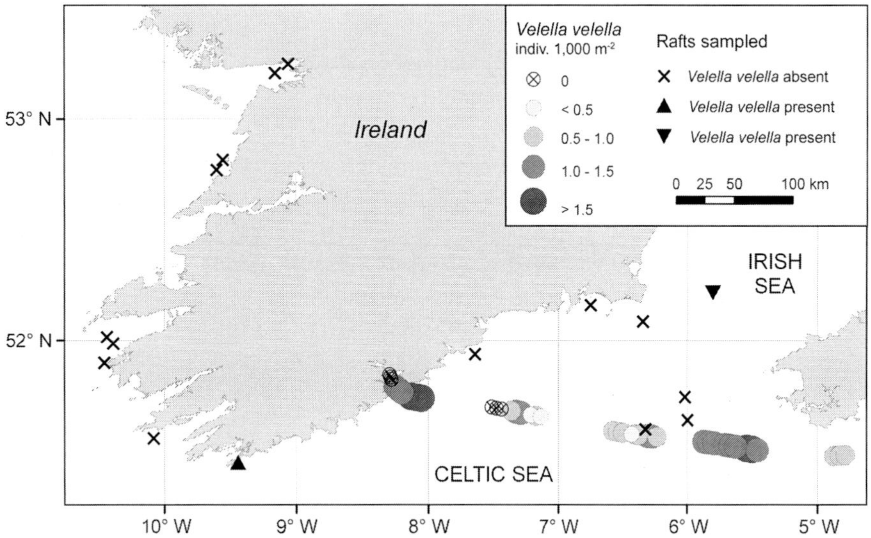

Fig. 1 Locations of *V. velella* collected in algal rafts on 14 and 17 June 2009 (*upward* and *downward triangles*, respectively). Locations in Irish waters where rafts lacking *V. velella* were collected 13–22 June 2009 are marked by ×. Numbers of large *V. velella* seen during a ferry survey in the Celtic Sea 17 September 2004

10% formalin–seawater solution for later analysis in the laboratory. We estimate the time between removal of the raft from the water and preservation was ~15 min. In the laboratory, all specimens were analyzed by first removing the sail from the surrounding tissue and measuring its maximum length with the aid of an optical micrometer or a ruler placed on the stage of a dissecting microscope. Then the central stomach was opened and all food remains identified to the lowest possible taxon, counted, and length measured, if possible. All surrounding gonozooids were split lengthwise and prey identified, counted, and measured.

Zooplankton

The zooplankton available in the water column was collected near the raft in the Celtic Sea (51.444°N, −9.453°W) with a 1-m-diameter ring net hauled vertically from 45 m to the surface (35.325 m^3 filtered) (Table 1). The concentrated sample was preserved immediately in 5% formalin–seawater solution for later analysis in the laboratory. Rare taxa (e.g., fish eggs and larvae) were identified and counted from the entire sample and standardized to number m^{-3}. Three subsamples were taken from the whole sample with a 5-ml stempel pipette and all organisms identified to general taxon (e.g., calanoid copepods and barnacle nauplii) and counted from each. Each taxon in the subsamples were averaged and standardized to number m^{-3}.

Selection

We tested the null hypothesis that *V. velella* consumed all available prey equally by calculating prey selection. Pearre's (1982) electivity index, "C", was used to test for significant differences in consumption as compared with available prey in the raft and in the zooplankton sample using χ^2 analyses. Because the source (raft or plankton) was uncertain for some prey types (i.e., gastropods, decapods), the most abundant prey source (plankton) was used to calculate a conservative estimate.

Ferry survey

Surface counts for jellyfish were conducted from a passenger ferry crossing the Celtic Sea from Cork Harbour to Swansea Port on 17 September 2004. The methodology was described by Doyle et al. (2007). In brief, an observer on the outside deck of the ferry (~17 m above the water surface) visually identified jellyfish to species (including large *V. velella*) and estimated their numbers according to six categories of abundance: 0, 1–10, 11–50, 51–100, 101–500, and >500. During each 15-min sampling period, the numbers of each species was tallied every 5-min (three 5-min bins) to permit finer resolution of jellyfish densities. Five minutes separated successive 15-min periods and 30–60-min breaks were taken periodically to avoid observer fatigue. GPS positions were taken every 5 min using a handheld GPS. Sea state (Beaufort scale) and glare (as defined by Houghton et al., 2006a) were recorded every 15 min at the beginning and end of each sampling period. The area sampled visually during each 5-min count was assigned according to previous estimates (Doyle et al., 2007) and organisms standardized to numbers km^{-2}.

Results

Rafts were collected from 18 locations and *V. velella* was found in rafts from two locations in the Celtic Sea (Fig. 1). Four small rafts collected off Clear Island contained 1, 8, 30, and 188 specimens; *V. velella* in those rafts were 4.9, 7.6 ± 1.5, 5.7 ± 1.7, and 5.7 ± 1.7 mm (mean ± standard deviation) in sail length, respectively. The large raft from the northern location contained 136 *V. velella* (12.1 ± 8.1 mm long) that were dispersed throughout the raft (Fig. 2). Of those, 98 were preserved for gut analysis. All of the *V. velella* were in undamaged, healthy condition. Comparison between the specimens used in gut analysis (12.6 ± 9.0 mm) and 38 remaining in the raft (11.0 ± 5.0 mm) did not differ significantly in size (Mann–Whitney Rank Sum Test; $U = 1733.500$; $P = 0.880$). Therefore, the gut contents specimens were representative of the whole group.

The largest raft was comprised of several algal species and many associated animals. The algae from the raft totaled 2,319 g wet weight of the following species: *Ulva* sp. (66.4%), *Fucus* spp. (23.6%), *Himanthalia elongata* (Linnaeus) (4.0%), *Ascophylum nodosum* (Linnaeus) (2.5%), *Chorda filum* (Linnaeus) (1.7%), *Pelvetia caniculata* (Linnaeus) (0.4%), *Polysiphonia lanosa* (Linnaeus) (<0.1%). Many types

Table 1 Prey organisms in the gut contents of 98 *V. velella* colonies, potential prey in an algal raft and in the water column in the Celtic Sea 17 June 2009

	In guts		In raft		In water		Selection	
	No.	%	No.	%	No. m^{-3}	%	*C*	X^2
Raft prey								
Brown harpacticoid	85	26.0	984	74.8	0	0	−513,219	456.5***
Blue harpacticoid	107	32.7	73	5.6	0	0	+336,393	339.4***
Other harpacticoids	0	0	5	0.4	0	0	NT	NT
Isopods	0	0	127	9.7	0	0	NT	NT
Flies + larvae	4	0.3	117	8.7	0	0	−278,080	178.9***
Cumaceans	27	8.3	0	0	0	0	NT	NT
Amphipods	1	0.3	3	0.2	0	0	NT	NT
Fish	17	5.2	2	0.15	0	0	+115,152	141.6***
Raft totals	241	73.7	1,315	100	0	0		
Pelagic prey								
Calanoid copepods	11	3.4	0	0	783.2	87.1	−317,118	495.0***
Cladocerans	10	3.1	0	0	0	0	NT	NT
Cirripede larvae	14	4.3	0	0	0	0	NT	NT
Polychaete larvae	23	7.0	0	0	1.4	0.2	+112,595	122.8***
Gastropods	12	3.7	1	0.08	58.5	6.5	+154,645	84.4***
Decapods	12	3.7	3	0.2	38.2	4.2	+187,615	45.6***
Chaetognaths	0	0	0	0	3.8	0.4	NT	NT
Appendicularia	0	0	0	0	7.6	0.8	NT	NT
Jellies	0	0	0	0	2.3	0.25	NT	NT
Fish eggs	4	1.2	0	0	3.3	0.4	+139,975	112.0***
Fish larvae	0	0	0	0	0.5	0.5	NT	NT
Pelagic totals	86	26.3			898.8	100		

Prey selection index (*C*) was tested by Chi-square (X^2) analysis, according to Pearre (1982)

NT not tested

*** $P < 0.001$

of mobile invertebrates and one fish species (*Ciliata mustela* (Linnaeus)) that were potential prey for *V. velella* were living in the raft (Table 1). The most numerous were harpacticoid copepods (~81% of total), predominantly of three types: short brown (~1 mm long), light with a blue eye spot and blue-edged or red-edged appendages and white eggs (1–2 mm long), and slender brown (1–2 mm long excluding tail rami); unfortunately, species identification for the harpacticoids was not possible. Isopods and flies and their larvae also were abundant (~9–10%). Two small fish (~30 mm in length) were collected with the raft.

Zooplankton and ichthyoplankton also were potential prey for *V. velella* (Table 1). Calanoid copepods predominated in the zooplankton (~87%), with crab and gastropod larvae common (4–6%). Fish eggs occurred at a density of 3.3 m^{-3} above 45 m depth.

Gut contents

A total of 327 prey items were identified from the gut contents of *V. velella* (Table 1). Most of the prey remains in the central stomach and gonozooids were very digested; a minimum of 538 additional prey (crustaceans) could not be identified further. The predominant identifiable prey items were harpacticoid copepods (58.7%), cumaceans (8.3%), and fish (5.2%). At least 74% of all prey taxa were from the raft. Gastropods and decapods were found both on the

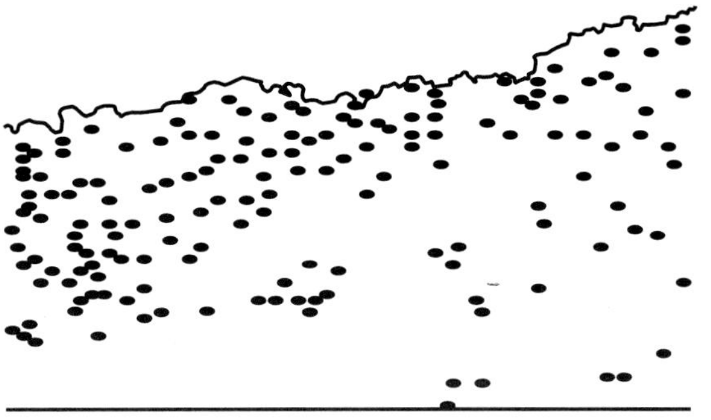

Fig. 2 Locations of *V. velella* (*black ovals*) in a large algal raft in the Celtic Sea collected 17 June 2009. The *uneven line* is the raft edge and the *straight line* is the side of the ship. The *V. velella* are ~12 mm in length (*scale* for the figure). Colony locations were mapped from a photograph; however, the orientation of each *oval* does not indicate colony orientation

raft and in the plankton, but we attributed them to the source where they were most numerous (plankton). The decapods in the gut contents were too digested to identify more precisely. Relatively few prey, including calanoid copepods, cladocerans, barnacle and polychaete larvae were from the plankton (total 26%). Seventeen fish >6 mm in length from the raft were found in the central stomach; four fish eggs (1.0–1.3 mm diameter) from the plankton also were eaten. Thus, *V. velella* eat both raft and pelagic prey.

Selection

Although the short brown harpacticoid was most numerous in the raft, they were second in the gut contents and selection by *V. velella* was negative (Table 1). The blue-eyed harpacticoids were less numerous in the raft, but were the most frequent prey and selection was positive. Several prey types were too rare to be tested. Fish from the raft were large and also could be counted from their hard eye lenses; selection for fish was positive. Among planktonic prey, polychaete, gastropod, and decapod larvae all had positive selection indices. Selection for fish eggs by *V. velella* also was positive. Thus, we rejected the null hypothesis that *V. velella* consumed all available prey equally.

Ferry survey

A total of 39 five-minute counts (190 min of sea surface observations) for jellyfish were made across the Celtic Sea 17 September 2004. *V. velella* was recorded in 84.6% of the counts. Their mean density along the transect was 1.3 ind. 1,000 m^{-2} and their highest density was 5.1 ind. 1,000 m^{-2}.

Discussion

V. velella in Irish seas

Beach strandings of *V. velella* have been reported around the entire coastline of Ireland, with the exception of the Irish Sea (reviewed in McGrath, 1985, 1994). McGrath (1985) concluded that the consistently large colonies in winter were at the end of the asexual stage. The small size of the specimens we found in the spring suggests new colonies that recently arrived at the surface (Fig. 3). Our data from Ireland suggest that newly produced colonies grow over the summer and release medusae that reproduce at depth. Bieri (1977) combined data from the North Atlantic Ocean, where small specimens occurred in June, and the Mediterranean Sea, where small specimens were found in February through April, and concluded that *V. velella* produced two generations per year. Although two generations may occur in the warm waters of the Mediterranean Sea (15–30°C), in the cool Irish waters (10–15°C) one main cycle of production per year seems more likely. Although small specimens (mainly <20 mm) were found on Irish beaches in October–December, survival and acquisition of enough food during the cold, rough North Atlantic winter to complete the two-stage life cycle before spring seems unlikely. It is possible that release and maturation of *V. velella* medusae is continuous. The seasonal patterns of *V. velella* require further study.

V. velella may play important roles in Irish coastal waters when it occurs in high abundance, as on 17 September 2004 (Fig. 1) when it was observed across the entire Celtic Sea. Although only large *V. velella* could be counted with certainty, such widespread blooms may be common but unreported, except as beach strandings (i.e., McGrath, 1985, 1994). Large numbers of the European storm petrel, *Hydrobates pelagicus* (Linnaeus), forage in the Celtic Sea during the summer months and may take from the sea surface small fish, plankton, mollusks, crustaceans, and *V. velella*, as do Leach's storm-petrels in the northeast

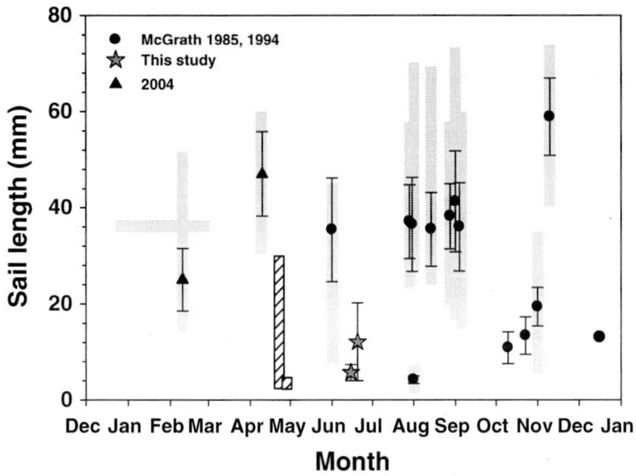

Fig. 3 Sizes of *V. velella* collected in algal rafts in the Celtic sea (*stars*, this study, June 2009) and washed ashore in Ireland throughout the year in 1976–1984 and 1992 (*circles*, from McGrath, 1985, 1994), and previously unpublished strandings data from 2004 (*triangles*) and 2011 (*hatched bars*). *Symbols with error bars* means ± s.d.; *vertical bars* ranges; *horizontal bar* size of specimens in January–March

Pacific (Vermeer & Devito, 1988). Other probable predators of *V. velella* in Irish waters include the ocean sunfish, *Mola mola* (Linnaeus), and leatherback turtles, *Dermochelys coriacea* Vandelli, which feed on surface aggregations of gelatinous prey (Houghton et al., 2006b).

Feeding of *V. velella*

Our results show for the first time that *V. velella* colonies are predators in algal raft communities. They were actively consuming raft epifauna, primarily harpacticoid copepods, fish, and dipteran larvae and adults. They also consumed components of the plankton community, including calanoid copepods, cladocerans, barnacle and polychaete larvae, and fish eggs. Our study confirmed data on fish eggs in the diet, as found off California by Bieri (1961); however, the predominant prey items in *V. velella* off California were euphausiid eggs, which were not available in the Celtic Sea. As in our study, *V. velella* off California also contained barnacle larvae and a cumacean (Wickham, 1979). Diets within species differ among locations depending on the prey available (e.g., Purcell et al., 2001). Therefore, the diets of *V. velella* in our study reflect the availability of raft fauna and differ from the foods found in open water by Bieri (1961).

In the raft and plankton food webs, *V. velella* is a predator of fish eggs and young, and it also consumes the crustacean foods of fish. Other pelagic cnidarians also are known to be important predators on early stages of fish and also potential competitors for zooplankton foods (reviewed in Purcell & Arai, 2001). The diet of another pleustonic hydrozoan, *Physalia physalis* (Linnaeus), consists mainly of small fish (Purcell, 1984). Both of these surface-dwelling species accumulate in convergences, where other floating and positively buoyant organisms, such as some fish eggs, also accumulate. Small fishes often seek out floating objects, presumably for protection from predators (e.g., Kingsford & Choat, 1985). Thus, both hydrozoans are well adapted for predation on fish eggs and small fish. Although, *V. velella* probably was not recognized as a threat to ichthyoplankton, the extensive bloom reported by García et al. (2003) may have been an important cause of mortality for the tuna eggs and larvae sampled in their study in the Mediterranean Sea.

Although we rejected the null hypothesis that *V. velella* consumed all available prey equally, there are many uncertainties in our analysis of prey selection. Some of the prey types were in the gut contents but were missing or scarce among raft dwellers (i.e., cumaceans, small fish). We believe it likely that those prey escaped during collection of the raft by dip net. Second, plankton was collected from 45 m to the surface; however, prey capture by *V. velella* colonies is only at the water surface. Some fish eggs are neustonic and their near-surface availability may have been greater than estimated. These two potential sampling errors would reduce apparent prey availability and incorrectly inflate selection indices. Conversely, if calanoid copepods were more abundant at depth than near-surface, their availability would have been overestimated and selection underestimated. Third, different durations of prey in the guts could have biased their apparent importance in the diet of *V. velella*. Some prey types (i.e., shelled gastropods, large fish) could take longer to digest and be recognizable longer than small copepods. Because we had no data on gut clearance times, which vary by prey type and size in other pelagic cnidarians (e.g., Martinussen & Båmstedt, 1999), no adjustments were made. This potential analysis error would increase apparent predation and, again, incorrectly inflate selection indices. Other uncertainties include the large amounts of unidentifiable crustacean exoskeleton in the gut contents, isopods being absent in the gut contents but common in the raft, and the source (plankton or raft) of some prey (i.e., gastropods and decapods).

The importance of rafting

The extent that *V. velella* can be classified as a rafter remains to be fully determined because it has been studied little. They occurred in 11% of the locations rafts were collected. *V. velella* are known to be passively associated with drifting macroalgae and now gut contents show they feed on raft-associated organisms. Gut content analysis of ctenophores (*Pleurobranchia pileus* Müller) revealed portunid larvae (Davenport & Rees, 1993), which are found on rafting material; therefore, other pelagic predators also take advantage of rafting material when available (Thiel & Gutow, 2005).

Previous rafting studies classified self-buoyant hydrozoans, such as *V. velella* and *Physalia* spp. as 'associated organisms', because they are in drifting seaweed when caught in the same current, front, or eddy (Adams, 1960; Ingólfsson, 1995; Thiel & Gutow, 2005). Our study indicates the importance of such interactions for rafting communities, significantly affecting survival, and hence transport, of many species, and apparently advantageous for pelagic predators because of the presence and attraction of prey. In addition to enhanced prey availability of prey in the raft, *V. velella* may gain protection from its vertebrate predators while in algal rafts.

V. velella is an abundant predator of zooplankton and ichthyoplankton in vast regions of the oceans. It also gains nutrition from symbiosis with zooxanthellae. It is eaten by a variety of oceanic vertebrate predators, some of which are commercially valuable species, and serves as transport for megalopae of commercial crabs. Not only it is of importance in the trophic web of the open ocean, it is blown ashore and deposits large amounts of pelagic production along coastlines. Because of its great abundance and widespread occurrence, *V. velella* may be important as a predator of fish and fish eggs and in carbon transport and should be targeted for study rather than merely noted as a curiosity.

Acknowledgments Funding for ship time was granted through the "Integrated marine exploration programme" by the Marine Institute, Galway. Cruise Co-ordinator was Prof. Mark Johnson and Chief Scientist was Dr. Louise Allcock from the National University of Ireland, Galway. TKD was supported by Grant-Aid Agreement No PBA/AF/08/002(01) under the Sea Change strategy with the support of the Marine Institute and the Marine Research Sub-Programme of the National Development Plan 2007–2013, co-financed under the European Regional Development Fund. TKD and JEP were supported by the European Union Regional Development Fund INTERREG 4A (EcoJel). Funding for EC was provided by the Department for Employment and Learning (DEL) Northern Ireland. We thank T. Bastian for preparation of Fig. 1.

References

Adams, J. A., 1960. A contribution to the biology and postlarval development of the *Sargassum* fish, *Histro histro* (Linneaus), with a discussion of the *Sargassum* complex. Bulletin of Marine Science of the Gulf and Caribbean 10: 55–82.

Bieri, R., 1961. Post-larval food of the pelagic coelenterate, *Velella lata*. Pacific Science 15: 553–556.

Bieri, R., 1977. The ecological significance of seasonal occurrence and growth rate of *Velella* (Hydrozoa). Publication of the Seto Mar. Biological Laboratory 24: 63–76.

Daniel, R., 1976. Chondrophora of the Indian Ocean. Journal of the Marine Biological Association of India 18: 110–121.

Davenport, J. & E. I. S. Rees, 1993. Observations on neuston and floating weed patches in the Irish Sea. Estuarine, Coastal and Shelf Science 36: 395–411.

Doyle, T. K., J. D. R. Houghton, S. M. Buckley, G. C. Hays & J. Davenport, 2007. The broad-scale distribution of five jellyfish species across a temperate coastal environment. Hydrobiologia 579: 29–39.

Emmett, R. L. & G. K. Krutzkowsky, 2008. Nocturnal feeding of Pacific hake and juvenile mackerel off the mouth of the Columbia River, 1998–2004: implications for juvenile salmon predation. Transactions of the American Fisheries Society 137: 657–676.

Evans, F., 1986. *Velella velella* (L.), the 'by-the-wind-sailor', in the North Pacific Ocean in 1985. Marine Observer 56: 196–200.

Fedoryako, B. I., 1989. A comparative characteristic of the oceanic fish assemblage associated with floating debris. Journal of Ichthyology 29: 128–137.

Flux, J. E. C., 2008. First mass stranding of *Velella velella* in New Zealand. Marine Biodiversity Records 1: 1–2.

García, A., F. Alemany, P. Velez-Belchí, J. M. Rodríguez, J. L. López Jurado, C. González Pola & J. M. de la Serna, 2003. Bluefin and frigate tuna spawning off the Balearic archipelago in the environmental conditions observed during the 2002 spawning season. Collective Volume of Scientifics Papers ICCAT 55: 1261–1270.

Houghton, J. D. R., T. K. Doyle, J. Davenport & G. C. Hays, 2006a. Developing a simple, rapid method for identifying and monitoring jellyfish aggregations from the air. Marine Ecology Progress Series 314: 159–170.

Houghton, J. D. R., T. K. Doyle, J. Davenport & G. C. Hays, 2006b. The ocean sunfish *Mola mola*: insights into distribution, abundance and behaviour in the Irish and Celtic

Seas. Journal of the Marine Biological Association of the United Kingdom 86: 1237–1243.

Ingólfsson, A., 1995. Floating clumps of seaweed around Iceland: natural microcosms and a means of dispersal for shore fauna. Marine Biology 122: 13–21.

Ingólfsson, A., 1998. Dynamics of macrofaunal communities of floating seaweed clumps off western Iceland: a study of patches on the surface of the sea. Journal of Experimental Marine Biology and Ecology 231: 119–137.

Kemp, P. F., 1986. Deposition of organic matter on a high-energy sand beach by a mass stranding of the cnidarian *Velella velella* (L.). Estuarine, Coastal and Shelf Science 23: 575–579.

Kingsford, M. J., 1995. Drift algae: a contribution to near-shore habitat complexity in the pelagic environment and an attractant for fish. Marine Ecology Progress Series 116: 297–301.

Kingsford, M. J. & J. H. Choat, 1985. The fauna associated with drift algae captured with a plankton-mesh purse seine net. Limnology and Oceanography 30: 618–630.

Kirkpatrick, P. A. & P. R. Pugh, 1984. Siphonophores and Velellids. In Kermack, D. M. & R. S. K. Barnes (eds), Synopsis of the British Fauna no. 29. Pittman Press, Bath: 1–155.

Komatsu, T., K. Tatsukawa, J. B. Filippa, T. Sagawa, D. Matsunaga, A. Mikami, K. Ishida, T. Ajisaka, K. Tanaka, M. Aoki, W.-D. Wang, H.-F. Liu, S.-D. Zhang, M.-D. Zhou & T. Sugimoto, 2007. Distribution of drifting seaweeds in eastern East China Sea. Journal of Marine Systems 67: 245–252.

Larson, R. J., 1980. The medusa of *Velella velella* (Linnaeus, 1758) (Hydrozoa, Chondrophorae). Journal of Plankton Research 2: 183–186.

Martinussen, M. B. & U. Båmstedt, 1999. Nutritional ecology of gelatinous planktonic predators. Digestion rate in relation to type and amount of prey. Journal of Experimental Marine Biology and Ecology 232: 61–84.

Matthew, C. H. & R. J. Orth, 2002. Long-distance dispersal potential in a marine macrophyte. Ecology 83: 3319–3330.

McCormick, T. B., L. M. Buckley, J. Brogan & L. M. Perry, 2008. Drift macroalgae as a potential dispersal mechanism for the white abalone *Haliotis sorenseni*. Marine Ecology Progress Series 362: 225–232.

McGrath, D., 1985. The by-the-wind sailor *Velella velella* (L.) (Coelenterata: Hydrozoa) in Irish waters 1976–1984. The Irish Naturalists' Journal 21: 479–484.

McGrath, D., 1994. Extraordinary occurrences of the by-the-wind sailor *Velella velella* (L.) (Cnidaria) in Irish waters in 1992. The Irish Naturalists' Journal 24: 383–388.

Mianzan, H. W. & C. V. Girola, 1990. The pleustonic coelenterates *Physalia physalis* (Linne, 1758), *Velella velella* (Linne, 1758) and *Porpita umbella* Muller, 1776 in southwestern Atlantic waters. Investigaciones Marinas CICIMAR 5: 97–98.

Ó Foighil, D., B. A. Marshall, T. J. Hilbish & M. A. Pino, 1999. Trans-Pacific range extension by rafting is inferred for the flat oyster *Ostrea chilensis*. Biological Bulletin 196: 122–126.

Parker, D. M., W. J. Cooke & G. H. Balazs, 2005. Diet of oceanic loggerhead sea turtles (*Caretta caretta*) in the Central North Pacific. Fishery Bulletin 103: 142–152.

Pearre, S. Jr., 1982. Estimating prey preference by predators: uses of various indices, and a proposal of another based on χ^2. Canadian Journal of Fisheries and Aquatic Science 39: 914–923.

Purcell, J. E., 1984. Predation on fish larvae by *Physalia physalis*, the Portuguese man of war. Marine Ecology Progress Series 19: 189–191.

Purcell, J. E. & M. N. Arai, 2001. Interactions of pelagic cnidarians and ctenophores with fish: a review. Hydrobiologia 451: 27–44.

Purcell, J. E., T. A. Shiganova, M. B. Decker & E. D. Houde, 2001. The ctenophore *Mnemiopsis* in native and exotic habitats: U.S. estuaries versus the Black Sea basin. Hydrobiologia 451: 145–176.

Shenker, J. M., 1988. Oceanographic associations of neustonic larval and juvenile fishes and Dungeness crab megalopae off Oregon. Fishery Bulletin 86: 299–317.

Thiel, M. & L. Gutow, 2005. The ecology of rafting in the marine environment. II. The rafting organisms and community. Oceanography and Marine Biology: An Annual Review 43: 279–418.

Vandendriessche, S., M. Vincx & S. Degraer, 2006. Floating seaweed in the neustonic environment: a case study from Belgian coastal waters. Journal of Sea Research 55: 103–112.

Vandendriessche, S., M. Messiaen, S. O'Flynn, M. Vincx & S. Degraer, 2007. Hiding and feeding in seaweed: floating seaweed clumps as possible refuges or feeding grounds for fishes. Estuarine, Coastal and Shelf Science 71: 691–703.

Vermeer, K. & K. Devito, 1988. The importance of *Paracallisoma coecus* and myctophid fishes to nesting fork-tailed and Leach's storm-petrels in the Queen Charlotte Islands, British Columbia. Journal of Plankton Research 10: 63–75.

Waters, J. M., 2008. Driven by the West Wind Drift? A synthesis of southern temperate marine biogeography, with new directions for dispersalism. Journal of Biogeography 35: 417–427.

Wickham, D. E., 1979. The relationship between megalopae of the Dungeness crab, *Cancer magister*, and hydroid, *Velella velella*, and its influence on abundance estimates of *C. magister* megalopae. California Fish and Game 65: 184–186.

Woltereck, R., 1904. Ueber die Entwicklung der *Velella* aus einer in der tiefe vorkommonden Larvae. Zoologische Jahrbuch Supplement 7, Festschrift A. Weissmann: 347–372.

Hydrobiologia (2012) 690:57–67
DOI 10.1007/s10750-012-1048-6

Review Paper

Do Staurozoa bloom? A review of stauromedusan population biology

Lucília S. Miranda · André C. Morandini · Antonio C. Marques

Published online: 15 March 2012
© Springer Science+Business Media B.V. 2012

Abstract The study of "jellyfish blooms" provides important data toward determining the causes and consequences of these phenomena; however, the definition of "bloom" remains controversial and different concepts have been adopted in recent works. By addressing the biological and convenience definitions, this study tested the adequacy of the different concepts of "blooms" for the Class Staurozoa (Cnidaria). From seasonal monitoring data of some species of Staurozoa, we concluded that stauromedusae bloom if we used the biological concept of "bloom", which considers the life cycle and resulting changes in the abundances of these animals. By contrast, the small, benthic, inconspicuous, and non-harmful stauromedusae do not bloom if we use the convenience concept of "bloom", which constrains the events to those that humans can observe and that cause damage to human activities. In other words, the same group of organisms either is or is not capable of blooming depending on which concept of "bloom" is used. In fact, previous literature has suggested that Staurozoa could not bloom, which indicates that the study of "jellyfish blooms" can be biased, considering convenience rather than biological reasoning.

Keywords Stauromedusae · Seasonality · Life cycle · Evolution

Introduction

Increases in the abundance of medusae are a phenomena of great interest, especially because of the negative impacts on ecosystem dynamics (Morandini & Marques, 2010) and human activities, such as fisheries, tourism, mariculture, and power production (Purcell et al., 2007). Several studies on "jellyfish blooms" (regarding only the medusoid stages of Phylum Cnidaria) have been carried out using different approaches, all contributing to a better understanding of the "bloom" phenomenon. Such studies have focused on methods of detection (Houghton et al., 2006; Bayha & Graham, 2009; Straehler-Pohl & Jarms, 2010), importance of different life stages (Purcell, 2007; Boero et al., 2008; Willcox et al., 2008; Arai, 2009; Bayha & Graham, 2009; Hoover & Purcell, 2009; Straehler-Pohl & Jarms, 2010), influence of climate (Dawson et al., 2001; Mills, 2001; Parsons & Lalli, 2002; Purcell, 2005; Hong et al., 2008), anthropogenic causes and subsequent consequences to

Electronic supplementary material The online version of this article (doi:10.1007/s10750-012-1048-6) contains supplementary material, which is available to authorized users.

Guest editors: J. E. Purcell, H. Mianzan & J. R. Frost /Jellyfish Blooms: Interactions with Humans and Fisheries

L. S. Miranda (✉) · A. C. Morandini · A. C. Marques
Instituto de Biociências, Departamento de Zoologia, Universidade de São Paulo, Rua do Matão, travessa 14, 101. Cidade Universitária, São Paulo 05508-090, SP, Brazil
e-mail: mirandals@ib.usp.br

the economy (Dawson et al., 2001; Purcell et al., 2007; Hong et al., 2008; Mariottini et al., 2008; Richardson et al., 2009), ecological (Pitt et al., 2009), and evolutionary (Dawson & Hamner, 2009; Hamner & Dawson, 2009) aspects of "blooms".

Despite the importance of these studies, the concept of "bloom" remains controversial (Table 1). Some even dispute the term "bloom", which is taken from plant ecology where it is applied to the flourishing appearance and high densities of plants, algae, and general phytoplankton in the marine environment (Smayda, 1997). The animal equivalent would be the term "outbreak", considered by some to be more appropriate to describe "…exceptional abundances (usually sudden and monospecific) of zooplankton, which can be seasonal or non-seasonal" (CIESM, 2001). Recently, an interesting perception of "bloom" emerged in the literature: the concept of "true bloom" and "apparent bloom" (Graham et al., 2001), later incorporated into evolutionary studies (Dawson & Hamner, 2009; Hamner & Dawson, 2009). In the "true bloom", there is a rapid population growth due to reproduction, and in the "apparent bloom", a stable population is re-distributed or re-dispersed as a consequence of physical, chemical, or behavioral factors (Graham et al., 2001). Thus, the term "jellyfish bloom" is common in the jellyfish literature (Table 1) and also accepted (CIESM, 2001); therefore, we will use both terms synonymously.

The study of "blooms" is often hampered by the need for appropriate methods of sampling and long-term monitoring (Genzano et al., 2008; Kogovšek et al., 2010). In addition, it is important to consider the different stages of the life cycle, such as benthic polyps, since they are crucial to understanding "bloom" dynamics (Purcell, 2007; Willcox et al., 2008; Bayha & Graham, 2009; Dawson & Hamner, 2009; Hoover & Purcell, 2009). Because "blooms" of

Table 1 Compilation of definitions of "bloom"/"outbreak" published during the last decade

References	Remarks
CIESM (2001)	"Bloom" synonymous with "outbreak", including "abnormal" and "normal (seasonal) blooms"
Benović & Lučić (2001)	"Bloom" as an abnormal event
Graham et al. (2001)	Differences between "apparent bloom" and "true bloom"
Mills (2001)	"Normal" and "abnormal blooms" ("true bloom")
Nival & Gorsky (2001)	Convenience concept of "bloom"
Parsons & Lalli (2002)	Differences between "apparent bloom" and "true bloom"
Purcell et al. (2007)	"Bloom" related to reproduction, i.e., "true bloom"
Boero et al. (2008)	"Bloom" (multispecific) is different from "outbreak" (monospecific); seasonal ("normal bloom") or unexpected ("abnormal bloom")
Daryanabard & Dawson (2008)	Differences between "apparent bloom" (aggregations) and "true bloom"
Genzano et al. (2008)	"Bloom" as an abnormal event; discussion on the convenience concept of "bloom", and differences between "apparent" and "true bloom"
Albert (2009)	Swarm and aggregation ("apparent bloom")
Dawson & Hamner (2009)	Definition of mass occurrence, "normal" and "abnormal blooms" ("true blooms"), and swarm ("apparent bloom")
Gibbons & Richardson (2009)	"Normal" and "abnormal blooms" ("true bloom") and "apparent bloom"
Hamner & Dawson (2009)	Definition of "true bloom" ("normal" and "abnormal"), "apparent bloom" (aggregations and swarms), mass occurrence, and accumulations
Lilley et al. (2009)	"Bloom" synonym of large aggregations
Purcell et al. (2009)	"Bloom" synonym of "outbreak"
Licandro et al. (2010)	"Bloom" as a synonym of "outbreak", related to the density of organisms
Purcell (2012)	"Bloom" as unusually high abundance (i.e., "abnormal bloom") synonym of "outbreak"
Brotz et al. (2012)	"Bloom" defined as increase in presence, numbers, or biomass

large medusae primarily are responsible for the negative impacts reported (Uye, 2008; Richardson et al., 2009), a biased view in the characterization of "blooms" neglects some inconspicuous groups, as the hydromedusae (Hamner & Dawson, 2009). Nevertheless, examples of hydromedusan "blooms" are in the literature (e.g., Purcell & Grover, 1990; Mianzan et al., 2000; Raskoff, 2001; Stefani et al., 2010; Purcell, 2012 electronic supplementary material).

Within the Phylum Cnidaria, the Class Staurozoa (Marques & Collins, 2004; Van Iten et al., 2006) comprises the small and inconspicuous benthic stauromedusae, some of which have a differentiated life cycle (Wietrzykowski, 1912; Kikinger & von Salvini-Plawen, 1995; Miranda et al., 2010). There are no reports of direct impacts on human activities by these animals: they do not prey on fish, and they are not harmful to humans (Davenport, 1998; Zagal, 2004a). These features make them interesting to contrast against the bias often introduced for "bloom" studies. Fortunately, long-term monitoring exists for some species (Corbin, 1979), which facilitates the study of "blooms" in this group. Thus, in this study, we test the different concepts of "bloom" for Staurozoa, discussing different applications of the term in relation to biological phenomena or human convenience.

Staurozoan population biology

Stauromedusae are small, stalked, benthic medusae that live mainly in the intertidal zone, attached to algae or rocks (Mayer, 1910; Mills & Hirano, 2007; S1. 1). They belong to the Class Staurozoa, the most recently proposed class for the Phylum Cnidaria (Marques & Collins, 2004; Collins et al., 2006; Van Iten et al., 2006; Daly et al., 2007). Although metagenetic, the general life cycle of Staurozoa is distinctive in that a planula larva attaches to the substrate and develops into a stauropolyp, which subsequently undergoes an apical metamorphosis into an adult stauromedusa. Consequently, the species do not produce a pelagic medusa, because the transformation to adult takes place without fission or budding. This developmental pattern results in a mosaic individual, in which the structures of the oral part are similar to those of an adult medusa (particularly, scyphozoans and cubozoans), whereas the basal part retains characteristics of the sessile polyp. The dioecious adult stauromedusa reproduces sexually and produces new planulae (Wietrzykowski, 1912; Kikinger & von Salvini-Plawen, 1995; Mills & Hirano, 2007).

The general life cycle described for Staurozoa states that one polyp produces directly only one adult medusa; however, they also possess a capacity for asexual reproduction, which occurs by lateral budding of the calyx (upper part of the animal) and at the distal portions of some special tentacles of the polyp (Kikinger & von Salvini-Plawen, 1995); and in the early stage of attached larvae ("microhydrula" stage) through frustulation (Jarms & Tiemann, 1996; Miranda et al., 2010). Asexual reproduction greatly increases the ability of a single individual to generate many clones. This ability was corroborated by molecular markers (16S, ITS1, and ITS2), which showed a unique haplotype for ten individuals of each population (from King George Island, Antarctica and Valdivia, Chile) of *Haliclystus antarcticus* Pfeffer 1889 analyzed (Miranda et al., 2010).

Although stauromedusae are widely distributed (Kramp, 1961), limited information exists on their biology and ecology, especially concerning their life cycle (Miranda et al., 2010). As a result, long-term monitoring data are rare, with a few exceptions (Table 2). Periods of high abundances of stauromedusae were recorded during the end of spring and summer for different species in both hemispheres (Fig. 1a; Table 2). In fact, high abundances of species of *Haliclystus* were found in the Northern Hemisphere summer by Uchida (1927), Ling (1937), Berrill (1962), and Corbin (1979), and in the Southern Hemisphere summer by Amor (1962), Davenport (1998), Zagal (2004a, b), and Miranda et al. (2009). According to Ling (1937) these animals "...occur most abundantly during the first part of August, (...) becoming quite rare by the end of October" in the Northern Hemisphere. Similarly, the population of *H. antarcticus* reached its highest density (1,405 ind m^{-2}) during the summer period in January in the Southern Hemisphere (Table 2) and then decreased drastically during the winter months (Zagal, 2004b). Additional records of stauromedusae have peak abundance in the summer, including the genera *Sasakiella* (Ling, 1937), *Manania*, and *Lucernaria* (Berrill, 1962); however, there are also exceptions, such as species of *Lucernariopsis* that peak in abundance during the autumn/winter (Corbin, 1979; Fig. 1b; Table 2).

Table 2 Life cycles and population abundances of some species of Staurozoa

Species	Generations year^{-1}	Peak abundance	Hemisphere	Season	Reference
Craterolophus convolvulus (Johnston 1835)	2	April (1st generation) September (2nd generation)	North	Spring & summer-autumn	Corbin (1979)
Haliclystus antarcticus Pfeffer 1889	1	January	South	Summer	Zagal (2004b)
Haliclystus auricula (Rathke 1806)	1	July	North	Summer	Corbin (1979)
Haliclystus salpinx Clark 1863	1	July	North	Summer	Berrill (1962)
Lucernaria quadricornis Müller 1776	1	July	North	Summer	Berrill (1962)
Lucernariopsis campanulata (Lamouroux 1815)	1	October	North	Autumn	Corbin (1979)
Lucernariopsis cruxmelitensis Corbin 1978	1	February	North	Winter	Corbin (1979)
Manania atlantica (Berrill, 1962)	1	July	North	Summer	Berrill (1962)
Sasakiella cruciformis Okubo 1917	1?	August	North	Summer	Ling (1937)

A single peak in the monitoring curve of abundance indicates that several species of stauromedusae have an annual cycle with one generation year^{-1} (Fig. 1a, b) (Berrill, 1962; Corbin, 1979; Zagal, 2004b). Corbin (1979), who monitored the variation in abundance of stauromedusae for 23 years, suggested that an annual life cycle is a common trait of Stauromedusae, with the exception of *Craterolophus convolvulus* (Johnston 1835), which has two generations year^{-1} (Fig. 1c; Table 2).

Although we do not know the exact environmental signals that synchronize these annual events (Mills & Hirano, 2007), the seasonality of stauromedusae was hypothesized to be related to the availability of suitable algal substrates; specifically, the periods of highest abundance of stauromedusae and greatest algal cover coincide, indicating optimal conditions for growth and nutrition of these animals (Zagal, 2004b). Environmental conditions are probably less favorable outside of these peak periods (Zagal, 2004b); however, species develop slightly differently from one another, and those that co-occur at one location may emerge, age, and disappear at different times of the year (Mills & Hirano, 2007).

How can stauromedusae appear in great numbers during specific seasons and then suddenly disappear? Stauromedusae are hard to find because of the relatively small size of some species, especially if they are camouflaged in color and texture against the background of macrophytes or coralline algae (Corbin, 1979; Larson & Fautin, 1989; Zagal, 2004b; S1). Moreover, the life cycle of stauromedusae includes inconspicuous creeping planula larvae that lack cilia (Otto, 1976, 1978) and stauropolyps, which are also small and difficult to observe (Wietrzykowski, 1912; Kikinger & von Salvini-Plawen, 1995). Many individuals of these "hidden" life cycle stages may occur with adult stauromedusae in the intertidal, but stauropolyps (~0.3–0.8 mm; Hirano, 1986; Kikinger & von Salvini-Plawen, 1995) and planulae (20–100 μm; Otto, 1978) are even more difficult to find than the camouflaged stauromedusae (usually varying from 1 to 4 cm; Mills & Hirano, 2007; S1). In fact, planulae from only five species (Kowalevsky, 1884; Wietrzykowski, 1912; Hanaoka, 1934; Otto, 1976, 1978) and polyps from only nine species (Wietrzykowski, 1912; Hirano, 1986; von Salvini-Plawen, 1987; Kikinger & von Salvini-Plawen, 1995) have been recorded from a total of 51 known species of Staurozoa (Daly et al., 2007; Mills, 2011).

Migration, in which the stauromedusae move from the littoral zone to deep waters, is another possibility for their absence during some seasons of the year (Sars, 1846; Corbin, 1979; Zagal, 2004b). Literature suggests that not only the medusoid but also other staurozoan life stages migrate. Gwilliam (1956) hypothesized that "...there exists in the vicinity [of the intertidal population of stauromedusae] a permanent subtidal population which is continually supplying planulae...", and when conditions were suitable, these planulae would settle, grow, and mature on algae. The "microhydrula" stage, which occurs after planula attachment, of *H. antarcticus* was collected at a depth of about 31 m attached to bivalve shells (Jarms & Tiemann, 1996; Miranda et al., 2010), also suggesting a migratory behavior. This could be indirect evidence that intertidal

Fig. 1 Seasonal patterns of abundance of Staurozoa. **a** Spring/summer peak (annual cycle); **b** autumn/winter peak (annual cycle); **c** a biannual cycle. Adapted from Corbin (1979)

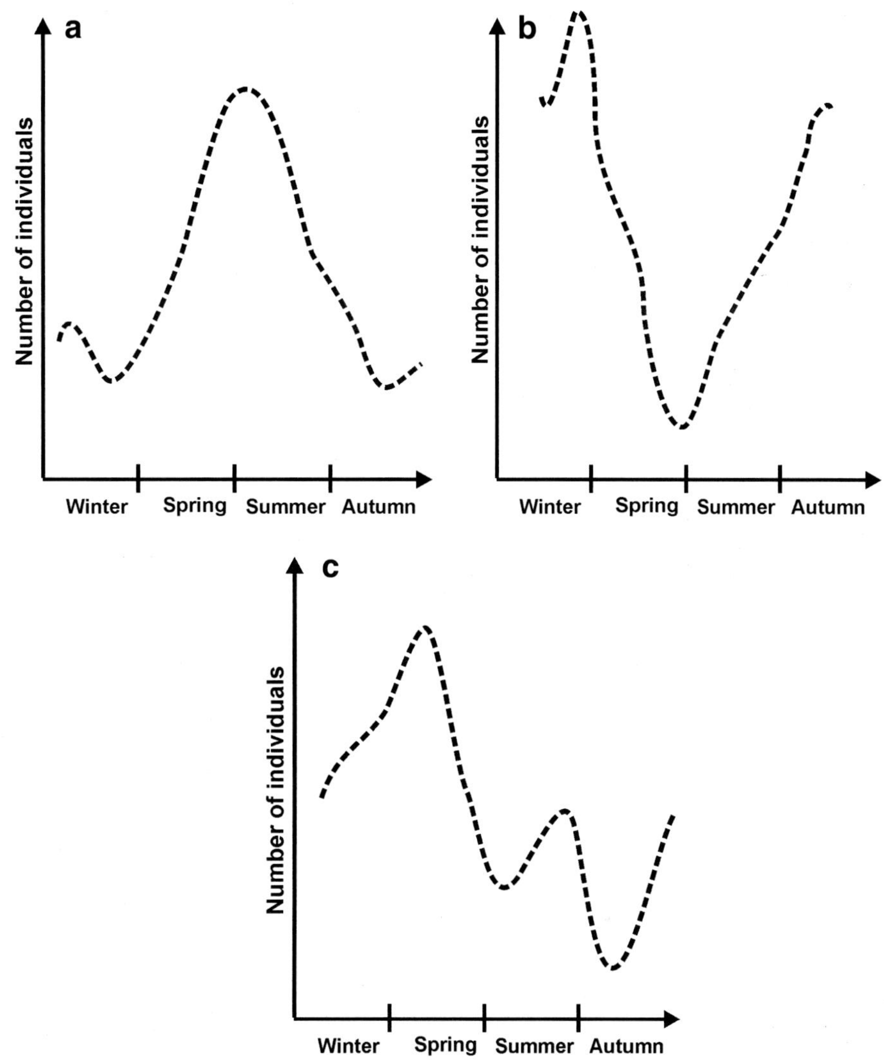

adult stauromedusae of *H. antarcticus* could co-exist with at least one other life history stage surviving in a different habitat than the adult, perhaps developing and migrating when conditions become favorable. It remains unknown whether this migration occurs actively or passively.

In addition to the planula and "microhydrula" stages, stauropolyps might also perform migration. The stauropolyp of *Haliclystus octoradiatus* (Lamarck 1816) tightly adheres to the substrate until the 4-tentacle stage, when the polyp spontaneously detaches itself, even when water conditions are calm. This was observed consistently in laboratory cultures and similar events may occur under natural conditions (Wietrzykowski, 1912), providing an opportunity for polyps to migrate from a deeper region, where the "microhydrula" stage was located, to shallow waters, where stauromedusae are observed. However, there are no other reports to corroborate that planulae and stauropolyps of Staurozoa can migrate seasonally from shallow to deep waters or vice versa (Gwilliam, 1956; Corbin, 1979; Zagal, 2004b), and there is no evidence of migratory behavior during the stauromedusa stage.

Information on the duration of each stage of the life cycle of Staurozoa, which would help in understanding the appearance of great numbers of stauromedusae, is also scarce. However, the migratory hypothesis has some support from scattered information on the life cycle of many species. The "microhydrula" stage of *H. antarcticus* was found in December at King George Island (Jarms & Tiemann, 1996) and could have been the source of the large number of adult stauromedusae recorded in late February–early March (Miranda et al., 2009). Although no further information exists for this species, some data in the literature help to deduce the life cycle. The planula stage of *H. octoradiatus* develops into a fully developed stauropolyp in about 15 days (Wietrzykowski, 1912) and the stauropolyp of *Stylocoronella* takes 2 months to become a mature stauromedusae under laboratory conditions (Kikinger & von Salvini-Plawen, 1995). Finally, Wietrzykowski (1912) observed young polyps of *H. octoradiatus* in the field during April and mature stauromedusae during July. Thus, about 3 months between December and March would be enough for planulae to become mature stauromedusae for *H. antarcticus*.

Field observations confirm that most species of stauromedusae disappear for several months before their young stages reappear, so encystment of the larvae also seems likely to occur in the field (Mills & Hirano, 2007). Although encystment or resistance stages have not been observed, when planulae of *Haliclystus salpinx* Clark 1863 settle in the laboratory, they can aggregate and become surrounded by an amorphous substance that forms packed subunits enclosing the planulae, which could be an overwintering stage (Otto, 1978). Otto (1978) suggests that the planulae of *Haliclystus* spp. probably do not develop directly after settlement in nature, but form cysts during unfavorable conditions. Interestingly, the "microhydrula" stage of *H. antarcticus* also aggregates (Miranda et al., 2010), and could represent the "resistant" stage suggested by Otto (1978). In addition, this "microhydrula" stage can produce frustules (Jarms & Tiemann, 1996; Miranda et al., 2010). In the genus *Stylocoronella*, many frustules further divide into two smaller ones during the development of some specimens, and this stage encysts, showing no further development (Kikinger & von Salvini-Plawen, 1995). These frustules could act as resting stages, later developing into stauropolyps under favorable conditions (Kikinger & von Salvini-Plawen, 1995). Probably due to their resistance, Otto (1978) was unable to trigger further development of staurozoan planulae or frustules in the laboratory. In fact, the "microhydrula" stage of *H. antarcticus* has been kept in continuous culture for nearly 20 years without further development (Jarms & Tiemann, 1996; Miranda et al., 2010).

In addition to a seasonal peak, which occurs once or twice a year depending on the species (Table 2), Corbin (1979) reported exceptional increases in abundances of certain species during specific years. On average, *Lucernariopsis cruxmelitensis* Corbin 1978, for instance, produced about 200 individuals during its peak abundance during winter. In 1968, it was estimated that this number reached 2,000 individuals in February (Northern Hemisphere winter). Corbin (1979) also detailed an event of considerable increase in abundance for different species (*Lucernariopsis campanulata* (Lamouroux 1815) and *C. convolvulus*) in 1974, which was maintained for a considerable period of time and gave "every indication of the occurrence of a population explosion of these two species" (Corbin, 1979). Corbin (1979) suggested that certain factors were especially favorable to these species at the time of spawning, prior to maturity, and continuing through the development and growth of the resulting "explosion generations". A combination of extrinsic and intrinsic factors could lead to an unusually high number of specimens.

Despite these high densities ("seasonal" or "exceptional") documented for Staurozoa, evolutionarily these animals are not considered able to bloom (either "apparent" or "true bloom"; Dawson & Hamner, 2009; Hamner & Dawson, 2009). In order to discuss the suitability of considering staurozoan "blooms", we will address different concepts of "bloom" in the literature, verifying whether or not they are applicable to Staurozoa.

Stauromedusae and the concepts of "bloom"

Biological concept

"Apparent bloom"

An "apparent bloom" is an event caused by environmental (physical and chemical) and/or biological (swimming) factors. These factors may lead to aggregations or swarms that could affect local perception of large numbers of jellyfish as a real population increase (Graham et al., 2001; Table 1).

Stauromedusae have limited capacity of movement and spend most of their lives attached to rocks or seaweeds by their peduncle; however, they can use their tentacles or anchors (adhesive structures) to restrict locomotion (Hyman, 1940; Mills & Hirano, 2007; personal observations). Their planulae do not possess cilia and so also have limited movement (Otto, 1976, 1978). Because traversing physical barriers depends on swimming (Graham et al., 2001), stauromedusae are susceptible to accumulation due to environmental variables. Currents or winds can accumulate their substrata or stauromedusae when they reach a barrier, thereby creating an "apparent bloom" of Staurozoa. Nevertheless, evolutionary analysis of "jellyfish blooms" (Dawson & Hamner, 2009; Hamner & Dawson, 2009) concluded that "apparent blooms" were "rare or absent" in the class. Contrary to this conclusion, there are records of dense stauromedusa aggregations (Collins & Daly, 2005), but insufficient information was available to conclude if these events were "apparent blooms" (Table 3).

"True blooms"

A "true bloom" can be a regular event ("normal bloom"; Graham et al., 2001). Jellyfish populations in temperate climates usually vary seasonally in abundance in a somewhat predictable manner, with their peak following the regular sequence of the phytoplankton spring pulse (CIESM, 2001). Some authors considered these seasonal, natural population phenomena to be a "true (demographic) normal bloom" (Mills, 2001; Dawson & Hamner, 2009; Hamner & Dawson, 2009; Table 1). According to those definitions, the annual pattern of abundance of a staurozoan population, with one or two annual peaks (Fig. 1) should be classified as a "normal true bloom".

"Blooms" are one possible consequence of a metagenetic life cycle because all metagenetic cnidarians have the potential to bloom (Mills, 2001; Hamner & Dawson, 2009). Among Scyphozoa, each benthic polyp may produce many ephyrae through strobilation (asexual reproduction), and each of the ephyrae can develop into an adult scyphomedusa. Thus, one polyp can produce many medusae, contributing decisively to the formation of "blooms" (Purcell, 2007). On the other hand, one stauropolyp only produces one adult stauromedusa because it does not strobilate, but metamorphoses directly into a benthic medusa (Wietrzykowski, 1912; Kikinger & von Salvini-Plawen, 1995). Similarly in cubozoans, each cubopolyp metamorphoses directly into a single cubomedusa (Stangl et al., 2002); this was one reason to conclude that this taxon cannot form "extraordinary blooms" (Hamner & Dawson, 2009). Nevertheless, cubozoans (Werner et al., 1971; Arneson & Cutress, 1976; Yamaguchi & Hartwick, 1980) and at least some, if not all, Staurozoa can asexually reproduce in the early life cycle stages (settled planulae—"microhydrula"—and stauropolyps; Wietrzykowski, 1912; Kikinger & von Salvini-Plawen, 1995; Miranda et al., 2010). This asexual reproduction would create many new polyps, which could each metamorphose into one new medusa. Thus, even without strobilation, there could still be rapid population growth in Staurozoa due to asexual reproduction. Constraining a species' ability to bloom to only those capable of strobilation is an oversimplification of the biological question. Also, these "blooms" are not biologically homologous (e.g., strobilation in Scyphozoa and frustulation in Staurozoa); therefore, evolutionary reconstructions of the "bloom" capacity may need to consider additional biological processes/features.

Evidence indicates that Staurozoa may occur periodically at high densities. During 23 years of monitoring (Corbin, 1979), stauromedusae presented a seasonal, predictable, "normal true bloom". This also was observed during 1 year of monitoring by Zagal (2004b). In conclusion, although data on the life cycle of Staurozoa are rare, the population dynamics of some species allows us to reject the conclusion that the Class Staurozoa does not have "normal true blooms" (Hamner & Dawson, 2009; Table 3).

"True blooms" also can be unpredictable events ("abnormal blooms"; Graham et al., 2001). Some authors suggest that the normal, seasonal increases in the abundance of individuals of a population (herein called "normal true bloom") should not be considered a "bloom"; however, in some years, the expected occurrence of a species may exceed the usual abundance level and only then should be considered a "bloom" (CIESM,

Table 3 The adequacy of considering a "Staurozoa bloom", according to the different concepts of "bloom"

Concepts of "bloom"	Do Staurozoa bloom?
Biological	
"Apparent bloom"	?
"True bloom"	
"Normal" (Regular)	Yes
"Abnormal" (Irregular)	Yes
Convenience	
Size/conspicuity	No
Plankton	No
Impact on human activities	No

2001; Genzano et al., 2008; Table 1). This was called an "abnormal true bloom" because it resulted from a population increase following asexual reproduction ("true bloom"; Dawson & Hamner, 2009; Hamner & Dawson, 2009) and "abnormal" because it surpassed the usual levels for the species.

In order to recognize such events as unusual, there must be good knowledge of what is "normal" abundance or biomass. Answering this question is only possible when proper baseline information, such as long-term monitoring, is available (Genzano et al., 2008). Corbin (1979) noted some years of "exceptionally high" abundances of stauromedusae. In 1968, the number of specimens of *L. cruxmelitensis* was 10-times higher than the maximum mean observed in other years. In 1974, the population of *C. convolvulus* and *L. campanulata* reached a high density, an event with "every indication of the occurrence of a population explosion" (Corbin, 1979). Zagal (2004b) also recorded remarkably high densities of *H. antarcticus*, reaching 1,405 ind m^{-2} (vs. a mean density of 385 ind m^{-2}) in summer 2002. Unfortunately in this case, the species was monitored for only 1 year, so we cannot know if those numbers were abnormal. Thus, the limited available data have shown "abnormal true blooms" in Staurozoa (Table 3), contrary to earlier conclusions (Dawson & Hamner, 2009; Hamner & Dawson, 2009).

Literature highlights different reasons for causes of "abnormal true blooms" of jellyfish, including anthropogenic sources (eutrophication), fishing pressure (overfishing), aquaculture, construction (human modification of aquatic habitats, altering coastal waters and circulation), climate change, and invasions (translocations) (Purcell et al., 2007; Richardson et al., 2009; Purcell, 2012). In Staurozoa, attempts to identify the events would be purely speculative, but likely include extrinsic factors, such as food availability, suitability of substrate, environmental water temperature, exposure, and concentration of predators; and intrinsic factors affecting more intense gametogenesis in the parent generation, higher levels of fertilization, better survival of larvae and cysts, more successful settlement, and optimal survival of post-larvae stages (Corbin, 1979).

The "convenience concept" ("biased bloom")

Convenience concepts are those not based on biological patterns and, unfortunately, they are spread throughout many biological examples including Medusozoans (Marques, 2001). Likewise, the concept of "bloom" should be based on biological aspects related to evolution or life history of the species. Nevertheless, the terms "bloom" or "outbreak" are sometimes used without considering the biological processes involved in the phenomenon.

Some authors consider a species to be blooming "...when it becomes conspicuous in the sea and when it is harmful to humans" (Nival & Gorsky, 2001; Table 1), incorporating high levels of anthropocentrism, because "...this condition depends on the size of the sample and on the sampling method used" (Nival & Gorsky, 2001). "Normal (annual) blooms" and absence of noxious species for humans do not draw attention (Benović & Lučić, 2001); therefore, most of the times they are not noticed or reported. Usually, "blooms" are observed when massive appearances of conspicuous, stinging jellyfish occur near coastal areas, significantly and visibly affecting human activities, like fisheries, tourism, and power production (Genzano et al., 2008). Hence, there is a focus on "macromedusae", mainly scyphozoans, which are individually large and cause negative impacts on various economic activities (Bayha & Graham, 2009; Richardson et al., 2009). We use the term "convenience concept of bloom" to refer to these events because they constrain "bloom" events to those observed by humans and that interfere with human activities. In other words, it is a "biased" concept.

Staurozoa are inconspicuous, often camouflaged animals (S1), in all stages of their life cycle (Wietrzykowski, 1912; Hanaoka, 1934; Otto, 1976, 1978; Hirano, 1986; Mills & Hirano, 2007). Their cnidae are not harmful to humans (personal observations), and there is no reported negative impact on human activities. Moreover, they are benthic animals, and the term "bloom" only has been applied to planktonic animals (CIESM, 2001), even though biologically they are medusae (Marques & Collins, 2004; Collins et al., 2006). Consequently, according to the convenience concept, Staurozoa do not bloom (Table 2).

Conclusion

Whether or not the Staurozoa have the capacity to bloom depends on the concept of "bloom" used (Table 1). When we considered the biological concept of "bloom", the Staurozoa do bloom according to the concepts of "true bloom" (normal and abnormal) and probably bloom according to the "apparent bloom" concept. In contrast, Staurozoa do not bloom if we use the convenience concept of "bloom", which refers to planktonic, conspicuous, and harmful to humans. Although an evolutionary analysis considered the Staurozoa to be a non-blooming clade (Dawson & Hamner, 2009), biological data reviewed here indicate that they can bloom. Consequently, the evolution of "bloom" (Dawson & Hamner, 2009) should be reviewed, as well as their phylogenetic implications.

It usually is difficult to identify the processes involved in high abundances of organisms, i.e., re-distribution of a stable population, real population growth, or both (Graham et al., 2001; Hamner & Dawson, 2009). Monitoring data are fundamental to defining such events (Genzano et al., 2008). It is also important to consider the various stages of the life cycle, such as planulae and polyps, which are often neglected in "bloom" studies (but see Arai, 2009; Dawson & Hamner, 2009; and Htun et al., 2012 on podocysts; Bayha & Graham, 2009; Astorga et al., 2012; Holst, 2012; and Purcell et al., 2012 on polyps; Straehler-Pohl & Jarms, 2010 for ephyrae detection). The biological concept of "bloom" also has weaknesses because there may be different, non-homologous processes for the origin of the "blooms". A "jellyfish bloom" is an interesting event, and if we want to be able to predict it, we need to study its whole complexity, with efforts to minimize possible biases introduced by the human point of view.

Acknowledgments We are grateful to the reviewers, C. E. Mills, M. N. Dawson, and editors, whose comments improved this paper and A. G. Collins for his comments on a previous abstract. We also thank the organizers of the 3rd International Jellyfish Blooms Symposium. L. S. Miranda was supported by CNPq (142270/2010-5) and FAPESP (2010/07362-7). A. C. Morandini was supported by CNPq (490158/2009-9) and FAPESP (2010/50174-7). A. C. Marques was supported by CNPq (557333/2005-9; 490348/2006-8; 474672/2007-7; 304720/2009-7; 563106/2010-7; 562143/2010-6) and FAPESP (2003/02432-3; 2004/09961-4; 2010/52324-6).

References

Albert, D. J., 2009. *Aurelia labiata* (Scyphozoa) jellyfish in Roscoe Bay: their spatial distribution varies with population size and their behaviour changes with water depth. Journal of Sea Research 61: 140–143.

Amor, A., 1962. Sobre "Stauromedusae" do litoral Patagónico, *Haliclystus auricula* (Rathke). Notas del Museo de la Universidad Nacional de La Plata 20: 89–96.

Arai, M. N., 2009. The potential importance of podocysts to the formation of scyphozoan blooms: a review. Hydrobiologia 616: 241–246.

Arneson, A. C. & C. E. Cutress, 1976. Life history of *Carybdea alata* Reynaud, 1830 (Cubomedusae). In Mackie, G. O. (ed.), Coelenterate Ecology and Behavior. Plenum Press, New York: 227–236.

Astorga, D., J. Ruiz & L. Prieto, 2012. Ecological aspects of early life stages of *Cotylorhiza tuberculata* (Macri, 1778) affecting its pelagic population success. Hydrobiologia. doi:10.1007/s10750-012-1036-x.

Bayha, K. M. & W. M. Graham, 2009. A new Taqman© PCR-based method for the detection and identification of scyphozoan jellyfish polyps. Hydrobiologia 616: 217–228.

Benović, A. & D. Lučić, 2001. Jellyfish outbreaks: natural cycle or stress response effect? CIESM Workshop Series 14: 59–62.

Berrill, M., 1962. The biology of three New England stauromedusae, with a description of a new species. Canadian Journal of Zoology 40: 1249–1262.

Boero, F., J. Bouillon, C. Gravili, M. P. Miglietta, T. Parsons & S. Piraino, 2008. Gelatinous plankton: irregularities rule the world (sometimes). Marine Ecology Progress Series 356: 299–310.

Brotz, L., W. W. L. Cheung, K. Kleisner, E. Pakhomov & D. Pauly, 2012. Increasing jellyfish populations: trends in large marine ecosystems. Hydrobiologia. doi:10.1007/s10750-012-1039-7.

CIESM (The Mediterranean Science Commission), 2001. Gelatinous zooplankton outbreaks: theory and practice. CIESM Workshop Series 14: 112 pp. http://www.ciesm.org/online/monographs/Naples.html. Accessed 05 Nov 2011

Collins, A. G. & M. Daly, 2005. A new deepwater species of stauromedusae, *Lucernaria janetae* (Cnidaria, Staurozoa, Lucernariidae), and a preliminary investigation of stauromedusan phylogeny based on nuclear and mitochondrial rDNA data. Biological Bulletin 208: 221–230.

Collins, A. G., P. Schuchert, A. C. Marques, T. Jankowski, M. Medina & B. Schierwater, 2006. Medusozoan phylogeny and character evolution clarified by new large and small subunit rDNA data and an assessment of the utility of phylogenetic mixture models. Systematic Biology 55: 97–115.

Corbin, P. G., 1979. The seasonal abundance of four species of stauromedusae (Coelenterata: Scyphomedusae) at Plymouth. Journal of the Marine Biological Association of the United Kingdom 59: 385–391.

Daly, M., M. R. Brugler, P. Cartwright, A. G. Collins, M. N. Dawson, D. G. Fautin, S. C. France, C. S. McFadden, D. M. Opresko, E. Rodriguez, S. L. Romano & J. L. Stake, 2007. The phylum Cnidaria: a review of phylogenetic patterns and diversity 300 years after Linnaeus. Zootaxa 1668: 127–182.

Daryanabard, R. & M. N. Dawson, 2008. Jellyfish blooms: *Crambionella orsini* (Scyphozoa: Rhizostomeae) in the Gulf of Oman, Iran, 2002–2003. Journal of the Marine Biological Association of the United Kingdom 88: 477–483.

Davenport, J., 1998. Note on the trophic relationships of the stauromedusa *Haliclystus antarcticus* from subantarctic South Georgia. Journal of the Marine Biological Association of the United Kingdom 78: 663–664.

Dawson, M. N. & W. M. Hamner, 2009. A character-based analysis of the evolution of jellyfish blooms: adaptation and exaptation. Hydrobiologia 616: 193–215.

Dawson, M. N., L. E. Martin & L. K. Penland, 2001. Jellyfish swarms, tourists, and the Christ-child. Hydrobiologia 451: 131–144.

Genzano, G., H. Mianzan, L. Diaz-Briz & C. Rodriguez, 2008. On the occurrence of *Obelia* medusa blooms and empirical evidence of unusual massive accumulations of *Obelia* and *Amphisbetia* hydroids on the Argentina shoreline. Latin American Journal of Aquatic Research 36: 301–307.

Gibbons, M. J. & A. J. Richardson, 2009. Patterns of jellyfish abundance in the North Atlantic. Hydrobiologia 616: 51–65.

Graham, W. M., F. Pagès & W. M. Hamner, 2001. A physical context for gelatinous zooplankton aggregations: a review. Hydrobiologia 451: 199–212.

Gwilliam, G. F., 1956. Studies on west coast stauromedusae. PhD Dissertation, University of California, Berkeley: 192 pp

Hamner, W. M. & M. N. Dawson, 2009. A review and synthesis on the systematics and evolution of jellyfish blooms: advantageous aggregations and adaptive assemblages. Hydrobiologia 616: 161–191.

Hanaoka, K.-I., 1934. Notes on the early development of a stalked medusa. Proceedings of the Imperial Academy (Tokyo) 10: 117–120.

Hirano, Y. M., 1986. Species of Stauromedusae from Hokkaido, with notes on their metamorphosis. Journal of the Faculty of Science, Hokkaido University, Series IV, Zoology 24: 182–201.

Holst, S., 2012. Effects of climate warming on strobilation and ephyra production of North Sea scyphozoan jellyfish. Hydrobiologia. doi:10.1007/s10750-012-1043-y.

Hong, J., C. He-Qin, X. Hai-Gen, F. Arreguín-Sánchez, M. J. Zetina-Rejón, P. D. M. Luna & W. J. F. L. Quesne, 2008. Trophic controls of jellyfish blooms and links with fisheries in the East China Sea. Ecological Modelling 212: 492–503.

Hoover, R. A. & J. E. Purcell, 2009. Substrate preferences of scyphozoan *Aurelia labiata* polyps among common dock-building materials. Hydrobiologia 616: 259–267.

Houghton, J. D. R., T. K. Doyle, J. Davenport & G. C. Hays, 2006. Developing a simple, rapid method for identifying and monitoring jellyfish aggregations from the air. Marine Ecology Progress Series 314: 159–170.

Htun, T., H. Ikeda & S. I. Uye, 2012. The potential role of podocysts in perpetuation of the common jellyfish *Aurelia aurita* s.l. (Cnidaria: Scyphozoa) in anthropogenically perturbed coastal waters. Hydrobiologia. doi:10.1007/s10750-012-1045-9.

Hyman, L. H., 1940. Observations and experiments on the physiology of medusae. Biological Bulletin 79: 282–296.

Jarms, G. & H. Tiemann, 1996. On a new hydropolyp without tentacles, *Microhydrula limopsicola* n. sp., epibiotic on bivalve shells from the Antarctic. Scientia Marina 60: 109–115.

Kikinger, R. & L. von Salvini-Plawen, 1995. Development from polyp to stauromedusa in *Stylocoronella* (Cnidaria: Scyphozoa). Journal of the Marine Biological Association of the United Kingdom 75: 899–912.

Kogovšek, T., B. Bogunović & A. Malej, 2010. Recurrence of bloom-forming scyphomedusae: wavelet analysis of a 200-year time series. Hydrobiologia 645: 81–96.

Kowalevsky, A. O., 1884. Zur Entwicklungsgeschichte der Lucernaria. Zoologische Anzeiger 7: 712–719.

Kramp, P. L., 1961. Synopsis of the medusae of the world. Journal of the Marine Biological Association of the United Kingdom 40: 7–469.

Larson, J. L. & D. G. Fautin, 1989. Stauromedusae of the genus *Manania* (=*Thaumatoscyphus*) (Cnidaria, Scyphozoa) in the northeast Pacific, including descriptions of new species *Manania gwilliami* and *Manania handi*. Canadian Journal of Zoology 67: 1543–1549.

Licandro, P., D. V. P. Conway, M. N. Daly Yahia, M. L. Fernandez de Puelles, S. Gasparini, J. H. Hecq, P. Tranter & R. R. Kirby, 2010. A blooming jellyfish in the northeast Atlantic and Mediterranean. Biology Letters 6: 688–691.

Lilley, M. K. S., J. D. R. Houghton & G. C. Hays, 2009. Distribution, extent of inter-annual variability and diet of the bloom-forming jellyfish *Rhizostoma* in European waters. Journal of the Marine Biological Association of the United Kingdom 89: 39–48.

Ling, S. W., 1937. Studies on Chinese stauromedusae. I. Stauromedusae from Tsingtao. Amoy Marine Biology Bulletin 3: 1–35.

Mariottini, G. L., E. Giacco & L. Pane, 2008. The mauve stinger *Pelagia noctiluca* (Forsskål [SIC], 1775). Distribution, ecology, toxicity and epidemiology of stings. A review. Marine Drugs 6: 496–513.

Marques, A. C., 2001. Simplifying hydrozoan classification: inappropriateness of the group Hydroidomedusae in a phylogenetic context. Contributions to Zoology 70: 175–179.

Marques, A. C. & A. G. Collins, 2004. Cladistic analysis of Medusozoa and cnidarian evolution. Invertebrate Biology 123: 23–42.

Mayer, A. G., 1910. The Medusae of the World, Vol. III. The Scyphomedusae. Publication No. 109. Carnegie Institution of Washington, Washington, DC: 499–735

Mianzan, H., D. Sorarrain, J. W. Burnett & L. L. Lutz, 2000. Mucocutaneous junctional and flexural paresthesias caused by the holoplanktonic Trachymedusa *Liriope tetraphylla*. Dermatology 201: 46–48.

Mills, C. E., 2001. Jellyfish blooms: are populations increasing globally in response to changing ocean conditions? Hydrobiologia 451: 55–68.

Mills, C. E., 2011. Stauromedusae: list of all valid species names. http://faculty.washington.edu/cemills/Staurolist.html. Accessed 05 Nov 2011

Mills, C. E. & Y. M. Hirano, 2007. Stauromedusae. In Denny, M. W. & S. D. Gaines (eds), Encyclopedia of Tidepools and Rocky Shores. University of California Press, Berkeley: 541–543.

Miranda, L. S., A. C. Morandini & A. C. Marques, 2009. Taxonomic review of *Haliclystus antarcticus* Pfeffer, 1889 (Stauromedusae, Staurozoa, Cnidaria), with remarks on the genus *Haliclystus* Clark, 1863. Polar Biology 32: 1507–1519.

Miranda, L. S., A. G. Collins & A. C. Marques, 2010. Molecules clarify a cnidarian life cycle—the ''hydrozoan''

Microhydrula limopsicola is an early life stage of the staurozoan *Haliclystus antarcticus*. Public Library of Science ONE 5: e10182.
Morandini, A. C. & A. C. Marques, 2010. Revision of the genus *Chrysaora* Péron & Lesueur, 1810 (Cnidaria, Scyphozoa). Zootaxa 2464: 1–97.
Nival, P. & G. Gorsky, 2001. Conditions for blooming of pelagic species. CIESM Workshop Series 14: 23–27.
Otto, J. J., 1976. Early development and planula movement in *Haliclystus* (Scyphozoa, Stauromedusae). In Mackie, G. O. (ed.), Coelenterate Ecology and Behavior. Plenum Press, New York: 319–329.
Otto, J. J., 1978. The settlement of *Haliclystus* planulae. In Chia, F. S. & M. E. Rice (eds), Settlement and Metamorphosis of Marine Invertebrate Larvae. Elsevier/North-Holland Biomedical Press, New York: 13–22.
Parsons, T. R. & C. M. Lalli, 2002. Jellyfish population explosions: revisiting a hypothesis of possible causes. La Mer 40: 111–121.
Pitt, K. A., D. T. Welsh & R. H. Condon, 2009. Influence of jellyfish blooms on carbon, nitrogen and phosphorus cycling and plankton production. Hydrobiologia 616: 133–149.
Purcell, J. E., 2005. Climate effects on formation of jellyfish and ctenophore blooms: a review. Journal of the Marine Biological Association of the United Kingdom 85: 461–476.
Purcell, J. E., 2007. Environmental effects on asexual reproduction rates of the scyphozoan *Aurelia labiata*. Marine Ecology Progress Series 348: 183–196.
Purcell, J. E., 2012. Jellyfish and ctenophore blooms coincide with human proliferations and environmental perturbations. Annual Review of Marine Science 4: 209–235.
Purcell, J. E. & J. J. Grover, 1990. Predation and food limitation as causes of mortality in larval herring at a spawning ground in British Columbia. Marine Ecology Progress Series 59: 55–67.
Purcell, J. E., S. Uye & W.-T. Lo, 2007. Anthropogenic causes of jellyfish blooms and their direct consequences for humans: a review. Marine Ecology Progress Series 350: 153–174.
Purcell, J. E., R. A. Hoover & N. T. Schwarck, 2009. Interannual variation of strobilation by the scyphozoan *Aurelia labiata* in relation to polyp density, temperature, salinity, and light conditions in situ. Marine Ecology Progress Series 375: 139–149.
Purcell, J. E., D. Atienza, V. Fuentes, A. Olariaga, U. Tilves, C. Colahan & J.-M. Gili, 2012. Temperature effects on asexual reproduction rates of scyphozoan polyps from the NW Mediterranean Sea. Hydrobiologia. doi:10.1007/s10750-012-1047-7.
Raskoff, K. A., 2001. The impact of El Niño events on populations of mesopelagic hydromedusae. Hydrobiologia 451: 121–129.
Richardson, A. J., A. Bakun, G. C. Hays & M. J. Gibbons, 2009. The jellyfish joyride: causes, consequences and management responses to a more gelatinous future. Trends in Ecology & Evolution 24: 312–322.
Sars, M., 1846. Fauna littoralis Norvegiae. Hefte 1: 1–94.
Smayda, T., 1997. What is a bloom? A commentary. Limnology and Oceanography 42: 1132–1136.
Stangl, K., L. von Salvini-Plawen & T. W. Holstein, 2002. Staging and induction of medusa metamorphosis in *Carybdea marsupialis* (Cnidaria, Cubozoa). Vie Milieu 52: 131–140.
Stefani, F., B. Leoni, A. Marieni & L. Garibaldi, 2010. A new record of *Craspedacusta sowerbii*, Lankester 1880 (Cnidaria, Limnomedusae) in Northern Italy. Journal of Limnology 69: 189–192.
Straehler-Pohl, I. & G. Jarms, 2010. Identification key for young ephyrae: a first step for early detection of jellyfish blooms. Hydrobiologia 645: 3–21.
Uchida, T., 1927. Report of the biological survey of Mutsu Bay. 2. Medusae of Mutsu Bay. Science Reports of the Tohoku University 2: 215–238.
Uye, S., 2008. Blooms of the giant jellyfish *Nemopilema nomurai*: a threat to the fisheries sustainability of the East Asian Marginal Seas. Plankton & Benthos Research 3: 125–131.
Van Iten, H., J. M. Leme, M. G. Simões, A. C. Marques & A. G. Collins, 2006. Reassessment of the phylogenetic position of conulariids (?Ediacaran-Triassic) within the subphylum Medusozoa (phylum Cnidaria). Journal of Systematic Palaeontology 4: 109–118.
von Salvini-Plawen, L., 1987. Mesopsammic Cnidaria from Plymouth (with systematic notes). Journal of the Marine Biological Association of the United Kingdom 67: 623–637.
Werner, B., C. Cutress & J. P. Studebaker, 1971. Life cycle of *Tripedalia cystophora* Conant (Cubomedusae). Nature 232: 582–583.
Wietrzykowski, W., 1912. Recherches sur le développement des Lucernaires. Archives de Zoologie Expérimentale et Générale 5e Série 10: 1–95
Willcox, S., N. A. Moltschaniwskyj & C. M. Crawford, 2008. Population dynamics of natural colonies of *Aurelia* sp. scyphistomae in Tasmania. Australia. Marine Biology 154: 661–670.
Yamaguchi, M. & R. Hartwick, 1980. Early life history of the sea wasp, *Chironex fleckeri* (class Cubozoa). In Tardent, P. & R. Tardent (eds), Developmental and Cellular Biology of Coelenterates. Elsevier/North-Holland Biomedical Press, Amsterdam: 11–16.
Zagal, C. J., 2004a. Diet of the stauromedusa *Haliclystus auricula* from southern Chile. Journal of the Marine Biological Association of the United Kingdom 84: 337–340.
Zagal, C. J., 2004b. Population biology and habitat of the stauromedusa *Haliclystus auricula* in southern Chile. Journal of the Marine Biological Association of the United Kingdom 84: 331–336.

Hydrobiologia (2012) 690:69–80
DOI 10.1007/s10750-012-1053-9

JELLYFISH BLOOMS

Comparative phylogeography of meroplanktonic species, *Aurelia* spp. and *Rhizostoma pulmo* (Cnidaria: Scyphozoa) in European Seas

Andreja Ramšak · Katja Stopar · Alenka Malej

Published online: 9 March 2012
© Springer Science+Business Media B.V. 2012

Abstract Mass occurrences of scypozoan medusae have become increasingly common in recent decades in European seas, including species in the genera *Aurelia* and *Rhizostoma.* We inferred the phylogeographic patterns of metagenetic scyphozoa *Aurelia* spp. and *Rhizostoma pulmo* from mitochondrial COI and nuclear ITS regions. No genetic structure was detected in *R. pulmo* over the Mediterranean Sea. By contrast, the phylogeographic analyses confirmed the separation of *Aurelia* spp. to several proposed cryptic species. Our results do not support the null hypothesis that both genera have concordant phylogeographic patterns. The resolvable parsimony network of haplotypes was retrieved for *Aurelia aurita*, *Aurelia* sp. 5, and *Aurelia* sp. 8 without connectivity between them and no genetic structure were found within those groups. Even though evidence of hybridization was found between *A. aurita* and *Aurelia* sp. 5, that did not break down the phylogenetic separation among them. The lowest haplotype and nucleotide diversity were found in samples of *Aurelia* sp. 8 and *R. pulmo* from the northern Adriatic, which acts as a sink area due to strong genetic drift. These new findings will facilitate linking the phenotype of the organism and its ability to survive in a particular environment—which shapes phylogeographic patterns.

Keywords Phylogeny · Genetic markers · mtDNA · ITS region

Electronic supplementary material The online version of this article (doi:10.1007/s10750-012-1053-9) contains supplementary material, which is available to authorized users.

Guest editors: J. E. Purcell, H. Mianzan & J. R. Frost / Jellyfish Blooms: Interactions with Humans and Fisheries

A. Ramšak (✉) · K. Stopar · A. Malej
National Institute of Biology, Marine Biology Station, Fornače 41, 6330 Piran, Slovenia
e-mail: ramsak@mbss.org

Introduction

The spatial distribution of genetic lineages within and among closely related species through use of genetic markers is the core of phylogeography. Through this approach, it is possible to resolve the consequences of historical demographic processes in populations and geological history on present distribution of genealogical lineages within species and between species. Phylogeographic studies benefit from a wealth of availability of genetic markers (Avise, 2000, 2009). The marine environment is very challenging for phylogeographic work because it does not have obvious geographic barriers like the terrestrial environment. In spite of this, researchers have been able to detect phylogeographic breaks in the marine environment and phylogeographic patterns in many taxa, thereby providing evidence of major barriers to gene flow in the sea (Kuo & Avise, 2005). There are many possible scenarios, and much evidence for and against such scenarios, if a particular geographical barrier is potentially responsible for phylogeographic breaks (Dawson, 2001; Patarnello et al., 2007; Pelc et al., 2009). The Straight of Gibraltar between the Mediterranean Sea and Atlantic Ocean sometimes provides a phylogeographic break for species on both sides (Patarnello et al., 2007). Authors also have argued that organism determinism does not have a significant role in divergence (Patarnello et al., 2007; Ayre et al., 2009). To the contrary, evidence from the southeast and southwest United States coasts indicates that dispersal ability is an important determinant of phylogeographic patterns in marine species. The phylogeographic structure in species with planktonic larvae may reflect contemporary oceanographic conditions, while species with restricted dispersion may be more likely to reflect historical processes (Pelc et al., 2009).

In the light of these findings, scyphozoan species are interesting because typically they have a metagenetic life history with a benthic phase and planktonic medusae (meroplanktonic species), while some others lack a benthic phase (holoplanktonic). Scyphozoan polyps and medusae have limited morphological characters, and ecological plasticity is characteristic for the whole group (Hamner & Jenssen, 1974; Greenberg et al., 1996; Dawson, 2003), leading to much confusion and rearrangement in taxonomy (Mayer, 1910; Kramp, 1961; Russell, 1970). Use of modern approaches in taxonomy, such as molecular markers to infer phylogenetic relationships, uncover many cryptic species, and challenging phylogeographic patterns among some scyphozoans. Phylogenetic studies on *Aurelia aurita* (Linnaeus, 1758) based on mitochondrial (COI) and nuclear ribosomal DNA (nrDNA) revealed ten cryptic species hidden in "*A. aurita*" sampled worldwide (designated as *Aurelia* sp. 1 to *Aurelia* sp. 10; Dawson and Jacobs, 2001; Dawson et al., 2005). A study of speciation and phylogeography of *A. aurita*, using an alternative set of markers, mitochondrial 16S rDNA, and the nuclear marker ITS 1/5.8 S rDNA, revealed six phylogroups and evidence of recent hybridization and introgression (Schroth et al., 2002). These groups agreed with the previous study where sampling regions overlapped (Dawson and Jacobs, 2001). In contrast to a previously described pattern, we confirmed that holoplanktonic *Pelagia noctiluca* (Forskåll, 1775), has no geographically supported genetic structure nor any distinct groups across European seas (Stopar et al., 2010). No previous phylogeographic study is devoted to a meroplanktonic *Rhizostoma pulmo* (Macri, 1778). A common characteristic of these scyphozoa genera is mass occurrences in the Mediterranean and adjacent seas (Kogovšek et al., 2010; Fuentes et al., 2011; Lilley et al., 2011).

Based on the meroplanktonic life cycle and a similarity in the distribution pattern, we tested the null hypothesis that *Aurelia* and *Rhizostoma* have concordant phylogeographic patterns over the European seas. The relationships between haplotypes from sampled localities were represented by a haplotype network in *Aurelia* spp. and *R. pulmo*, respectively. Analyses were made using sequences of the mitochondrial COI gene and nuclear ITS1 and ITS2 regions. A spatial distribution of haplotypes in *Aurelia* spp. was inferred with new sequences generated as part of this study and supplemented with existing data from previous studies (Dawson et al., 2005). Phylogeographic analyses of *R. pulmo* were conducted on sequences derived in this study.

Materials and methods

Sample collection

In total, 60 individuals of the *Aurelia* spp. and 26 individuals of *R. pulmo* were collected in the Mediterranean, North, and South Adriatic (Mljet Lake), Black, North, and Baltic seas during 2005–2010 (Fig. 1). Individuals were collected during local

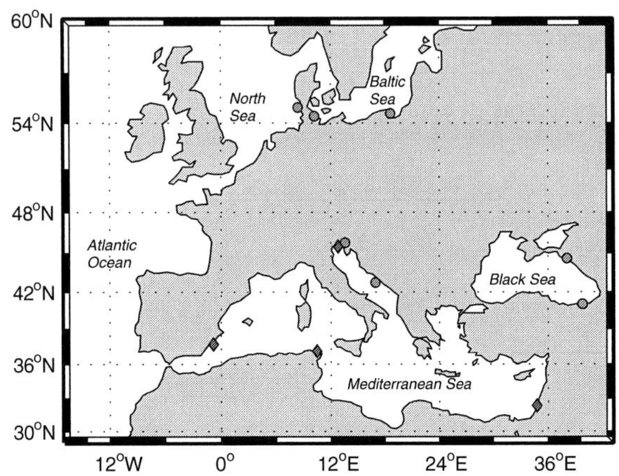

Fig. 1 Localities of *Aurelia* spp. (*circles*) and *R. pulmo* (*rhombs*) sample collection for the phylogenetic analysis

blooms. Medusa tissue (bell margin or gonads) was stored in 96% ethanol or in salt-saturated dimethyl sulfoxide (DMSO), and then frozen at −80°C until DNA extraction. Genomic DNA was extracted with a CTAB-based extraction or the DNeasy Blood and Tissue Kit (QIAGEN).

PCR amplification, sequencing, and alignment of genetic markers

Amplification of COI fragments from *Aurelia* spp. were done by using primer pair Au-LCO (5′-TCTC AACTAACCACAAAGATATAGG) and Au-HCO (5′-TACACTTCTGGGTGTCCAAAAAACCA), which are slightly modified universal invertebrate COI primers. The original universal invertebrate primers were used for amplification of COI in *R. pulmo* (Folmer et al., 1994). The internal transcribed spacers ITS1 and ITS2 of *Aurelia* spp. and *R. pulmo* were amplified using primer pairs ITS1/ITS2 and ITS3/ITS4 (White et al., 1990). Detailed PCR conditions for all primer pairs were described previously (Stopar et al., 2010). PCR products were purified and sequenced in both directions by Macrogen (Seoul, Korea) using the same primer pairs as used for PCR amplification. The amplified internal transcribed spacers ITS1 and ITS2 were not cloned and they gave clear sequences. Contigs were assembled and edited using ChromasPro 1.34 (Technelysium Pty). The identity of the sequences was confirmed by BLAST. New sequences were deposited in GenBank and are listed in Appendix A—Supplementary Material. Alignments were made with CLUSTALX (Thompson et al., 1997) under selected pairwise alignment conditions (gap open—10 and extension penalties—0, 1/0, 2). The correctness of the COI alignment was verified at the amino acid level using the Coelenterate mitochondrial code.

The alignment dataset for the *Aurelia* spp. consisted of 44 COI sequences retrieved in this study and 30 sequences taken from GenBank. The total length of the COI alignment was 655 base pairs (bp). Sequences from GenBank belong to the ten species of *Aurelia* sp. 1 to *Aurelia* sp. 10 and to *A. aurita* (Dawson et al., 2005). Sequences of *A. limbata* (Brandt, 1835) (AY903189 and AY935215) and *A. labiata* (Chamisso & Eysenhardt, 1821) (AY903073 and AY935202) were used as an outgroup for the phylogenetic tree. The alignment of ITS1 and ITS2 regions was made of sequences from 44 medusae in total length 783 bp. Twenty-eight ITS1 and ITS2 sequences come from this study and 16 sequences from GenBank.

The alignment dataset for *R. pulmo* consisted of COI from 13 medusae (length 655 bp) and of ITS1 from 26 medusae (length 417 bp). Sequences from ITS2 region were not used for further analyses because they were invariant. Sequences of *Nemopilema nomurai* (Kishinouye, 1922) (EU373728 and AB377548) and *Rhopilema esculentum* (Kishinouye, 1891) (EU373723 and AB377589) were used as outgroups for the phylogenetic tree. A detailed list of sequences used in this study is in Appendix A—Supplementary Material. Phylogenetic analyses in investigated species were made on COI, concatenated ITS1-ITS2, and COI-ITS1-ITS2 datasets.

Phylogenetic and population analyses

The optimal model of evolution was estimated in Modelgenerator (Keane et al., 2006) for subsequent use for the phylogenetic analyses. The most appropriate models under the AIC conditions (Akaike, 1973) were in *Aurelia* HYK + I + G ($\alpha = 1.48$, for COI) and, TVMef + G ($\alpha = 0.51$, for the ITS region), in *Rhizostoma* K81uf + G ($\alpha = 0.16$, for COI), and K80 (for the ITS region). A Bayesian approach was used for phylogenetic analyses on each dataset (COI, ITS, and COI-ITS) using MrBayes 3.1 (Ronquist & Huelsenbeck, 2003). For all datasets, two parallel Markov chain Monte Carlo (MCMC) searches were run with four chains for 2×10^6 generations, and sampled every 100 generations. With TRACER 1.4.1 (Rambaut and Drummond, 2009) the stationarity of values was checked. According to obtained results from TRACER, 75% of the trees were discarded as burn-in. The remaining 5,000 trees were used to calculate majority rule consensus tree and to estimate posterior probabilities. A phylogenetic relationship between haplotypes was visualized with FigTree 1.2.2 (Rambaut, 2006). The relationships in phylogroups were visualized and networks constructed using the program TCS 1.21 (Clement et al., 2000). Statistical parsimony haplotype networks were calculated using default settings.

Analyses of genetic differentiation were performed separately for COI and ITS datasets due to different models of inheritance in particular genetic markers. Estimates of genetic variation were obtained for the species as a whole, and for samples grouped according to the geographic region and to the phylogroups recovered in phylogenetic analyses (see "Results" section). Nucleotide (π) and haplotype (h) diversity were estimated using DnaSP ver 5.1 (Librado & Rozas, 2009). The program ARLEQUIN 3.1 (Excoffier et al., 1992) was used for analyses of molecular variance (AMOVA) to determine the relative partitioning of variation within and between populations, and to calculate genetic differentiation by means of pairwise Φ_{ST} values using 10,000 permutations, and TrN model for COI or Kimura 2-parameters for ITS with appropriate gamma distributed rate heterogeneity.

To test the null hypothesis that *Aurelia* and *Rhizostoma* have concordant phylogeographic patterns over the European seas, isolation by distance analyses were calculated for *Aurelia* spp. to describe patterns of population structuring based on the correlation coefficient between two pairwise matrices of genetic differentiation and geographic distances among groups recognized by phylogenetic analyses. The significance of the correlation was tested with the Mantel test (1,000 permutations) in ARLEQUIN 3.1. Further, we tested climatic influences on distribution pattern. In order to calculate the Mantel test, we grouped nine location sites into two climatic zones defined with minimum sea temperature in winter time: temperate 8–15°C (Mediterranean Sea) and boreal 0–8°C (North Atlantic Ocean and Black, North, and Baltic seas).

Results

Genetic diversity and phylogeography of *Aurelia* spp.

We analyzed 60 individuals of *Aurelia* spp. from seven localities and found 17 COI haplotypes, 14 of which were found only at one geographic region. Some geographic regions shared COI haplotypes; one haplotype was shared between the Baltic and North seas, one between Mljet, the North and Black seas, and one between Mljet and the Black Sea. ITS haplotypes were amplified from 31 individuals of *Aurelia* spp. from five locations. Altogether, 14 ITS1 haplotypes were retrieved; one haplotype was shared between the Baltic and North seas, and one between the Baltic, North, and Black seas. ITS2 sequences yielded 16 haplotypes, among which one was shared between the North and Black seas, and two were shared between Mljet and the Black Sea. No COI and ITS haplotypes were shared between individuals collected in the Adriatic Sea and other regions nor in the adjoining Mljet Lake.

All COI sequences had uniform length (655 bp) with 184 polymorphic sites, of which 173 were parsimony informative. Measures of genetic diversity (haplotype diversity (h), nucleotide diversity (π), and number of private haplotypes) were calculated across geographic regions and indicated higher nuclear (ITS) than mitochondrial (COI) diversity (Table 1). Phylogenetic analyses revealed several phylogroups restricted to geographic regions. Length polymorphism and nucleotide diversity (π) of ITS regions were higher among than within phylogroups: *Aurelia* sp. 8 from the Adriatic Sea (ITS1 328–336 bp, ITS2 438–451 bp; $\pi = 0.63\%$; 11 parsimony informative sites), *Aurelia* sp. 5 from Mljet Lake (ITS1 332–338 bp, ITS2 499–504 bp; $\pi = 4.49\%$; 8 parsimony informative sites), and *A. aurita* (ITS1 329–332 bp, ITS2 447–450 bp; $\pi = 4.84\%$; 8 parsimony informative sites). Overall, the ITS length polymorphism (due to many small indels) was 767–842 bp with 205 polymorphic sites, of which 169 were parsimony informative, and overall nucleotide diversity was very high ($\pi = 13.44\%$). ITS haplotype diversity was higher in *A. aurita* and Mljet Lake (0.956 and 1, respectively) and lowest in the *Aurelia* sp. 8 (0.583).

The tree of combined COI–ITS sequences (Fig. 2) revealed six main phylogroups which are restricted to geographic areas: *Aurelia* sp. 8, *Aurelia* sp. 5, (Adriatic Sea), *Aurelia* sp. 1 (Pacific Ocean), *Aurelia* sp. 2, and *Aurelia* sp. 9 (West Atlantic), *Aurelia* sp. 3, *Aurelia* sp. 4, and *Aurelia* sp. 6 (Palau lakes). The last was *A. aurita*, which consisted of specimens from the North, Baltic, and Black seas. Names of the phylogroups were taken from previous studies (Dawson & Jacobs, 2001; Schroth et al., 2002; Dawson et al., 2005). The major tree topologies with six phylogroups were consistent with all datasets (separated COI, ITS, and combined COI–ITS). The combined COI–ITS tree (Fig. 2) is the same as the tree made of ITS regions (not shown). Due to low within-group diversity in COI, all the COI sequences strictly belonged to the

Table 1 Genetic diversity of mitochondrial COI and nuclear ITS sequences in *Aurelia* spp. and *R. pulmo* according to geographic region

Geographic area	Specimens (no.)		Haplotypes[a] (no.)		Haplotype diversity *h*		Nucleotide diversity π (%)	
	COI	ITS	COI	ITS	COI	ITS	COI	ITS
Aurelia spp.								
Adriatic Sea (*Aurelia* sp. 8)	13	9	5(5)	5(5)	0.756	0.583	0.54	0.63
Mljet Lake (*Aurelia* sp. 5)	8	6	6(4)	6(6)	0.893	1.000	4.55	4.49
Black Sea (*A. aurita*)	21	9	9(7)	7(7)	0.653	0.917	2.42	9.63
Baltic Sea (*A. aurita*)	4	3	2(1)	1(1)	0.500	1.000	0.78	0
North Sea (*A. aurita*)	6	4	5(3)	4(4)	0.900	1.000	0.85	0.69
NE Atlantic (*A. aurita*)	5	–	4(4)	–	0.900	–	0.56	–
NW Atlantic (*A. aurita*)	3	–	3(3)	–	1.000	–	0.93	–
All *A. aurita*	39	16	12(11)	12(12)	0.862	0.956	1.55	4.84
All *Aurelia* spp.	60	31	30	24	0.919	0.952	9.40	13.44
R. pulmo								
North Adriatic	8	11	5(5)	5(3)	0.893	0.345	0.27	0.14
South Mediterranean	5	9	2(2)	3(2)	0.600	0	0.37	0
West Mediterranean	–	5	–	3(2)	–	0.733	–	0.52
All *R. pulmo*	13	26	7		0.910	0.385	0.48	0.18

[a] Number of private haplotypes are shown in parentheses

phylogroup related with geographic region. All the COI haplotypes from the North Atlantic, North, Baltic, and Black seas were together in the same homogenous group *A. aurita*. Moreover, *Aurelia* sp. 5 and *Aurelia* sp. 8 here were placed closely together, but in separated groups. There were differences in the support values for the deeper nodes, all shallower nodes in combined COI–ITS, and ITS tree were moderate to highly supported (posterior probabilities 75–100%), while those nodes in the COI tree had low support (50–58%).

A detailed view of the COI–ITS tree revealed that few haplotypes had different positions than in the COI tree. One haplotype, AA2210 (Mljet Lake), was separated from the main phylogroup *Aurelia* sp. 5 and positioned closer to a haplotype from the SW Atlantic (AA2501). A second haplotype, AA2721 from the Black Sea, was positioned at the base of phylogroup *Aurelia* sp. 5. A third difference concerned two haplotypes originating from the Black Sea (AA2711 and AA2714), which also were separated from the phylogroup *A. aurita* and positioned at the base of phylogroup west Atlantic (consisting of sequences from *Aurelia* sp. 2 and *Aurelia* sp. 9).

Comparisons of COI and ITS haplotypes of *Aurelia* spp. (Table 2) revealed significant values for several pairs: Mljet/Adriatic, Mljet/North Sea, Mljet/Baltic Sea, Mljet/Black Sea, Adriatic/North Sea, Adriatic/Baltic Sea, and Adriatic/Black Sea. All pairwise comparisons among geographic regions that were grouped by phylogenetic analyses into *A. aurita* (NE and NW Atlantic, Baltic, North, and Black seas) were low and not significant. The highest COI and ITS pairwise Φ_{ST} values were found between phylogroup *Aurelia* sp. 8 (Adriatic Sea) and localities included in *A. aurita* (North, Baltic, and Black seas). In concordance with pairwise Φ_{ST} values and phylogeographic analyses, the Mantel test failed to identify any evidence of isolation by distance. The correlation between genetic and geographic distances was not significant and slightly negative ($rY = -0.217$, $P = 0.723$). A Mantel correlation coefficient among genetic distance and minimum winter temperatures at sampling localities showed positive and significant correlation ($rY = 0.903$, $P = 0.043$). Relationships between COI haplotypes within *Aurelia* spp. were represented on parsimony networks. A clear single network was impossible to reconstruct for all species haplotypes because there were too many mutational steps between them. Haplotype networks revealed three separated groups: *A. aurita*, *Aurelia* sp. 5, and *Aurelia* sp. 8, without any sign of gene flow

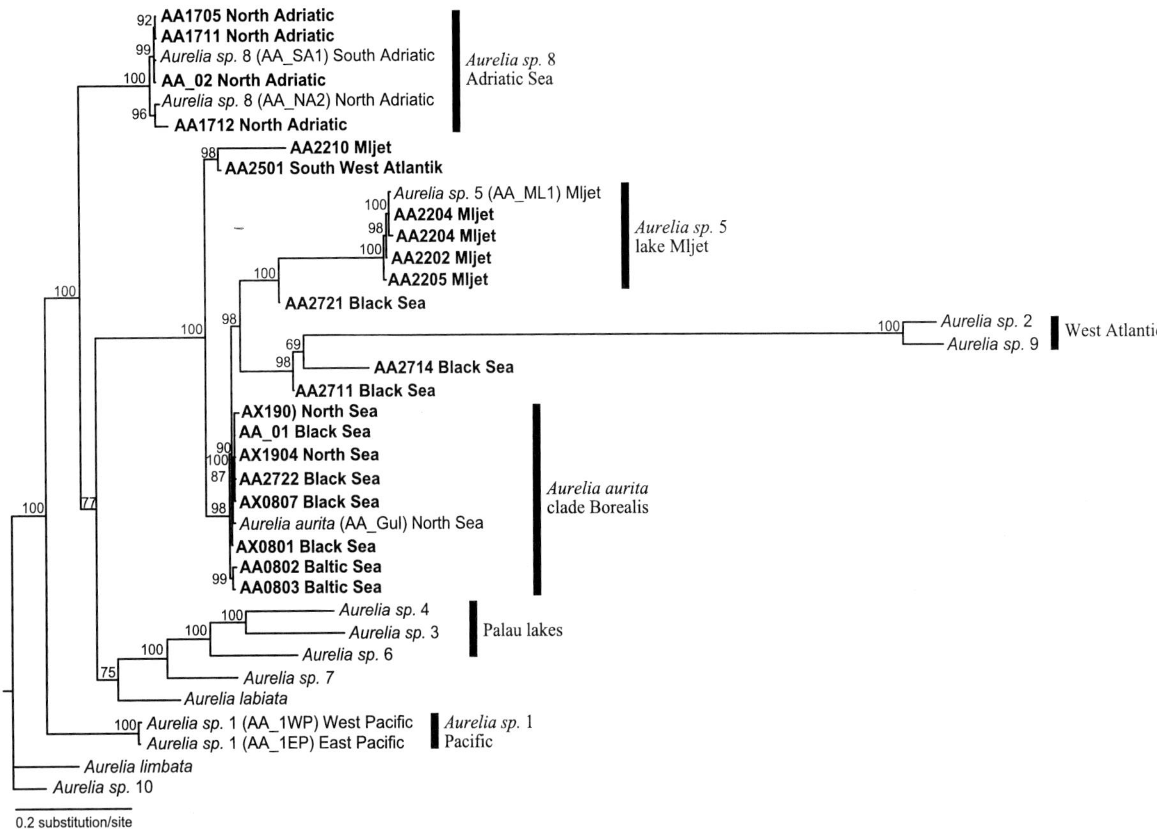

Fig. 2 Phylogenetic relationships within *Aurelia* spp. derived by Bayesian inference based on combined COI-ITS1-ITS2 sequences under a HKY + I + G (COI) and TVMef + G (ITS) model of evolution. *Numbers at nodes* indicate posterior probabilities. *Bolded* haplotypes were retrieved during this study, and the others were from GenBank

Table 2 Pairwise Φ_{ST} values in *Aurelia* spp. between geographic regions based on 10,000 permutations

	Adriatic Sea	Lake Mljet	Black Sea[a]	Baltic Sea[a]	North Sea[a]	NE Atlantic[a]	NW Atlantic[a]
Adriatic Sea	–	0.899*	0.799*	0.982*	0.977*	–	–
Mljet Lake	0.885*	–	0.653*	0.856*	0.861*	–	–
Black Sea[a]	0.933*	0.825*	–	0.029	0.030	–	–
Baltic Sea[a]	0.976*	0.797*	−0.169	–	0.444	–	–
North Sea[a]	0.948*	0.821*	0.143	−0.169	–	–	–
NE Atlantic[a]	0.978*	0.807*	−0.057	0.461	0.389	–	–
NW Atlantic[a]	0.977*	0.766*	0.029	0.489	0.460	0.172	–

Below diagonal: Φ_{ST} values for COI; above diagonal: Φ_{ST} values for ITS

NE northeast, *NW* northwest

* Significant differentiation $P < 0.01$

[a] Individuals from these geographical regions belong into phylogroup *A. aurita*

among them (Fig. 3). The most frequent haplotype (AA_CO2) found in the Black Sea also was shared with the North Sea and was inferred to be the ancestral haplotype. This haplotype had direct connections to other haplotypes from the Black Sea and through several mutational steps was connected with the haplotypes from the North and Baltic seas and the North Atlantic Ocean.

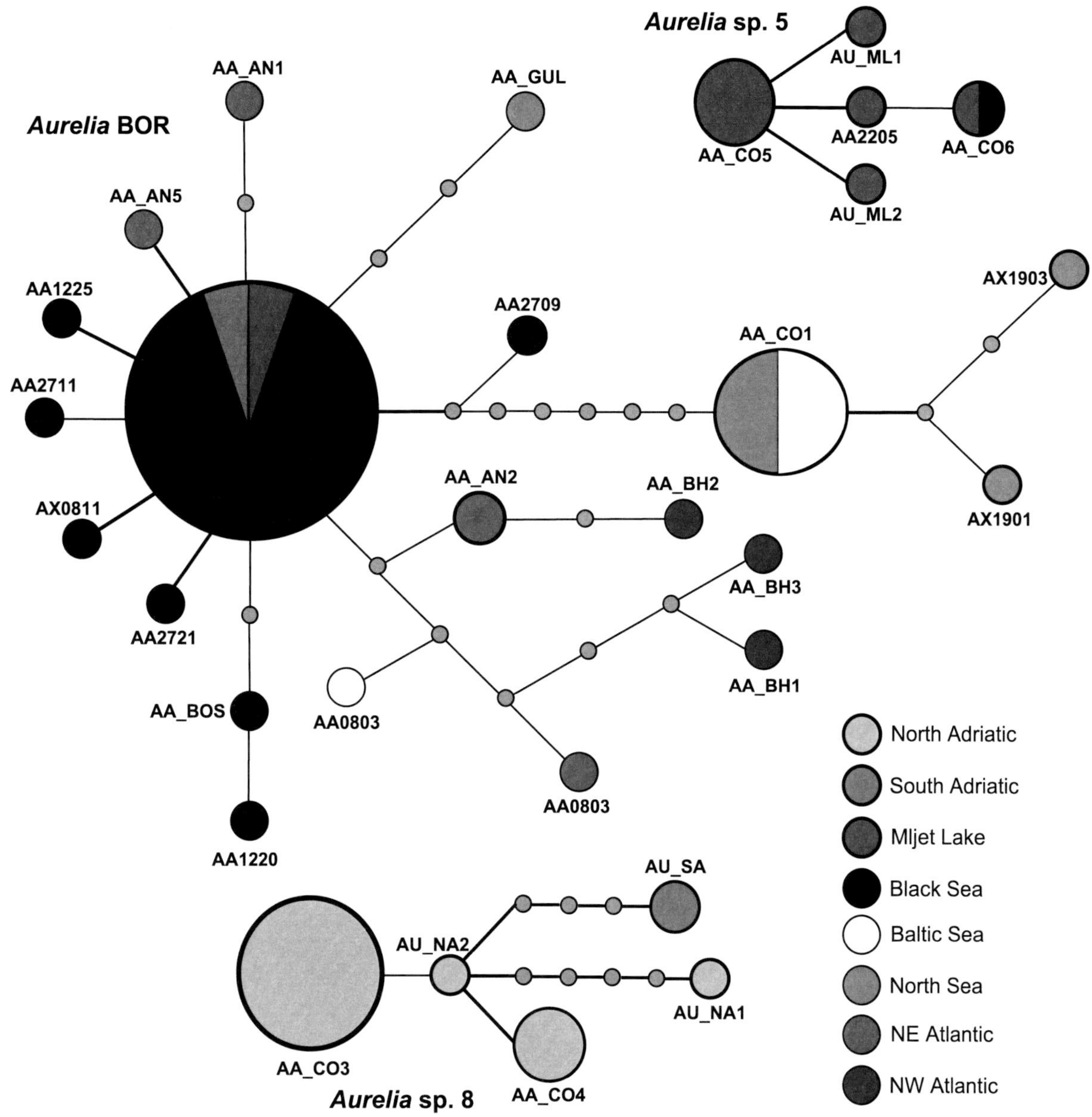

Fig. 3 Parsimony network of COI haplotypes from *Aurelia* spp. The *size of circle* is proportional to the haplotype frequency and the *color of circle* indicates the geographic origin. Each *branch* represents a one-nucleotide mutation. *Small circles* symbolize median vectors that represent hypothetical ancestral haplotypes

Genetic diversity and phylogeography of *R. pulmo*

Seven unique COI haplotypes were found in 13 individuals of *R. pulmo* sampled in the Adriatic Sea and the south and western Mediterranean Sea. Nine ITS1 haplotypes were found in 26 individuals of *R. pulmo* sampled at five localities in the Mediterranean Sea. Among them were seven unique ITS1 haplotypes, one shared between the Adriatic and southern Mediterranean seas (RP1_02) and one shared between the Adriatic, western and south Mediterranean seas (RP1_01). Only one ITS2 haplotype was found in 24 individuals from five localities in the Mediterranean Sea. In COI-ITS phylogenetic tree haplotype RP_04 was shared between the North Adriatic and two locations (eastern and western) in the Mediterranean Sea (Fig. 4).

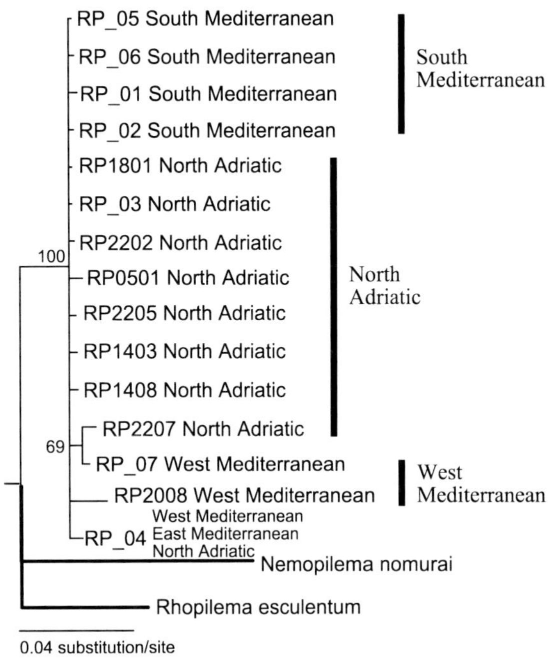

Fig. 4 Phylogenetic relationships within *R. pulmo* derived by Bayesian inference based on combined COI-ITS1 sequences under a K81uf + G (COI) and K80 (ITS1) model of evolution. *Numbers at nodes* indicate posterior probabilities

All measures of genetic diversity (h, π, and number of private haplotypes) were calculated across geographic regions and indicated higher mitochondrial (COI) than nuclear (ITS) diversity (Table 1). Sequence analyses of the COI region revealed no indels or length polymorphism (655 bp). COI haplotypes had nine polymorphic sites, of which eight were parsimony informative. Overall haplotype diversity of the COI region was high (0.910), but nucleotide diversity was low (0.48%). Length polymorphism and nucleotide diversity of the ITS region were only due to the ITS1 region. The ITS2 region in *R. pulmo* was invariable (total length 499 bp). Alignment of ITS1 sequences only from *R. pulmo* revealed variations in length from 328 to 344 bp, due to insertion of 10 bp in some haplotypes and deletion of 38 bp in the southern Mediterranean haplotype. Overall, the ITS1 region had six polymorphic sites, of which one was parsimony informative. Overall nucleotide and haplotype diversity of the ITS1 region were very low ($\pi = 0.18\%$ and $h = 0.385$, respectively). The highest ITS1 genetic diversity was in the western Mediterranean ($\pi = 0.52\%$; $h = 0.733$; one parsimony informative site) and lower in the North Adriatic ($\pi = 0.14\%$; $h = 0.345$; one parsimony informative site). No diversity was detected among individuals in the southern Mediterranean.

Phylogenetic analyses of the COI and ITS regions did not support geographically restricted groups among samples of *R. pulmo* in the Mediterranean Sea (Fig. 4). Pairwise comparisons of COI and ITS haplotypes were significant for the Adriatic and southern Mediterranean seas (COI, 0.534). Differences between the southern and western Mediterranean in the ITS region were significant (ITS1, 0.303). Low, not significant differences occurred between the Adriatic and southern Mediterranean seas (ITS1 0.048) and between the Adriatic and western Mediterranean seas (ITS1, 0.032). A parsimony network of relationships between COI haplotypes in *R. pulmo* revealed connections between haplotypes from the North Adriatic and the southern Mediterranean seas through several mutational steps (network not shown).

Discussion

The comparison of phylogeographic patterns between meroplanktonic *Aurelia* spp. and *R. pulmo* was made using data from mitochondrial COI and nuclear ITS regions. Even though both genera have metagenetic life histories and inhabit similar coastal environments, they have different phylogeographic patterns: *Aurelia* spp. were divided into several cryptic species, meanwhile *R. pulmo* showed no genetic structuring or separation into phylogenetic groups related to the investigated area. The tested null hypothesis that both meroplanktonic taxa have concordant phylogeographic patterns was rejected. The phylogenetic analyses confirm several clades of *Aurelia* in the surveyed area correspond to cryptic species of *Aurelia* sp. 8, *Aurelia* sp. 5, and *A. aurita*. Analyses of sequences from *Aurelia* sp. 8 showed only one clade of that cryptic species in the northern and southern Adriatic Sea without phylogeographic pattern over the investigated area. Meanwhile, *Aurelia* sp. 5 was restricted to Mljet Lake and no further structuring was expected. The large group consisted of sequences from the North, Baltic, and Black seas and represent the phylogroup *A. aurita* without any genetic differentiation correlated with the geographic distance as confirmed with the model of isolation by distance. Few haplotypes were shared between *Aurelia* sp. 5 and *A. aurita* clade and could result from hybridization between ancestral lineages. Some extraordinary genetic lineages found previously by Schroth et al. (2002) were explained by recent hybridization events rather than ancestral polymorphisms. Both *A. aurita* and *Aurelia* sp. 5 (Mljet Lake) represent lineages adapted to temperate-boreal temperatures for strobilation (Schroth et al., 2002), which might be reason that the *Aurelia* sp. 5 is confined to Mljet Lake, even though there is no physical barrier to prevent water exchange or the mixing of planktonic sexually reproducing medusae.

The distribution of cryptic species reflects past geological processes and the spreading of ancestor species to new environments after the last glaciations period (Dawson & Jacobs, 2001; Schroth et al., 2002). Nowadays, the distribution map can be confounded as well with anthropogenic translocation, as for *Aurelia* sp. 1 (Dawson et al., 2005). Both genetically different entities in adjacent localities (Mljet Lake and Adriatic Sea) are adapted to particular climatic factors (minimum winter temperature for strobilation) and strobilation in the *Aurelia* sp. 8 region occurs during the cold part of the year, while in the Mljet Lake all year round (Malej et al., 2012). The strong Mantel test correlation between minimum temperature and genetic distance ($r = 0.903$) confirmed this observation. Ecological adaptations, phenotypic, and ecological differences by divergent selection may be a reasonable explanation for the deep branching and phylogenetic patterns found in *Aurelia* spp. Also, peripatric processes can justify speciation in *Aurelia* lineages because there is not sufficient isolation by distance (Schroth et al., 2002).

Interestingly, many mutational steps connected two haplotypes in *Aurelia* sp. 8, one from the South Adriatic and one from the North Adriatic. Strong genetic drift could have caused a separation of lineages in the North Adriatic population, which lives at the edge of its environmental tolerance and had the lowest *Aurelia* spp. diversity in COI and ITS regions. A similar situation was found in the European sprat from the North Adriatic (Debes et al., 2008). The network made of *A. aurita* was star shaped with haplotypes from the Black Sea and the ancient haplotype was connected with several mutational steps with haplotypes from the North Sea, Baltic Sea, and North Atlantic Ocean. An explanation for this network pattern and high diversity in the COI and ITS regions is that the population in the Black Sea is stable and originated from an ancestral population in the North Atlantic. Many mutational sites between the population in the Black Sea and in the North Atlantic in modern times can be explained by reduced gene flow.

In contrast, phylogeographic analyses of *R. pulmo* specimens from locations around the Mediterranean Sea (southern Mediterranean and the North Adriatic) showed phylogenetic homogeneity and no geographical subdivisions. A few haplotypes from different locations formed a weakly supported clade, in which no genetic differentiation was observed in specimens from the Mediterranean Sea, similar to the holoplanktonic scyphomedusa *P. noctiluca* (Stopar et al., 2010). The obtained phylogeographic pattern was consistent regardless of which marker was used.

Several mechanisms shape genomes and as well as leave fingerprints on them. This is evident also in adaptive mutations in mitochondrial DNA and are important in the determination of the life history of an organism. Even more, mitochondrial phenotypes (mostly enzymatic processes in mitochondria) have a huge impact on their ability to survive in a particular environment, and consequently limit biogeographic distribution of organisms (reviewed in Ballard & Melvin, 2010). In this sense, COI is a very useful marker in uncovering phylogeographic patterns and ecological adaptations. *Aurelia* spp. has the most prominent phylogenetic structuring coupled with extensive morphological and ecological plasticity (e.g., biphasic life cycle, brooding larvae, temperature-dependent strobilation of polyps). *R. pulmo* has lower diversity of COI haplotypes in the Mediterranean Sea, lacking a structured phylogeographic pattern even though it has a biphasic life cycle. On the basis of the phylogeographic analysis, we concluded that *R. pulmo* have more relaxed habitats specifities. This is in contrast to *Aurelia* spp. where three phylogroups within the same geographic regions were found. The life cycle of *R. octopus* and *R. pulmo* was described in laboratory conditions including planula settlement and strobilation, but this has not been observed in nature (Holst et al., 2007; Fuentes et al., 2011). Can some parts of the life cycle, such as brooding of larvae, shape phylogeographic structuring? This is characteristic of *A. aurita* (in Ishii & Båmstedt, 1998), as well as in other rhizostomes, including *Catostylus mosaicus* (Quoy & Gaimard, 1824), *Mastigias papua* (Lesson, 1830) (in Sugiura, 1965), *Phyllorhiza punctata* (von Lendenfeld, 1884) (in Rippingale & Kelly, 1995), *Cotylorhiza tuberculata* (Macri, 1778) (in Kikinger, 1992), *Cephea cephea* (Forskål, 1775) (in Sugiura, 1966), and *Cassiopea andromeda* (Forskål, 1775) (in Gohar & Eisawy, 1960) with short time before larval settlement (Holst et al., 2007 and references therein). A phylogeographic pattern was confirmed in *C. mosaicus* (Quoy & Gaimard, 1824) (Dawson, 2005) and in *C. andromeda* (Forskål, 1775), with multiple introductions to the Hawaiian Islands (Holland et al., 2004).

Phylogeographic pattern in *R. pulmo* was similar to the scyphozoan species *P. noctiluca*, which lacks a polyp stage and does not show any obvious structured populations between Mediterranean Sea and Atlantic Ocean (Stopar et al., 2010). This suggests that recent oceanographic characteristics of European seas enable good mixing and high gene flow, and that the Straight of Gibraltar and other straights within Mediterranean Sea do not represent a geographic barrier (Stopar et al., 2010).

A much more comprehensive phylogeographic pattern is possible by use of many different genetic markers, usually mitochondrial and nuclear genes such as rDNA operons (e.g., ITS 1 and ITS 2). The suitability of ITS regions in inferring population differentiation, phylogeographic patterns, and in phylogenetic studies is well known (Schlötterer & Tautz, 1994). In general, both markers are recognized as good markers, but some studies reveal different features of these markers that make them unsuitable for phylogeographic studies for some taxa (Shearer et al., 2002; Erpenbeck et al., 2005; Wörheide et al., 2004; Calderón et al., 2006). Use of concatenated sequences COI-ITS regions in scyphozoan species does not jeopardize inference of a phylogeographic pattern. Comparison of haplotype and nucleotide diversity in COI and ITS markers between *Aurelia* spp. and *R. pulmo* revealed comparable diversity in COI, while ITS diversity was lower in *R. pulmo*. Comparison of ITS1 and ITS2 regions revealed that ITS1 region was more diverse than ITS2 region in all species examined here. Moreover, the ITS2 region in *R. pulmo* did not express nucleotide diversity. Similarly, the ITS markers were also more uniform and uninformative in *P. noctiluca* (Stopar et al., 2010). Very uniform ITS regions are probably result of concerted evolution, which homogenizes sequences to ensure production of homogenous transcripts for functional ribosomes. The molecular mechanisms for concerted evolution are well known (e.g., unequal crossing over and gene conversion) and are important for the differentiation and discrimination of natural populations (Elder & Turner, 1995).

Some significant differences between geographic regions were found in pairwise comparisons of COI and ITS haplotypes. All examined species of meroplanktonic *Aurelia* spp. and *R. pulmo* as well as holoplanktonic *P. noctiluca* had the lowest haplotype and nucleotide diversity in the North Adriatic. This suggests that the northern Adriatic may act as sink area for all three scyphozan species. These observations raise many questions: how strong is the impact of ecological factors to sort out haplotypes from this region with particular features? How are COI co-adapted to the local environment? There is evidence that COI can co-adapt to the local environment with regard to the local temperature (Blier & Lemieux, 2001; Somero, 2002). Moreover, discussion about COI must be consider the functioning of the whole mitochondria and processes within it. There is increasing evidence that adaptive mitochondrial mutations determine the life history of an organism (Ballard & Melvin, 2010); as they proposed, these new findings will lead to linking the mitochondrial genotype with the phenotype of the organism and its ability to survive in a particular environment—which ultimately shapes phylogeographic patterns. This concept can be useful in helping to explain some adaptive traits, in particular scyphozoan species and the dispersion of mitochondrial haplotypes.

Acknowledgments We are very grateful to the people who helped us with the samples: Dror Angel (Israel), Diana Astorga Garcia (Spain), Reinhold Hanel (Germany), Manuela Krakau (Germany), Bojan Marčeta (Slovenia), Hermes Mianzan (Argentina), Mohamed Néjib Daly Yahia (Tunisia), Tamara A. Shiganova (Russia), and Maciej Wołowicz and Michal Olenycz (Poland). Further thanks are also given to Peter Trontelj for helpful comments on phylogenetic analysis. This research was supported by the Slovenian Research Agency Program P1-0237, the European Commission 6th Framework Programme: MarBEF NoE (Contract No. GOCET-2003-505446) and SESAME project (Contract No. GOCE-036949).

References

Akaike, H., 1973. Information theory as an extension of the maximum-likelihood principle. In Petrov, B. N. & F. Csaki (eds), Second International Symposium on Information Theory. Akademiai Kiado, Budapest: 267–281.

Avise, J. C., 2000. Phylogeography: the history and formation of species. Harvard University Press, Cambridge.

Avise, J. C., 2009. Phylogeography: retrospect and prospect. Journal of Biogeography 36: 3–15.

Ayre, D. J., T. E. Minchinton & C. Perrin, 2009. Does life history predict past and current connectivity for rocky intertidal invertebrates across a marine biogeographic barrier? Molecular Ecology 18: 1887–1903.

Ballard, J. W. O. & R. G. Melvin, 2010. Linking the mitochondrial genotype to the organismal phenotype. Molecular Ecology 19: 1523–1539.

Blier, P. U. & H. Lemieux, 2001. The impact of the thermal sensitivity of cytochrome c oxidase on the respiration of Arctic charr red muscle mitochondria. Journal of Comparative Physiology B 171: 247–253.

Calderón, I., J. Garrabou & D. Aurelle, 2006. Evaluation of the utility of COI and ITS markers as tools for population genetic studies of temperate gorgonians. Journal of Experimental Marine Biology and Ecology 336: 184–197.

Clement, M., D. Posada & K. A. Crandall, 2000. TCS: a computer program to estimate gene genealogies. Molecular Ecology 9: 1657–1659.

Dawson, M. N., 2001. Phylogeography in coastal marine animals: a solution from California? Journal of Biogeography 28: 723–736.

Dawson, M. N., 2003. Macro-morphological variation among cryptic species of the moon jellyfish, *Aurelia* (Cnidaria: Scyphozoa). Marine Biology 143: 369–379.

Dawson, M. N., 2005. Incipient speciation of *Catostylus mosaicus* (Syphozoa, Rhizostomeae, Catostylidae), comparative phylogeography and biogeography in south-east Australia. Journal of Biogeography 32: 515–533.

Dawson, M. N. & D. K. Jacobs, 2001. Molecular evidence for cryptic species of *Aurelia aurita* (Cnidaria, Scyphozoa). Biological Bulletin 200: 92–96.

Dawson, M. N., A. S. Gupta & M. H. England, 2005. Coupled biophysical global ocean model and molecular genetic analyses identify multiple introductions of cryptogenic species. Proceedings of the National Academy of Sciences of the United States of America 102: 11968–11973.

Debes, P. V., F. E. Zachos & R. Hanel, 2008. Mitochondrial phylogeography of the European sprat (*Sprattus sprattus* L., Clupeide) reveals isolated climatically vulnerable populations in the Mediterranean Sea and range expansion in the northeast Atlantic. Molecular Ecology 17: 3873–3888.

Elder, J. F. & B. J. Turner, 1995. Concerted evolution of repetitive DNA sequences in eukaryotes. The Quarterly Review of Biology 70(3): 297–320.

Erpenbeck, D., J. N. A. Hooper & G. Wörheide, 2005. CO1 phylogenies in diploblasts and the 'Barcoding of Life' – are we sequencing a suboptimal partition? Molecular Ecology Notes 6: 550–553.

Excoffier, L., P. E. Smouse & J. M. Quattro, 1992. Analysis of molecular variance inferred from metric distances among DNA haplotypes: application to mitochondrial restriction data. Genetics 131: 479–491.

Folmer, O., M. Black, W. Hoeh, R. Lutz & R. Vrijenhoek, 1994. DNA primers for amplification of mitochondrial cytochrome c oxidase I subunit from diverse metazoan invertebrates. Molecular Marine Biology and Biotechnology 3: 294–299.

Fuentes, V., I. Straehler-Pohl, D. Atienza, I. Franco, U. Tilves, M. Gentile, M. Acevedo, A. Olariaga & J. M. Gili, 2011. Life cycle of the jellyfish *Rhizostoma pulmo* (Scyphozoa: Rhizostomeae) and its distribution, seasonality and inter-annual variability along Catalan coast and the Mar Menor (Spain, NW Mediterranean). Marine Biology 158: 2247–2266.

Gohar, H. A. F. & A. M. Eisawy, 1960. The development of *Casiopea andromeda* (Scyphomedusae). Publications of the Marine Biology Station of Al Ghardaq̄a 11: 147–190.

Greenberg, N., R. L. Garthwaite & D. C. Potts, 1996. Allozyme and morphological evidence for a newly introduced species of *Aurelia aurita* in San Francisco Bay, California. Marine Biology 125: 401–410.

Hamner, W. M. & R. M. Jenssen, 1974. Growth, degrowth, and irreversible cell differentiation in *Aurelia aurita*. Animal Zoology 14: 833–849.

Holland, B. S., M. N. Dawson, G. L. Crow & D. K. Hofmann, 2004. Global phylogeography of *Cassiopea* (Scyphozoa: Rhizostomeae): molecular evidence for cryptic species and multiple invasions of the Hawaiian Islands. Marine Biology 145: 1119–1128.

Holst, S., I. Sötje, H. Tiemann & G. Jarms, 2007. Life cycle of the rhizostome jellyfish *Rhizostoma octopus* (L.) (Scyphozoa, Rhizostomeae), with the studies on cnidocysts and statoliths. Marine Biology 151: 1695–1710.

Ishii, H. & U. Båmstedt, 1998. Food regulation of growth and maturation in a natural population of *Aurelia aurita* (L.). Journal of Plankton Research 20: 805–816.

Keane, T. M., C. J. Creevey, M. M. Pentony, T. J. Naughton & O. J. McInerney, 2006. Assessment of methods for amino acid matrix selection and their use on empirical data shows that *ad hoc* assumptions for choice of matrix are justified. BMC Evolutionary Biology 6: 29.

Kikinger, R., 1992. *Cotylorhiza tuberculata* (Cnidaria: Scyphozoa) – life history of a stationary population. Marine Ecology 13: 333–362.

Kogovšek, T., B. Bogunovič & A. Malej, 2010. Recurrence of bloom-forming scyphomedusae: wavelet analysis of a 200 year time series. Hydrobiologia 217: 81–96.

Kramp, P. L., 1961. Synopsis of the medusae of the world. Journal of the Marine Biology Association of the United Kingdom 40: 5–469.

Kuo, C. H. & J. C. Avise, 2005. Phylogeographic breaks in low-dispersal species: the emergence of concordance across gene trees. Genetics 124: 179–186.

Librado, P. & J. Rozas, 2009. DnaSP v5: a software for comprehensive analysis of DNA polymorphism data. Bioinformatics 25: 1451–1452.

Lilley, M. S. K., S. E. Beggs, T. K. Doyle, V. J. Hobson, K. H. P. Stromberg & G. C. Hays, 2011. Global patterns of epipelagic gelatinous zooplankton biomass. Marine Biology. doi:10.1007/s00227-011-1744-1.

Malej, A., T. Kogovšek, A. Ramšak & L. Catenacci, 2012. Blooms and population dynamics of moon jellyfish in the Northern Adriatic. Cahiers de Biologie Marine 53: 3.

Mayer, A. G., 1910. Medusae of the World III: The Scyphomedusae. Carnegie Institution of Washington, Washington.

Patarnello, T., F. A. M. J. Volckaert & R. Castilho, 2007. Pillars of Hercules: is the Atlantic–Mediterranean transition a phylogeographical break? Molecular Ecology 16: 4426–4444.

Pelc, R. A., R. R. Warner & S. D. Gaines, 2009. Geographical patterns of genetic structure in marine species with contrasting life histories. Journal of Biogeography 36: 1881–1890.

Rambaut, A., 2006. FigTree: Tree Figure Drawing Tool, version 1.2.2. Institute of Evolutionary Biology, University of Edinburgh, Edinburgh.

Rambaut, A. & A. J. Drummond, 2009. TRACER: MCMC Trace Analysis Tool. 1.4.1 edn. Institute of Evolution Biology, Edinburgh. http://tree.bio.ed.ac.uk/software/tracer. Accessed 18 Jan 2010.

Rippingale, R. J. & S. J. Kelly, 1995. Reproduction and survival of *Phyllorhiza punctata* (Cnidaria: Rhizostomeae) in a seasonally fluctuating salinity regime in Western Australia. Marine & Freshwater Research 46: 1145–1151.

Ronquist, F. & J. P. Huelsenbeck, 2003. MRBAYES 3: Bayesian phylogenetic inference under mixed models. Bioinformatics 19: 1572–1574.

Russell, F. S., 1970. The Medusae of the British Isles: II. Pelagic Scyphozoa with a Supplement to the First Volume on Hydromedusae. Cambridge University Press, London.

Schlötterer, C. & D. Tautz, 1994. Chromosomal homogeneity of Drosophila ribosomal DNA arrays suggests intrachromosomal exchanges drive concerted evolution. Current Biology 4: 777–783.

Schroth, W., G. Jarms, B. Streit & B. Schierwater, 2002. Speciation and phylogeography in the cosmopolitan marine moon jelly, *Aurelia* sp. BMC Evolutionary Biology 2: 1–10.

Shearer, T. L., M. J. H. van Oppen, S. L. Romano & G. Wörheide, 2002. Slow mitochondrial DNA sequence evolution in the Anthozoa (Cnidaria). Molecular Ecology 11: 2475–2487.

Somero, G. N., 2002. Thermal physiology and vertical zonation of intertidal animals: optima, limits, and costs of living. Integrative Comparative Biology 42: 780–789.

Stopar, K., A. Ramšak, P. Trontelj & A. Malej, 2010. Lack of genetic structure in the jellyfish *Pelagia noctiluca* (Cnidaria: Scyphozoa: Semaestomeae) across European seas. Molecular Phylogenetics and Evolution 57: 417–428.

Sugiura, Y., 1965. On the life-history of rhizostome medusae. III. On the effects of temperature on the strobilation of *Mastigias papua*. Biological Bulletin 128: 493–496.

Sugiura, Y., 1966. On the life-history of rhizostome medusae IV, *Cephea cephea*. Embryologia 9: 105–122.

Thompson, J. D., T. J. Gibson, F. Plewniak, F. Jeanmougin & D. G. Higgins, 1997. The CLUSTAL_X windows interface: flexible strategies for multiple sequence alignment aided by quality analysis tools. Nucleic Acids Research 25: 4876–4882.

White, T. J., T. Bruns, S. Lee & J. Taylor, 1990. Amplification and direct sequencing of fungal ribosomal RNA genes for phylogenetics. In Innis, M. A., G. H. Gelfand, J. J. Sninsky & T. J. White (eds.), PCR Protocols: A Guide to Methods and Applications. Academic Press, San Diego: 315–332.

Wörheide, G., S. A. Nichols & J. Goldberg, 2004. Intragenomic variation of the rDNA internal transcribed spacers in sponges (Phylum Porifera): implications for phylogenetic studies. Molecular Phylogenetics and Evolution 33: 816–830.

Hydrobiologia (2012) 690:81–96
DOI 10.1007/s10750-012-1057-5

JELLYFISH BLOOMS

Associations of large jellyfish distributions with temperature and salinity in the Yellow Sea and East China Sea

Fang Zhang · Song Sun · Xianshi Jin · Chaolun Li

Published online: 6 March 2012
© Springer Science+Business Media B.V. 2012

Abstract Climate change may contribute to the increasing frequency and intensity of jellyfish blooms around the world. To test the null hypotheses that distributions did not differ among species of jellyfish or according to temperature salinity, we sampled large jellyfishes using bottom trawl surveys during 2006–2007 in the Yellow Sea (YS) and East China Sea (ECS). The total biomass of large jellyfish in the YS was low in April 2006 in cool waters, increased with warming waters, peaked in early September 2006 (22,891 ± 25,888 kg km^{-2}), and then decreased with cooling to minimal biomass during March 2007. During its peak early September 2006, *Nemopilema nomurai* was relatively eurythermal and distributed throughout the YS. *Cyanea* spp. occurred in warmer waters and attained maximum biomass in May 2007 in the ECS. Ulmaridae, which preferred colder temperatures, reached maximum biomass in October 2006 and occurred mainly in the central YS. *Aequorea* spp. usually occurred in colder waters, with maximum biomass in May 2007 mainly north of 30°N. Our analyses suggest that environmental preferences of the large jellyfish may enable prediction of jellyfish population sizes and distributions in Chinese waters, which is essential in order to address ecological problems caused by large jellyfish blooms in East Asia Waters.

Keywords *Nemopilema nomurai* · *Cyanea nozakii* · Jellyfish bloom · Species composition · Medusae · Biomass

Guest editors: J. E. Purcell, H. Mianzan & J. R. Frost / Jellyfish Blooms: Interactions with Humans and Fisheries

F. Zhang · S. Sun (✉) · C. Li
Institute of Oceanology, Chinese Academy of Sciences, 7 Nanhai Road, Qingdao 266071, People's Republic of China
e-mail: sunsong@ms.qdio.ac.cn

F. Zhang
e-mail: fzhang2008@gmail.com

X. Jin
Yellow Sea Fisheries Research Institute, Chinese Academy of Fishery Sciences, 106 Nanjing Road, Qingdao 266071, People's Republic of China

Introduction

In recent years, many studies have demonstrated that jellyfish and ctenophores play a critical role in structuring coastal marine and estuarine ecosystems, and their high production within such ecosystems has increased global concern (e.g., Arai, 2001; Lynam et al., 2005; Purcell, 2005; Doyle et al., 2007; Purcell et al., 2007). This attention is largely the result of jellyfish and ctenophore blooms that have caused ecological and economic losses, which are linked intrinsically to overfishing, eutrophication, climate change, and species invasions (reviewed in Purcell et al., 2007). Thus, climate change has been reported to be one of the possible reasons for the increasing frequency and intensity of jellyfish blooms, which has been locally indicated in many sea areas. For example, climate-related increases of jellyfish frequency suggest a more gelatinous future for the North Sea (Attrill et al., 2007). Nevertheless, increases in jellyfish and ctenophores provide valuable insights into the bloom phenomenon (e.g., Shiganova, 1998, 2005; Uye et al., 2003; Uye & Ueta, 2004).

Blooms of large jellyfish have formed increasingly in the southern Yellow Sea (YS) and northern East China Sea (ECS) during summer and fall since the end of the 1990s (Yan et al., 2004; Cheng et al., 2004, 2005; Ding & Cheng, 2005; Dong et al., 2005, 2006a, b), bringing severe economic loss to Chinese people. Large jellyfish are slimy, fragile, hard to process, unsuitable for fish forage, and thus usually are not studied by Chinese scientists. Consequently, jellyfish are considered a nuisance when collected in trawls and thrown directly back into the sea. Because little scientific attention has been devoted to them, there is a paucity of knowledge and data of species composition, geographical distribution, and general natural history in the YS and ECS. Recently, however, the occurrence of jellyfish, their biomass, and preliminary analysis on bloom causes has been locally documented for the YS and ECS region (Cheng et al., 2004, 2005; Ding & Cheng, 2005, 2007). In addition, that "the population of *Nemopilema nomurai*, the bloom forming species in the Sea of Japan, is thought transported from the China coastal seas" was assumed in previous studies (e.g., Kawahara et al., 2006; Uye, 2011). Nevertheless, data on this species over larger spatial and temporal scale in the YS and ECS have been unavailable until our study.

To provide quantitative data on large jellyfish in the YS and ECS, this study documents the annual geographical abundance and biomass of large jellyfish using data collected as a part of the China-Integrated Marine Biogeochemistry and Ecosystem Research (IMBER), National Basic Research Program (China Jellyfish Project 973 Program), National Natural Sciences Foundation of P. R. China program (NSFC), and Public Science and Technology Research Funds Projects of Ocean. Our objective was to provide synchronously extensive data on species composition, spatial and temporal distribution patterns, and relationships of the predominant species to environmental factors. We test the null hypotheses that the spatial and temporal distributions of jellyfish, and their temperature and salinity preferences, do not differ among species, which may elucidate the relationships between jellyfish blooms and climate change, and ecological problems caused by large jellyfish blooms in the East Asia Waters.

Materials and methods

Site description

The YS and ECS are temperate marginal seas in the northwest Pacific Ocean, both being semi-enclosed by the contiguous lands of China in the west and Korea in the east. They are connected topographically, but divided subjectively by a line from the Yangtze River mouth to Cheju Island (Fig. 1). The two seas are highly dynamic regions with widely varying water properties, which are impacted by seasonal variation of environmental factors like wind, river runoff, and air temperature (Liu et al., 1992; Hur et al., 1999).

A few typical hydrological features are present in the YS, such as a pronounced tidal front, the Yellow Sea cold bottom water (YSCBW; located in the middle part of the YS) and the Yellow Sea warm current (YSWC) (Figs. 1, 2). Thermal stratification begins at the end of spring, peaks in summer, and then breaks down at the end of fall. This stratification occurs simultaneously with mixing of the shallow region by high-energy tidal waves. As a result, tidal fronts form between the stratified and mixed areas (Wei et al., 2003). The stratified area is primarily located in the YSCBW mass, which is entrenched deep below the thermocline and acts as a bottom nutrient pool (Wei et al., 2002). The thermocline is a barrier to the vertical distribution and migration of zooplankton by offering a bottom, over-summering site to temperature species that cannot tolerate high temperatures in the upper layer and in the coastal area (Wang et al., 2003). In winter, monsoon conditions prevail with a strong north wind that drives a surface, southward flow along both the Chinese and Korean coasts, which is also balanced by a northward flow along the YS trough (Jacobs et al., 2000; Teague & Jacobs, 2000). The YSWC advects warmer and saltier water into the YS, which greatly changes the biological community structure.

The hydrology of ECS is mainly influenced by the Yangtze River runoff and the embranchment or main body of the Kuroshio Current. Kuroshio Current is

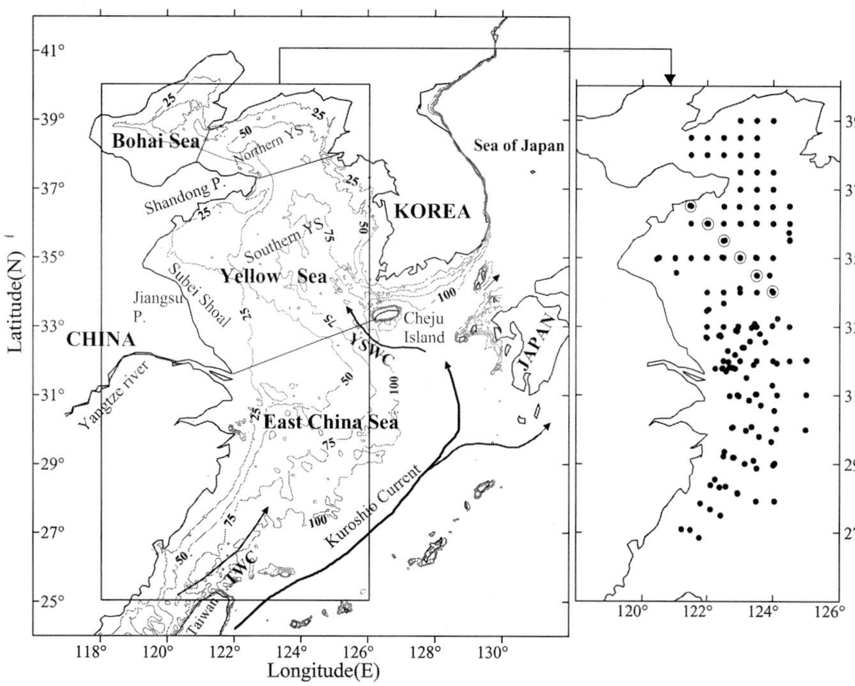

Fig. 1 Map of study area with boundary of Bohai Sea, YS, southern YS, northern YS, and ECS. Isobaths (*gray line*) of 25, 50, 75, and 100 m in coastal China Sea. Positions of Kuroshio Current, Taiwan current (*TWC*), and Yellow Sea warm current (*YSWC*) are shown (*arrows*). Sampling stations (*dots*), and 6 stations (*circles*) at one transect are also indicated

characterized by warm, saline waters. The extent of influence by them is related to the climate phenomenon of the North Pacific Ocean such as El Niño/La Niña-Southern Oscillation (He & Hong, 1999).

Sampling

Ten bottom trawl surveys were conducted in the YS and ECS aboard the R/V *Beidou* from April 2006 to August 2007 (Table 1). Different sampling strategies consisted of 5 grid surveys (GS), spanning ca. 4 longitude degrees and 5–11 latitude degrees, and 5 transect surveys (TS). Survey stations varied somewhat from cruise to cruise due to different study objectives of the program (Fig. 1). All surveys were carried out over a 24 h period. At each station, we deployed a bottom trawl net having a total length of 83.2 m, opening circumference of 167.2 m (836 mesh × 20 cm), opening height ca. 7 m, opening width ca. 22 m, and 10-cm mesh size cod-end with a 2.4-cm mesh liner. The towing speed was approximately 3 kt and the duration of trawl hauls at each station was 0.3–1 h. Haul duration was shortened when masses of jellyfish were present so that the net was not damaged or the towing capacity of the research vessel exceeded. An average tow filtered ca. 840,000 m^3. A Seabird SBE-25 CTD was deployed to determinate temperature and salinity on every station during all cruises.

Catches at each station were enumerated as soon as the trawl was hauled onboard. Each specimen was identified, species were weighed in aggregate (wet weight), and individuals of each species were enumerated. Bell diameter (BD) and wet weight (WW) of each specimen for dominant species were measured at random during earlier cruises to obtain the species-specific exponential fitting formula between BD (cm) and WW (g) (Fig. 3). Thus, the WW of individual jellyfish from later sampling could be calculated using their measured BD in the appropriate equation. Because variation of biomass or density from different catches was large for big catches, BD of a subsample of each species was measured. WW of each species in each catch was calculated from the total number of this species multiplied by the average WW, as estimated from the equation of WW versus BD of individuals measured from subsamples.

Calculations of abundance and biomass

Biomass density (ρ) for each species was expressed as kg km^{-2} of area swept by the trawl using the equation:

Fig. 2 Isoclines of surface and bottom temperatures during June and August & September 2006 in the YS and ECS

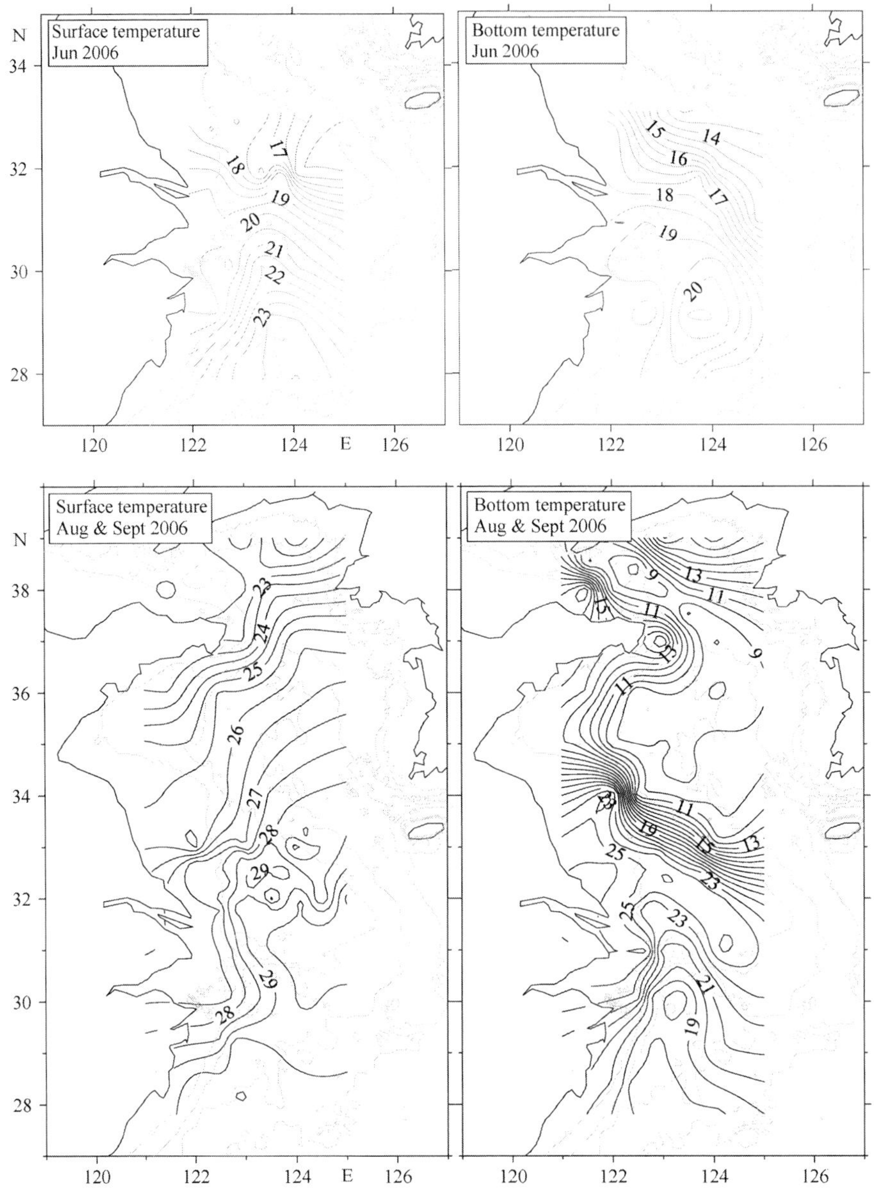

$$\rho = 10^3 c/(b\,l),$$

where c (kg) is the WW of each or total species in a trawl, l (km) is the integral distance of a trawl (read from a Simrad EK500 echo sounder integral system), and b (m) is the width of the net mouth, considering the influence for the trawl mouth area to fluctuate under variable hydrodynamic pressure. To account for this fluctuation, Tang et al. (2006) determined a fitting formula ($b = 0.12\ d + 15.7$) through an experiment with this trawl, where d (m) is the average water depth of hauling at each station. When c was changed into n (number of jellyfish in one catch), a (abundance density) was calculated. Variation from the mean of all units in this paper is shown as ±SD.

Statistical analyses

To test the null hypotheses that the spatial and temporal distributions of jellyfish did not differ among

Table 1 Dates, areas, and type of surveys for jellyfish in the YS and the ECS

Year	Date	Sea area	Abbreviation	Region covered	Type
2006	10–29 Apr	Southern YS	Apr 2006	31.9–36.5°N, 121.0–125.0°E	Grid
	01–13 June	Mainly ECS	Jun 2006	27.9–33.1°N, 122.4–125.0°E	Grid
	19 Aug–12 Sep	YS and ECS	Aug and Sep 2006	27.8–39.0°N, 120.5–125.0°E	Grid
	18-27 Sep	Southern YS	Sep 2006	32.0–36.5°N, 120.5–124.0°E	Transect
	02 Oct–03 Nov	Southern YS and ECS	Oct and Nov 2006	27.8–36.5°N, 121.0–125.0°E	Grid
2006/2007	25 Dec–02 Jan	Southern YS	Dec 2006	32.0–36.5°N, 120.5–124.0°E	Transect
2007	14–23 Mar	Southern YS	Mar 2007	32.0–36.5°N, 120.5–124.0°E	Transect
	06–15 May	ECS	May 2007	26.9–32.0°N, 121.2–124.0°E	Grid
	16–29 May	Southern YS	May 2007	32.0–36.5°N, 120.5–124.0°E	Transect
	03–09 Aug	Southern YS	Aug 2007	34.0–36.5°N, 120.5–124.0°E	Transect

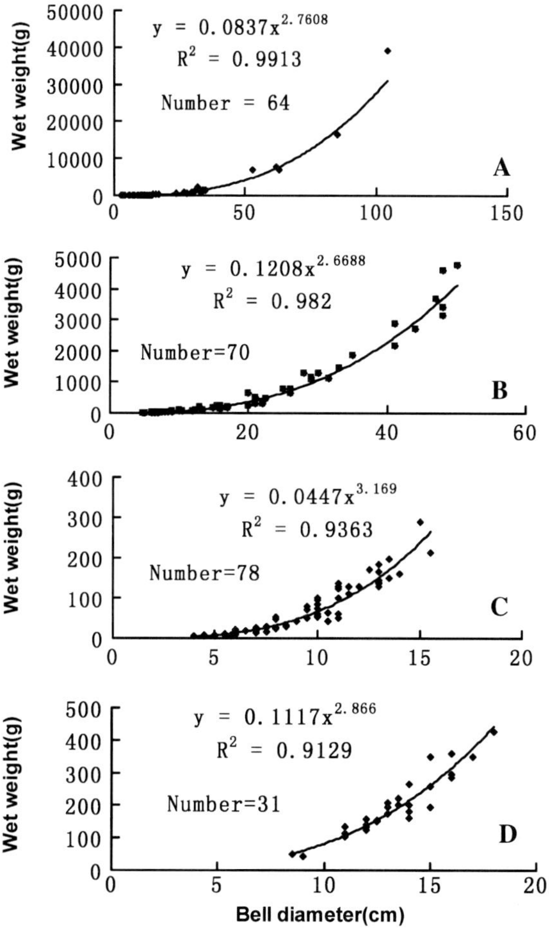

Fig. 3 The relationships between bell diameters and wet weights of jellyfish species in Chinese waters. **A** *Nemopilema nomurai*, **B** *Cyanea* spp., **C** *Aequorea* spp., and **D** Ulmaridae (genus and species undetermined)

species, biomass data from all cruises in the YS and ECS were log-transformed and subjected to a q-type cluster analysis based on the Bray–Curtis similarity and group average linkage classification (Bray & Curtis, 1957; Field et al., 1982). Combined with cluster analysis, non-metric multidimensional scaling (MDS) was also performed to check adequacy and mutual consistency of both representations (Clarke & Warwick, 2001). As a complimentary method, the cluster and MDS analyses were also performed separately for each season, except in winter due to fewer samples and small jellyfish biomass.

To test the null hypotheses that the temperature and salinity preferences of jellyfish did not differ among predominant species, a series of frequency plots and a three-dimensional scatter plot indicated the bottom temperature and salinity at each station where predominant species appeared.

Results

Hydrography

Water depth varied from 13 to 94 m for the study area. Temperature and salinity for all cruises from April 2006 to August 2007 ranged between 4.9 and 30.2°C and 27.1 and 34.6, respectively (Table 2). The Yangtze River discharge and the Subei shoal had lower salinities in the western part than the eastern part of the study area. Temperature varied mainly over seasons and latitude, but also was affected by the fronts, YSCBW, warm saline Kuroshio-ECS water, and

Table 2 Surface and bottom temperature (*T* in °C) and salinity (*S*) at 6 stations representing a transect (Fig. 1) from northwest to southeast during 2006–2007 cruises in Chinese waters

	St.	2006								2007					
		April		Late September		October		December		March		May		August	
		T (°C)	*S*	*T* (°C)	*S*	*T* (°C)	*S*	*T* (°C)	*S*	*T* (°C)	*S*	*T* (°C)	*S*	*T* (°C)	S
Surface northwest	1	7.0	31.3	22.8	31.0	19.8	31.2	7.6	31.4	6.8	31.6	15.0	31.6	26.0	30.8
↓	2	7.9	31.5	23.4	31.7	20.1	31.1	9.8	31.5	6.2	31.7	15.2	31.7	25.7	30.2
	3	8.3	32.1	23.8	31.5	19.5	31.2	11.1	31.8	8.7	32.6	15.2	32.3	26.1	31.8
	4	10.6	34.2	23.9	30.9	21.1	31.2	12.0	31.9	11.0	33.8	15.4	33.2	27.2	31.8
	5	10.5	33.8	23.4	30.6	22.8	31.0	12.4	32.2	10.1	33.5	15.5	33.0	26.7	31.6
Southeast	6	10.6	29.8	22.9	30.9	22.7	30.9	14.2	32.9	10.5	33.4	16.2	33.3	26.9	31.2
Bottom northwest	1	6.9	31.3	22.2	31.1	19.8	31.1	7.6	31.5	6.5	31.7	11.2	31.6	19.8	31.7
↓	2	5.0	31.6	9.1	32.3	19.2	31.2	9.8	31.4	5.9	31.7	9.4	32.8	9.3	32.0
	3	9.1	33.7	8.6	33.2	8.4	32.8	10.0	32.6	10.0	33.4	11.2	34.0	10.2	33.4
	4	10.1	34.3	9.0	33.5	8.8	33.2	9.6	33.4	11.1	33.8	11.5	33.9	11.5	33.9
	5	10.1	34.2	9.9	34.0	9.6	33.8	9.9	33.7	10.1	33.8	10.6	33.5	11.6	33.8
Southeast	6	10.1	34.0	9.4	33.7	9.6	33.7	15.7	33.6	11.5	33.9	11.5	33.7	11.9	33.7

YSWC (Fig. 1). In June 2006 in the ECS, a weak thermocline, as indicated by surface and bottom temperatures, was present. In August and September 2006, a strong tidal front, here the YSCBW, was observed (Fig. 2).

Species composition

Eleven species of jellyfish, *Nemopilema nomurai* Kishinouye, *Cyanea nozakii* Kishinouye, and *Cyanea purpurea* Kishinouye, less prevalent *Cyanea* sp., Ulmaridae (genus and species undetermined), *Aequorea coerulescens* Brandt, *Aequorea* sp.1, *Aequorea* sp.2, *Aurelia aurita* Linnaeus, *Liriope tetraphylla* (Chamisso & Eysenhardt), and *Pelagia noctiluca* (Forskal) were identified during the survey from April 2006 to August 2007. *Cyanea* spp. and *Aequorea* spp. were treated at the generic level for simplicity. In the YS, *N. nomurai*, Ulmaridae, and *Aequorea* spp. had the widest occurrence of the 11 species for all cruises (Table 3) and greatest biomass (97.8, 0.97, and 0.93%, respectively) of all jellyfish. By comparison, *N. nomurai* only occurred in a narrow area of the northern ECS (representing 82.8% of all jellyfish there), while *Cyanea* spp. were the predominant jellyfish in the ECS (representing 14.4% of total jellyfish).

Jellyfish assemblage in the YS

During spring sampling (April 2006, March and May 2007; Fig. 4A, F, G), *Aequorea* spp. and Ulmaridae were the predominant species. Ulmaridae occurred in the central YS where the temperature and salinity were directly affected by YSCBW, while *Aequorea* spp. generally occurred at coastal stations. In the summer (Fig. 4B, B', C, H), large jellyfish were widely distributed throughout the study region, occurring in 86.7 ± 23.2% of the biomass trawls. In the southern YS (32–36°N), *N. nomurai* represented 96.7 ± 8.9% of all large jellyfish and 86.1 ± 24.4% of the total catch (average biomass was 20,446 kg km^{-2}), and these jellyfish were widespread over the coastal area of the YS, forming blooms that overwhelmed other gelatinous species during August and September 2006 sampling. Ulmaridae was second in numerical predominance, and on average, the abundance of this species accounted for 40% (0.3–95%), 30% (5.0–80%), and 18% (7–47%) of total jellyfish abundance at stations where they occurred during early September 2006, late September 2006, and August 2007, respectively.

During fall and winter sampling (October and November 2006, December 2006) total jellyfish biomass declined (Fig. 4D, D', E), *N. nomurai* and

Table 3 Occurrences of large jellyfish in the YS and ECS from different surveys

Species	Yellow Sea area (mainly southern YS)								East China Sea area (mainly northern ECS)			
	Apr 2006	Early Sep 2006	Late Sep 2006	Oct and Nov 2006	Dec 2006	Mar 2007	May 2007	Aug 2007	Jun 2006	Aug 2006	Oct 2006	May 2007
Nemopilema nomurai	+	+	+	+	+		+	+	+	+	+	+
Cyanea spp.		+	+				+	+	+	+	+	+
Ulmaridae.	+	+	+	+	+	+	+	+				
Aequorea spp.	+	+			+	+	+	+	+			+
Aurilia aurita	+											
Pelagia noctiluca									+			
Liriope tetraphylla							+					+

+ indicates occurrence, blank indicates no occurrence

Ulmaridae were still the predominant species. In October and November 2006, the biomass distribution of *N. nomurai* shrank toward the northern area, being mainly distributed north of 33°N. The biomass of this species decreased, with the maximum biomass (1.6×10^4 kg km^{-2}) about one-sixth of that in August and September 2006 (Fig. 4D'). Ulmaridae became the most numerous species at 21–2,780 ind. km^{-2}, with a biomass of 7.7–1,807 kg km^{-2}. On average, abundances of Ulmaridae accounted for 68% (2.3–100%) of total jellyfish in October and November 2006 and 88% (50–100%) in December 2006. In summary, abundance percentages of Ulmaridae gradually increased from summer to winter sampling and they became the most abundant species during October, November, and December 2006.

To test the null hypotheses that the spatial and temporal distributions of jellyfish did not differ among species, we used the Bray–Curtis similarity and group average linkage classification. The clustering and ordination analysis of all surveys in the YS identified three major seasonal groupings of stations that corresponded to spring, spring–autumn–winter, and summer–autumn samples. The spring community was located at the south shore of Shandong Peninsula and the central YS, and was distinguished by one indicator genus, *Aequorea* spp. The spring–autumn–winter community was found at the middle part of the YS and was mainly characterized by a higher biomass of Ulmaridae. The summer–autumn community was located throughout the YS and had a notably high biomass of *N. nomurai*. Results of the ordination analysis of MDS were consistent with the cluster analysis (all stress coefficients <0.1).

For spring sampling, cluster analysis indicated two groupings. One grouping was located in waters deeper than the 50 m isobath, predominated by Ulmaridae, and the other grouping was shallower than the 50 m isobath, predominated by *Aequorea* spp. For summer sampling, three major groupings were identified, which were named large group A, small groups B and C. Large group A was located at the YS coast, in waters between the 25-m and 50-m isobaths, and was characterized by *N. nomurai*. Small group B was located at the YS coast, shallower than the 25-m isobath and was predominated by *N. nomurai* and *Cyanea* spp. Group C was in the central part of the YS, in waters at or beyond the 50 m isobath, and was predominated by Ulmaridae and *N. nomurai*. For fall sampling, two groupings were identified. One was located at the central part of the YS characterized by Ulmaridae. The other grouping was in the central YS and predominated by a higher biomass of *N. nomurai* with some Ulmaridae.

Jellyfish assemblage in the East China Sea

Due to the more southern location of the ECS, the predominant species and their distributional patterns differed some from the YS. In general, the most numerous species were *Cyanea* spp. along the southern transects and *Aequorea* spp. along the northern transect. *N. nomurai* and *Pelagia noctiluca* occurred sporadically in the northern transect during June 2006 (Fig. 4I). In August 2006, *N. nomurai* was as numerous in the northern region as in June 2006, *Cyanea* spp. had a low average biomass of 10.6 kg km^{-2}, abundance of 2.3 ind. km^{-2}, and was distributed sporadically in the ECS (Fig. 4B, B'). In October 2006, jellyfish were present at only 26% of the stations, with *Cyanea* spp. occurring at three stations on the most southern transect. *N. nomurai* occurred sporadically in the northern part of the ECS study area (Fig. 4D, D'). In May 2007, *Liriope tetraphylla* occurred sporadically in the northern transect (Fig. 4J). *Cyanea* spp. occurred at almost all southern transect stations (96% frequency of occurrence), had a maximum average biomass of 380 kg km^{-2} and abundance of 512 ind. km^{-2}, which accounted for 82.1 ± 34.9% of the large jellyfish assemblage in May 2007.

To better understand the species composition in the ECS, hierarchical cluster analysis results identified three groupings with indicator species N (*N. nomurai*), C (*Cyanea* spp.), and A (*Aequorea* spp.). The N grouping was located north of 30°N and sampled mainly during June, August, and October 2006. The C grouping was south of 31°N and found mainly during June 2006 and May 2007. The A grouping was north of 30°N and found mainly during June 2006 and May 2007. Therefore, we rejected the null hypothesis that the distributions of jellyfish did not differ among species in both the YS and ECS.

Temperature and salinity preferences of predominant species

Total jellyfish biomass was low during the April 2006 survey (6.8 ± 15.0 kg km^{-2}), coinciding with cooler sea surface temperatures (9.3 ± 1.5°C), but biomass then increased with warming temperatures, reaching the highest observed value of 22,891 ± 25,888 kg km^{-2} in early September 2006 (Fig. 5). Afterward, biomass decreased with cooling sea surface temperatures, reaching the lowest observed value in March 2007 (4.6 ± 9.4 kg km^{-2}; 9.1 ± 1.7°C).

Overall, temperature and salinity ranges of species were more variable than were their frequencies of occurrence (Figs. 6, 7). For *N. nomurai*, frequency of occurrence was highest at the bottom temperature range of 7–10°C, while they occurred in waters from 7.7 to 26.1°C. The salinity range of 29.8–34.2 for *N. nomurai* matched their frequency of occurrence, with notable peaks at bottom salinities of 32 and 33.

Fig. 4 Distribution, biomass, and biomass composition of the ▶ jellyfish assemblage during April 2006 to August 2007 surveys in Chinese waters. **A–J** *Pies* indicate total jellyfish biomass (*pie* size: biomass, kg km^{-2}. Biomass increases linearly with the pie area at each station), different *shadings* indicate different jellyfish species, with proportions as percentages of the pie area. Note that in (**B**) and (**D**), biomass of *Nemopilema nomurai* was excluded to show biomasses of other species, which otherwise would be masked by *N. nomurai* due to their huge biomass. The biomass of *N. nomurai* was shown instead in (**B'**) and (**D'**). Different *scale bars* used for different surveys. *A. aurita* = *Aurelia aurita*; *L. tetraphylla* = *Liriope tetraphylla*; *P. noctiluca* = *Pelagia noctiluca*

Cyanea spp. occurred most frequently when bottom temperatures ranged from 17 to 20°C and bottom salinity was 33–34. Their distribution indicated that *Cyanea* spp. could occur in waters of 14.8–25.8°C with bottom salinities of 29.8–34.6. *N. nomurai* seemed to tolerate a wider range bottom temperature (frequency of occurrence 6–25°C) than *Cyanea* spp. (frequency of occurrence 17–20°C). The frequency of occurrence for *Aequorea* spp. spanned bottom temperatures from 8 to 20°C and peaked when bottom salinity was 33. They were distributed throughout bottom temperatures ranging from 8.7 to 19.4°C and bottom salinities 30.3 to 34.6. The frequency of occurrence for Ulmaridae indicated a peak at bottom temperatures of 9–10°C and bottom salinities of 33–34. The range of temperature and salinity preferences was most narrow for Ulmaridae, ranging from 7.7 to 11.9°C and 31 to 34, respectively. The frequencies of occurrence ranked from widest to narrowest were *N. nomurai*, *Aequorea* spp., *Cyanea* spp., and Ulmaridae (Fig. 6). Therefore, we rejected the null hypotheses that the temperature and salinity preferences of jellyfish did not differ among species. The differences among species may indicate that particular environmental conditions promote their increased production and could allow blooms of certain jellyfish species to be forecast.

Discussion

Large-scale distributions of jellyfish in Chinese waters

In this study, bottom trawl surveys provided the first opportunity to sample large jellyfish in the YS and

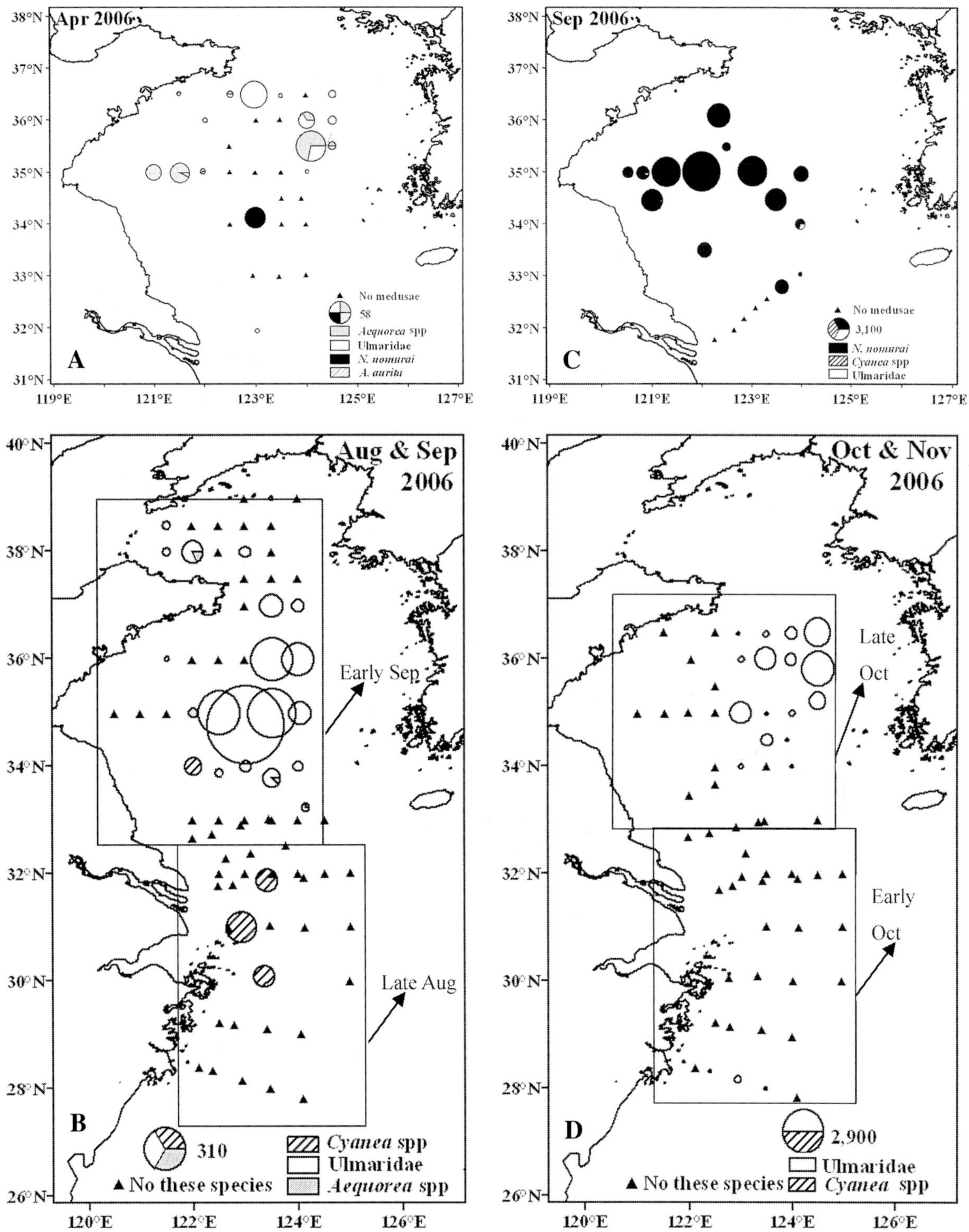

ECS to assess population biomass and abundance over large areas and in different seasons. This was the most practical sampling method due to available concomitant fishery resources and hydrographic data sets of the coastal China Sea. Furthermore, because consistent trawling methods were employed, comparisons among

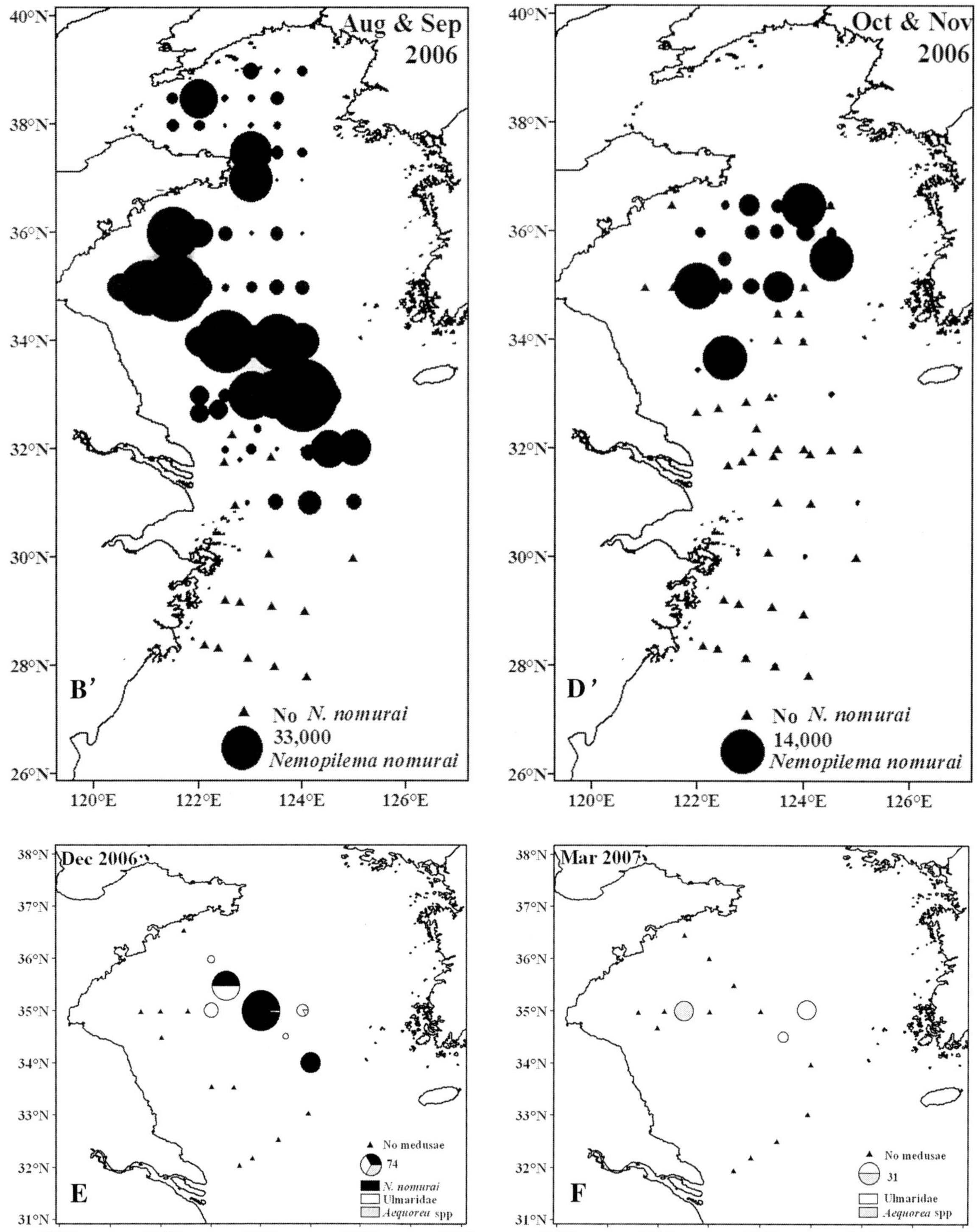

Fig. 4 continued

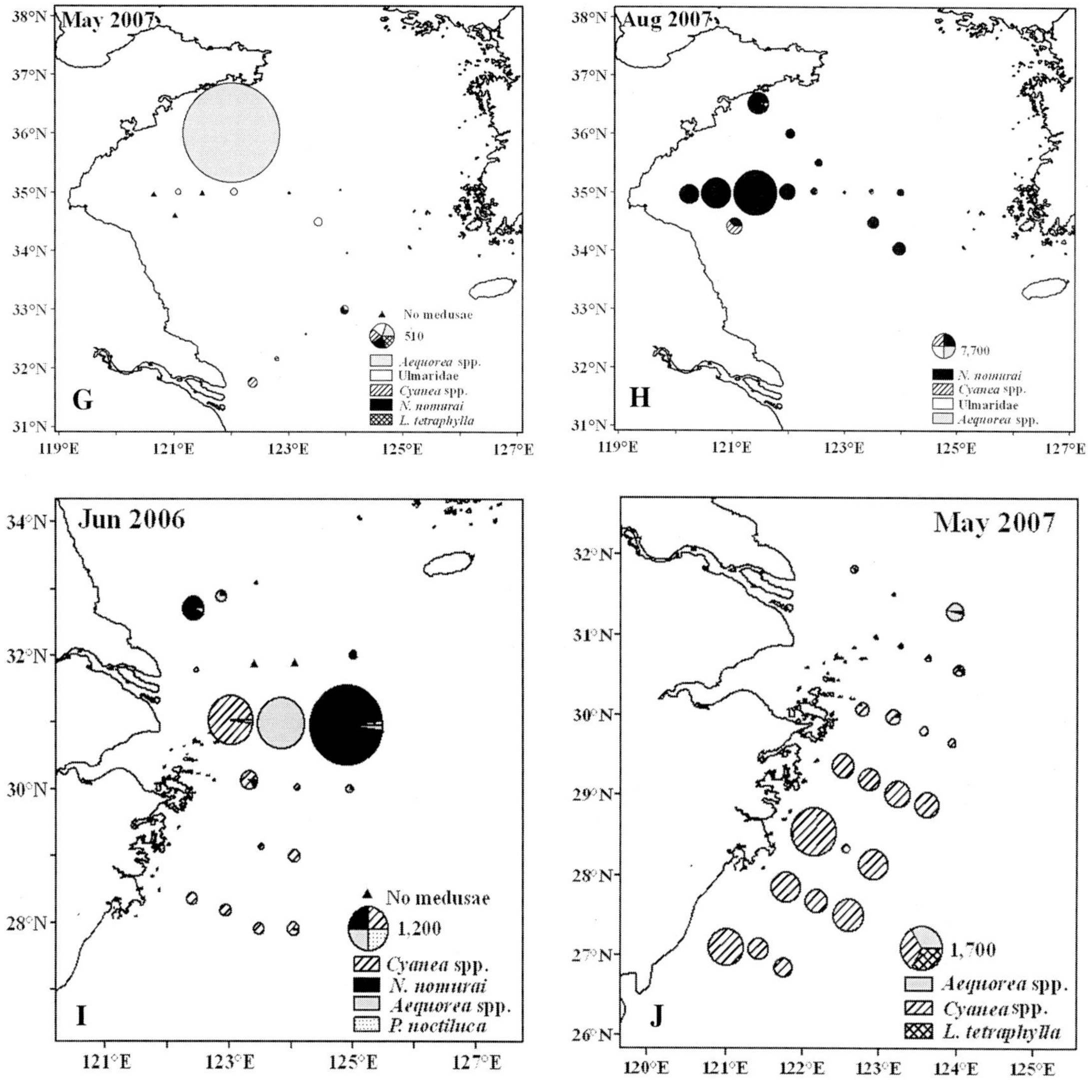

Fig. 4 continued

trawls and seasons were possible. Our analyses showed clear differences in temporal and spatial distributions of predominant species.

Quantification of large jellyfish is notoriously difficult and rife with problems unique to their watery composition, large size, and localized concentration (Hamner et al., 1975; Omori & Hamner, 1982; Graham et al., 2003). Large jellyfish also are generally too patchy to allow good sampling by standard zooplankton nets (Hay et al., 1990). Previous studies have used fishing gears to catch jellyfish (e.g., Mironov, 1971; Shenker, 1984; Brodeur et al., 1999, 2002, 2008a; Graham et al., 2001; Lynam et al., 2005), as well as large-scale, pelagic nets (e.g., Mianzan & Guerrero, 2000; Suchman & Brodeur, 2005; Brodeur et al., 2008b).

The source of *N. nomurai* in Japanese waters is still undetermined. *N. nomurai* in the YS and ECS previously was identified as *Stomolophus meleagris* (namely "Sha haizhe" in Chinese) by Chinese researchers (e.g., Hong & Zhang, 1982, Gao et al., 2002, Cheng et al., 2004), while this species in the Sea of Japan was reassigned to the genus *Nemopilema* by Omori & Kitamura (2004) after going through their confusing phase of identification (Kawahara et al., 2006). *N. nomurai* was a common species in China coastal sea and the Sea of Japan, however, the population could be different for the largest bell diameter (135 cm) of this

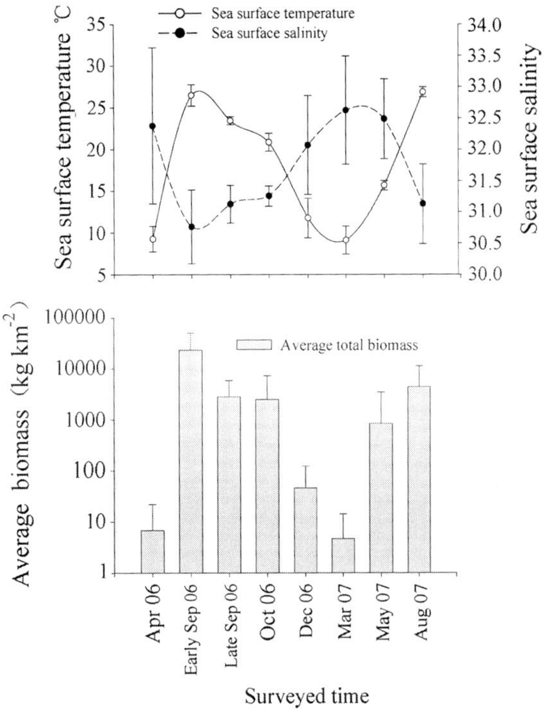

Fig. 5 Seasonal variation of sea surface temperature and salinity and total jellyfish biomass in the YS in 2006 and 2007

species ever sampled in the YS and ECS is far smaller than the 200 cm sampled in the Japan Sea. Furthermore, no scyphistomae or ephyrae were found in China coastal seas, so the assumption that "*N. nomurai* are released as ephyrae from benthic scyphistomae residing in China coastal seas" (Uye, 2011) lacks direct evidence. Further studies on the life cycle and population dynamics of this species in China coastal seas are necessary in East Asia Waters.

Latitudinal patterns of distribution have been observed for species in other regions, as in the Celtic and Irish seas (Doyle et al., 2007). *N. nomurai* and *Cyanea* spp. have increasingly formed blooms in summer and autumn in the southern YS and northern ECS since the end of the 1990s (Cheng et al., 2004, 2005; Yan et al., 2004; Ding & Cheng, 2005), especially in 2003, although the intensity decreased a little in 2004 and 2005 (Ding & Cheng, 2007; Li et al., 2007). Our research indicated that the two species bloomed in 2006 with different distributions; *N. nomurai* was abundant in the northern area (i.e., YS and adjacent northern ECS) and *Cyanea* spp. in the southern area (i.e., ECS). The boundary line was about 31°N in June, 2006; however, the species boundary was further north in June, 2008 (authors' unpublished data). Their distributional differences could be related to the annual variation of the convergence of YSCBW and Kuroshio Current, which is forced by climate changes.

Previous studies have indicated that hydrographic features affect jellyfish accumulation. For example, high jellyfish abundance was associated with physical discontinuities in the Celtic and Irish seas (Doyle et al., 2007), numerous tidal fronts affected populations of large medusae in the eastern Bering Sea (Brodeur et al., 1997), and very dense *Crambionella orsini* (Vanhöffen) aggregations occurred at frontal systems in the Arabian Sea (Billett et al., 2006). During the August and September 2006 grid survey, *N. nomurai* medusae formed a bloom and occurred at tidal front areas (Fig. 2, 4B'). During the June 2006 survey, high biomasses of *N. nomurai* and *Cyanea* spp. were found within temperature discontinuities (Fig. 2, 4I). The mechanism of how aggregations of *N. nomurai* respond to the tidal front in the YS was undetermined. Their increased abundance could reflect the locally high chlorophyll *a* concentration and zooplankton biomass (Liu et al., 2003), or the interaction between frontal circulation and their swimming behavior (Purcell et al., 2000).

Temperature and salinity preferences of predominant species

The importance of environmental factors in determining jellyfish distributions is well known (e.g., Graham et al., 2001). Analysis of temperature and salinity preferences of the predominant species allowed us to determine environmental variables that could influence jellyfish frequency of occurrence and distribution in the YS and the ECS.

Because our results showed that *Cyanea* spp. were mainly distributed throughout the ECS at 17–20°C, we considered them to be warm-water species. In comparison, *N. nomurai* medusae in this study seemed to be eurythermal rather than a cold-water species, as was suggested by previous studies (Gao et al., 2002; Cheng et al., 2005). To understand why *N. nomurai* occurred only in the YS and the northern part of the ECS, rather than the southern part, requires study of their benthic phase. Temperature and salinity ranges were narrowest for Ulmaridae (9–10°C, salinity 33–34), which occurred in the middle part of the YS and we considered to be a saline, cold-water species.

Fig. 6 The frequencies of occurrence of jellyfish according to bottom temperatures and salinities in Chinese waters in 2006 and 2007. Expected distribution curves of the predominant species/taxa are also shown

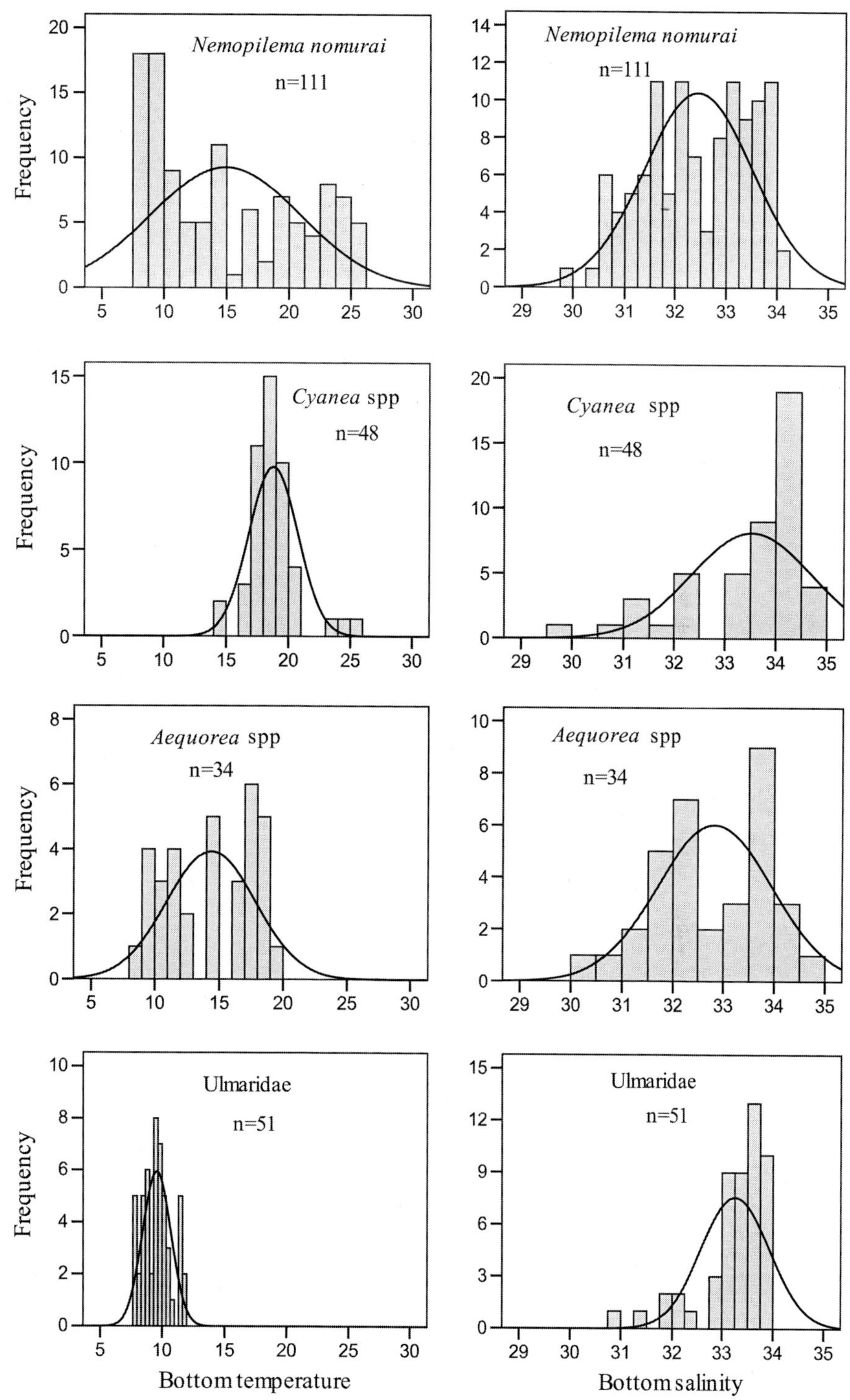

Aequorea spp. were distributed mainly in the YS and northern ECS, had a wider temperature range than Ulmaridae, but we also considered to be cold-water species.

Like other zooplankton, distributional patterns of jellyfish can be influenced by sea currents or water masses (Gao, 1982; Graham et al., 2001). Thus, jellyfish can be good indicators of specific water

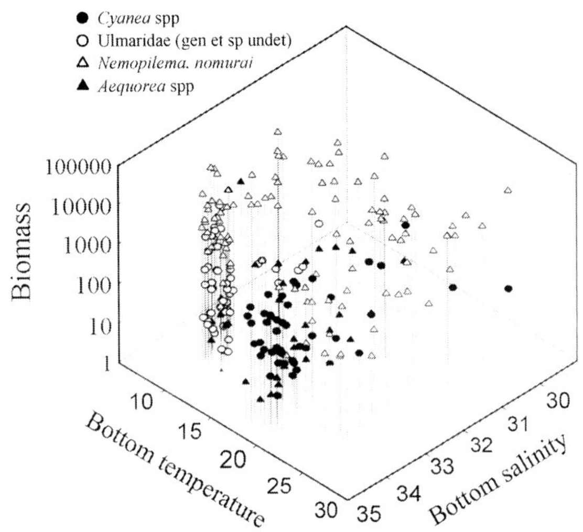

Fig. 7 Three-dimensional scatter plot with biomass (kg km^{-2}) of *Nemopilema nomurai*, *Cyanea* spp., Ulmaridae (genus and species undetermined), and *Aequorea* spp. and sea bottom water temperature (°C) and salinity at investigated stations in Chinese waters in 2006 and 2007

currents and masses because they exist within certain limits of temperature and salinity. For instance, Alvarino (1971) characterized the siphonophore *Chelophyes contorta* Lens & van Reimsdijk as an indicator of warm water flowing to California. In our study, Ulmaridae was clearly adapted to cool temperatures and higher salinities (7.7–11.9°C and salinity 30.9–33.9).

The YSCBW occupies ~30% of the total volume of the YS (Su & Weng, 1994), as demarcated in the temperature field (Hur et al., 1999; Chu et al., 2005; Zhang et al., 2008). Previous studies have documented the variations in range of temperature and salinity of the YSCBW (e.g., 4.03–11.73°C, salinity 31.34–33.92; Weng et al., 1989). Because Ulmaridae was adapted to temperatures and salinities consistent with the character of the YSCBW, and their occurrence exactly corresponded with the area of YSCBW, even in different seasons (Fig. 4); therefore, this species can be used as an indicator of YSCBW. YSCBW, a conservative water masses in the YS, is likely to contain the long-term signals for understanding climatological evolutions of the YS. Park et al. (2011) reported that the inter-annual and inter-decadal variability in the YSCBW related to climate changes, including the Pacific Decadal Oscillation, Arctic Oscillation, increased SST in the Kuroshio and others. Therefore, the annual and seasonal distribution pattern and biomass of the indicator species Ulmaridae could respond to the spread and retreat of the YSCBW forced by climate change.

In summary, wide-scale trawl net sampling provided information on annual geographical occurrences and biomass of large jellyfish assemblages in the YS and ECS. This study presented synchronously extensive spatial and temporal distribution patterns by species. Preferred temperature and salinity regimes of the predominant species were analyzed. The paucity of historic or current data has made it difficult to relate recent jellyfish blooms to anthropogenic activity and climate change in the coastal China Sea. This study provided fundamental knowledge that which will be essential to understand problems related to the causes and effects of jellyfish blooms in the YS, ECS, and in East Asia waters.

Acknowledgments We thank Dr. D. Huang and Dr. H. Wei for contributing CTD data. We are grateful to the captain and crew of the RV "Beidou" for their efforts in the field and to the people who provided support in our sampling. We also thank the reviewers and editors for their suggested improvements to the manuscript. Funding for this research was provided by the national key fundamental developing projects Nos. 2011CB403601 and 2006CB400606 to S. Sun, NSFC project No. 40606036 to F. Zhang, and public science and technology research funds projects of ocean No. 201005018 to C. Li.

References

Alvarino, A., 1971. Siphonophores of the pacific with a review of the world distribution. Bulletin of the Scripps Institution of Oceanography 16: 1–432.

Arai, M. N., 2001. Pelagic coelenterates and eutrophication: review. Hydrobiologia 451: 69–87.

Attrill, M. J., J. Wright & M. Edwards, 2007. Climate-related increases in jellyfish frequency suggest a more gelatinous future for the North Sea. Limnology and Oceanography 52: 480–485.

Billett, D. S. M., B. J. Bett, C. L. Jacobs, I. P. Rouse & B. D. Wigham, 2006. Mass deposition of jellyfish in the deep Arabian Sea. Limnology and Oceanography 51: 2077–2083.

Bray, J. R. & J. T. Curtis, 1957. An ordination of the upland forest communities of southern Wisconsin. Ecological Monographs 27: 325–349.

Brodeur, R. D., M. T. Wilson & J. M. Napp, 1997. Distribution of juvenile pollock relative to frontal structure near the Pribilof Islands, Bering Sea. In Forage Fishes in Marine Ecosystems, American Fisheries Society, Lowell Wakefield Fisheries Symposium Series, no. 14: 573–589.

Brodeur, R. D., C. E. Mills, J. E. Overland, G. E. Walters & J. D. Schumacher, 1999. Evidence for a substantial increase in gelatinous zooplankton in the Bering Sea, with possible

links to climate change. Fisheries Oceanography 8: 296–306.
Brodeur, R. D., H. Sugisaki & G. L. Hunt Jr, 2002. Increases in jellyfish biomass in the Bering Sea: implications for the ecosystem. Marine Ecology Progress Series 233: 89–103.
Brodeur, R. D., M. B. Decker, L. Ciannelli, J. E. Purcell, N. A. Bond, P. J. Stabeno, E. Acuna & G. L. Hunt Jr, 2008a. Rise and fall of jellyfish in the eastern Bering Sea in relation to climate regime shifts. Progress in Oceanography 77: 103–111.
Brodeur, R. D., C. L. Suchman, D. C. Reese, T. W. Miller & E. A. Daly, 2008b. Spatial overlap and trophic interactions between pelagic fish and large jellyfish in the northern California Current. Marine Biology 154: 649–659.
Cheng, J., S. Li, F. Ding & L. Yan, 2004. Primary analysis on the jellyfish blooms and its cause in the East China Sea and the Yellow Sea. Journal of Modern Fisheries Information 19: 10–12 (in Chinese).
Cheng, J., F. Ding, S. Li, L. Yan, J. Ling, J. Li & Y. Liu, 2005. A study on the quantity distribution of macro-jellyfish and its relationship to sea water temperature and salinity in the East China Sea Region. Acta Ecologica Sinica 25: 440–445 (in Chinese with English Abstract).
Chu, P. C., Y. C. Chen & A. Kuninaka, 2005. Seasonal variability of the East China/Yellow Sea surface buoyancy flux and thermohaline structure. Advances in Atmospheric Sciences 22: 1–20.
Clarke, K. R. & R. M. Warwick, 2001. Change in Marine Communities: An Approach to Statistical Analysis and Interpretation, 2nd ed. PRIMER-E, Plymouth.
Ding, F. & J. Cheng, 2005. The analysis on fish stock characteristics in the distribution areas of large jellyfish during summer and autumn in the East China Sea region. Marine Fisheries 27: 120–128 (in Chinese with English Abstract).
Ding, F. & J. Cheng, 2007. Dynamic distribution of *Stomolophus meleagris* in the East China Sea region. Journal of Fishery Sciences of China 14: 83–89 (in Chinese with English Abstract).
Dong, J., C. Liu, W. Li, H. Yu, B. Wang & Y. Wang, 2005. The morphology and structure of jellyfish (*Cyanea nozakii* Kishinouye). Fisheries Science 24: 22–23 (in Chinese with English Abstract).
Dong, J., C. Liu, Y. Wang & B. Wang, 2006a. Laboratory observations on the life cycle of *Cyanea nozakii* (Semeostomida, Scyphozoa). Acta Zoologica Sinica 52: 389–395.
Dong, J., B. Wang & C. Liu, 2006b. Morphology of *Cyanea nozozakii* in different developmental stages. Journal of Fisheries of China 30: 761–766 (in Chinese with English Abstract).
Doyle, T. K., J. D. R. Houghton, S. M. Buckley, G. C. Hays & J. Davenport, 2007. The broad-scale distribution of five jellyfish species across a temperate coastal environment. Hydrobiologia 579: 29–39.
Field, J. G., K. R. Clarke & R. M. Warwick, 1982. A practical strategy for analyzing multispecies distribution patterns. Marine Ecology Progress Series 8: 37–52.
Gao, S., 1982. The medusae of the East China Sea. Studia Marina Sinica 19: 33–42 (in Chinese).
Gao, S., H. Hong & S. Zhang, 2002. Fauna Sinica invertebrate 27 Phylum Cnidaria. Science Press, Beijing (in Chinese).
Graham, W. M., F. Pages & W. M. Hamner, 2001. A physical context for gelatinous zooplankton aggregations: a review. Hydrobiologia 451: 199–212.
Graham, W. M., D. L. Martin & J. C. Martin, 2003. In situ quantification and analysis of large jellyfish using a novel video profiler. Marine Ecology Progress Series 254: 129–140.
Hamner, W. M., L. P. Madin, A. L. Alldredge, R. W. Gilmer & P. P. Hamner, 1975. Underwater observations of gelatinous zooplankton: sampling problems, feeding biology, and behavior. Limnology and Oceanography 20: 907–917.
Hay, S. J., J. R. G. Hislop & A. M. Shanks, 1990. North Sea scyphomedusae; summer distribution, estimated biomass and significance particularly for 0-group gadoid fish. Netherlands Journal of Sea Research 25: 113–130.
He, F. & H. Hong, 1999. Variation of hydrologic structure and circulation in relation to ENSO events in the Kuroshio area in the East China Sea and its adjacent seas. Transactions of Oceanology and limnology 4: 16–24 (in Chinese with English Abstract).
Hong, H. X. & S. M. Zhang, 1982. Edible jellyfish in the coasts of China. Journal of Jimei University 1: 12–17 (in Chinese with English Abstract).
Hur, H. B., G. A. Jacobs & W. J. Teague, 1999. Monthly variations of water masses in the Yellow and East China Seas, November 6, 1998. Journal of Oceanography 55: 171–184.
Jacobs, G. A., H. B. Hur & S. K. Riedlinger, 2000. Yellow and East China Seas response to winds and currents. Journal of Geophysical Research (C Oceans) 105: 21947–21968.
Kawahara, M., S.-I. Uye, K. Ohtsu & H. Iizumi, 2006. Unusual population explosion of the giant jellyfish *Nemopilema nomurai* (Scyphozoa: Rhizostomeae) in East Asian waters. Marine Ecology Progress Series 307: 161–173.
Li, H., J. Li, F. Ding & J. Cheng, 2007. Distribution characteristics of *Nemopilema nomurai* (Scyphozoa: Rhizostomeae) and zooplankton in East China Sea. Chinese Journal of Ecology 26: 1974–1980 (in Chinese with English Abstract).
Liu, S. X., X. Q. Shen, Y. Q. Wang & S. X. Han, 1992. Preliminary analysis of distribution and variation of perennial monthly mean water masses in the Bohai Sea, the Huanghai Sea and the East China Sea. Acta Oceanologica Sinica 11: 483–498.
Liu, G. M., S. Sun, H. Wang, Y. Zhang, B. Yang & P. Ji, 2003. Abundance of *Calanus sinicus* across the tidal front in the Yellow Sea, China. Fisheries Oceanography 12: 291–298.
Lynam, C. P., S. J. Hay & A. S. Brierley, 2005. Jellyfish abundance and climatic variation: contrasting responses in oceanographically distinct regions of the North Sea, and possible implications for fisheries. Journal of the Marine Biological Association of the United Kingdom 85: 435–450.
Mianzan, H. W. & R. A. Guerrero, 2000. Environmental patterns and biomass distribution of gelatinous macrozooplankton. Three study cases in the South-western Atlantic Ocean. Scientia Marina 64: 215–224.
Mironov, G. N., 1971. Biomass and distribution of the jellyfish *Aurelia aurita* (L.) based on data of trawling catches in 1949–1962 in the Black Sea. Biologiyamorya 24: 49–69.
Omori, M. & W. M. Hamner, 1982. Patchy distribution of zooplankton: behavior, population assessment and sampling problems. Marine Biology 72: 193–200.

Omori, M. & M. Kitamura, 2004. Taxonomic review of three Japanese species of edible jellyfish (Scyphozoa: Rhizostomeae). Plankton Biology and Ecology 51: 36–51.

Park, S., P. C. Chu & J. H. Lee, 2011. Interannual-to-interdecadal variability of the Yellow Sea Cold Water Mass in 1967–2008: characteristics and seasonal forcings. Journal of Marine Systems 87: 177–193.

Purcell, J. E., 2005. Climate effects on formation of jellyfish and ctenophore blooms: a review. Journal of the Marine Biological Association of the United Kingdom 85: 461–476.

Purcell, J. E., E. D. Brown, K. D. E. Stokesbury, L. H. Haldorson & T. C. Shirley, 2000. Aggregations of the jellyfish *Aurelia labiata*: abundance, distribution, association with age-0 walleye pollock, and behaviors promoting aggregation in Prince William Sound, Alaska, USA. Marine Ecology Progress Series 195: 145–158.

Purcell, J. E., S. Uye & W. T. Lo, 2007. Anthropogenic causes of jellyfish blooms and their direct consequences for humans: a review. Marine Ecology Progress Series 350: 153–174.

Shenker, J. M., 1984. Scyphomedusae in surface waters near the Oregon coast, May-August 1981. Estuarine, Coastal and Shelf Science 19: 619–632.

Shiganova, T. A., 1998. Invasion of the Black Sea by the ctenophore *Mnemiopsis leidyi* and recent changes in pelagic community structure. Fisheries Oceanography 7: 305–310.

Shiganova, T. A., 2005. Changes in appendicularian *Oikopleura dioia* abundance caused by invasion of alien ctenophores in the Black Sea. Journal of the Marine Biological Association of the United Kingdom 85: 477–494.

Su, Y. S. & X. C. Weng, 1994. Watermasses in China Seas. In Zhou, D., Y. B. Liang & C. K. Tseng (eds), Oceanology of China Seas, Vol. 1. Kluwer Academic Publishers, Dordrecht: 3–16.

Suchman, C. L. & R. D. Brodeur, 2005. Abundance and distribution of large medusae in surface waters of the northern California Current. Deep-Sea Research Part II 52: 51–72.

Tang, Q., X. Jin, Y. Zheng & T. Meng, 2006. Marine living resources and inhabited environment of Chinese exclusive economic zones. Science Press, Beijing: 3–28 (in Chinese).

Teague, W. J. & G. A. Jacobs, 2000. Current observations on the development of the Yellow Sea Warm Current. Journal of Geophysical Research (C Oceans) 105: 3401–3411.

Uye, S.-I., 2011. Human forcing of the copepod–fish–jellyfish triangular trophic relationship. Hydrobiologia 666: 71–83.

Uye, S. & Y. Ueta, 2004. Recent increase of jellyfish populations and their nuisance to fisheries in the Inland Sea of Japan. Bulletin of the Japanese Society of Fisheries Oceanography 68: 9–19 (in Japanese with English Abstract).

Uye, S., N. Fujii & H. Takeoka, 2003. Unusual aggregations of the scyphomedusa *Aurelia aurita* in coastal waters along western Shikoku, Japan. Plankton Biology and Ecology 50: 17–21.

Wang, R., T. Zuo & K. Wang, 2003. The Yellow Sea cold bottom water – an oversummering site for *Calanus sinicus* (Copepoda, Crustacea). Journal of Plankton Research 25: 169–185.

Wei, H., L. Wang, Y. A. Lin & C. S. Chung, 2002. Nutrient transport across the thermocline in the Central Yellow Sea. Advances in Marine Science 20: 15–20 (in Chinese with English abstract).

Wei, H., J. Su, R. Wan, L. Wang & Y. Lin, 2003. Tidal front and the convergence of anchovy (*Engraulis japonicus*) eggs in the Yellow Sea. Fisheries Oceanography 12: 434–442.

Weng, X., Y. Zhang, C. Wang & Q. Zhang, 1989. The variational characteristics of the Huanghai Sea (Yellow Sea) cold water mass. Oceanologia et Limnologia Sinica 19: 368–379 (in Chinese with English Abstract).

Yan, L., S. Li & F. Ding, 2004. The preliminary studies on the dynamics of macro-jellyfish resources and their relationship with fisheries in the East China Sea and Yellow Sea. Marine Fisheries 26: 9–12 (in Chinese with English Abstract).

Zhang, S. W., Q. Y. Wang, Y. Lu, H. Cui & Y. L. Yuan, 2008. Observation of the seasonal evolution of the Yellow Sea cold water mass in 1996–1998. Continental Shelf Research 28: 442–457.

Hydrobiologia (2012) 690:97–112
DOI 10.1007/s10750-012-1054-8

JELLYFISH BLOOMS

Limnocnida tanganyicae medusae (Cnidaria: Hydrozoa): a semiautonomous microcosm in the food web of Lake Tanganyika

Kalevi Salonen · Pia Högmander · Victor Langenberg · Hannu Mölsä · Jouko Sarvala · Anne Tarvainen · Marja Tiirola

Published online: 6 March 2012
© Springer Science+Business Media B.V. 2012

Abstract Medusae are important members of marine food webs, but are rare in lakes. In one of the largest lakes in the world, Lake Tanganyika, a small medusa (*Limnocnida tanganyicae*) is a prominent component of zooplankton. We used field and laboratory methods to study the ecological role of Lake Tanganyika medusae, which occasionally reached high local densities in the whole epilimnion. The largest individuals showed low amplitude, diel vertical migration which minimized their exposure to harmful UV radiation and also may be important for picocyanobacteria regularly present in the medusae. The endosymbiotic picocyanobacteria differed morphologically among medusae and were predominantly one Lake Biwa type *Cyanobium* sp. that typically was abundant in the water column. Under light, some medusae were net primary producers. Although nitrogen stable isotopic ratios indicated that the free-living cyanobacteria were nitrogen-fixers, the picocyanobacteria in medusae obtained nitrogen predominantly from their host. Stable isotopic ratios of carbon and nitrogen further suggested that copepods were the most likely prey for the medusae. Lake Tanganyika medusae apparently base their metabolism both on animal and plant sources, with possible internal cycling of nutrients; however, the role of picocyanobacteria gardening in the Lake Tanganyika ecosystem and its medusae requires quantification.

Keywords Genetics · Jellyfish · Hydromedusa · Stable isotopes · Symbiosis · UV light

Guest editors: J. E. Purcell, H. Mianzan & J. R. Frost / Jellyfish Blooms: Interactions with Humans and Fisheries

K. Salonen (✉) · P. Högmander · A. Tarvainen · M. Tiirola
Department of Biological and Environmental Science, University of Jyväskylä, P.O. Box 35 (YAC), 40014 Jyväskylä, Finland
e-mail: kalevi.salonen@jyu.fi

V. Langenberg
Department of Water Quality and Ecology, DELTARES, P.O. Box 177, 2600 MH Delft, The Netherlands

H. Mölsä
Fish Innovation Centre, Lohitie 701, 72210 Tervo, Finland

J. Sarvala
Department of Biology, University of Turku, 20014 Turku, Finland

Introduction

Lake Tanganyika is the second deepest and oldest lake on the earth. Its numerous endemic biota reflect the history of about ten million years under rather stable conditions prevailing near the equator (Tiercelin & Mondeguer, 1990; Cohen et al., 1993). The total biodiversity of the lake, one of the highest in the world (Coulter, 1994), is largely confined to the littoral zone. Its pelagic biodiversity, by contrast, is low and leads to a rather simple food web. In Lake Tanganyika, a hydromedusa, *Limnocnida tanganyicae* Böhm, 1883 (Hydrozoa, Limnomedusae), is a prominent component in zooplankton (Sarvala et al., 1999; Langenberg et al., 2008). Of the two most common freshwater medusa genera, *Craspedacusta*, has colonized all continents, while *Limnocnida* is restricted to Asian and African tropics and subtropics (Dumont, 1994a; Jankowski, 2001). In Africa, *L. tanganyicae* seems to be the only species (Goy, 1977). In Lake Tanganyika, the medusa stage is predominant and in fact, the tiny (<0.5 mm) hydroid stage of the hydromedusa was discovered later because of its small size and cryptic life style (Bouillon, 1954).

The ecology of freshwater medusae is poorly known and their taxonomy is still debated. All non-parasitic Cnidaria are predators, but due to the absence of knowledge of their food and feeding, the trophic position of freshwater medusae remains obscure (Rayner & Appleton, 1989; Dumont, 1994b). In Lake Tanganyika, medusae up to 25-mm diameter are abundant (Kurki et al., 1999), and their biomass is of the same order as that of predatory crustacean zooplankton (Sarvala et al., 1999).

The predators of freshwater medusae are unknown (Dumont, 1994a), but it has been hypothesized, albeit contradicted by the observations of Viherluoto (1999), that they might be consumed by benthic decapods. There is no evidence that pelagic fish feed on them (Coulter, 1991). Consequently, *Limnocnida* may be considered as a dead end in the food web.

Examination of *L. tanganyicae* medusae with a high resolution epifluorescence microscope during a cruise in 1996 surprisingly showed a multitude of picocyanobacteria inside them. This led us to hypothesize that these animals might be able to garden picocyanobacteria and partly base their metabolism on that. To better understand the role of possible gardening by *L. tanganyicae* medusae in Lake Tanganyika, we studied several aspects of their ecology during several expeditions covering the whole lake, utilizing field abundance data, laboratory experiments, as well as genetic and stable isotope analyses.

Materials and methods

This study was performed during 1994–2001. To estimate medusa biomass, a regression was established between the umbrella diameter, dry mass (DM), and ash-free dry mass (AFDM) of medusae. Individuals were selected over the range of sizes. After measurement of the umbrella diameters, individuals were dried on pre-weighed aluminum foil cups. These samples were taken to Finland, dried again at 60°C, and weighed on a Cahn Electrobalance. AFDM was obtained by difference from DM after re-weighing following combustion at 500°C.

In the pelagic waters off Kigoma, Tanzania (4°51.00′S, 29°35.00′E), quantitative samples of *L. tanganyicae* medusae were taken at 10-m-vertical hauls from 120 m to the surface using a 500-μm-mesh closing net. In the vicinity of Mpulungu, Zambia (08°43.98′S, 31°02.43′E), medusa bloom samples were taken with a 7-l tube sampler (Limnos Ltd., Finland) and from the immediate surface by scooping with a 10-l PVC container. Qualitative epifluorescence microscopic observations were made of medusae collected throughout the lake. To avoid possible damage to medusae from excessive light, individuals for the experiments were collected at dusk when they ascended to the surface. Generally, single large animals were caught by hand-scooping into a 0.5-l beaker from the surface or with a tube water sampler from slightly deeper layers. In 1998, medusae also were collected at night or in dim light by divers in ≤2 m water depth using 50-mm-diameter acrylic tubes with 300-μm-mesh plankton netting covering one end. When a medusa was captured in the tube, the open end was closed with a stopper and immediately taken to the laboratory aboard the research vessel.

In some cases, the umbrella diameters of animals were measured with the aid of a dissecting microscope. The presence of protozoans inside the animals also was recorded. The occurrence of internal algae was checked with an epifluorescence microscope (Nikon Optiphot) at 1,250-power magnification. Eukaryotic algae were observed with blue excitation and prokaryotic picocyanobacteria with green excitation.

To investigate the effect of UV light on medusa survival, an experiment was conducted on board the R/V Tanganyika Explorer. Twelve or thirteen animals were placed in each of three 2-l quartz bottles, which were put in a water bath in an open, white polystyrene box. One bottle was exposed to direct sunlight, another was kept under a UV-protected polyethylene film, and one wrapped in aluminum foil was used as the dark control. Water temperature was kept similar to ambient by pumping lake water through the box. During the experiment (beginning at 11:45), spectral sunlight radiation was measured every 15 min with a Macam SR 991 spectroradiometer (planar cosine light collector). Spectral penetration of light into water of Lake Tanganyika was also measured using a 4-m quartz light cable. Without the UV-protected polyethylene film, measurable radiation was observed down to 300-nm wavelength; with the film, the limit was about 350 nm. At wavelengths longer than 400 nm, the film absorbed ca. 1/3 of the radiation. Medusae pulsing their swimming umbrella were considered alive and were counted every 10 min. In the dark bottle, animals were counted only at the end of the experiment. To avoid even short exposure to high UV radiation, the bottle under the UV screen was counted only once before the end of the experiment; the bottle was placed in a black cotton bag and transferred to the laboratory of the ship for counting. The experiment was terminated when all animals had died in the bottle kept under direct sunlight. We tested the null hypotheses that UV radiation does not adversely affect *L. tanganyicae* medusae and that their vertical distribution shows no avoidance of surface water in bright light.

Fluorescent beads (Polysciences Inc.) were used to qualitatively study the ingestion of picoplankton-sized organisms. To remove possible bead aggregations, a small volume of stock suspension of beads was filtered through a 5-μm Nuclepore filter. The final concentration of beads in lake water offered to medusae in a 50-ml water bottle was adjusted roughly to the same density as picoplankton abundance in lake water (10^5 cells ml^{-1}). We tested the hypothesis that picocyanobacteria were taken up by medusae from the water.

Bacterial composition of medusae was studied from 11 large (>10 mm) individuals sampled during December 2001 from the lake surface (0–2 m) off Kigoma harbor at dusk and then stored in 70% ethanol. Water samples (1 l) were taken with a Limnos sampler at the same time at 10-m depth intervals from the surface to 60 m. For DNA analyses, 0.5 l of water was screened through 50-μm-mesh plankton netting and then filtered onto Filtropur acetylacetate filter units (0.2-μm pore size). Bacterial DNA extractions were performed using the combined enzymatic and bead-beating method, and the length heterogeneity-PCR (LH-PCR) targeted V1–V3 variable regions of the 16S rRNA (area 8-534, *Escherichia coli* numbering), as described by Tiirola et al. (2002a). Direct sequencing of the heterogeneous PCR products was performed bi-directionally using the ABI BigDye kit and ABI 3100 DNA sequencer (Applied Biosystems). Sequences were compared against the EMBL database using the BLAST algorithm (Altschul et al., 1997). A boot-strapped neighbor-joining tree was calculated using Jukes-Cantor correction with the MEGA 4 software. Reference sequences for inferring the tree were the following (from top to bottom in the tree): AF330249, DQ463712 (Lake Tanganyika clone), AF330250, AF448063, AF216955, AF317074, AF330247, AB015058, AF330252, AF001477, AY151249, AF098370, AF330251, AY172819, AY172811, AY172810, AY172801 AF001479, AY172833, AF245618, S000388727, AF053398, S000628344, and AY946243. We tested the null hypothesis that the picocyanobacterial assemblage in the medusae is similar to that in the water.

To measure oxygen consumption, medusae were transferred individually into 50-ml bottles filled with lake water and sealed with ground glass stoppers. Bottles with the same water, but without a medusa, served as controls. After filling, oxygen concentration was measured and the bottles were placed in an incubator. Some of the bottles were darkened with aluminum foil. In the incubator, the bottles were kept under an illumination of 511 μmol m^{-2} s^{-1} supplied by 6 daylight type fluorescent tubes. Water temperature was maintained within 1°C of the lake temperature (around 27°C) by pumping water from the lake through the incubator. Temperature was monitored with a thermometer during the oxygen measurements. Oxygen concentration was measured with an YSI BOD bottle probe with a stirrer at the upper part of the bottle. It was placed in the bottle and mixing was kept on for 30 s before reading the value. After the measurement, the bottle was re-stoppered and returned to the incubator. Oxygen consumption by medusa was calculated as the difference between the initial and final concentrations taking into account the incubation time and the bottle volume. Differences in control bottles were subtracted from the results of bottles with medusae. We tested the null hypothesis that the photosynthesis of picocyanobacteria does not affect the oxygen budget of the medusa-picocyanobacteria microcosm.

In one oxygen production/consumption experiment, lake water was amended with autoclaved stock solutions of KH_2PO and NH_4Cl to final concentrations of 0.8 μmol P l^{-1} and 12.5 μmol N l^{-1}, roughly in accordance with the highest concentrations (phosphate 0.1–0.6 μmol, nitrate 1.6–3.7 μmol) reported by De Wever et al. (2008a) for the epilimnion of Lake Tanganyika. Ammonium nitrogen was used because algae preferentially take up ammonium and medusae excrete ammonium. NH_4-N is typically very low (0–0.05 g m^{-3}) down to roughly 100-m depth in the lake (Plisnier et al., 1999). If nutrients released from the digestion of invertebrate food of medusae were sufficient for endosymbiotic picocyanobacteria, then nutrient addition would not affect their photosynthesis. Therefore, we tested the null hypothesis that external nutrients do not affect the photosynthesis by the medusae.

To clarify the trophic position of the medusa, samples for stable carbon and nitrogen isotope determinations were collected off Kigoma (4°51′S, 29°35′E), in late November to early December 2001. Medusae were sampled on Dec 4–10, 2001 either with vertical net hauls (100- or 250-μm mesh) from 120 m to the surface (or from 50 m after sunset), or by scooping individual medusae from surface water. In the latter case, the abundance of associated picoplankton was estimated visually by color, and the medusae were sorted into five groups accordingly (colorless = no or very few picocyanobacteria, slight pink hue = few picocyanobacteria, medium pink hue = picocyanobacteria moderately abundant, entire medusa intensely pink = picocyanobacteria very abundant overall, and pink color only around the marginal ring of the medusa). The correlation between medusa color and abundance of internal picocyanobacteria was confirmed with the epifluorescence microscope. Medusae were stored in carbon- and nitrogen-free alkaline Lugol's iodine. Later, medusae were rinsed with deionized water and placed in tin cups as groups of small, similar individuals or as pieces of large individuals (total sample dry mass 1–4 mg). The cups were sealed, dried at 60°C, and sent for analysis by an Europa Scientific Hydra 20/20 isotope ratio mass spectrometer at the Stable Isotope Facility, University of California-Davis, California, USA. The results are given using the δ notation, where $\delta = [(R_{sample}/R_{reference}) - 1] \times 1000$, expressed in units per thousand (‰), and where $R = {}^{13}C/{}^{12}C$ or ${}^{15}N/{}^{14}N$. Reference materials were PeeDee belemnite for carbon and atmospheric N_2 for nitrogen. Nitrogen-fixing algae have very low $\delta^{15}N$ values (around 2‰ or less; Vuorio et al., 2006). If nitrogen fixation was important for the internal picocyanobacteria, nitrogen isotope signatures would be lower in the medusae that had higher abundance of picocyanobacteria. Therefore, we tested the null hypothesis that the nitrogen isotope signatures of the medusae do not reflect the abundance of picocyanobacteria.

Zooplankton, shrimps, and fish larvae from the same net hauls were fixed with carbon- and nitrogen-free alkaline Lugol's iodine immediately after sampling and later sorted by species and size groups in the laboratory. After rinsing with deionized water, groups of 1 to ~3,000 individuals were transferred to tin cups, sealed, dried, weighed, and sent for stable isotope analysis. Water was sampled from different depths (0–100 m) with a tube sampler (Limnos Ltd., Finland) from Nov 19–Dec 10, 2001. The samples (4–20 l) were pre-screened through 50-μm-mesh netting to remove zooplankton and large phytoplankton, and the filtrates were then filtered through pre-combusted (at 500°C overnight) glass fiber filters (Whatman) using a low vacuum (<20 kPa), first through a GF/D filter (median pore size 2.8 μm, which retained mainly eukaryotic nano- and microplankton) and then through a GF/F filter (median pore size 0.7 μm, which retained mainly picocyano- and heterotrophic bacteria). Larger phytoplankton (mainly cyanobacteria) were collected on Dec 6–10, 2001 in net hauls with 50-μm mesh from 5–10 m depth to the surface, which then were concentrated on GF/D filters. The filters were put on pre-combusted aluminum foil and dried at 60°C. In Finland, the dried filters were weighed and 16–18 (GF/D) or 10 (GF/F) 3-mm-diameter subsample disks were punctured from the filters and placed in pre-weighed tin cups, weighed, and sent for stable isotope analysis.

The samples of the most important pelagic planktivore, the clupeid fish, *Stolothrissa tanganicae* Regan, were obtained on Nov 22 and 26, 2001 directly from the fishermen as they came ashore. The fish were measured for length and a tissue sample of the dorsal white muscle was cut from behind the dorsal fin. The tissue samples were put on aluminum foil and dried in an oven at 60°C. In Finland, the tissue samples were ground to a fine powder and ca. 0.8 mg from each sample was transferred into pre-weighed tin cups and sent for stable isotope analysis.

Linear mixing models were applied to the isotope signatures to quantify the contributions of potential food sources to the diet of the medusae (program IsoSource; Phillips & Gregg, 2003).

Isotope signatures were adjusted for the stepwise enrichment in the heavier isotopes from one trophic level to the next, using steps of 0.5 and 1‰ for ^{13}C (France & Peters, 1997), and 2 and 3.4 ‰ for ^{15}N (McCutchan et al., 2003).

Results

Although *L. tanganyicae* are so large that they easily attract visual attention, their individual biomasses were low, with AFDM ca. 2 mg for a 12-mm-diameter medusa (Fig. 1). During several cruises, *L. tanganyicae* medusae were found at the surface in locally high densities. In early September 1995, we examined their detailed horizontal and vertical distributions by sampling a medusa bloom covering nine locations near Mpulungu in the southern basin of Lake Tanganyika (Fig. 2). Medusa density was roughly 3,000 ind m^{-3} nearest to the coast, but it was an order of magnitude lower at >15 km offshore.

Vertical day and night distributions of *L. tanganyicae* were studied in April 1998 off Kigoma, when the epilimnion was <15-m thick (Fig. 3). Medusae were present throughout the oxygenated water column to ~100 m deep. Vertical distributions of medusa abundances near noon and midnight both were maximum within the 10- to 20-m zone; however, at night their median depth of occurrence was 13 m, while at noon it was deeper at 21 m. The difference between day and night vertical distributions was highly significant (Smirnov test, $D = 0.219, P < 0.001, n_1 = 261, n_2 = 367$), and we rejected the null hypothesis that the vertical distribution of jellyfish was even throughout the water column. Small medusae were evenly

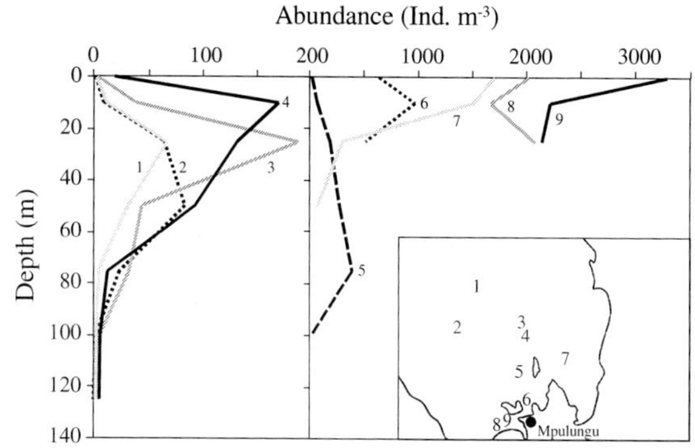

Fig. 2 Abundance of *L. tanganyicae* medusae in the southern basin of Lake Tanganyika near the town Mpulungu, Zambia Sept 2–4, 1995. Numbers *1–9* represent sampling stations

distributed but large (>10-mm diameter) ones more frequently occurred in the upper 30 m. Visual observations also showed that, at sunset, medusae appeared near the surface when photosynthetically active radiation (PAR) above water reached ca. 200 μmol m^{-2} s^{-1} (roughly 20% of the daytime average; Sarvala et al., 1999) and large individuals arrived first. During daytime, medusae were observed only occasionally near the surface. Although they were often alive, in agreement with Dumont (1994a, b), when used in the experiments they did not survive long, suggesting that they were somehow damaged.

Because UV radiation can be damaging in clear water lakes, such as Tanganyika, we studied whether UV might explain the absence of or damage to *L. tanganyicae*. The most harmful UV-B radiation was restricted to the top 5 m of the water column (Fig. 4). In a survival experiment performed near solar noon on the deck of the research vessel, the accumulation of UV radiation developed linearly (Fig. 4). Under a UV-screen, UV-B radiation (290–320 nm) was virtually

Fig. 1 Correlation of dry and ash-free dry mass of *L. tanganyicae* medusae with umbrella diameter. *Dashed line* for dry mass of *C. sowerbii* medusae (Jankowski, 2000) is shown for comparison

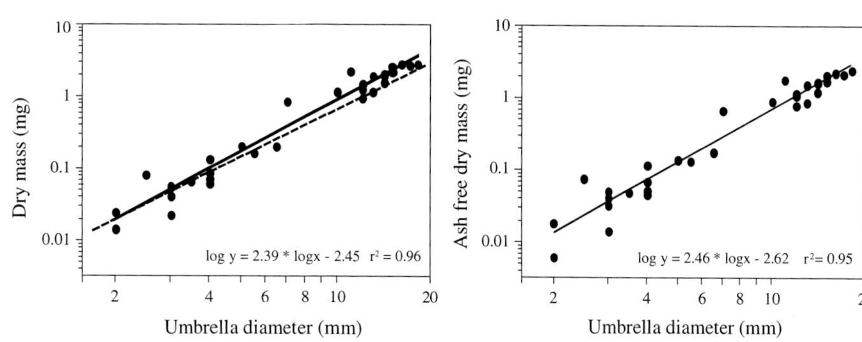

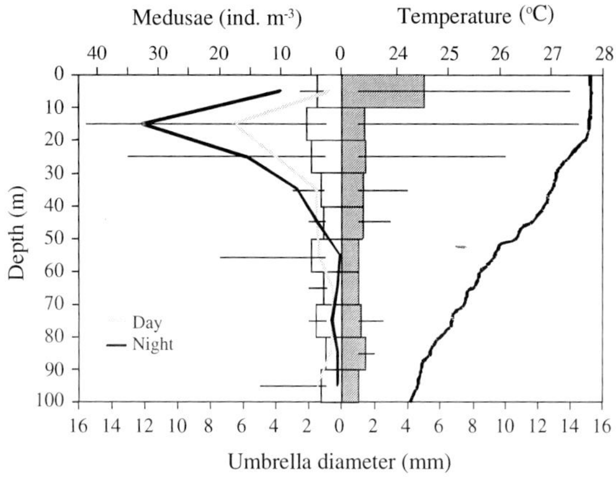

Fig. 3 Abundances (*left side lines*; day *gray line*, night *black line*) and mean umbrella diameters (day *white bars*; night *gray bars*) with ranges (*horizontal lines*) of *L. tanganyicae* medusae in the epi- and metalimnion of Lake Tanganyika off Kigoma Harbor during April 7, 1998. The temperature versus depth curve is shown at the *right*

eliminated. UV-A radiation was detectable at wavelengths of 350–400 nm, but the experiment was too short to cause significant mortality of medusae. By contrast, medusae exposed to natural solar radiation died within 1 h (Fig. 5). We rejected the null hypothesis that the UV radiation was harmless for medusae.

Observations with an epifluorescence microscope of living *L. tanganyicae* without any staining only occasionally showed cells that fluoresced under blue excitation, and only in the gut region. By contrast, under green excitation, the field of view was often full of generally <1-μm-diameter orange fluorescing cells, which indicated that they contained phycoerythrine pigment of picocyanobacteria. Some of the cells moved freely in the body fluid of the medusae and some were stationary. Often it was difficult to judge whether stationary cells were on surface or inside the medusae. Sometimes picocyanobacteria were in different types of colonies (Fig. 6). Generally, the cells were globular or very short rods. When the colonies formed a felt-like structure, the picocyanobacteria were rods arranged in filaments and the medusae had a pink color visible at distance in the lake. Although picocyanobacteria were nearly always present in medusae caught during all expeditions, their abundance differed remarkably among individuals. A sample collected by net from the entire 100-m water column off Kigoma contained about 20% of 123 medusae with so many picocyanobacteria that their enumeration at 1,250× magnification was impossible; however, 56% of the medusae had only scattered single cells or small groups of picocyanobacteria.

In addition to picocyanobacteria, 21% of the studied 705 medusa guts carried wheel-like *Trichodina* type ciliates, similar to those found for *Craspedacusta sowerbyi* (Lankester) by Green (1998). When fluorescent beads 0.2 to 2 μm-diameter were offered, ciliates ingested large numbers of all bead sizes, indicating that they can consume both bacteria and picocyanobacteria; however, we found no uptake of beads by medusae. Thus, we rejected the hypothesis that medusae ingested picocyanobacteria from the water.

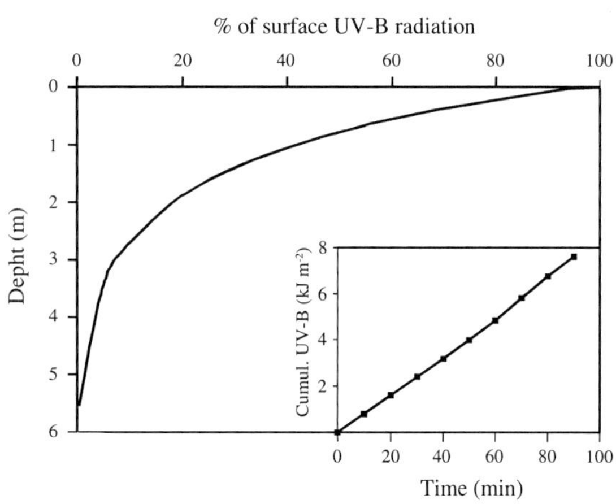

Fig. 4 Penetration of UV-B radiation into water of Lake Tanganyika (April 1998) and cumulative UV-B radiation during the UV-exposure experiment (Fig. 5) on ship board in January 2001

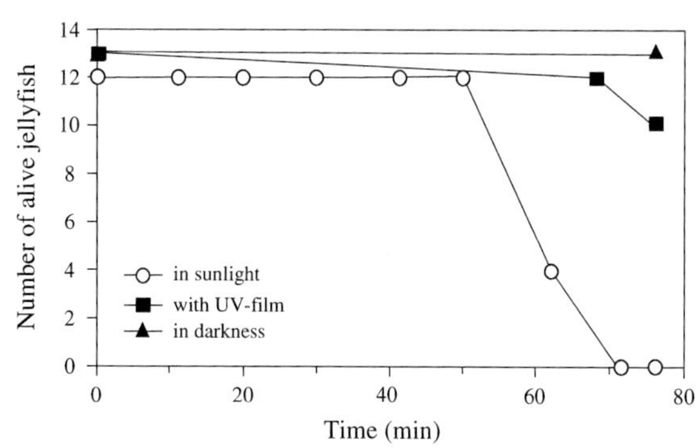

Fig. 5 Survival of >10-mm-diameter *L. tanganyicae* medusae collected off Kigoma Harbor and kept in darkened, UV-film-covered, and uncovered (sunlight) quartz bottles in a water bath on ship board in December 2001

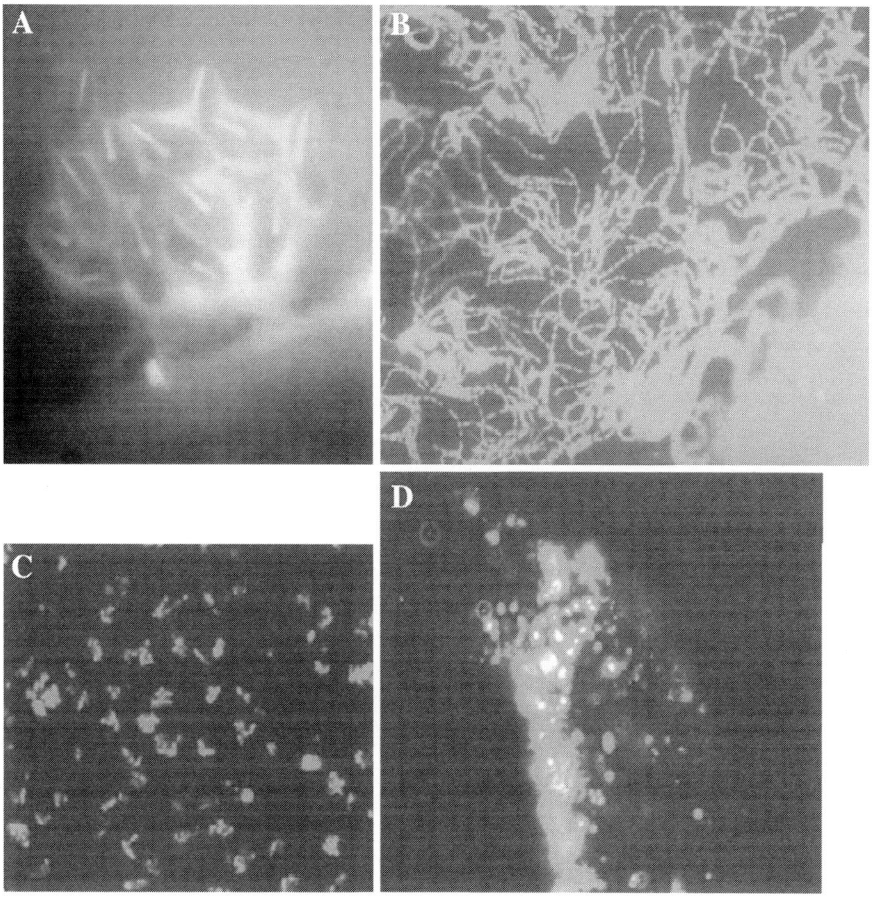

Fig. 6 A trichocyst battery (**A**), filamentous (**B**), and globular colonies (**C**) as well as picocyanobacteria in a tentacle (**D**) of a *L. tanganyicae* medusa

Dozens of individually caught medusae observed immediately by microscopy never had zooplankton in their guts. In contrast, medusae inspected immediately after collection by plankton net had immobilized copepods in the corners at the bottom of the gut, which we believe were caught in the concentrated sample.

LH-PCR analysis of the 16S rRNA genes showed that a single LH-PCR fragment size (470 bp) was present in all medusa samples (Fig. 7). This fragment was highly dominant, covering >60–90% of the total fragment intensity in 7 out of the 11 studied medusae. Picocyanobacterial biomarker (470 bp) of medusae also was very abundant in the epilimnion of Lake Tanganyika, contributing up to 55% of the bacterial abundance at 0–10 m sampling depth (Fig. 8). Another fragment size (472 bp), probably belonging to another picocyanobacterial group, comprised >15% of the bacterial abundance in the water (0–50 m), but was undetectable from medusae. This indicates that the microbial diversity of the medusae did not mirror the diversity of the environment, and we rejected the null hypothesis that the picocyanobacterial assemblage in the medusae was similar to that in water.

Five of the 16S rRNA gene PCR products of the medusae (numbers 1–5) were subjected to bi-directional sequencing, which revealed that the consensus sequences were 100% identical (387-bp overlap). The BLAST search of all available nucleotide sequences revealed that this sequence was identical to the cyanobacterial sequence TK-SE6 (EMBL accession number DQ463712, length 1,432 bp) dominating the oxic epilimnion of Lake Tanganyika (De Wever et al., 2008b) and to sequences obtained from the Chinese MiYu reservoir water (GU305743) and freshwater lakes (GU323646, GU323613, and GU323608). Similar sequences (99% identity) were obtained from

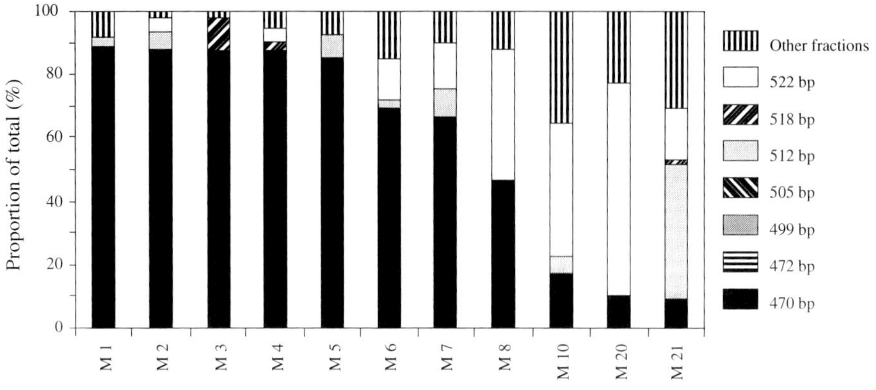

Fig. 7 Diversity of bacteria in 11 *L. tanganyicae* medusae according to LH-PCR analysis of the 16S rRNA gene. The biomarker size 470 bp was affiliated to the *Cyanobium*-type picocyanobacteria

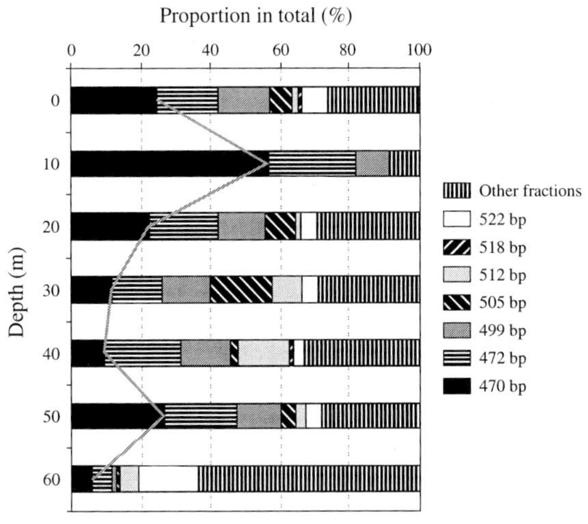

Fig. 8 Depth distributions of bacteria in Lake Tanganyika water column according to LH-PCR analysis of the 16S rRNA gene. The *gray line* shows that the biomarker 470 bp had its relative maximum at 10 m depth

Synechococcus sp. strains isolated from tropical and boreal freshwater environments, such as lakes Biwa (strain BS2, HM346183), Taihu (strain LBG2, AF330249), and Tuusulanjärvi (strain 0tu30s01, AM259220), as well as from marine environments (strain CCMP839, AY946244; strain KORDI-28, FJ497720). In the cluster analysis, we used the longer Lake Tanganyika sequence (DQ463712) to reveal the detailed phylogenetic affiliation of the endosymbiont in comparison to well-established picocyanobacterial clusters (Fig. 9). Lake Tanganyika type sequence clustered with the Lake Biwa strains, which form the freshwater picocyanobacteria group E designated by Crosbie et al. (2003). This cluster contains strains recently re-classified from *Synechococcus* to *Cyanobium*, and their closest described species is *Cyanobium gracile*. One of the other medusae (#10) had a high dominance of the LH-PCR peak 522 bp. The sequence of this peak indicated the dominance of betaproteobacterial heterotroph closest to the type strain of *Vogesella indigofera* (strain ATCC 19706^{T}, AB021385, 95% identity).

The potential role of picocyanobacteria in the metabolism of the medusa-picocyanobacteria microcosm was examined in experiments. In the first light incubation experiment, medusae >10 mm in diameter sometimes showed net oxygen production (Fig. 10). In subsequent experiments, we studied how oxygen consumption/production would develop when the same medusae were kept sequentially in light and dark. The results corroborated our initial finding, specifically, that under light, oxygen consumption decreased or even switched to production (Fig. 11). The same trend was observed in a similar experiment, although the medusae generally were net oxygen consumers (Fig. 12). When inorganic nutrients were added late in the light phase of the experiment, however, three of four individuals had remarkably increased oxygen production. Finally, after the light was switched off, no medusa produced oxygen. These results suggested high variation among individuals, both in consumption and production of oxygen by medusae. Oxygen consumption increased in relation to the diameter of medusa in the dark (Fig. 13); however, in light there was no clear trend, and oxygen production rates in the smallest size classes were surprisingly high. We rejected the hypothesis that photosynthesis of picocyanobacteria does not affect the oxygen budget of medusa-picocyanobacteria microcosm. We could not reject the null hypothesis that external nutrients are unimportant for the photosynthesis of the picocyanobacteria in the medusae.

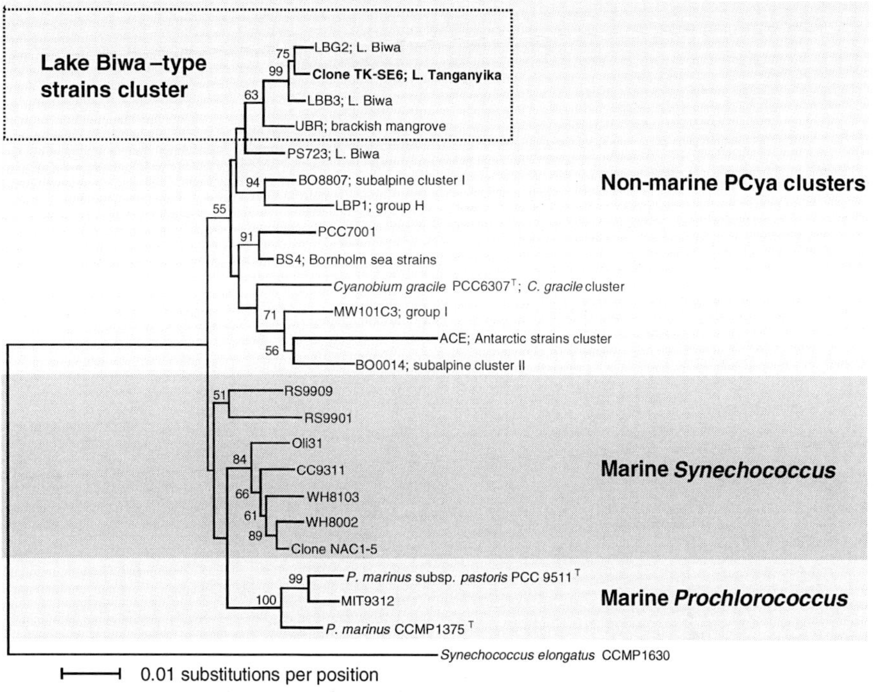

Fig. 9 A neighbor-joining tree of the 16S rRNA gene sequence of the *Cyanobium* (represented by clone TK-SE6) dominating in *L. tanganyicae* medusae, with reference sequences of non-marine and marine picocyanobacteria. Terminal branches display strain code and place of isolation (for Lake Biwa cluster), or the affiliated picocyanobacteria cluster (for strains of other non-marine clusters), based on the clusters designated by Ernst et al. (2003) and Crosbie et al. (2003). *Numbers* at nodes indicate the percent frequency (if >50%) obtained from the bootstrap analysis of 1,273 nt positions of the tree

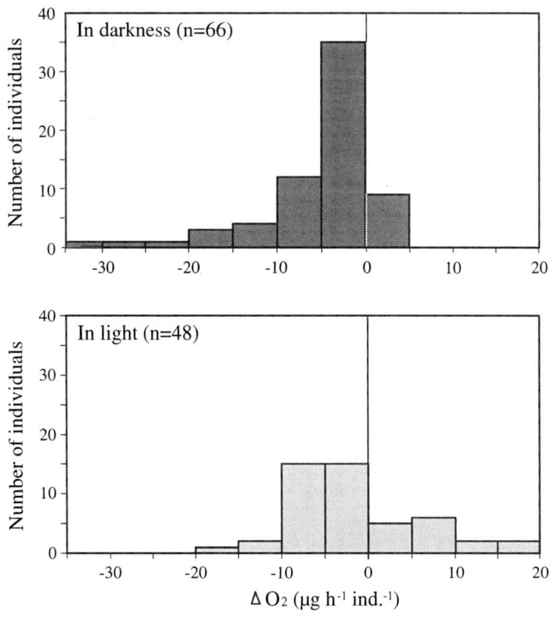

Fig. 10 Oxygen consumption or production by individual *L. tanganyicae* medusae kept for 2 h in light and dark bottles at ~27°C during November 1996 off Mpulungu, Zambia. Medusa diameters ranged from 10 to 20 mm

The stable carbon isotope signatures of *L. tanganyicae* were similar to those of crustacean zooplankton and fish larvae, but lower than in big shrimps or in the planktivorous clupeid fish *S. tanganicae*, slightly higher than in pico-, nano-, and microphytoplankton, or copepod nauplii, and much higher than in small shrimps or in the net phytoplankton mainly consisting of cyanobacteria (Fig. 14). The nitrogen signatures of *L. tanganyicae* were similar to those of adult *S. tanganyicae*, but higher than those of all other groups except big shrimps. Neither the carbon nor nitrogen isotope signatures of *L. tanganyicae* were significantly affected by the abundance of internal picocyanobacteria (ANOVA, $F_{4,6} = 0.587$, $P = 0.34$ for carbon, $F_{4,6} = 0.055$, $P = 0.90$ for nitrogen), and we rejected the null hypothesis that nitrogen isotope signatures of the medusae would reflect the abundance of picocyanobacteria. This implies that nitrogen fixation was not important for the internal picocyanobacteria of the medusae.

Isotope mixing models produced broad distributions of calculated diet proportions, which were sensitive to the choice and grouping of potential food sources, as well as to the trophic step adjustment. The diffuse results were consistent with the fact that the isotope polygon defined by the potential food sources was rather narrow (Fig. 14). No feasible solutions were obtained with a nitrogen step of 3.4‰; however, use of steps of 0.5‰ for carbon and 2.0‰ for nitrogen, achieved isotope mass balance for numerous

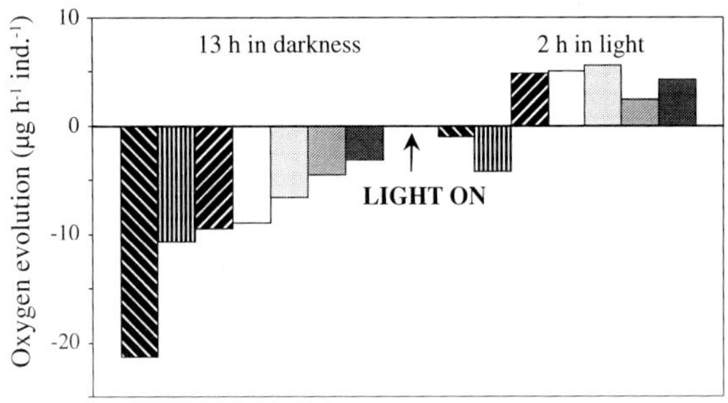

Fig. 11 Oxygen consumption or production by seven *L. tanganyicae* medusae kept first in darkness and then in light at ~27°C during November 1996. The bell diameters of medusae ranged from 10 to 20 mm. Each individual is represented by a *uniquely shaded bar*

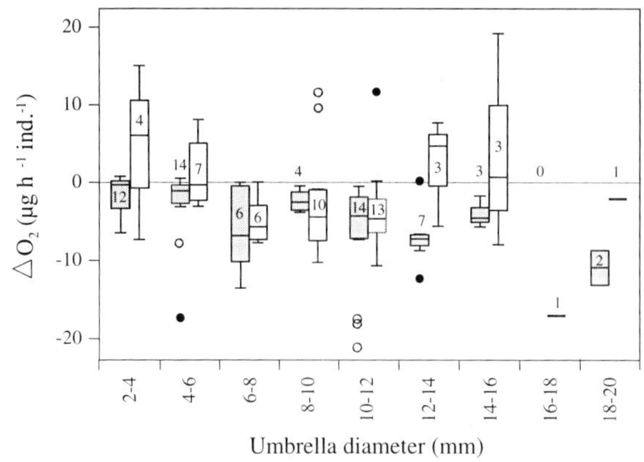

Fig. 13 Oxygen consumption or production of *L. tanganyicae* medusae in relation to their size (umbrella diameter) in light (*white boxes*) and darkness (*gray boxes*) at ~28°C during March 1998. The *box* represents the middle interquartile range of 50% of the observed values. The *whiskers* show the highest and lowest values, excluding outliers (*black dots* 1.5–3 box heights; *circles* >3 box heights from the edge of the box). *Horizontal lines* across the boxes indicate medians. The *numbers* denote the number of animals in each size class

combinations. When all major groups except adult and larval fish were included as potential food sources for *L. tanganyicae*, the calculated contributions of all sources remained variable and low (all included zero contributions); the 1–99th percentiles varied from 0–18 to 0–46%, with picocyanobacteria, *Tropocyclops tenellus* (Sars), *Tropodiaptomus simplex* (Sars), and copepod nauplii showing slightly higher maximum values than other groups. Net phytoplankton, nano-, and microplankton, as well as small shrimps, always showed low contributions. The results remained qualitatively similar if cyclopoid or all copepod adults and copepodids were one group, but the importance of copepods and picocyanobacteria then became more pronounced (1–99th percentiles 0–58 and 0–46%, respectively). In such cases, the big shrimps appeared as a moderately important diet component (1–99th percentiles 14–38%), provided that phytoplankton with low carbon and nitrogen signatures was simultaneously consumed at a higher extent. Similarly, mixtures of several food sources also could include fish larvae. If copepod nauplii were grouped with adults and copepodids, no feasible solutions were found. The most clear-cut model solution was obtained when all phytoplankton groups, grouped cyclopoid adults and copepodids, *T. simplex*, copepod

Fig. 12 Oxygen consumption or production at ~27°C of four *L. tanganyicae* medusae kept first in darkness, then in light, and finally with additional phosphorus and nitrogen during November 1996. The bell diameter of medusae ranged from 10 to 20 mm. Different individuals are represented by *differently shaded bars*

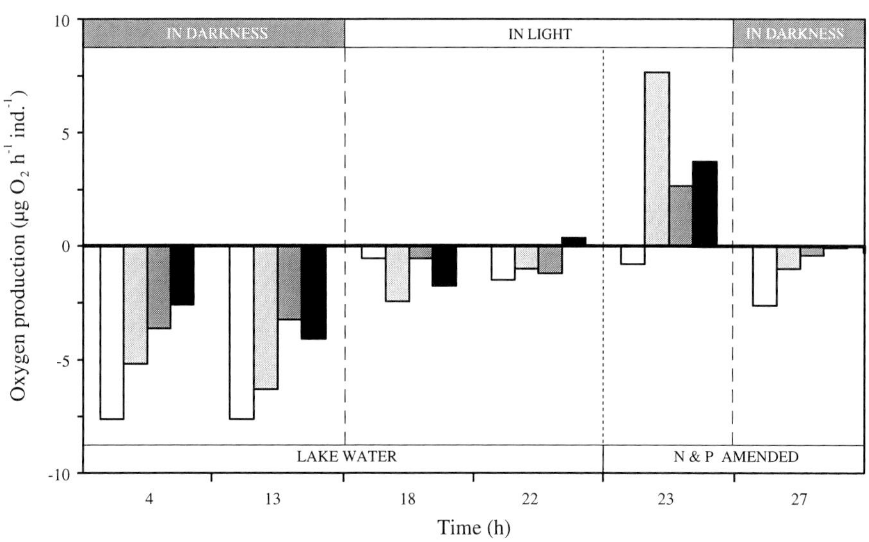

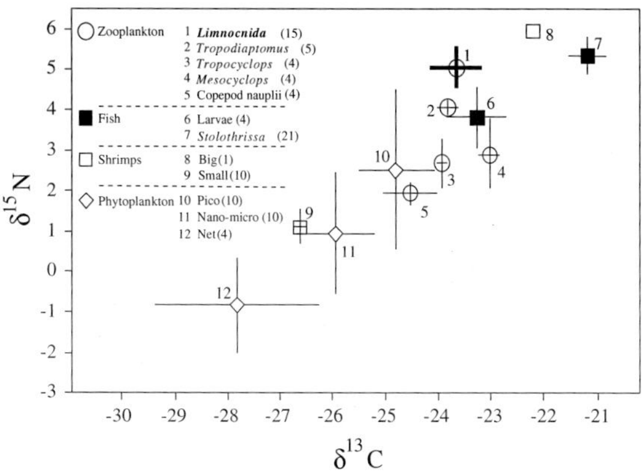

Fig. 14 Mean stable isotopic composition (with standard deviations) of *L. tanganyicae* medusae and other major groups of organisms in the plankton of Lake Tanganyika. Numbers of replicate determinations are in *parentheses*

nauplii, and small shrimps were included as potential food sources. The highest contributions then were shown by cyclopoids, *T. simplex*, copepod nauplii, and picocyanobacteria (1–99th percentiles 2–62, 24–56, 0–40, and 0–30%, respectively).

Discussion

Our results verified that *L. tanganyicae* medusae have very low dry and ash-free biomass relative to their diameter, and the regression between diameter and dry mass was similar to that for *C. sowerbii* (Jankowski, 2000). Medusae are an interesting product of evolution which, during the 600 million years of their existence, may have expanded their prey catching machinery as large as possible with minimum material cost.

Kurki et al. (1999) sampled monthly for 2 years at three sites and found typical abundances of *L. tanganyicae* between 10 and 100 ind m^{-3} in the upper 100 m, with the highest abundances (>500 ind m^{-3}) in the northern part of Lake Tanganyika. By contrast, four lake-wide cruises indicated highest abundances in the southern basin (Kurki, 1998). In our study, the observed maximal abundances were up to 5–6 times higher than those observed by Kurki et al. (1999), and even higher concentrations were recorded in December 1994 in surface waters at Kigoma (personal observations). It seems clear that high abundances of medusae are temporally and spatially variable (Kurki et al., 1999; Langenberg et al., 2008). The factors behind high density blooms are poorly known. The large bloom in this study appeared in the southern basin at the end of the windy upwelling season (Coulter & Spigel, 1991), during a transition from a 3-month period of increased availability of nutrients and primary production in cool, well-mixed waters to warmer water, increased epilimnetic stratification, and oligotrophy (Langenberg et al., 2003). Although the blooms generally seem regionally limited, this bloom covered a large area of ca. 400 km^2. The medusa blooms interfere with the fishery because nearby fishing communities in Zambia and Tanzania must stop fishing for the duration of the bloom (I. Kimirei, Tanzanian Fisheries Research Institute, personal communication).

UV radiation, particularly UV-B, is harmful to zooplankton (e.g., Williamson, 1995; Hylander & Hansson, 2010). Vulnerability to radiation differs among different aquatic organisms (Rhode et al., 2001) and environments, but there is no information about UV-B radiation effects on freshwater medusae. Although *L. tanganyicae* was present throughout the whole water column, their abundance was clearly highest in the epilimnion. Similar to marine medusae (Schuyler & Sullivan, 1997), *L. tanganyicae* actively avoided water layers with high illumination. This avoidance, combined with the results of our light incubation experiment (Fig. 5) and the fact that animals found near the surface at daytime did not survive long, suggests that daytime UV-B radiation is harmful to these highly transparent *L. tanganyicae* medusae. UV-B radiation started to cause mortality of copepods at doses of ca. 0.1 W m^{-2} (Zagarese et al., 1997), which suggests that in the upper 3–5 m of the water column, UV-B is excessive for zooplankton in Lake Tanganyika. In fact, at high radiation levels prevailing around noon at the surface, our experimental results show clear harmful effects of UV-B radiation on *L. tanganyicae* medusae (Fig. 5). For more realistic estimates of the effect, lower exposures over longer periods of time are needed. It seems probable that vertical migration of *L. tanganyicae* reflects avoidance of UV radiation in the upper epilimnion, although other factors may be involved as well.

The LH-PCR results indicated that bacterial groups in medusae did not match the diversity in the surrounding water. Instead, individual medusae had their own bacterial communities more or less dominated by a certain biomarker size, which also was at a maximum in the water column at 10 m depth. Because several cyanobacterial and alphaproteobacterial genera have the same LH-PCR size (Tiirola et al., 2003), further sequencing was needed to identify the microbes represented by the biomarker. All the sequenced medusae carried certain *Cyanobium*-type picocyanobacteria. In other than pathogenic situations (e.g., Rantakokko-Jalava et al., 2000; Tiirola et al., 2002b), it is unusual for a PCR fragment from environmental samples amplified by universal bacterial primers to be directly sequenced without a cloning step. This is usually possible only if some rRNA gene template contributes to >60–70% of all templates (Tiirola et al., unpublished). The prevailing 16S rRNA sequence of the medusae *Cyanobium* belonged to the same taxonomic unit that predominates in Lake Tanganyika (De Wever et al., 2008b). The characteristic *Cyanobium* sp. sequence belonged to the Lake Biwa-type freshwater cluster, but it also was closely related to sequences obtained from both freshwater and marine samples ranging from tropical to boreal environments, showing the cosmopolitan nature of the cluster (see Crosbie et al., 2003).

Among marine invertebrates, including medusae (e.g., Hofmann & Kremer, 1981; Hamner et al., 1982; Muscatine & Marian, 1982; Kremer et al., 1990), endosymbiotic algae are rather common including algae from at least five different classes and animal hosts ranging from protozoans to tunicates (e.g., Trench, 1993). By contrast, endosymbiotic cyanobacteria seem to be less common. Nevertheless, they have been found in invertebrates other than *L. tanganyicae* medusae. Erwin & Thacker (2008) found *Synechococcus*-type cyanobacteria in marine poriferans, while Lesser et al. (2004) found endosymbiotic cyanobacteria to coexist with the symbiotic dinoflagellates (zooxanthellae) in a coral and to express the nitrogen-fixing enzyme, nitrogenase. The presence of this prokaryotic symbiont in a nitrogen-limited zooxanthellate coral suggests that nitrogen fixation may be an important source of that limiting element for the symbiotic association.

Medusae capture photosynthetic organisms from water (Rumpho et al., 2011) and form a symbiotic association with them. Carbohydrates and low molecular weight lipids are produced for the host, which probably return nutrients to the microorganisms. Thus, each generation has to acquire its photosynthetic partner from the surrounding environment. The picocyanobacteria, that are predominant in *L. tanganyicae* medusae, are very abundant in Lake Tanganyika water (Vuorio et al., 2003; De Wever et al., 2008b; Stenuite et al., 2009). Stenuite et al. (2009) estimated that autotrophic picoplankton with "*Synechococcus*-type pigment" accounted for 41–99% of the total phytoplankton biomass. Boulenger (1911) suggested that the typical function of the stomach, which is reduced in *L. tanganyicae*, is accomplished by the canal system, and that the medusae live upon unicellular algae and protozoa driven into the radial canals. Although we saw picocyanobacteria flowing in the canals, ingestion of pico-sized particles was not confirmed by uptake of fluorescent beads, which suggests an endosymbiotic association of the microbiota inside the medusae.

A well-known example of medusae with endosymbiotic algae is the scyphozoan *Mastigias* sp., which relies up to 100% on photosynthesis of endosymbiotic algae and could contribute substantially (16%) to primary production of Eil Malk Jellyfish Lake (McCloskey et al., 1994). *Mastigias* sp. medusae even orient their umbrella to obtain optimum illumination (Hamner et al., 1982). The vertical migration of *L. tanganyicae* medusae also could provide improved growth conditions for its picocyanobacteria in Lake Tanganyika. If the medusae adjust their vertical position according to the optimum illumination, they could maximize photosynthesis of their endosymbionts. This possibility seems plausible because medusae arrive at the surface at dusk too early when light intensity is too high to capture vertically migrating zooplankton. Previous sampling was too coarse to reveal layers of medusae at 100–200 μmol m^{-2} s^{-2} PAR illumination throughout the day, but the daytime vertical distribution of large individuals supports this speculation.

In low-nutrient environments, heterotrophic–autotrophic associations of organisms offer a competitive advantage through the mutual transfer of otherwise limiting resources (Yellowlees et al., 2008). The results of our oxygen consumption/production experiment with medusae and additional nutrients suggest that Lake Tanganyika medusae can have net nutrient uptake similar to that of other medusae (Muscatine & Marian, 1982; Pitt et al., 2005; Todd et al., 2006). Thus, *L. tanganyicae* medusa may acquire key nutrients both from the water and from their prey; however, if they have a high phosphorus level, similar to

C. sowerbii (Jankowski, 2000), then while fulfilling their phosphorus requirements, the medusae obtain excess nitrogen that has to be removed.

Because many photosynthetic organisms can store nutrients, internal picocyanobacteria might be able to effectively harvest nutrients released by medusae so that the organisms form a closed semiautonomous microecosystem where nutrients (at least nitrogen) are recycled internally (Pitt et al., 2009), as was supported by our stable isotope results. The nitrogen signature of the medusae was relatively high and independent of the abundance of internal picocyanobacteria. Because the concurrent free-living picocyanobacteria showed low nitrogen signatures indicative of nitrogen fixation (Vuorio et al., 2006) and thus a shortage of inorganic nitrogen in the environment, the internal picocyanobacteria most likely obtained their nitrogen from *L. tanganyicae* medusae.

The oxygen production of endosymbiotic algae may be so high that it exceeds the oxygen consumption of its host (Kremer et al., 1990). Factors affecting their oxygen production are similar to those of free-living algae (Muscatine & Marian, 1982). Contrary to our results showing high variation in *L. tanganyicae* (Fig. 13), in *Linuche uniquiculata* medusae (Kremer et al., 1990), oxygen production depended linearly on the size of the medusa, on light intensity, and on the behavior of medusae containing endosymbiotic algae (Muscatine & Marian, 1982). Although we found that *L. tanganyicae* medusae sometimes can be net producers, the high variation of the abundance of associated picocyanobacteria complicates understanding their role in medusa metabolism, as well as in the open water ecosystem. The observed picocyanobacteria abundance in the medusae may vary according to their nutritional status. When the status is poor, medusae may deplete the picocyanobacteria, while in the opposite case, picocyanobacteria flourish. Thus, both feeding and life histories may affect the picocyanobacteria abundance. Variation in picocyanobacteria abundance was high even among large animals collected from the surface.

In Eil Malk Jellyfish Lake, *Mastigias* sp. medusae maximized their exposure to inorganic nutrients by swimming 15 m to the chemocline at night (Muscatine & Marian, 1982). It might be possible that such migration to the darker part of the water column also contributes to variation in the abundance of picocyanobacteria. Although the distance to the chemocline in Lake Tanganyika is roughly 100 m, *L. tanganyicae* may harvest nutrients there. The results of our nutrient addition experiment suggest that nutrients from digested prey are not always sufficient. If so, photosynthesis by picocyanobacteria may not be a steady source of food but may extend the resources obtained from prey over a longer duration. As for algae in *Mastigias* sp. medusae (Muscatine et al., 1986), there is no direct evidence of digestion of picocyanobacteria by *L. tanganyicae* medusae. They may obtain photosynthetic products directly through the cell walls of the picocyanobacteria as in the coral–zooxanthellae relationship (Lesser et al., 2004; Woolridge, 2010).

The stable isotope results helped to clarify the trophic position of *L. tanganyicae* in the Lake Tanganyika food web. The isotope composition of an organism reflects its diet, with a stepwise enrichment in the heavier isotopes from one trophic level to the next. Enrichment in ^{13}C is small (0–1‰), permitting identification of carbon sources (France & Peters, 1997), while the enrichment in ^{15}N is higher (2–4‰; empirical average 3.4‰), allowing estimation of the trophic position of consumers (McCutchan et al., 2003). In the pelagic food web of Lake Tanganyika, the ^{15}N enrichment per trophic step seems to be rather low, around 2‰ (Sarvala et al., 2003), as is common for ammonotelic freshwater organisms (Vanderklift & Ponsard, 2003). The similarity of carbon signatures and the mean difference of 1.8‰ in nitrogen signatures between the medusae and the copepodid and adult copepods are consistent with feeding on mixed copepods by the medusae, which was supported by the results of the mixing model application. The mixing model results indicated that medusae also might feed on picocyanobacteria, but that feeding on small shrimps was unlikely. Although feeding on big shrimps in combination with low-signature phytoplankton could balance the isotope equilibrium, the big shrimps are too scarce in the lake (Bosma et al., 1998) to be a realistic food source for the medusae. Thus, the mixing model can be used only to quantify the contributions of known food items in the diet, but not used to identify the true diet components. In general, food items with strong isotopic signatures in opposite directions appear important in the calculated mixture whether or not they are actually eaten and may conceal the importance of items with moderate signatures.

It was suggested that *L. tanganyicae* fed on fish eggs (Dumont, 1994b), but on the basis of stable isotope signatures, it is unlikely that fish formed substantial portions of the diet. The nitrogen signature of the medusae was only 1.3‰ higher than that of fish larvae and did not differ from the signature in adult *Stolothrissa* (Fig. 14). The nitrogen signature of fish eggs is normally very close to that of the female body; thus, medusae cannot have been feeding on fish eggs to any significant extent. The mixing model did show that some isotopically feasible diet combinations could include fish larvae. During five lake-wide cruises in Lake Tanganyika in 1995–1998, abundances of fish larvae and eggs were one or two magnitudes lower than medusae, shrimps, and copepods (Bosma et al., 1998); therefore, ichthyoplankton probably would have been only occasional prey of the medusae that could not be detected by the isotopic signatures.

Medusae can be powerful modifiers of the zooplankton community. In River Yamuna, India *L. indica* medusae removed cladocerans (*Moina* sp.), as well as the rotifer *Keratella* sp. from zooplankton; however, some rotifers (*Asplanchna* sp. and *Brachionus* sp.) and some copepods were not affected (Sharma & Chakrabarti, 2000). *C. sowerbii* medusae consume zooplankton up to 2 mm (Dodson & Cooper, 1983; Jankowski et al., 2005; Smith & Alexander, 2008; Stefani et al., 2010). Thus, *L. tanganyicae* medusae also may exert strong influence on zooplankton particularly during blooms in Lake Tanganyika. More information on the feeding of *L. tanganyicae* on different components of zooplankton is needed to better understand its trophic role in Lake Tanganyika.

Limnocnida tanganyicae medusae are a prominent component in the open water ecosystem of Lake Tanganyika with many metabolic and behavioral adaptations. Primarily, the medusae are likely predators of zooplankton and probably also can garden photosynthetic picocyanobacteria, which enables their survival in the oligotrophic conditions of Lake Tanganyika. This study provides novel insights into the ecology of freshwater medusae.

Acknowledgments This study was a part of the FAO/FINNIDA Lake Tanganyika Research Project GCP/RAF/271/FIN "Research for the Management of the Fisheries on Lake Tanganyika (LTR)". Additional funding was received from the Academy of Finland (Grants 44130, 52271 and 201414), the University of Turku Foundation and the Universities of Turku, Kuopio and Jyväskylä, Finland. We also thank Jitka Jezberova and David Fewer for commenting the picocyanobacterial systematics, and the reviewers and editors for their assistance.

References

Altschul, S. F., T. L. Madden, A. A. Schaffer, J. Zhang, Z. Zhang, W. Miller & D. J. Lipman, 1997. Gapped BLAST and PSI-BLAST: a new generation of protein database search programs. Nucleic Acids Research 25: 3389–3402.

Bosma E. M., S. Muhoza & I. Zulu, 1998. The gulf net sample results of five cruises with the *R/V Tanganyika Explorer*. FAO/FINNIDA Research for the Management of the Fisheries of Lake Tanganyika. GCP/RAF/271/FIN-TD/81 (En): 24.

Bouillon, J., 1954. A hydropolyp in the biological cycle of a freshwater jellyfish. Nature 174: 1112.

Boulenger, C. L., 1911. On some points in the anatomy and bud-formation of *Limnocnida tanganicæ*. Quarterly Journal of Microscopic Sciences 57: 83–106.

Cohen, A. S., M. J. Soreghan & C. A. Scholz, 1993. Estimating the age of formation of lakes: an example from Lake Tanganyika, East African Rift system. Geology 21: 511–514.

Coulter, G. W., 1991. Lake Tanganyika and its Life. British Museum, Natural History & Oxford University Press, London. 354.

Coulter, G. W., 1994. Lake Tanganyika. In Martens, K., B. Goddeeris & G. Coulter (eds), Speciation in Ancient Lakes. Advances in Limnology, 44: 13–18.

Coulter, G. W. & R. H. Spigel, 1991. Hydrodynamics. In Coulter, G. W. (ed.), Lake Tanganyika and its Life. Oxford University Press, London: 49–75.

Crosbie, N. D., M. Pockl & T. Weisse, 2003. Dispersal and phylogenetic diversity of nonmarine picocyanobacteria, inferred from 16S rRNA gene and cpcBA-intergenic spacer sequence analyses. Applied and Environmental Microbiology 69: 5716–5721.

De Wever, A., K. Muylaert, D. Langlet, L. Alleman, J.-P. Descy, L. Andre, C. Cocquyt & W. Vyverman, 2008a. Differential response of phytoplankton to additions of nitrogen, phosphorus and iron in Lake Tanganyika. Freshwater Biology 53: 264–277.

De Wever, A., K. Van der Gucht, K. Muylaert, S. Cousin & W. Vyverman, 2008b. Clone library analysis reveals an unusual composition and strong habitat partitioning of pelagic bacterial communities in Lake Tanganyika. Aquatic Microbial Ecology 50: 113–122.

Dodson, S. I. & S. D. Cooper, 1983. Trophic relationships of the freshwater jellyfish *Craspedacusta sowerbyi* Lankester 1880. Limnology and Oceanography 28: 345–351.

Dumont, H. J., 1994a. The distribution and ecology of fresh- and brackish-water medusae of the world. Hydrobiologia 272: 1–12.

Dumont, H. J., 1994b. Ancient lakes have simplified pelagic food webs. Archiv für Hydrobiologie Beihefte Ergebnisse der Limnologie 44: 223–234.

Ernst, A., S. Becker, U. I. A. Wollenzien & C. Postius, 2003. Ecosystem-dependent adaptive radiations of picocyanobacteria inferred from 16S rRNA and ITS-1 sequence analysis. Microbiology 149: 217–228.

Erwin, P. M. & R. W. Thacker, 2008. Phototrophic nutrition and symbiont diversity of two Caribbean sponge–cyanobacteria symbioses. Marine Ecology Progress Series 362: 139–147.

France, R. L. & R. H. Peters, 1997. Ecosystem differences in the trophic enrichment of ^{13}C in aquatic food webs. Canadian Journal of Fisheries and Aquatic Science 54: 1255–1258.
Goy, J., 1977. Sur les *Limnocnida* africaines (Cnidaires: Limnoméduses). Bulletin de l'Institut francais d'Afrique Noire, série A 39: 563–582.
Green, J., 1998. Plankton associated with medusae of the freshwater jellyfish *Craspedacusta sowerbyi* (Lankester) in a Thames backwater. Freshwater Forum 11: 69–76.
Hamner, W. H., R. W. Gilmer & P. P. Hamner, 1982. The physical, chemical and biological characteristics of a stratified, saline sulfide lake in Palau. Limnology and Oceanography 27: 896–909.
Hofmann, D. K. & B. P. Kremer, 1981. Carbon metabolism and strobilation of *Cassiopea andromeda* (Cnidaria: Scyphozoa): significance of endosymbiotic dinoflagellates. Marine Biology 65: 25–33.
Hylander, S. & L.-A. Hansson, 2010. Vertical migration mitigates UV effects on zooplankton community composition. Journal of Plankton Research 32: 971–980.
Jankowski, T., 2000. Chemical composition and biomass parameters of a population of *Craspedacusta sowerbii* Lank 1880 (Cnidaria: Limnomedusa). Journal of Plankton Research 22: 1329–1340.
Jankowski, T., 2001. The freshwater medusae of the world—a taxonomic and systematic literature study with some remarks on other inland water jellyfish. Hydrobiologia 462: 91–113.
Jankowski, T., T. Strauss & H. T. Ratte, 2005. Trophic interactions of the freshwater jellyfish *Craspedacusta sowerbii*. Journal of Plankton Research 27: 811–823.
Kremer, P., J. Costello, J. Kremer & M. Canino, 1990. Significance of photosynthetic endosymbionts to the carbon budget of the scyphomedusa *Linuche unquiculata*. Limnology and Oceanography 35: 609–624.
Kurki, H., 1998. Results of plankton net and torpedo sampling during cruises on board R/V Tanganyika Explorer. FAO/FINNIDA Research for the management of the fisheries on Lake Tanganyika, GCP/RAF/271/FIN-TD/85 (En): 33.
Kurki, H., P. Mannini, I. Vuorinen, E. Aro, H. Mölsä & O. V. Lindqvist, 1999. Macrozooplankton communities in Lake Tanganyika indicate food chain differences between the northern part and the main basins. Hydrobiologia 407: 123–129.
Langenberg, V. T., J. Sarvala & R. Roijackers, 2003. Effect of wind induced water movements on nutrients, chlorophyll-a, and primary production in Lake Tanganyika. Aquatic Ecosystem Health & Management 6: 279–288.
Langenberg, V. T., J.-M. Tumba, K. Tshibangu, C. Lukwesa, D. Chitamwebwa, D. Bwebwa, L. Makasa & R. Roijackers, 2008. Heterogeneity in physical, chemical and plankton-community structures in Lake Tanganyika. Aquatic Ecosystem Health & Management 11: 16–28.
Lesser, M. P., C. H. Mazel, M. Y. Gorbunov & P. Falkowski, 2004. Discovery of symbiotic nitrogen-fixing cyanobacteria in corals. Science 305: 997–1000.
McCloskey, L. R., L. Muscatine & F. P. Wilkerson, 1994. Daily photosynthesis, respiration, and carbon budgets in a tropical marine jellyfish (*Mastigias* sp.). Marine Biology 119: 13–22.
McCutchan, J. H., W. M. Lewis Jr., C. Kendall & C. C. McGarth, 2003. Variation in trophic shift for stable isotope ratios of carbon, nitrogen, and sulfur. Oikos 102: 378–390.
Muscatine, L. & R. E. Marian, 1982. Dissolved inorganic nitrogen flux in symbiotic and nonsymbiotic medusae. Limnology and Oceanography 27: 910–917.
Muscatine, L., F. P. Wilkerson & L. R. McCloskey, 1986. Regulation of population density of symbiotic algae in a tropical marine jellyfish (*Mastigias* sp.). Marine Ecology Progress Series 32: 279–290.
Phillips, D. L. & J. W. Gregg, 2003. Source partitioning using stable isotopes: coping with too many sources. Oecologia 136: 261–269.
Pitt, K. A., K. Koop & D. Rissik, 2005. Contrasting contributions to inorganic nutrient recycling by the co-occurring jellyfishes, *Catostylus mosaicus* and *Phyllorhiza punctata* (Scyphozoa, Rhizostomeae). Journal of Experimental Marine Biology and Ecology 315: 71–86.
Pitt, K. A., D. T. Welsh & R. H. Condon, 2009. Influence of jellyfish blooms on carbon, nitrogen and phosphorus cycling and plankton production. Hydrobiologia 616: 133–149.
Plisnier, P. D., D. Chitamwebwa, L. M. Mwape, K. Tshibangu, V. T. Langenberg & E. Coenen, 1999. Limnological annual cycle inferred from physico-chemical fluctuations at three stations of Lake Tanganyika. Hydrobiologia 407: 45–58.
Rantakokko-Jalava, K., S. Nikkari, J. Jalava, E. Eerola, M. Skurnik, O. Meurman, O. Ruuskanen, A. Alanen, E. Kotilainen, P. Toivanen & P. Kotilainen, 2000. Direct amplification of rRNA genes in diagnosis of bacterial infections. Journal of Clinical Microbiology 38: 32–39.
Rayner, N. A. & C. C. Appleton, 1989. Occurrence of introduced *Craspedacusta sowerbii* and indigenous *Limnocnida tanganjicae* (Cnidaria: Limnomedusae) in Southern Africa. Environmental Conservation 16: 267–270.
Rhode, S. C., M. Pawlowski & R. Tollrian, 2001. The impact of ultraviolet radiation on the vertical distribution of zooplankton of the genus *Daphnia*. Nature 412: 69–72.
Rumpho, M. E., K. N. Pelletreau, A. Moustafa & D. Bhattacharya, 2011. The making of a photosynthetic animal. Journal of Experimental Biology 214: 303–311.
Sarvala, J., K. Salonen, M. Järvinen, E. Aro, T. Huttula, P. Kotilainen, H. Kurki, V. Langenberg, P. Mannini, P.-D. Plisnier, I. Vuorinen, H. Mölsä & O. V. Lindqvist, 1999. Trophic structure of Lake Tanganyika: carbon flows in the pelagic food web. Hydrobiologia 407: 149–173.
Sarvala, J., S. Badende, D. Chitamwebwa, P. Juvonen, L. Mwape, H. Mölsä, N. Mulimbwa, K. Salonen, M. Tarvainen & K. Vuorio, 2003. Size-fractionated δ^{15}N and δ^{13}C isotope ratios elucidate the role of the microbial food web in the pelagial of Lake Tanganyika. Aquatic Ecosystem Health & Management 6: 241–250.
Schuyler, Q. & B. K. Sullivan, 1997. Light responses and diel migration of the scyphomedusa *Chrysaora quinquecirrha* in mesocosms. Journal of Plankton Research 19: 1417–1428.
Sharma, J. G. & R. Chakrabarti, 2000. Seasonal occurrence of freshwater medusa *Limnocnida indica annandale* (Cnidaria: Limnomedusae) in a lake associated with the river Yamuna, India. Aquatic Ecology 34: 205–207.

Smith, A. S. & J. E. Alexander Jr., 2008. Potential effects of the freshwater jellyfish *Craspedacusta sowerbii* on zooplankton community abundance. Journal of Plankton Research 30: 1323–1327.

Stefani, F., B. Leoni, A. Marieni & L. Garibaldi, 2010. A new record of *Craspedacusta sowerbii*, Lankester 1880 (Cnidaria, Limnomedusae) in Northern Italy. Journal of Limnology 69: 189–192.

Stenuite, S., A.-L. Tarbe, H. Sarmento, F. Unrein, S. Pirlot, D. Sinyinza, S. Thill, M. Lecomte, B. Leporcq, J. M. Gasol & J.-P. Descy, 2009. Photosynthetic picoplankton in Lake Tanganyika: biomass distribution patterns with depth, season and basin. Journal of Plankton Research 31: 1531–1544.

Tiercelin, J. & A. Mondeguer, 1990. The geology of the Tanganyika Trough. In Coulter, G. (ed.), Lake Tanganyika and its Life. Oxford University Press, London: 7–48.

Tiirola, M. A., M. K. Männistö, J. A. Puhakka & M. S. Kulomaa, 2002a. Isolation and characterization of *Novosphingobium* sp. MT1, a dominant polychlorophenol degrading strain in a groundwater bioremediation system. Applied and Environmental Microbiology 68: 173–180.

Tiirola, M., E. T. Valtonen, P. Rintamäki-Kinnunen & M. Kulomaa, 2002b. Diagnosis of flavobacteriosis by direct amplification of rRNA genes. Diseases of Aquatic Organisms 51: 93–100.

Tiirola, M. A., J. E. Suvilampi, M. S. Kulomaa & J. A. Rintala, 2003. Microbial diversity in a thermophilic aerobic biofilm process: analysis by length heterogeneity PCR (LH-PCR). Water Research 37: 2259–2268.

Todd, B. D., D. J. Thornhill & W. K. Fitt, 2006. Patterns of inorganic phosphate uptake in *Cassiopea xamachana*: a bioindicator species. Marine Pollution Bulletin 52: 515–521.

Trench, R. K., 1993. Microalgal-invertebrate symbiosis: a review. Endocytosis Cell Research 9: 135–175.

Vanderklift, M. A. & S. Ponsard, 2003. Sources of variation in consumer-diet δ^{15}N enrichment: a meta-analysis. Oecologia 136: 169–182.

Viherluoto, M., 1999. The food utilisation and diel feeding pattern of shrimps (Atyidea and Palaemonidea) in Lake Tanganyika. In Mölsä, H., K. Salonen & J. Sarvala (eds), Results of the LTR's 20th Multi-Disciplinary Cruise. FAO/FINNIDA Research for the Management of the Fisheries of Lake Tanganyika. GCP/RAF/271/FIN-TD/93 (En): 96.

Vuorio, K., M. Nuottajärvi, K. Salonen & J. Sarvala, 2003. Spatial distribution of phytoplankton and picocyanobacteria in Lake Tanganyika, in March–April 1998. Aquatic Ecosystem Health & Management 6: 268–278.

Vuorio, K., M. Meili & J. Sarvala, 2006. Taxon-specific variation in the stable isotopic signatures (δ^{13}C and δ^{15}N) of lake phytoplankton. Freshwater Biology 51: 807–822.

Williamson, C. E., 1995. What role does UV-B radiation play in freshwater ecosystems? Limnology and Oceanography 40: 386–392.

Woolridge, S. A., 2010. Is the coral-algae symbiosis really 'mutually beneficial' for the partners? BioEssays 32: 615–625.

Yellowlees, D., T. A. V. Rees & W. Leggat, 2008. Metabolic interactions between algal symbionts and invertebrate hosts. Plant, Cell and Environment 31: 679–694.

Zagarese, H. E., M. Feldman & C. E. Williamson, 1997. UV-B-induced damage and photoreactivation in three species of *Boeckella* (Copepoda, Calanoida). Journal of Plankton Research 19: 357–367.

Hydrobiologia (2012) 690:113–125
DOI 10.1007/s10750-012-1055-7

JELLYFISH BLOOMS

Large medusae in surface waters of the Northern California Current: variability in relation to environmental conditions

Cynthia L. Suchman · Richard D. Brodeur · Elizabeth A. Daly · Robert L. Emmett

Published online: 15 March 2012
© Springer Science+Business Media B.V. 2012

Abstract Blooms of jellyfish around the world have been correlated with climatic variables related to environmental causes. Sizeable populations of large medusae, primarily *Chrysaora fuscescens* and *Aequorea* sp., appear annually in shelf waters of the Northeast Pacific Ocean. Previous research has shown that *C. fuscescens* is abundant seasonally in the inner shelf and exhibits high feeding rates on zooplankton. We examined medusae caught in surface trawls over an 8-year period (2000–2007) using (1) mesoscale surveys sampling 8–10 transects in May, June, and September, and (2) biweekly surveys along two transects from April to August, relating abundance to environmental parameters. *C. fuscescens* abundances generally peaked in late summer, whereas *Aequorea* sp. peaked in May or June. General additive models of the mesoscale data indicated that station catches for both species correlated with latitude, temperature, salinity, and distance from shore (and chlorophyll *a* for *Aequorea* sp.). Analysis of interannual variability revealed that highest catches of medusae correlated with cool spring–summer conditions, or negative anomalies of the Pacific Decadal Oscillation, and low winter–summer runoff from the Columbia River. Results confirmed our hypothesis of connections between jellyfish populations and regional climate conditions in a region known for strong physical forcing of ecosystem processes.

Keywords Jellyfish · *Chrysaora* · *Aequorea* · Climate · Upwelling · California Current

Introduction

Gelatinous zooplankton remain understudied components of marine ecosystems, largely due to difficulty in collecting them with standard sampling gear (Purcell, 2009) and the perception that they represent trophic dead ends in marine food webs (but see Purcell & Arai, 2001; Houghton et al., 2006b; Pauly et al., 2009). Some studies have suggested that populations of these taxa may be increasing worldwide (reviewed by Mills, 2001; Purcell, 2005, 2012; Purcell et al., 2007). Consequently, fisheries and ecosystem scientists have identified a need to document existing populations, their role in food webs, and how they are affected by

Guest editors: J. E. Purcell, H. Mianzan & J. R. Frost / Jellyfish Blooms: Interactions with Humans and Fisheries

C. L. Suchman (✉)
North Pacific Research Board, Anchorage, AK 99501, USA
e-mail: cynthia.suchman@nprb.org

R. D. Brodeur · R. L. Emmett
Northwest Fisheries Science Center, National Marine Fisheries Service, National Oceanic and Atmospheric Administration, Newport, OR 97365, USA

E. A. Daly
Cooperative Institute for Marine Resources Studies, Oregon State University, Newport, OR 97365, USA

ecosystem stressors, such as eutrophication, overfishing, exotic species introductions, and climate change (Richardson et al., 2009).

Because they have fast growth rates, gelatinous zooplankton respond quickly to variability in local or regional environmental conditions, but general abundance patterns and the mechanisms responsible for those patterns have been difficult to discern. Although some literature suggests that with warming climate populations of large medusae will increase, the evidence is equivocal (reviewed in Purcell, 2005, 2012; Purcell et al., 2007). For example, in the Bering Sea, a dramatic increase in *Chrysaora melanaster* Brandt 1838 was followed by a steep decline; modeling studies suggest that these population changes best correlate with ice cover, spring and summer sea surface temperature, and wind mixing (Brodeur et al., 2008a). In the Irish Sea, abundance of *Aurelia aurita* Linnaeus 1758 and *Cyanea* spp. increased over a 16-year period and was positively correlated with sea surface temperature and negatively correlated with precipitation (Lynam et al., 2011). In contrast, in the North Sea, some scyphomedusan populations correlated with low temperature anomalies associated with the negative North Atlantic Oscillation Index (Lynam et al., 2004).

In this study, we examine the hypothesis that populations of medusae in coastal upwelling systems are closely tied to within-year environmental conditions. The shelf waters of the Northern California Current provide a particularly good system to look for the influence of the environment on plankton populations. First, biological productivity is closely tied to physical processes, in this case seasonal coastal upwelling (Hickey & Banas, 2003; Checkley & Barth, 2009). Moreover, because many regional-scale interdisciplinary studies have been supported in the area (e.g., US Global Ocean Ecosystem Dynamics and River Influences on Shelf Ecosystems Programs), there are rich data sets and synthetic activities, such as index development, to draw from and add to (e.g., Batchelder et al., 2002; Hickey et al., 2010). Large medusae are conspicuous in surface water in this system and coupled with their fast growth rates, may have potential to serve as an indicator species of changing ocean conditions. Because the annual growth period for these organisms is tied closely to the months with upwelling winds, we hypothesize that increases in populations may be more likely to have linear relationships with seasonal environmental variables than other plankton with shorter generation times (Hsieh & Ohman, 2006).

Large medusae appear annually off the coasts of Washington and Oregon in the Northern California Current (Suchman & Brodeur, 2005). Previous work showed that large jellyfish, particularly the scyphomedusa *Chrysaora fuscescens* Brandt 1835 and the hydromedusa *Aequorea* sp.,[1] are seasonally abundant in shelf waters (Shenker, 1984; Suchman & Brodeur, 2005) and that *C. fuscescens* can be an important predator on zooplankton, particularly early stages of euphausiids (Suchman et al., 2008). Because of overlap in diet with co-occurring planktivorous fish (Brodeur et al., 2008b), *C. fuscescens* may also be an important competitor with these taxa and in turn may indirectly impact other commercially important species such as salmon (Ruzicka et al., 2007) and other planktivorous fishes (Brodeur et al., 2011). Therefore, factors that influence abundance of gelatinous taxa have implications for broader ecosystem processes in this and similar upwelling ecosystems. In this study, analysis of a multi-year data set enabled us to examine abundance and distribution of medusae in relation to environmental variables in the Northern California Current and determine whether relationships exist between regional climate indices and interannual patterns of jellyfish abundance. We hypothesize, as demonstrated for other ecosystems (Lynam et al., 2005, 2010; Brodeur et al., 2008a), that large medusae are highly opportunistic and will therefore respond quickly to regional and local forcing factors.

Materials and methods

Data collection

Sampling locations were in surface, shelf waters of the Northern California Current between Tatoosh Island,

[1] Identity of this hydromedusan species in the Northern California Current is not clear. Recent studies in the Northeast Pacific have called it *Aequorea victoria* (Murbach and Shearer, 1902), but in here it will be called *Aequorea* sp. Claudia Mills describes the confusion on her website as "a big problem and has led to the use of at least three different species names (*Aequorea aequorea*, *A. forskalea*, and *A. victoria*) in the modern literature for what is very likely all the same animal in the NE Pacific" (http://faculty.washington.edu/cemills/).

Washington (48.4°N, 124.7°W) and Newport, Oregon (44.6°N, 124.0°W; Fig. 1). Data were collected during two types of cruises, 2000–2007: (1) mesoscale cruises in May, June, and September sampled a grid of 5–10 cross-shelf transects during daylight hours; and (2) 8–10 biweekly nighttime cruises from April to August collected data along only the Willapa Bay and Columbia River transects (Fig. 1). Each transect consisted of 6–8 stations, from the shallowest depth practical using the trawl (~30 m) to approximately 50 km from shore, often extending beyond the continental shelf.

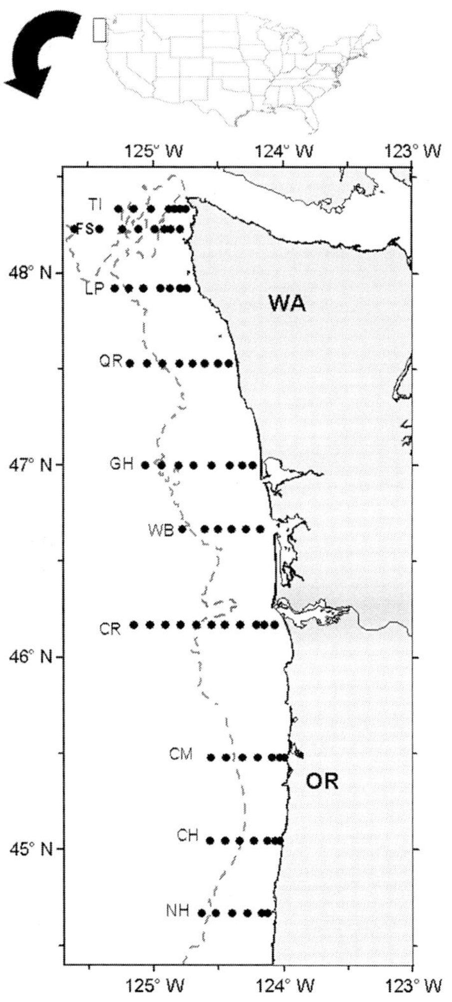

Fig. 1 Locations of stations sampled (*dots*) during 2000–2007 along transects off the Washington (*WA*) and Oregon (*OR*) coasts. The 200-m isobath is shown as a *dashed line*. Transect name abbreviations: *NP* Newport, *CH* Cascade Head, *CM* Cape Meares, *CR* Columbia River, *WB* Willapa Bay, *GH* Grays Harbor, *QR* Queets River, *LP* La Push, *FS* Father & Son, *TI* Tatoosh Island

In addition, paired surface and midwater trawls were conducted 21–22 June 2000 to compare catch of medusae from the surface (to approximately 18 m) to those from subsurface waters (approximately 20–40 m). Eight paired trawls were completed between 0515 PST and 2330 PST at a station 13 km from shore on the Columbia River transect (Emmett et al., 2004).

At each station, medusae were collected using a Nordic 264 rope trawl (30-m wide × 18-m deep) built by Nor-Eastern Trawl Systems, Inc., towed in surface waters for 30 min at 1.5–2.0 m s^{-1} (approximately 5–7 km h^{-1}). Start and end latitude and longitude were recorded using Global Positioning System. Mesh size of the trawl ranged from 162.6 cm at the throat to 8.9 cm at the cod end, with a 6.1-m long, 0.8-cm mesh liner sewn into the cod end. Medusae were identified, counted, and measured at sea immediately after capture. Each species was weighed in aggregate (wet weight), and individuals of each species were counted and bell diameters measured (to nearest mm). For exceptionally large hauls, a subsample (at least 50) of each species was measured, counted, and weighed, and the total number of medusae was calculated based on total species weight for the haul (total number = subsample number × total weight/subsample weight). Over the 8 years of the study, 1,746 trawls collected data for this analysis (Table 1; Fig. 2).

Area trawled was calculated as the distance towed (determined by start and end latitude and longitude, corrected for the curvature of the earth) multiplied by the mouth width of the trawl (0.03 km). Abundance of

Table 1 Annual FO and abundance (cumulative average) of medusae in all trawls (mesoscale and biweekly)

Year	FO	Medusae (km^{-2})	No. trawls
2000	0.54	3,633	170
2001	0.59	5,562	214
2002	0.66	3,272	220
2003	0.49	1,924	237
2004	0.73	3,245	230
2005	0.62	2,187	221
2006	0.67	1,667	252
2007	0.35	2,955	202

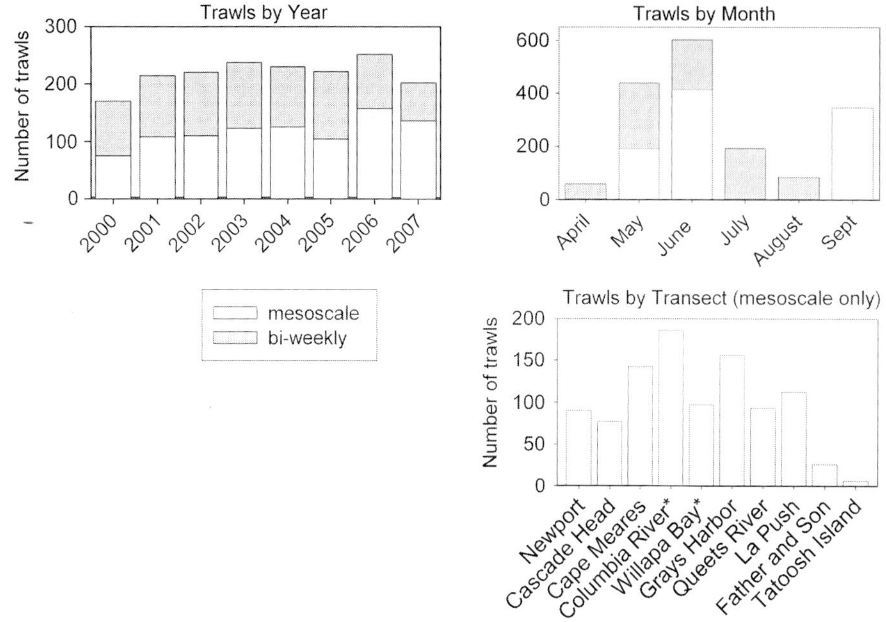

Fig. 2 Distribution of trawl sampling by year, month, and transect. Mesoscale cruises were conducted over 10 days in May, June, and September 2000–2007, covering 10 cross-shelf transects. Biweekly 2-day cruises were conducted April–August 2000–2007 along Columbia River and Willapa Bay transects

medusae was reported as number km^{-2}. Because of the large number of zero catches at individual stations, cumulative averages were used to report abundance (km^{-2}) by month or year unless otherwise noted.

Several environmental parameters were also measured at each station. The temperature and salinity throughout the water column were determined using a CTD deployed from 1 m below the surface to a depth of either 100 m or 5 m off the bottom (for shallower stations). In addition, during the mesoscale cruises, chlorophyll *a* samples were collected from 3-m depth using a Niskin bottle. Chlorophyll *a* values were determined in the laboratory using fluorescence measurements.

Statistical analysis

For the mesoscale cruises, we used generalized additive modeling (GAM) to test the hypotheses that both the occurrence and the abundances of the three dominant species (*Chrysaora fuscescens*, *Aequorea* sp., and *Aurelia labiata* Chamisso & Eysenhardt 1821) are related to environmental variables at the local (station) scale. GAM is a nonparametric regression technique that does not assume an explicit functional relationship between the dependent variable and the covariates. Relationships between the response variable and the covariates are modeled with nonparametric smooth functions (Hastie & Tibshirani, 1990). Because of the large number of stations with no jellyfish present (zero-inflated data), we used a two-stage approach, first modeling presence/absence data using a binomial distribution (Barry & Welsh, 2002). Then we modeled the standardized densities of dominant taxa assuming a normally distributed error (Gaussian family and identity link) on $\log_e(n + 1)$-transformed data excluding the zero catches. Initial covariates included: latitude (decimal degrees), bottom depth at station (m), sea surface temperature (°C), sea surface salinity, $\log_e$ of chlorophyll *a* concentration at 3 m (mg l^{-1}), and distance from the coast (km). The models also included two factors, month (May, June, or September) and year (2000 through 2007), to examine the significance of these temporal factors on jellyfish catch. To decide which covariates to retain in the final models, we applied a backward strategy based on the minimization of the generalized cross validation (GCV), a measure of the model prediction error (Wood & Augustin, 2002; Wood, 2006). Only stations with all variables available were used in GAM analyses.

We used linear regression to test interannual variability of dominant jellyfish taxa in relation to regional-scale environmental conditions. The monthly Pacific Decadal Oscillation (PDO) index was accessed from the University of Washington's Joint Institute for the Study of the Atmosphere and Ocean (http://jisao.washington.edu/pdo/), and monthly streamflow for the Columbia River (Beaver Army Terminal) was obtained from the US Geological Survey (http://waterdata.usgs.gov/nwis/uv?14246900). For both the dependent and the independent variables, annual anomalies were calculated as deviation from the means over the 8 years of the study.

Results

Abundance and size patterns

In this region, distribution and abundance of five species of medusae varied among the 8 years of the study (Table 1). In 2007, only 35% of trawls caught medusae (the lowest frequency of occurrence (FO)) compared with 73% in 2004. Magnitude of catch also varied, from a cumulative average of 1,667 km^{-2} in 2006 to 5,562 km^{-2} in 2001. FO and abundance were not correlated ($P > 0.05$), indicating variability in distribution, as well as number of medusae. The scyphomedusa *Chrysaora fuscescens* dominated, with catches an order of magnitude higher than the next most abundant species, the hydromedusa *Aequorea* sp. (Table 2). These two species were each caught in about 40% of trawls, when 66% contained at least one large medusa. *Aurelia labiata* was much less abundant and was present in nearly 20% of trawls. Repeated stratified trawls showed no significant difference in jellyfish catch between surface (to approximately 18 m) and subsurface (approximately 20–40 m) waters (Fig. 3).

Table 2 FO and abundance of medusae by species collected during mesoscale cruises, reported as cumulative averages and standard deviations across years (2000–2007) ($n = 951$)

Species	FO (±SD)	Medusae (km^{-2}) (±SD)
Chrysaora fuscescens	0.42 ± 0.15	2,930 ± 1,417
Aequorea sp.	0.40 ± 0.21	217 ± 177
Aurelia labiata	0.19 ± 0.18	20 ± 33
Other medusae	0.10 ± 0.11	2 ± 1
Total medusae	0.66 ± 0.19	

The "other" category of large medusae consisted of *Phacellophora camtschatica* (Brandt, 1835) and *Cyanea capillata* (Linnaeus, 1758)

Separating catch by month showed a seasonal pattern of abundance for the two most commonly caught species (Fig. 4). Over the 8 years of the study,

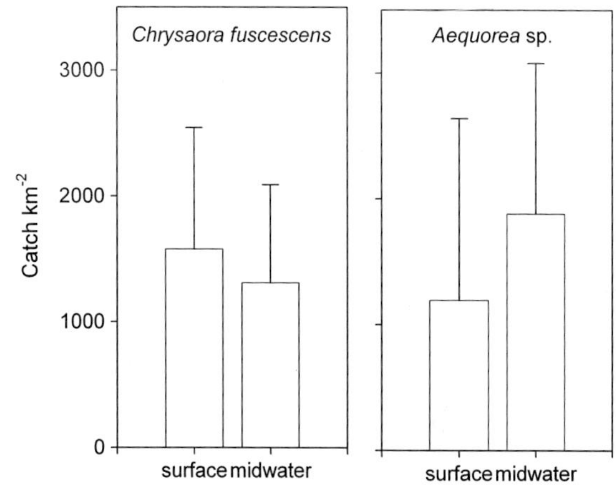

Fig. 3 Stratified tows showed no difference in catch of *Chrysaora fuscescens* or *Aequorea* sp. from surface versus subsurface trawls ($n = 8$, paired t-test $P > 0.05$). Sampling was conducted June 21–22, 2000 at a station 8 nm from shore on the Columbia River transect

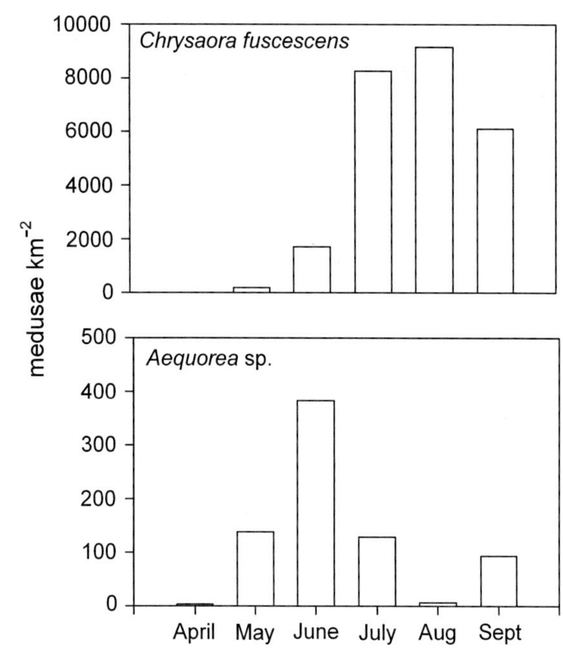

Fig. 4 *Chrysaora fuscescens* and *Aequorea* sp. caught in trawls by month for all years ($n = 1,746$)

Chrysaora fuscescens was not abundant in April or May and had relatively low abundances in June. Populations peaked in July and August, with relatively high numbers remaining in September at the end of the sampling season. *Aequorea* sp. showed a different pattern. Following low abundances in April, *Aequorea* sp. numbers peaked in June and declined in July and into late summer months. These seasonal patterns were relatively consistent, despite interannual variability in the magnitude of jellyfish populations (Fig. 5).

Bell diameters for all three species increased with month of sampling (Kolmogorov–Smirnov comparison of distribution, $P < 0.05$ for each species, each month) (Fig. 6). Average size of *C. fuscescens* increased from 12.6 cm in May to 17.3 cm in September, with most growth between May and June, earlier than the population increase (Figs. 4, 6). The two next most abundant species, *Aequorea* sp. and *Aurelia labiata*, increased in size later in the season (between June and September), although sample sizes for *A. labiata* were much smaller than for the other two species. *Aequorea* sp., though larger in September, were also much less abundant (Figs. 4, 6). Bell diameter analysis for this species was restricted to 2001, 2002, 2003, and 2005, as sample sizes were small in September during other years. In addition, a few very large *Aequorea* sp. individuals (>20 cm) were caught in September 2004, 2006, and 2007.

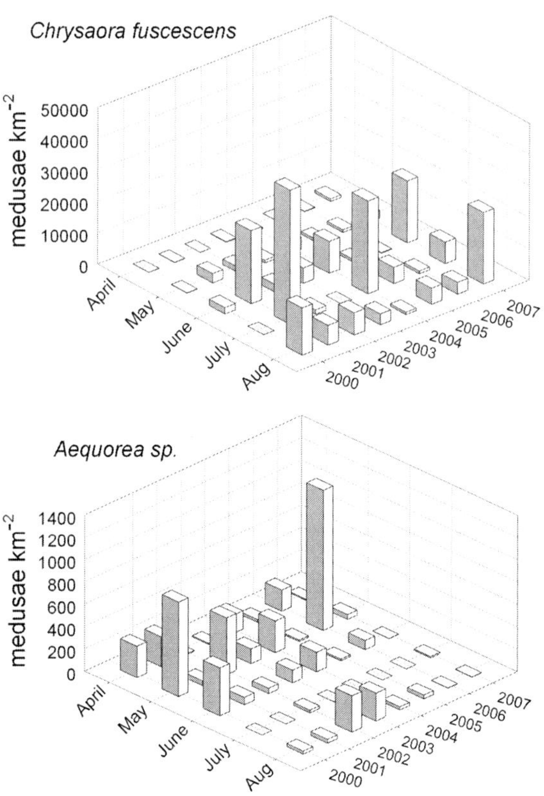

Fig. 5 *Chrysaora fuscescens* and *Aequorea* sp. caught in trawls by month and year. Number of trawls available in Fig. 2

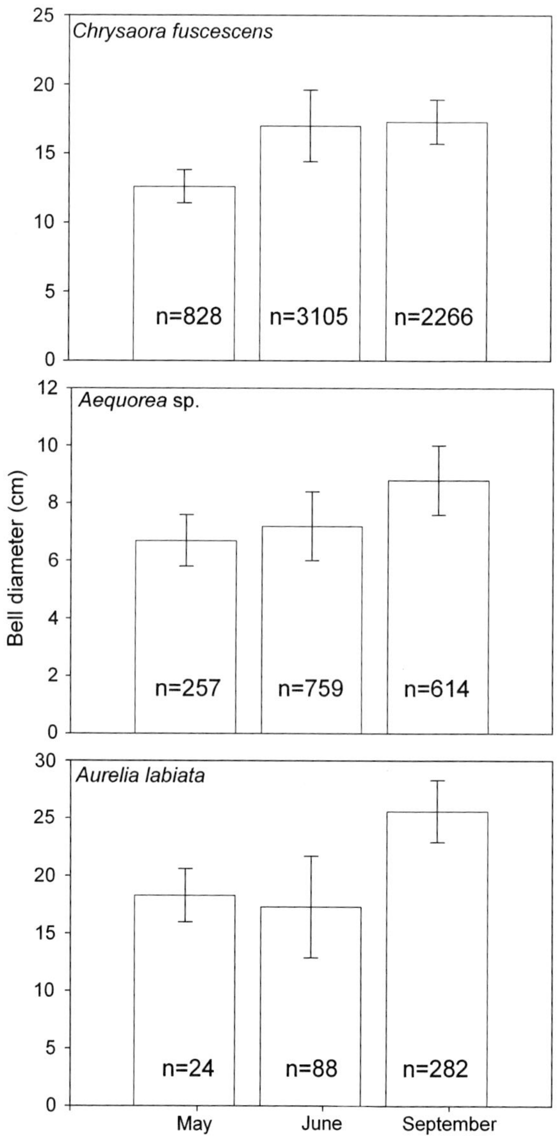

Fig. 6 Average bell diameters of medusae caught during mesoscale cruises, 2001–2007, by species and month. Averages and standard deviations are across years (*Aequorea* sp. data are for 2001, 2002, 2003, and 2005 only)

Maps of peak seasonal abundance in contrasting years provide snapshots of interannual variability, as well as differences in distribution between taxa. Figure 7 compares September 2001 and 2006 for *Chrysaora fuscescens* and Fig. 8 compares June 2001 and 2006 for *Aequorea* sp. In both years, the highest catches of *C. fuscescens* were at nearshore stations, whereas *Aequorea* sp. was more widely distributed across the shelf. More medusae were caught during 2001 than 2006 (Table 1). For *C. fuscescens*, a higher proportion of stations in the inner shelf had high catches in September 2001 (Fig. 7). In contrast, during June 2001, *Aequorea* sp, particularly along the northern transects, were caught at the shelf break. In June 2006, *Aequorea* sp. were more abundant and more widely distributed (Fig. 8).

Relationship to environmental variables

Results of GAM analysis for the mesoscale data showed station-specific nonparametric relationships with environmental conditions (Table 3). Presence–absence analysis of the full data set ($n = 876$) explained approximately 38% of *Chrysaora fuscescens* variance. The model included latitude, station depth, salinity, and distance offshore. The binomial models for *Aequorea* sp. and *Aurelia labiata* explained relatively little variance and showed few significant covariates or factors (Table 3). The models using only stations with medusae present proved much more useful. For *Chrysaora fuscescens*, 49% of the deviance in catch was explained for the 381 stations with complete data available for analysis. *Chrysaora fuscescens* abundance was nonlinearly related to latitude, with a peak in abundance around 47°N and smaller catches to the north and south in our sampling grid (Fig. 7). Abundance was highest at moderate salinities, and negatively correlated with both SST and distance from shore. Significant temporal factors also included all months and the majority of years (not 2000, 2003, or 2006). For *Aequorea* sp., the GAM explained 37% of the deviance ($n = 364$). Catches were lowest at mid-latitudes (46–47°N; Fig. 8), with higher abundances at higher salinity and SST. *Aequorea* sp. abundance showed negative relationships with other variables, including both distance from shore and chlorophyll *a*. Temporal factors were significant for May and September and some years (2001, 2004, 2005, and 2007). Finally, for *Aurelia labiata* ($n = 179$), 28% of the deviance was explained in the

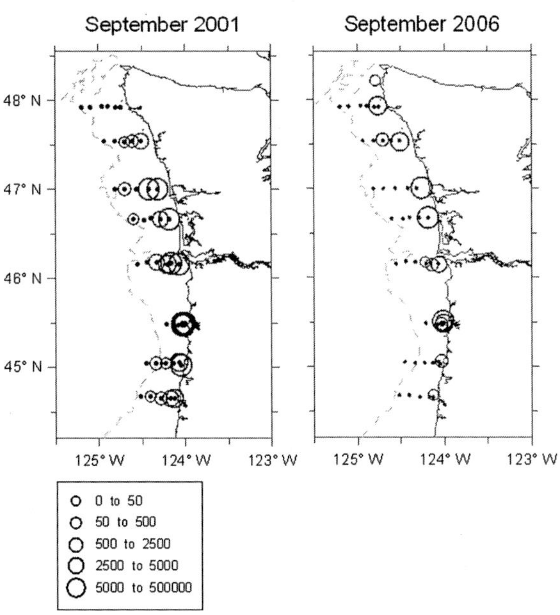

Fig. 7 *Chrysaora fuscescens* abundance and distribution (medusae km^{-2}) during mesoscale cruises in September 2001 and 2006. Stations sampled are represented by a *dot*

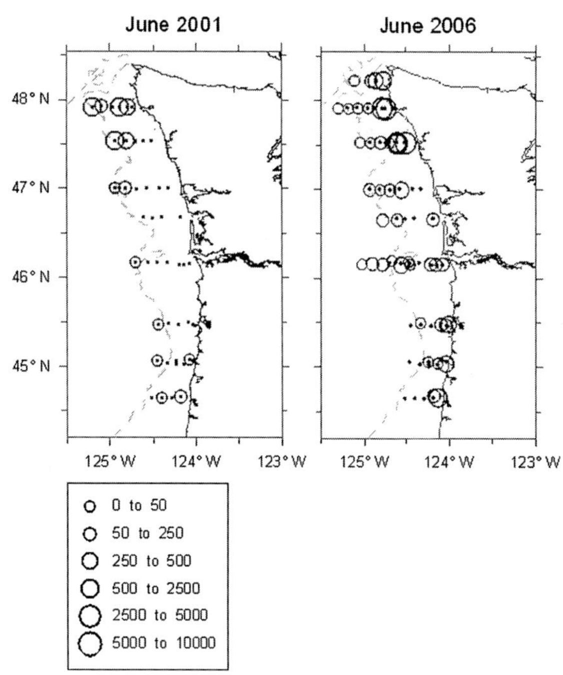

Fig. 8 *Aequorea* sp. abundance and distribution (medusae km^{-2}) during mesoscale cruises in June 2001 and 2006. Stations sampled are represented by a *dot*

Table 3 Results of GAM based on presence/absence (binomial model) and the natural log of positive catches (Gaussian model) for *Chrysaora fuscescens*, *Aequorea* sp., and *Aurelia labiata*

Covariates		Presence/absence			Natural log of positive catches		
		Chrysaora fuscescens	*Aequorea* sp.	*Aurelia labiata*	*Chrysaora fuscescens*	*Aequorea* sp.	*Aurelia labiata*
Latitude (decimal degrees)		(+/−)**	n.s.	(−)**	(+/−)**	(−/+)*	n.s.
Station bottom depth (m)		(−/+)**	n.s.	n.s.	n.s.	n.s.	n.s.
Sea surface temperature (°C)		n.s.	(+)**	n.s.	(−)**	(+)*	(−)*
Sea surface salinity		(+/−)*	n.s.	n.s.	(+/−)**	(+)**	(+/−)**
Chlorophyll *a* (mg l^{-1}, 3 m)		n.s.	(−)*	n.s.	n.s.	(−)**	(−/+)*
Distance from shore (km)		(−)*	n.s.	(−)*	(−)**	(−)*	(−/+)**
Factors	Levels						
Month	May	n.s.	n.s.	**	n.s.	**	**
	June	**	n.s.	n.s.	**	**	n.s.
	September	**	n.s.	**	**	*	*
Year	2000	n.s.	*	n.s.	n.s.	*	n.s.
	2001	*	n.s.	n.s.	**	*	n.s.
	2002	**	n.s.	n.s.	**	n.s.	n.s.
	2003	*	n.s.	*	n.s.	n.s.	n.s.
	2004	**	n.s.	n.s.	**	*	n.s.
	2005	*	n.s.	n.s.	**	n.s.	n.s.
	2006	n.s.	**	n.s.	n.s.	**	n.s.
	2007	**	n.s.	n.s.	**	n.s.	n.s.
Sample sizes		876	876	876	381	364	179
Deviance explained (%)		37.7	11.4	17.7	49.0	37.1	28.1

Significance values for each covariate and factor/level are coded as follows: * $P < 0.05$; ** $P < 0.01$; $^{n.s.}$ $P > 0.05$. In parentheses are relationships with covariates (+, increasing; −, decreasing; +/−, nonlinear concave; −/+, nonlinear convex). The bottom rows contain sample sizes and percentage of deviance explained by the final model

model, with higher abundances at moderate salinities, relatively low and high chlorophyll *a*, and inner shelf and shelf-break stations. Catches of *A. labiata* were significantly related to SST. May and September were significant temporal factors for this species, with no significant interannual variability (Table 3).

The aggregate catch of medusae in the region varied among years, with the highest abundances in 2001 and lowest in 2006 (Table 1). Multiple linear regression related the annual anomalies of medusan catch to the PDO (May–August), and to streamflow of the Columbia River (January–August) ($P < 0.05$, $r^2_{adj} = 0.93$), where

$$\text{Medusae km}^{-2} = 3309 - (328 \times \text{PDO anomaly}) - 0.002\,\text{streamflow anomaly}\left(\text{ft}^3\,\text{s}^{-1}\right).$$

PDO and Columbia River streamflow were not correlated ($P > 0.05$). Years with negative anomalies in PDO and streamflow were associated with highest catches of medusae (Fig. 9).

Chrysaora fuscescens, the most abundant species in the region, drove the overall relationship (multiple regression of anomalies with PDO and Columbia River streamflow: $P < 0.001$, $r^2_{adj} = 0.95$). The multiple regressions with anomalies of *Aequorea* sp. ($P = 0.06$, $r^2_{adj} = 0.54$) and *Aurelia labiata* ($P = 0.76$, $r^2_{adj} = 0$) were not significant.

Discussion

A number of scientific studies and reports in popular literature suggest that for a variety of reasons,

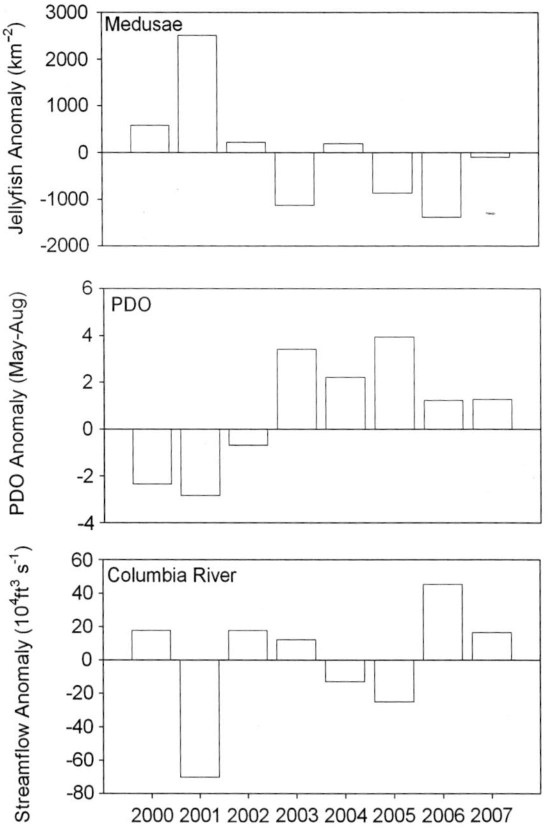

Fig. 9 Yearly anomalies of total jellyfish catch (all species), PDO index (May–August), and Columbia River streamflow (January–August, ft^3 s^{-1}). Anomalies were calculated as deviations from the means over the 8 years of the study

populations of gelatinous zooplankton are increasing, and these increases may lead to a shift in coastal or ocean food webs (Parsons & Lalli, 2002; Lynam et al., 2005; Richardson et al., 2009). Although some studies have linked changes in jellyfish populations with fluctuations in climate (reviewed in Purcell et al., 2007; Purcell, 2012), there remain relatively few published multi-year data sets to support generalized assertions (Pauly et al., 2009). Therefore, research to identify species- and region-specific population sizes and trends is sorely needed to establish baseline information and support or refute hypotheses about the roles of medusae in marine ecosystems and whether or how their population sizes are changing.

Several sampling challenges make quantifying populations of large medusae particularly difficult (Purcell, 2009). Smaller nets used for crustacean zooplankton often do not filter enough water to quantify larger gelatinous taxa and usually destroy more delicate organisms. Video from towed or remotely operated systems can be useful (Raskoff, 2001; Graham et al., 2003), particularly to determine fine-scale horizontal or vertical distribution, but these methods often do not cover large spatial scales effectively. Other noninvasive methods, such as acoustics (Brierley et al., 2005; Graham et al., 2010) or surface counts via observers or aircraft (Houghton et al., 2006a), must be compared with other collection methods for quantitative accuracy and species composition. A growing number of studies use data collected with a continuous plankton recorder (CPR) to correlate time series for zooplankton presence or absence with climate indices (e.g., Edwards & Richardson, 2004); however, the CPR was designed to capture much smaller plankton, and identification or quantification of gelatinous taxa, usually from preserved tentacles, must be interpreted with caution (Haddock, 2008; Baxter et al., 2010; Lynam et al., 2010).

Some of the best time series data have come from trawls deployed for fisheries-independent surveys, when fisheries scientists have identified and counted medusae caught incidentally (Lynam et al., 2004, 2011; Brodeur et al., 2008a); however, these data are best thought of as semi-quantitative, as the nets are selective with respect to size and behavior of the biota. The surface trawl used in this study sampled only the top ~20 m, and the mesh size near the trawl opening was large enough so that some of the smaller medusae may not have been caught. Another challenge is that, like other marine organisms, medusae are patchy in space and time. One methodological concern with using surface trawls is that medusae may be aggregated below the sampled depth or may migrate beyond the surface layer in response to light or temperature cues (Schuyler & Sullivan, 1997; Sparks et al., 2005). More directed studies might reveal fine-scale detail in vertical distribution patterns, but preliminary analysis of diel patterns (Suchman & Brodeur, 2005) and stratified trawls (present study) suggest that surface trawls in this region are not biased across time of day or across the upper 40 m of the water column.

Furthermore, even when interannual variability in medusa populations has been measured, mechanisms responsible for that variability remain difficult to identify. One reason for this is that scyphozoans alternate between benthic, asexual polyps and planktonic, medusan forms. Therefore, changes in abundance of medusae in the water column could be caused by factors influencing either or both stages. Identifying correlations with environmental or ecosystem variables allow scientists to form hypotheses and begin to test potential mechanisms underlying variability in populations. Understanding these mechanisms will allow us to distinguish between true "blooms" of jellyfish, which reflect rapid reproduction and growth rates and apparent bloom conditions caused by physical aggregations of organisms (Graham et al., 2001). Various hypotheses have been tested to link medusa abundance with environmental and ecosystem variables. These include the influence of temperature, salinity, and light on asexual reproduction or mortality of the polyp stages of scyphomedusae (Purcell, 2007; Purcell et al., 2007, 2012; Liu et al., 2009; Astorga et al., 2012; Holst, 2012; Thein et al., 2012); advection of ephyrae or medusae (Suchman & Brodeur, 2005; Lynam et al., 2010); physiological challenges associated with low salinity (Wright & Purcell, 1997); and availability of prey or abundance of competitors (Lynam et al., 2011).

We observed some correlations with local environmental variables by species and station within the study area. Because only 28–49% of the deviance in species distribution was explained through GAM analysis, it is clear that fine-scale distributions are difficult to predict using only the variables included. *Chrysaora fuscescens* was the dominant medusan species in the region, and its distribution was the one best explained by GAM analysis. As in the shelf region immediately to the south (Suchman & Brodeur, 2005), *C. fuscescens* catch was highest close to shore (distance covariate was significant). Suchman & Brodeur (2005) postulated that nearshore distribution minimized downstream transport in this highly advective system. In general, primary productivity in this study region is higher to the north, off the coast of Washington (Hickey & Banas, 2003; Hickey et al., 2010), and yet *C. fuscescens* abundances were higher off the Columbia River and, during years with highest abundances (e.g., 2001), to the south (Fig. 7). Catches of *C. fuscescens* were also correlated with latitude (Table 3), with highest abundances at mid-latitude transects near the Columbia River. Other studies have suggested that fronts created by the Columbia River plume could aggregate plankton and serve as areas of enhanced feeding potential for juvenile salmon (Morgan et al., 2005; Peterson & Peterson, 2008). Similar processes could lead to local retention of large medusae.

When medusa catch data were aggregated by year, variability among the 8 years of the study showed a strong negative correlation with the PDO and Columbia River streamflow (Fig. 9). The PDO represents sea surface temperature variability poleward of 20°N, with negative anomalies associated with the relatively cool conditions associated with stronger upwelling. In the Northern California Current, upwelling occurs when surface water is pushed offshore by northerly winds via Ekman transport and colder, nutrient-rich deeper water replaces it. Studies in the region have demonstrated a correlation between negative PDO, or cool conditions, and "good years" for coho salmon and northern copepod species (Peterson, 2009). This study demonstrates that in the Northern California Current, large medusae had a similar response to the PDO as other marine organisms: cool years corresponded with periods of high productivity. A similar pattern has also been demonstrated or suggested for medusae in coastal upwelling areas in other regions of the world (e.g., Buecher & Gibbons, 2000; Miglietta et al., 2008; Quiñones et al., 2010). These results in upwelling systems are important in the context of global jellyfish populations because they run counter to the prevailing trend for temperate species that warm temperatures lead to increased numbers. Large-scale environmental indices that we did not include such as the El Niño/Southern Oscillation (ENSO) Index may negatively affect jellyfish populations through decreased plankton productivity or positively affect biomass in terms of increased growth rates through higher temperatures. We did not include ENSO in our model because none of the values during the years we included (2000–2007) were highly anomalous. Longer time series would be required to reveal larger-scale climate and population trends such as ENSO.

The negative correlation between Columbia River streamflow and catches of large medusae in the study area was surprising to us. The relationship was most obvious when comparing 2000 versus 2001 or 2006 versus 2007. In each of these 2-year pairs, the PDO anomalies were similar, yet Columbia River streamflow was substantially different. In 2001, an anomalously dry year, catch of medusae was higher than in 2000; in 2006, a relatively wet year, catch was lower than in 2007. Despite the potential for aggregation within fronts created by the plume, a larger area of stratification of surface waters during high flow years may lead to changes in vertical distribution of medusae (but see Fig. 3; Peterson & Peterson, 2008) or increase transport away from nearshore retention areas. Conversely, years with low Columbia River streamflow may allow *Chrysaora fuscescens* to remain closer to shore and less prone to downstream transport. A few other studies have also shown negative relationships between abundance of medusae and precipitation or streamflow (Cargo & King, 1990; Purcell & Decker, 2005; Lynam et al., 2011), with a number of possible mechanisms presented to explain these correlations.

Despite our lack of full understanding of the underlying mechanisms at work, the strong relationships we found in this 8-year data set supports the hypothesis of the importance of physical forcing in regulating medusan biomass at a regional scale. Moreover, we demonstrate that these seasonally increasing populations have linear relationships with environmental variables on a similar time scale, as predicted by the "linear tracking window" hypothesis (Hsieh & Ohman, 2006). The availability of these indices by the middle of the summer of each year could allow us to make near real-time predictions of interannual variability of medusae in the Northern California Current and frame future research to explain causal relationships.

Conclusions

In this study, we analyzed surface trawls conducted during spring and summer in the Northern California Current from 2000 through 2007, reporting seasonal and interannual patterns of abundant large medusae in relation to small- and regional-scale environmental variables. Our results indicate that these populations respond to regional climate parameters, the PDO and Columbia River streamflow, with highest aggregate abundances associated with cool and relatively dry conditions. These findings run counter to the hypothesis that resident jellyfish populations would increase with warm conditions in the Northern California Current. Instead, conditions supporting high regional productivity (years with strong upwelling) lead to high abundances of gelatinous zooplankton as well as other taxa. These data, when considered with what we know about fish communities (Brodeur et al., 2005) and trophic interactions (Miller & Brodeur, 2007; Brodeur et al., 2008b; Suchman et al., 2008), provide a better understanding of ecosystem processes and the potential effects of climate variability in this highly productive region. Finally, we suggest that jellyfish biomass can be one of the most sensitive indicators of changing ecosystem status as shown for other ecosystems worldwide (Richardson et al., 2009; Samhouri et al., 2009) and should be continued to be monitored in this and other ecosystems.

Acknowledgments Thanks to the captains and crew of the FV *Frosti*, FV *Ocean Harvester*, FV *Sea Eagle*, FV *Piky*, RV *Miller Freeman*, and RV *W.E. Ricker*, as well as the many seagoing scientists who participated in the fieldwork component of this project. G. Krutzikowsky, C. Morgan, S. Pool, and C. Bucher helped with database management. E. Casillas, J. Field, C. Rice, J. Purcell and two anonymous reviewers provided helpful comments on earlier drafts of the manuscript. This study was funded by the Bonneville Power Administration, the US GLOBEC Northeast Pacific Program, and the Northwest Fisheries Science Center (NOAA). Completion of this manuscript was supported by the National Science Foundation, while the lead author was working at the Foundation. Any opinion, finding, and conclusions or recommendations expressed in this material are those of the authors and do not necessarily reflect the views of the National Science Foundation.

References

Astorga, D., J. Ruiz & L. Prieto, 2012. Ecological aspects of early life stages of *Cotylorhiza tuberculata* (Scyphozoa: Rhizostomae) affecting its pelagic population success. Hydrobiologia. doi:10.1007/s10750-012-1036-x.

Barry, S. C. & A. H. Welsh, 2002. Generalized additive modeling and zero inflated count data. Ecological Modeling 157: 179–188.

Batchelder, H. P., J. A. Barth, M. P. Kosro, P. T. Strub, R. D. Brodeur, W. T. Peterson, C. T. Tynan, M. D. Ohman, L. W. Botsford, T. M. Powell, F. B. Schwing, D. G. Ainley, D. L. Mackas, B. M. Hickey & S. R. Ramp, 2002. The GLOBEC Northeast Pacific California Current System Program. Oceanography 15: 36–47.

Baxter, E. J., A. W. Walne, J. E. Purcell, R. McAllen & T. K. Doyle, 2010. Identification of jellyfish from Continuous Plankton Recorder samples. Hydrobiologia 645: 193–201.

Brierley, A. S., D. C. Boyer, B. E. Axelsen, C. P. Lynam, C. A. Sparks, H. J. Boyer & M. J. Gibbons, 2005. Towards the acoustic estimation of jellyfish abundance. Marine Ecology Progress Series 295: 105–111.

Brodeur, R. D., J. P. Fisher, C. A. Morgan, R. L. Emmett & E. Casillas, 2005. Species composition and community structure of pelagic nekton off Oregon and Washington under variable oceanographic conditions. Marine Ecology Progress Series 298: 41–57.

Brodeur, R. D., M. B. Decker, L. Ciannelli, J. E. Purcell, N. A. Bond, P. J. Stabeno, E. Acuna & G. L. Hunt, 2008a. Rise

and fall of jellyfish in the eastern Bering Sea in relation to climate regime shifts. Progress in Oceanography 77: 103–111.
Brodeur, R. D., C. L. Suchman, D. Reese, T. Miller & E. Daly, 2008b. Spatial overlap and trophic interactions between fish and large jellyfish in the Northern California Current. Marine Biology 154: 649–659.
Brodeur, R. D., J. J. Ruzicka & J. H. Steele, 2011. Investigating alternate trophic pathways through gelatinous zooplankton and planktivorous fishes in an upwelling ecosystem using end-to-end models. In Omori, K., X. Guo, N. Yoshie, N. Fujii, I. C. Handoh, A. Isobe & S. Tanabe (eds), Interdisciplinary Studies on Environmental Chemistry – Marine Environmental Modeling & Analysis. TERRAPUB, Tokyo: 57–63.
Buecher, E. & M. J. Gibbons, 2000. Interannual variation in the composition of the assemblages of medusae and ctenophores in St. Helena Bay. Southern Benguela ecosystem. Scientia Marina 64: 123–134.
Cargo, D. G. & D. R. King, 1990. Forecasting the abundance of the sea nettle, *Chrysaora quinquecirrha*, in Chesapeake Bay. Estuaries 13: 486–491.
Checkley, D. M. & J. A. Barth, 2009. Patterns and processes in the California Current System. Progress in Oceanography 83: 49–64.
Edwards, M. & A. J. Richardson, 2004. Impact of climate change on marine pelagic phenology and trophic mismatch. Nature 430: 881–884.
Emmett, R. L., R. D. Brodeur & P. M. Orton, 2004. The vertical distribution of juvenile salmon (*Oncorhynchus* spp.) and associated fishes in the Columbia River plume. Fisheries Oceanography 13: 392–402.
Graham, W. M., F. Pagès & W. M. Hamner, 2001. A physical context for gelatinous zooplankton aggregations: a review. Hydrobiologia 451: 199–212.
Graham, W. M., D. L. Martin & J. C. Martin, 2003. In situ quantification and analysis of large jellyfish using a novel video profiler. Marine Ecology Progress Series 354: 129–140.
Graham, T. R., J. T. Harvey, S. R. Benson, J. S. Renfree & D. A. Demer, 2010. The acoustic identification and enumeration of scyphozoan jellyfish, prey for leatherback sea turtles (*Dermochelys coriacea*), off central California. ICES Journal of Marine Science 67: 1739–1748.
Haddock, S. H. D., 2008. Reconsidering evidence for potential climate-related increases in jellyfish. Limnology and Oceanography 53: 2759–2762.
Hastie, T. J. & R. J. Tibshirani, 1990. Generalized Additive Models. Chapman and Hall, New York: 335 pp.
Hickey, B. M. & N. S. Banas, 2003. Oceanography of the U.S. Pacific Northwest coastal ocean and estuaries with application to coastal ecology. Estuaries 26: 1010–1031.
Hickey, B. M., et al., 2010. River influences on shelf ecosystems: introduction and synthesis. Journal of Geophysical Research 115: C00B17.
Holst, S., 2012. Effects of climate warming on strobilation and ephyra production of North Sea scyphozoan jellyfish. Hydrobiologia. doi:10.1007/s10750-012-1043-y.
Houghton, J. D. R., T. K. Doyle, J. Davenport & G. C. Hays, 2006a. Developing a simple, rapid method for identifying and monitoring jellyfish aggregations from the air. Marine Ecology Progress Series 314: 159–170.
Houghton, J. D. R., T. K. Doyle, M. W. Wilson, J. Davenport & G. C. Hays, 2006b. Jellyfish aggregations and leatherback turtle foraging patterns in a temperate coastal environment. Ecology 87: 1967–1972.
Hsieh, C. & M. D. Ohman, 2006. Biological responses to environmental forcing: the linear tracking window hypothesis. Ecology 87: 1932–1938.
Liu, W. C., W. T. Lo, J. E. Purcell & H. H. Chang, 2009. Effects of temperature and light intensity on asexual reproduction of the scyphozoan, *Aurelia aurita* (L.) in Taiwan. Hydrobiologia 616: 247–258.
Lynam, C. P., S. J. Hay & A. S. Brierley, 2004. Interannual variability in abundance of North Sea jellyfish and links to the North Atlantic Oscillation. Limnology and Oceanography 49: 637–643.
Lynam, C. P., M. R. Heath, S. J. Hay & A. S. Brierley, 2005. Evidence for impacts by jellyfish on North Sea herring recruitment. Marine Ecology Progress Series 298: 157–167.
Lynam, C. P., M. J. Attrill & M. D. Skogen, 2010. Climatic and oceanic influences on the abundance of gelatinous zooplankton in the North Sea. Journal of the Marine Biological Association of the United Kingdom 90: 1153–1159.
Lynam, C. P., M. K. S. Lilley, T. Bastian, T. K. Doyle, S. E. Beggs & G. C. Hays, 2011. Have jellyfish in the Irish Sea benefited from climate change and overfishing? Global Change Biology 17: 767–782.
Miglietta, M. P., M. Rossi & R. Collin, 2008. Hydromedusa blooms and upwelling events in the Bay of Panama, Tropical East Pacific. Journal of Plankton Research 30: 783–793.
Miller, T. W. & R. D. Brodeur, 2007. Diets of and trophic relationships among dominant marine nekton within the Northern California Current ecosystem. Fishery Bulletin 105: 548–559.
Mills, C. E., 2001. Jellyfish blooms: are populations increasing globally in response to changing ocean conditions? Hydrobiologia 451: 55–68.
Morgan, C. A., A. De Robertis & R. W. Zabel, 2005. Columbia River plume fronts. I. Hydrography, zooplankton distribution, and community composition. Marine Ecology Progress Series 299: 19–31.
Parsons, T. R. & C. M. Lalli, 2002. Jellyfish population explosions: revisiting a hypothesis of possible causes. La Mer 40: 111–121.
Pauly, D., W. Graham, S. Libralato, L. Morissette & M. L. Deng Palomares, 2009. Jellyfish in ecosystems, online databases, and ecosystem models. Hydrobiologia 616: 67–85.
Peterson, W. T., 2009. Copepod species richness as an indicator of long-term changes in the coastal ecosystem of the Northern California Current. CalCOFI Reports 50: 73–81.
Peterson, J. O. & W. T. Peterson, 2008. Influence of the Columbia River plume (USA) on the vertical and horizontal distribution of mesozooplankton over the Washington and Oregon shelf. ICES Journal of Marine Science 65: 477–483.
Purcell, J. E., 2005. Climate effects on formation of jellyfish and ctenophore blooms: a review. Journal of the Marine Biological Association of the United Kingdom 85: 461–476.
Purcell, J. E., 2007. Environmental effects on asexual reproduction rates of the scyphozoan, *Aurelia labiata*. Marine Ecology Progress Series 348: 183–196.

Purcell, J. E., 2009. Extension of methods for jellyfish and ctenophore trophic ecology to large-scale research. Hydrobiologia 206: 23–50.

Purcell, J. E., 2012. Jellyfish and ctenophore blooms coincide with human proliferations and environmental perturbations. Annual Review of Marine Science 4: 209–235.

Purcell, J. E. & M. N. Arai, 2001. Interactions of pelagic cnidarians and ctenophores with fish: a review. Hydrobiologia 451: 27–44.

Purcell, J. E. & M. B. Decker, 2005. Effects of climate on relative predation by scyphomedusae and ctenophores on copepods in Chesapeake Bay during 1987–2000. Limnology and Oceanography 50: 376–387.

Purcell, J. E., S. I. Uye & W. S. Lo, 2007. Anthropogenic causes of jellyfish blooms and their direct consequences for humans: a review. Marine Ecology Progress Series 350: 153–174.

Purcell, J. E., D. Atienza, V. Fuentes, A. Olariaga, U. Tilves, C. Colahan & J.-M. Gili, 2012. Temperature effects on asexual reproduction rates of Scyphozoan species from the northwest Mediterranean Sea. Hydrobiologia. doi:10.1007/s10750-012-1047-7.

Quiñones, J., V. G. Carman, J. Zeballos, S. Purca & H. Mianzan, 2010. Effects of El Niño-driven environmental variability on black turtle migration to Peruvian foraging grounds. Hydrobiologia 645: 69–79.

Raskoff, K. A., 2001. The impact of El Niño events on populations of mesopelagic hydromedusae. Hydrobiologia 451: 121–129.

Richardson, A. J., A. Bakun, G. C. Hays & M. J. Gibbons, 2009. The jellyfish joyride: causes, consequences, and management responses to a more gelatinous future. Trends in Ecology & Evolution 24: 312–322.

Ruzicka, J. J., R. D. Brodeur & T. C. Wainwright, 2007. Seasonal food web models for the Oregon inner-shelf ecosystem: investigating the role of large jellyfish. CalCOFI Reports 48: 106–128.

Samhouri, J. F., P. S. Levin & C. J. Harvey, 2009. Quantitative evaluation of marine ecosystem indicator performance using food web models. Ecosystems 12: 1283–1298.

Schuyler, Q. & B. K. Sullivan, 1997. Light responses and diel migration of the scyphomedusa *Chrysaora quinquecirrha* in mesocosms. Journal of Plankton Research 19: 1417–1428.

Shenker, J. M., 1984. Scyphomedusae in surface waters near the Oregon coast, May–August 1981. Estuarine, Coastal and Shelf Science 19: 619–632.

Sparks, C., A. S. Brierley, E. Buecher, D. Boyer, B. Axelsen & M. J. Gibbons, 2005. Submersible observations on the daytime vertical distribution of *Aequorea forskalea* off the west coast of southern Africa. Journal of the Marine Biological Association of the UK 85: 519–522.

Suchman, C. L. & R. D. Brodeur, 2005. Abundance and distribution of large medusae in surface waters of an upwelling zone off coastal Oregon, USA. Deep Sea Research II 52: 51–72.

Suchman, C. L., E. A. Daly, J. E. Keister, W. T. Peterson & R. D. Brodeur, 2008. Feeding patterns and predation potential of scyphomedusae in a highly productive upwelling region. Marine Ecology Progress Series 358: 161–172.

Thein, H., H. Ikeda & S. I. Uye, 2012. The potential role of podocysts in perpetuation of the common jellyfish *Aurelia aurita* s.l. (Cnidaria: Scyphozoa) in anthropogenically perturbed coastal waters. Hydrobiologia. doi:10.1007/s10750-012-1045-9.

Wood, S. N., 2006. Generalized Additive Models: An Introduction with R. Chapman and Hall, New York: 391 pp.

Wood, S. N. & N. H. Augustin, 2002. GAMs with integrated model selection using penalized regression splines and applications to environmental modeling. Ecological Modeling 157: 157–177.

Wright, D. A. & J. E. Purcell, 1997. Effect of salinity on ionic shifts in mesohaline scyphomedusae, *Chrysaora quinquecirrha*. Biological Bulletin 192: 332–339.

Hydrobiologia (2012) 690:127–140
DOI 10.1007/s10750-012-1043-y

JELLYFISH BLOOMS

Effects of climate warming on strobilation and ephyra production of North Sea scyphozoan jellyfish

Sabine Holst

Published online: 9 March 2012
© Springer Science+Business Media B.V. 2012

Abstract Recent studies have correlated fluctuations in jellyfish abundances with climatic changes, leading to speculation that the warming trend in the North Sea will affect the strobilation activity of Scyphozoa. The present study provides long-term data (10–22 months) on temperature effects on the species *Aurelia aurita*, *Cyanea capillata*, *Cyanea lamarckii* and *Chrysaora hysoscella*. Strobilation at current winter temperature (5°C) in the German Bight was compared to strobilation at warmer winter temperatures. Simulated winter temperature of 10°C had several positive effects on strobilation, as compared to 5°C: 1. A longer strobilation period or higher ephyra production per polyp in *A. aurita*, *C. lamarckii* and *Ch. hysoscella*; 2. Higher percentages of polyps strobilating in *A. aurita* and *Ch. hysoscella*; 3. More ephyrae per strobila in *C. capillata* and *C. lamarckii*; 4. A shorter strobilation duration in *C. capillata* and *C. lamarckii*. Cold winter temperatures of 5°C promoted strobilation in *C. capillata*, but inhibited strobilation in *A. aurita* and reduced ephyra production in *C. lamarckii* and *Ch. hysoscella*. These results suggest that climate warming will benefit *A. aurita*, but not cold-water *C. capillata*. The distributions of *C. lamarckii* and *Ch. hysoscella* probably could expand to the north.

Keywords *Aurelia* · *Cyanea* · *Chrysaora* · Polyp · Temperature · Reproduction

Guest editors: J. E. Purcell, H. Mianzan & J. R. Frost / Jellyfish Blooms: Interactions with Humans and Fisheries

S. Holst (✉)
Senckenberg am Meer, German Center for Marine Biodiversity Research, c/o Biozentrum Grindel und Zoologisches Museum, Martin-Luther-King-Platz 3, 20146 Hamburg, Germany
e-mail: sabine.holst@senckenberg.de

Introduction

Reports of mass occurrences of large jellyfish (Scyphozoa) in many marine ecosystems worldwide have increased in recent decades (Purcell et al., 2007; Richardson et al., 2009). Negative impacts of such medusa blooms on ecosystems, fisheries, industries and tourism are obvious: medusae are food competitors of fish and feed on fish larvae and small fish (Barz & Hirche, 2007; Sabatés et al., 2010), the gelatinous bodies clog fishing nets and cooling systems of coastal industries, and jellyfish stinging swimmers have negative effects on the tourism industry (CIESM, 2001; Purcell et al., 2007). It is possible that consequences of anthropogenic activities, including overfishing, eutrophication, species invasions and especially climate change, have contributed to increased jellyfish abundances (Purcell et al., 2007; Richardson et al., 2009; Purcell, 2011).

Recent analyses of temperature data show a clear warming trend in global average air and ocean temperature (IPCC, 2007). A pronounced winter warming trend has been observed in the North and Baltic seas, which is predicted to continue in the future (HELCOM, 2007; Belkin, 2009). Recent rapid climate change has changed the abundances, population structures and biographical ranges of benthic and planktonic North Sea species (Mieszkowska et al., 2006; Wiltshire et al., 2010). Climate regime shifts also affect jellyfish populations. Changes in temperature, salinity, currents, predator–prey interactions and competition have measurable effects on jellyfish abundance and distribution (Purcell, 2005, 2009; Molinero et al., 2008; Lynam et al., 2010). These environmental changes will affect both the pelagic and the benthic stages of metagenetic medusae; however, the role of the polyps has been neglected in most analyses and models because of the lack of data on polyp ecology (Lynam et al., 2010).

The medusae of four semaeostome scyphozoans, *Aurelia aurita* (Linneaus, 1758), *Cyanea capillata* (Linneaus, 1758), *Cyanea lamarckii* Péron and Lesueur, 1809 and *Chrysaora hysoscella* (Linneaus, 1766), occur in the German Bight during the summer months, and mass occurrences have been documented in some years (Russell, 1970; Möller, 1980a; Hay et al., 1990; Barz & Hirche, 2007). Like most scyphozoans, the life cycle of these species includes planktonic medusa and benthic polyp generations. After settlement and metamorphosis of the planula larvae, the polyp populations increase by asexual reproduction. Each polyp seasonally produces one to several young medusae (ephyrae) in a process of transversal fission called strobilation. This process contributes to the development of jellyfish blooms because polydisc strobilation allows single polyps to produce many ephyrae (Boero et al., 2008).

Information on the environmental factors and stressors that determine the induction, timing and magnitude of the strobilation process is limited to a few species (e.g. *Aurelia* spp., reviewed in Purcell et al., 2012). Moreover, the benthic stage is completely undiscovered or undescribed for most scyphozoans (Tronolone et al., 2002; Jarms, 2010). Although many laboratory investigations on scyphopolyps have shown that temperature significantly affects the asexual polyp reproduction and strobilation (e.g. Holst et al., 2007; Purcell, 2007; Willcox et al., 2007; You et al., 2008; Liu et al., 2009), most of these studies were restricted to short periods of a few weeks and polyps were often exposed to extreme, abrupt changes, whereas changes in situ would be less severe (Purcell, 2005). Long-time investigations on the effects of natural seasonal temperature cycles on polyp ecology are unavailable for most species (but see Gröndahl, 1988; Brewer & Feingold, 1991; Miyake et al., 2002; Purcell et al., 2009; Di Camillo et al., 2010; Ishii & Katsukoshi, 2010). In the present study, I compare the long-term (10–22 months) effects of different temperature conditions on strobilation and ephyra production of four semaeostome jellyfish species from the North Sea to show the possible effects of increasing winter temperatures on the strobilation activity of Scyphozoa in the North Sea. Strobilation at temperature conditions similar to the current temperatures in the German Bight was compared to that at warmer winter temperatures. Additional experiments without a temperature change were conducted to determine the importance of temperature changes as inducers of strobilation in scyphozoan polyps. I tested the null hypotheses to ascertain whether the percentages of polyps strobilating and the numbers of ephyrae produced per strobila were independent of temperature; second, whether the percentages of strobilating polyps and the number of ephyrae produced per strobila were independent of polyp age; and third, whether the duration of strobilation was independent of the number of ephyrae per strobila and of temperature.

Materials and methods

Polyps of the species *A. aurita*, *C. capillata*, *C. lamarckii* and *Ch. hysoscella* were reared from planulae collected from female medusae in the summer periods of 2003 and 2004 around the island of Helgoland (details in Holst & Jarms, 2007, 2010). Watch glasses or polyethylene plates (for *C. lamarckii*) colonized by polyps in the laboratories of the Biologische Anstalt Helgoland were transported to the Zoological Institute of the University of Hamburg (Biocenter Grindel) 1–2 weeks after planula settlement. The substrates were transferred into 150-ml glass bowls filled with filtered North Sea water (salinity 35 ± 2). The bowls were kept in incubators at 15°C in darkness before the experiments. Only well-developed polyps with extended tentacles were used for the experiments, whereas small or contracted polyps were carefully removed from the substrates with a needle. Polyps were fed nauplii of *Artemia salina* for 1–2 h every

7–10 days, after which the seawater in the bowls was replaced with fresh filtered seawater of the same temperature. During strobilation, only half of the water was changed carefully, and uneaten food was removed from the bowls with a pipette to avoid disturbing the strobilation process. Polyps of each species were cultured in three different temperature groups (Table 1). In the 15°C group, the temperature was constant 15°C throughout the year. In the groups 15–10–15°C and 15–05–15°C, the temperature was 15°C until mid-October and lowered to 10°C in autumn and 5°C in winter, respectively. The temperatures were decreased in steps of 2.5°C per month (Fig. 1), and all cultures were kept in incubators in darkness. Polyps were monitored for 12 months (*A. aurita* from 15 September, and *C. capillata* from 15 October) or 10 months (*C. lamarckii* and *Ch. hysoscella* from 15 October) in the first year after settlement. Additional 10-month-long experiments (from 15 October) were conducted in the second year after settlement with *A. aurita* and *C. capillata* polyps. The polyps were counted monthly in all replicates, and all cultures were checked weekly to detect beginning strobilations. When strobilation appeared in the cultures, each strobila was monitored individually. The developing ephyrae in each strobila and the released ephyrae in the cultures were counted at least twice per week.

Calculations and statistics

The numbers of ephyrae produced monthly in each temperature group were summed and divided by the number of polyps to calculate ephyra production for each month. To account for production and mortality of polyps, the mean numbers of polyps (counted monthly) were calculated over the duration of the experiment (Table 1) and used in calculations and statistics. The total numbers of ephyrae produced per mean polyp numbers were calculated for each temperature group. The percentages of strobilating polyps in the first year were calculated for each replicate from the total numbers of strobilae divided by the mean numbers of polyps in each replicate. Percentages were arcsine square root transformed before statistical analysis to test the first null hypothesis, H_{01}: the percentages of polyps strobilating were independent of the temperature treatments. In *C. lamarckii*, all polyps strobilated once in all temperature groups, therefore the percentages of polyps that strobilated twice were compared to test H_{01}. Individually monitored strobilations in the first year were used to calculate the mean numbers of ephyrae per strobila to test H_{02}: the numbers of ephyrae produced per strobila were independent of the temperature treatments. Strobilations from replicates 1–3 were used for analysis in all species and temperature groups; additional strobilations were included from replicates 4–6 in *A. aurita* 15°C, *C. capillata* 15–10–15°C, *C. capillata* 15–05–15°C and *Ch. hysoscella* 15–05–15°C. For H_{01} and H_{02}, data with normal distributions were tested by one-way analysis of variance (ANOVA) followed by a Fisher's least significance difference (LSD) post-hoc test. Data not normally distributed were tested by a Kruskal–Wallis ANOVA on ranks and a Student–Newman–Keuls post-hoc test. Two experimental groups were compared with a Mann–Whitney Rank Sum Test.

Table 1 Replicates (*n*) in temperature groups used for analyses of strobilation in first year (1y) and second year (2y) after planula settlement in four species of scyphozoans

Species	Year	*n*	15°C		15–10–15°C		15–05–15°C	
			Start	Mean	Start	Mean	Start	Mean
A. aurita	1y	6	79	87.3 ± 4.9	78	102.6 ± 19.9	81	114.7 ± 32.5
	1y	3	29	37.1 ± 4.8	32	46.3 ± 11.9	36	54.0 ± 16.3
	2y	3	33	46.3 ± 15.2	32	47.2 ± 21.8	35	49.7 ± 22.0
C. capillata	1y	6	71	73.7 ± 4.0	75	86.3 ± 11.2	81	96.6 ± 17.3
	1y	3	25	26.0 ± 2.9	25	26.0 ± 2.9	26	28.0 ± 6.4
	2y	3	11	18.2 ± 8.3	20	28.9 ± 10.1	24	39.1 ± 13.2
C. lamarckii	1y	6	86	76.3 ± 7.4	86	64.5 ± 14.1	85	60.5 ± 17.0
Ch. hysoscella	1y	7	108	85.0 ± 22.3	108	92.9 ± 15.7	113	81.9 ± 28.8

The total polyp numbers at the beginning of the experiments (Start) and numbers of polyps in monthly counts (mean ± standard deviation) in different temperature groups (see Fig. 1) are listed

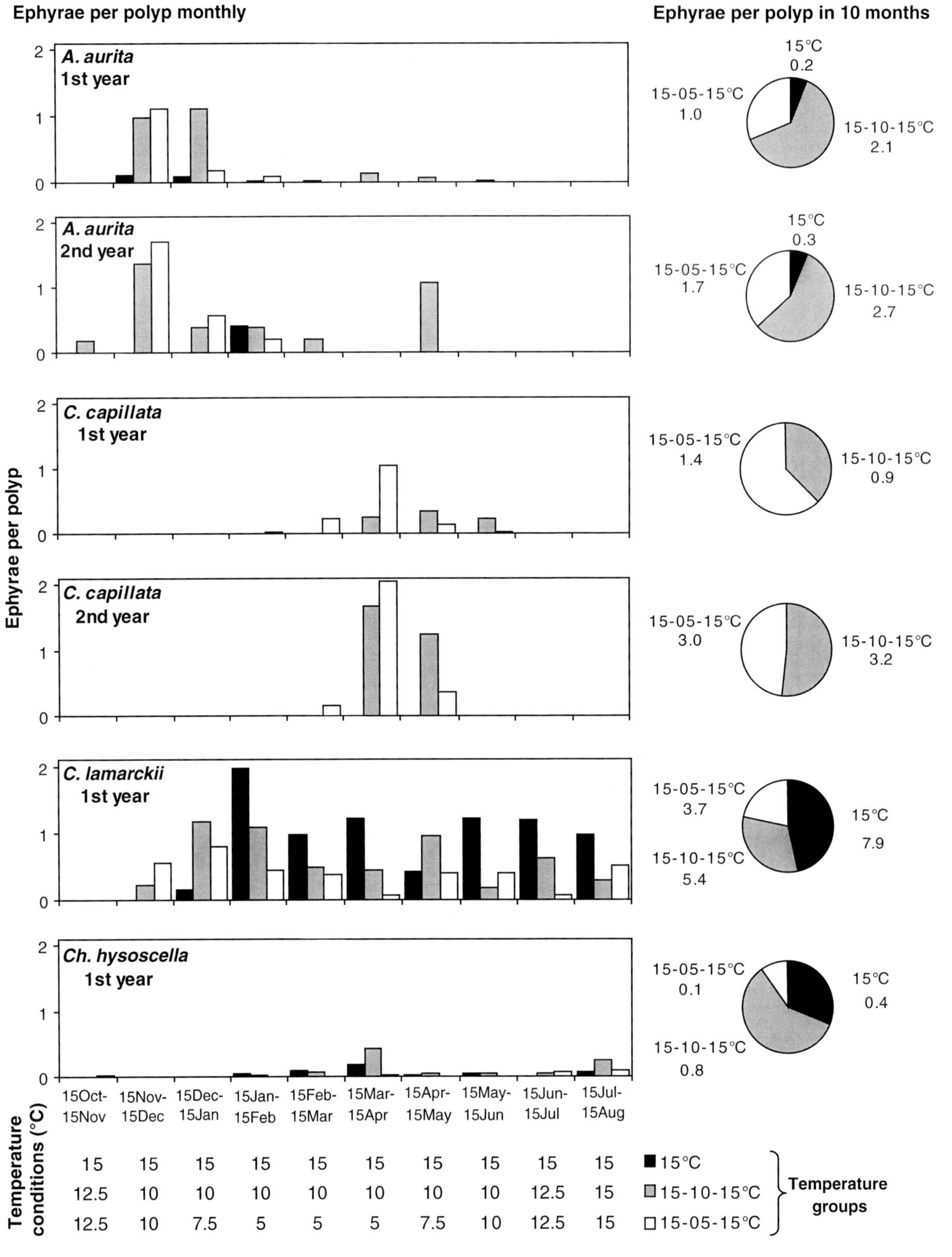

Ephyrae per polyp monthly
Ephyrae per polyp in 10 months
A. aurita 1st year
15°C 0.2
15-05-15°C 1.0
15-10-15°C 2.1
A. aurita 2nd year
15°C 0.3
15-05-15°C 1.7
15-10-15°C 2.7
C. capillata 1st year
15-05-15°C 1.4
15-10-15°C 0.9
C. capillata 2nd year
15-05-15°C 3.0
15-10-15°C 3.2
C. lamarckii 1st year
15-05-15°C 3.7
15°C 7.9
15-10-15°C 5.4
Ch. hysoscella 1st year
15-05-15°C 0.1
15°C 0.4
15-10-15°C 0.8
Ephyrae per polyp
0
1
2
15Oct-15Nov
15Nov-15Dec
15Dec-15Jan
15Jan-15Feb
15Feb-15Mar
15Mar-15Apr
15Apr-15May
15May-15Jun
15Jun-15Jul
15Jul-15Aug
Temperature conditions (°C)
15 15 15 15 15 15 15 15 15 15
12.5 10 10 10 10 10 10 10 12.5 15
12.5 10 7.5 5 5 5 7.5 10 12.5 15
15°C
15-10-15°C
15-05-15°C
Temperature groups

◀ **Fig. 1** Numbers of ephyrae per polyp produced monthly (*left side*) and total numbers of ephyrae produced per mean numbers of polyps counted monthly during 10 months of observation (15 Oct–15 Aug, *right side*) in different temperature groups of *A. aurita*, *C. capillata*, *C. lamarckii* and *Ch. hysoscella* polyps. The culture temperature was constant 15°C from 15th July to 15th October in all temperature groups (not shown in the figure). Polyp numbers in different temperature groups are shown in Table 1

For comparisons among strobilations 1 and 2 years after settlement, the same three replicates of each temperature group of *A. aurita* and *C. capillata* polyps were used in both years (Table 1). The same 10-month period (15 October–15 August) were analysed in both years. The percentages of strobilating polyps and the numbers of ephyrae produced per strobila were calculated for both years, as described above, and compared by a Mann–Whitney Rank Sum Test. Two null hypotheses were tested, H_{03}: The percentages of strobilating polyps were independent of polyp age and H_{04}: The number of ephyrae produced per strobila was independent of the polyp age.

The durations of strobilations (days) were determined from the number of days between the start of the strobilation, when first constrictions of the polyp body appeared, until the end the strobilation, when the last ephyra detached from the strobila. Only strobilations that began and ended at the same temperature were used in the analysis. Strobilae in the first year after polyp settlement were analysed in *C. lamarckii* and *Ch. hysoscella*, and in the first and second years combined in *A. aurita* and *C. capillata*. Strobilations were analysed at 15, 10 and 5°C culture temperatures. Two null hypotheses were tested by analysis of covariance (ANCOVA), H_{05}: The duration of strobilation was independent of the number of ephyrae per strobila and H_{06}: The duration of strobilation was independent of the temperature treatments. ANCOVA-tests were conducted separately for each species. In all tests, duration was the dependent variable; the number of ephyrae per strobila and the culture temperature were covariates.

Results

Ephyra production and strobilation rates

A. aurita

The first *A. aurita* strobilations occurred in autumn in all temperature groups in the first and second years after polyp settlement. Ephyra were produced until next May in both years in the 15–10–15°C group, whereas ephyra production stopped after a temperature decrease to 5°C in January in the 15–05–15°C groups (Fig. 1, left side). Few ephyrae per polyp were produced in the temperature group with constant 15°C and the most ephyrae per polyp were produced in the 15–10–15°C group in the first and in the second year after settlement (Fig. 1, right side). Significantly higher percentages of *A. aurita* polyps strobilated after a temperature decrease in autumn than at constant 15°C, and H_{01} was rejected for both years (Figs. 2, 3; Table 2). The numbers of ephyrae produced per strobila did not differ significantly among the temperature groups in the first year or in the second year and H_{02} was not rejected (Fig. 2; Table 2). The percentages of strobilating polyps were higher in the second than in the first year, but the differences were not significant, and H_{03} was not rejected (Fig. 3; Table 3). Although the maximum numbers of ephyrae per strobila were higher in the second year than in the first, the means differed only slightly among groups, and H_{04} was not rejected (Fig. 3; Table 3).

C. capillata

C. capillata polyps never strobilated in constant 15°C during 22 months of observation. The first ephyra appeared in February–March at the cold winter temperature (5°C) in both years, whereas the first ephyra appeared at least 1 month later at the 10°C winter temperature (Fig. 1, left side). Ephyra production continued until the beginning (second year) or end of May (first year). In the first year, production of ephyrae per polyp was higher in 5°C winter temperature, whereas it was slightly higher in 10°C winter temperature in the second year (Fig. 1, right side). In the first year, the percentages of polyps strobilating were significantly higher in the 15–05–15°C group than in the 15–10–15°C and 15°C groups and H_{01} was rejected (Fig. 2; Table 2). Although the percentages of polyps strobilating differed only slightly for the 15–10–15°C and the 15–05–15°C groups, no strobilation occurred at 15°C and H_{01} was rejected in the second year as well (Fig. 3; Table 2). The numbers of ephyrae per strobila were lower in groups with 5°C than in those with 10°C winter temperature, but differences were only significant in the first year (H_{02}

rejected, Fig. 2; Table 2). The percentages of strobilating polyps did not differ significantly in polyps of different ages and H_{03} was not rejected (Fig. 3; Table 3); however, the mean numbers of ephyrae per strobila were significantly higher in the second year in the 15–10–15°C group and the 15–05–15°C group (H_{04} rejected, Fig. 3; Table 3).

C. lamarckii

In *C. lamarckii* cultures, the first ephyra appeared after a temperature decrease to 10°C in groups 15–10–15°C and 15–05–15°C; however, the most ephyrae per polyp were produced in the constant 15°C group, whereas the fewest were in the coldest winter temperature (5°C;

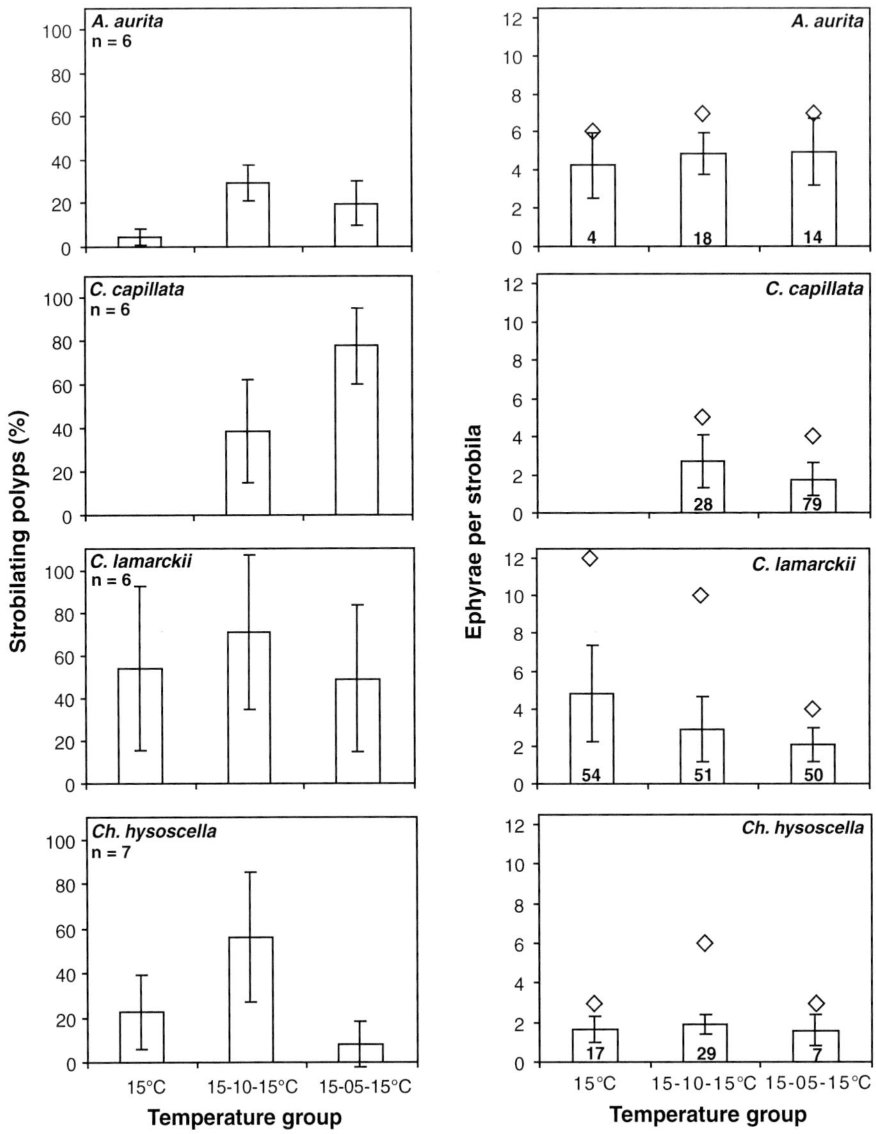

Fig. 2 Strobilating polyps (mean % ± SD) (*left side*) and ephyrae produced per strobila (mean number ± SD) (*right side*) in the first year after polyp settlement in different temperature groups of *A. aurita*, *C. capillata*, *C. lamarckii* and *Ch. hysoscella* polyps; *n* number of replicates (numbers of polyps as shown in Table 1). The numbers of analysed strobilae appear *inside the bars*. *Diamond* maximal number of ephyrae produced by one strobila. The observation time was 12 months for *A. aurita* (15 Sep–15 Sep) and *C. capillata* (15 Oct–15 Oct) and 10 months for *C. lamarckii* and *Ch. hysoscella* (15 Oct–15 Aug). Test statistics are shown in Table 2

Fig. 1). All *C. lamarckii* polyps strobilated at least once in all temperature groups (>100%). The percentages of polyps strobilating twice were the highest in the 15–10–15°C groups, but the differences were not significant, and H_{01} was not rejected (Fig. 2; Table 2). The number of ephyrae produced per strobila, however, was significantly higher at constant 15°C than in treatments with temperature decreases, and H_{02} was rejected (Fig. 2; Table 2).

Ch. hysoscella

The monthly ephyra production per polyp was relatively low in *Ch. hysoscella* (Fig. 1, left side). The most ephyrae per polyp were in the 15–10–15°C group, and the fewest ephyrae per polyp were produced in the 15–05–15°C group (Fig. 1, right side). The percentages of strobilating polyps were significantly higher in the 15–10–15°C group than in the other groups, and H_{01} was rejected (Fig. 2; Table 2). In the 15°C and 15–05–15°C groups, maxima of three ephyrae were produced per strobila, whereas as many as six ephyrae per strobila were produced in the 15–10–15°C group (Fig. 2); however, the production of more then three ephyrae occurred only in 7% of analysed strobilae. Consequently, differences in ephyra production per strobila were not significant among temperature groups and H_{02} was not rejected (Fig. 2; Table 2).

Temperature change as a strobilation inducer or inhibitor

Small percentages of polyps strobilated at constant 15°C in *A. aurita*, demonstrating that temperature decrease was an important strobilation inducer; however, ephyra production stopped after temperature decreased further to 5°C in winter. This result demonstrated that a low winter temperature of 5°C inhibited strobilation activity from February on whereas a higher winter temperature of 10°C lead to a longer strobilation period until spring (Fig. 1). Temperature increases from 5 to 10°C in spring and then to 15°C in summer did not induce strobilation in *A. aurita* (Fig. 1).

Strobilation never occurred at constant 15°C in *C. capillata* and ephyra production was lower when temperature rose in spring and summer following strobilation at 5 or 10°C winter temperatures (Fig. 1). No ephyra was produced at temperatures exceeding

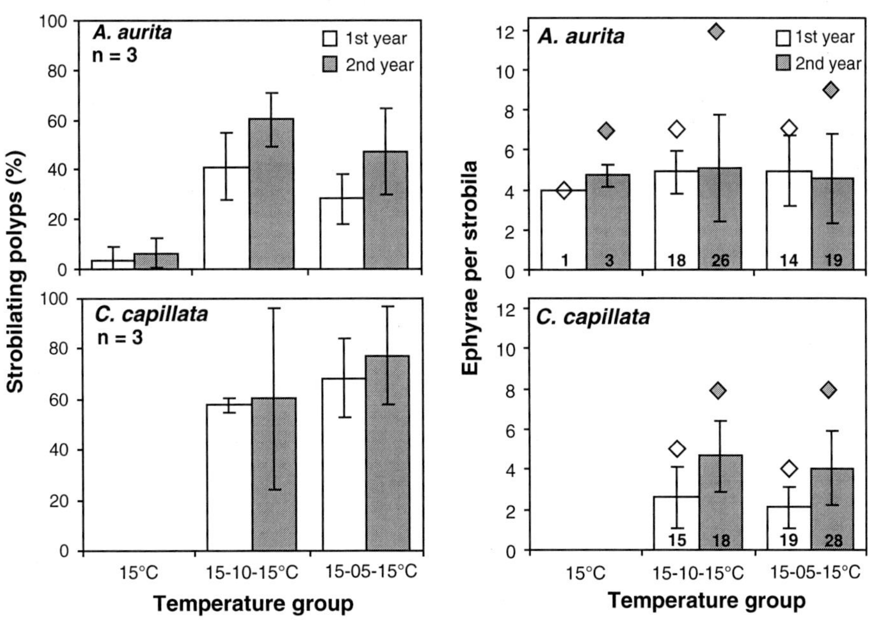

Fig. 3 Strobilating polyps (mean % ± SD) (*left side*) and ephyrae produced per strobila (mean number ± SD) (*right side*) in the first year and in the second year after polyp settlement in different temperature groups of *A. aurita*, *C. capillata*, *C. lamarckii* and *Ch. hysoscella* polyps, *n* number of replicates (numbers of polyps are shown in Table 1). The numbers of analysed strobilae appear *inside the bars*. *Diamond* maximal number of ephyrae produced by one strobila. In both analyses, the observation time was 10 months (15 Oct–15 Aug) in both years. Test statistics are shown in Table 3

10°C in summer, suggesting strobilation inhibition at warmer temperatures in *C. capillata*. In *C. lamarckii*, ephyra production started earlier after a temperature decrease to 10°C than at constant 15°C (Fig. 1), showing a positive effect of temperature decrease on strobilation induction, although the total ephyra production was the highest at constant 15°C. In *Ch. hysoscella*, strobilation activity was low in all temperature treatments without clear responses to temperature changes (Fig. 1).

Temperature effects on strobilation duration and ephyrae per strobila

A significant effect of the number of ephyrae per strobila on the strobilation duration was shown by ANCOVA for all tested species (Fig. 4), and H_{05} was rejected. Correlations among temperature treatments and strobilation duration were not possible for *A. aurita* because strobilation occurred mostly at 10°C; thus, data at other temperatures were insufficient. In all other tested species, the ANCOVA confirmed a significant effect of temperature treatment on the strobilation duration (Fig. 4), and H_{06} was rejected.

Discussion

Scyphopolyps are difficult to find and investigate in the field because of their small sizes and their preferences of colonizing the undersides of substrates and concealed habitats (Pierce, 2009; Di Camillo

Table 2 Test statistics on the effects of temperature treatments on the percentages of strobilating polyps (H_{01}) and ephyrae produced per strobila (H_{02}) in four species of scyphozoans

Species	Null hypothesis	Polyp age	Test statistic	P
A. aurita	H_{01}	1y	$F_{2,15} = 16.936$	0.001
		2y	$F_{2,6} = 14.779$	0.005
	H_{02}	1y	$H_2 = 0.573$	0.751
		2y	$H_2 = 0.315$	0.854
C. capillata	H_{01}	1y	$H_2 = 13.940$	<0.001
		2y	$H_2 = 6.161$	0.025
	H_{02}	1y	$T_{28,79} = 1936.000$	0.003
		2y	$T_{18,28} = 482.000$	0.182
C. lamarckii	H_{01}	1y	$F_{2,15} = 1.224$	0.322
	H_{02}	1y	$H_2 = 40.367$	<0.001
Ch. hysoscella	H_{01}	1y	$F_{2,18} = 9.486$	0.002
	H_{02}	1y	$H_2 = 0.477$	0.788

The temperature groups, numbers of replicates, mean values and observation times are shown in Figs. 2 and 3. Strobilations in the first year after polyp settlement (1y) were analysed in all species. Strobilations in the second year (2y) were analysed in *A. aurita* and *C. capillata*

Table 3 Test statistics of comparisons on strobilations in the first and second year after polyp settlement in two scyphozoan species

Species	Null hypothesis	15°C		15-10-15°C		15-5-15°C	
		Test statistics	P	Test statistics	P	Test statistics	P
A. aurita	H_{03}	$T_{3,3} = 9.0$	0.700	$T_{3,3} = 7.0$	0.200	$T_{3,3} = 7.0$	0.200
	H_{04}	No test, $n < 3$		$T_{18,26} = 424.5$	0.645	$T_{14,19} = 259.5$	0.439
C. capillata	H_{03}	No strobilae		$T_{3,3} = 8.0$	0.400	$T_{3,3} = 8.5$	0.400
	H_{04}	No strobilae		$T_{15,18} = 169.0$	0.002	$T_{19,28} = 292.5$	<0.001

Percentages of strobilating polyps (H_{03}) and ephyrae produced per strobila (H_{04}) in different temperature treatments were analysed. Replicates, mean values and observation times are shown in Fig. 3

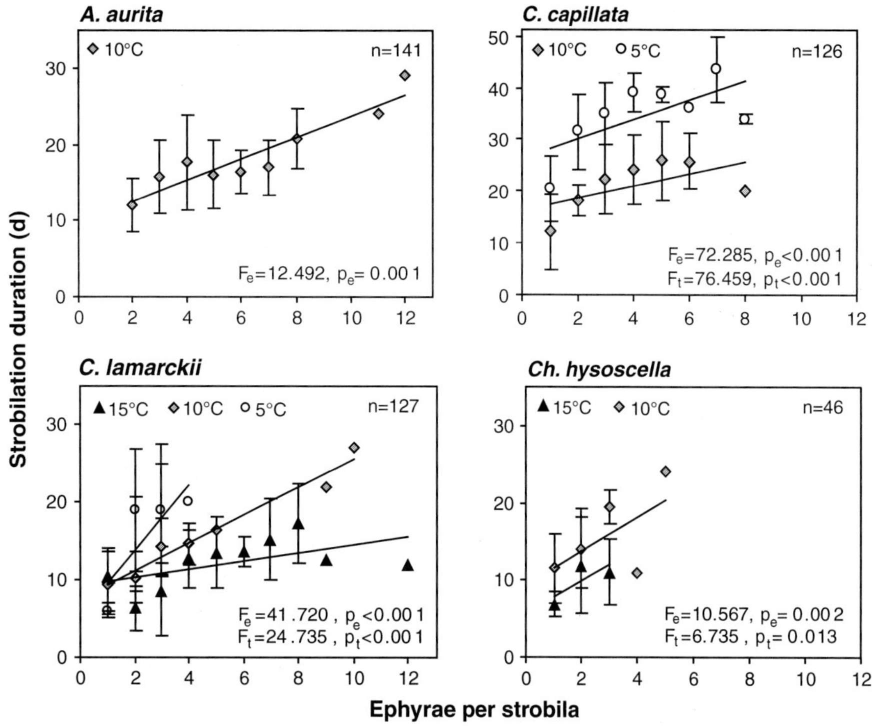

Fig. 4 Strobilation durations (mean days ± SD) in individually-monitored strobilae at different constant temperatures in relation to ephyrae numbers produced in the strobilae of *A. aurita*, *C. capillata*, *C. lamarckii* and *Ch. hysoscella*. The effect of ephyra numbers produced per strobila (e) on the strobilation duration and the effect of the temperature treatment (t) on the strobilation duration were tested by ANCOVA. *n* number of analysed strobilae. df = 1 for all performed tests. *F* and *P* values are shown in the figures

et al., 2010). Laboratory investigations allow quantification of polyp survival and reproduction and are therefore an important source of data (e.g. Purcell, 2007; Willcox et al., 2007; Liu et al., 2009; Sötje & Jarms, 2009; Holst & Jarms, 2010). Although the results from field observations may differ form laboratory experiments (Purcell et al., 2009), the results of this study demonstrated that simulation of natural temperature cycles in the laboratory can be used to test the effects of seasonal temperature changes on strobilation. The strobilation activity documented in two consecutive years in two species (*A. aurita* and *C. capillata*) showed that repeated temperature cycles in the laboratory experiments caused very similar responses of the polyps. The temperature cycles in the present experiments with 15°C maximum summer temperature and 5°C minimum winter temperature differed from the natural annual temperature cycles because of 2.5°C temperature changes per month. Nevertheless, the laboratory temperature conditions were similar to temperatures found recently in the German Bight (Wiltshire & Manly, 2004; www.bsh.de/en/Marine_data). The parallel experiments with warmer winter temperatures (10°C) indicated how warmer winter temperatures, as observed and predicted to progress in the North Sea (Belkin, 2009), could affect the strobilation activity of scyphopolyps in situ. In these experiments, warmer winter temperature (10°C in comparison to 5°C) positively affected strobilation in several ways: 1. A longer strobilation period or higher ephyra production per polyp (in *A. aurita*, *C. lamarckii* and *Ch. hysoscella*; Fig. 1); 2. Higher percentages of polyp strobilation (in *A. aurita* and *Ch. hysoscella*; Fig. 2); 3. More ephyrae per strobila (in *C. capillata*, *C. lamarckii*, see Fig. 2); 4. A shorter strobilation duration (in *C. capillata* and *C. lamarcki*, see Fig. 4). Many of the changes in species abundances, population structure and biogeographical ranges occur as a result of increased reproductive output and juvenile survival in response to increased warming (Mieszkowska et al., 2006). The results of this study suggest that this could also be true for the North Sea scyphozoans investigated.

In *A. aurita*, the experiments showed significantly higher percentages of strobilations after a temperature decrease in autumn compared to constant 15°C demonstrating the importance of a temperature decrease for strobilation induction in this species. Spring strobilation only occurred in experiments with 10°C winter temperature, but did not occur with rising temperature in spring after a cold winter period of 5°C. Perhaps under natural conditions, autumn strobilation in *A. aurita* polyps can happen only if the temperature decreases slowly and the period of moderate temperature of about 10°C lasts for several weeks, whereas a rapid temperature decrease may lead to inhibition of strobilation in autumn. In agreement with this idea, in situ autumn strobilations of *A. aurita* were observed in areas with moderate winter temperatures, such as in Sylt, German Bight (Thiel, 1938), Osterschelde, Netherlands (Korringa, 1953), and Gullmarfjord, western Sweden (Hernroth & Gröndahl, 1985, Gröndahl, 1988); however, in areas with a rapid temperature decrease in autumn, such as in several Baltic Sea areas, a later start of the strobilation activity in *A. aurita* was documented, and strobilation occurred mainly in winter and spring (Palmén, 1954; Kändler, 1961; Thiel, 1962; Rasmussen, 1973; Möller, 1980b). The ephyrae may survive the cold winter periods in deep water layers without further development, which could explain the main time of ephyra abundance in the plankton in autumn and spring (Rasmussen, 1973; Hernroth & Gröndahl, 1985; Gröndahl, 1988). In Southampton Water and Horsea Lake in southern England, ephyra production begins in December and lasts 7 months (Lucas, 1996; Lucas et al., 1997), confirming the results of this study of an extended strobilation phase in *A. aurita* at warmer winter temperatures (Fig. 1). Recent molecular genetic study detected that *A. aurita* is not a single species, but includes members of several molecular species (Dawson, 2003). *A. aurita* populations occurring in the North Atlantic and adjacent seas are probably adapted to the temperature regime in these areas (Dawson, 2003), and therefore, our results reflect the strobilation behaviour of polyps from this temperature regime only.

The tests of this study showed shorter strobilation duration and higher ephyra production per strobila of *C. capillata* polyps at warmer temperatures. On the other hand, cold winter temperatures had positive effects on strobilation in *C. capillata* polyps; strobilation started earlier at colder temperatures, and unlike *A. aurita*, ephyra production was not inhibited by the cold winter temperature (5°C). In these experiments, *C. capillata* was the only species without any strobilation during 22 months at the warm temperature of 15°C. Strobilation induction of *C. capillata* may strictly depend on a temperature decrease, which may explain why the distribution of the species is limited to northern boreal areas (Russell, 1970), whereas *Cyanea* medusae from the warmer U.S. Atlantic, the Gulf of Mexico and Australian waters may represent other species of *Cyanea* (Bayha, 2005; Dawson, 2005). In the North Sea, *C. capillata* medusae are rare in the southern part and do not appear in the English Channel. In the Irish Sea, their occurrence also is limited to the northern part (Russell, 1970; Hay et al., 1990; Doyle et al., 2007). The simulated annual seasons in present experiments induced ephyra production from February until June, which matched field observations on *C. capillata* strobilation and occurence of the ephyrae in areas with a similar temperature regime (Hartlaub, 1894; Verwey, 1942; Gröndahl, 1988). Although the polyps are obviously adapted to cold temperature conditions, they have not been reported from the southern or eastern Baltic Sea, where the *C. capillata* medusae appear each summer (Möller, 1980a; Barz et al. 2006). Recent studies demonstrated that *C. capillata* polyps are able to strobilate at a low salinity of 12 and may be more widespread in the Baltic Sea than previously thought (Holst & Jarms, 2010). In accordance with previous studies (Gröndahl, 1988), the results of this study confirm that *C. capillata* polyps are tolerant of cold temperatures, indicating that the low Baltic Sea winter temperatures probably do not limit their distribution. I therefore believe that the only reason why the polyps have not been found in this area to date is that there were too few efforts undertaken to find them.

The medusae of *C. lamarckii* have a more southern distribution in the North and Irish seas than do *C. capillata* medusae (Russell, 1970; Hay et al., 1990; Doyle et al., 2007). In the Baltic Sea, *C. lamarckii* medusae appear rarely in Danish waters (Rasmussen, 1973), whereas they occur periodically from March until June off the Swedish west coast (Gröndahl, 1988). *C. lamarckii* ephyrae were not found in 4 years of plankton sampling on the Swedish west coast and no strobilation was observed on settling plates in the field, leading to the conclusion that *C. lamarckii* polyps do not strobilate there (Gröndahl, 1988). *C. lamarckii* polyps have not yet been described in their natural habitat, but present laboratory studies indicate positive effects of warm temperature on strobilation: higher production of ephyrae per strobila and shorter strobilation duration. This may enable their distribution to expand to the northern North Sea with rising winter temperatures due to climate changes. From there, the polyps may also extend into the Baltic Sea because of their high tolerance of low salinity (Holst & Jarms, 2010). The present experiments demonstrated a long strobilation phase of *C. lamarckii* polyps from winter until the next summer in agreement with observations of abundant *C. lamarckii* medusae of various sizes and different developmental stages from spring until late summer in the German Bight and off the Dutch coast (Verwey, 1942; Künne, 1952).

Ch. hysoscella medusae also are distributed mainly in the southern North and Irish seas (Russell, 1970; Hay et al., 1990; Doyle et al., 2007). Verwey (1942), suggested strobilation of *Ch. hysoscella* with increasing temperature in spring and summer, but the polyps have not yet been found in their natural habitat in the North Sea. A mild winter in 1988 in the German Bight was followed by early appearance and high abundance of *Ch. hysoscella* medusae in the summer, leading to the conclusion that polyp's survival during the winter was higher at warm temperatures (Merck, 1990). These observations agree with experimental results of this study showing very low ephyra production at cold winter temperatures and shorter strobilation duration at warmer temperatures. *Ch. hysoscella* planulae are able to settle at salinities at least as low as 20 (Holst & Jarms, 2010), and thus *Ch. hysoscella* medusae and polyps may be able to spread from the North Sea into the Baltic Sea with continued climate warming.

The number of ephyrae produced by each strobila is affected by temperature, other abiotic factors, food supply and polyp size (Russell, 1970; Purcell et al., 1999). The results of this study agree with those of previous laboratory experiments on polyp cultures demonstrating that warm temperatures increase the strobilation rates of polyps and ephyra production, except at very high temperatures (Purcell et al., 1999, 2012; Purcell, 2007; Liu et al., 2009). The number of ephyrae in the strobila increases with polyp's size (Russell, 1970). This may explain why the *C. capillata* polyps in the present experiments were able to produce more ephyrae by the second year when they grew to a larger size. Salinity also is known to affect strobilation rates of North Sea Scyphozoa (Holst & Jarms, 2010) and other species (reviewed in Purcell et al., 2009). Purcell et al. (2009) concluded that the combined effects of temperature, salinity, light and food determined the amount and time of strobilation in situ.

Only a few recent studies have monitored polyps in situ (see Purcell et al., 2009; Di Camillo et al., 2010; Ishii & Katsukoshi, 2010). Climatic changes in the North Sea related to the North Atlantic Oscillation (NAO) affect jellyfish abundances and may also affect the strobilation of the benthic polyps in this area (Lynam et al., 2010); however, the effects of the NAO depends on the depth that macrozoobenthos animals occur, and therefore, the estimation of the NAO's effect on strobilation is not possible without knowing the locations and depths of polyp habitats (Lynam et al., 2005). The effect of the NAO on North Sea polyps is likely to be high because strobilation occurs mainly in winter and spring when the NAO's influence is the greatest in the North Sea (Lynam et al., 2010). The results of the present study support the idea that variable winter temperatures affect the strobilation activity of North Sea Scyphozoa.

Increasing winter temperature probably will affect the abundances and distributions of scyphozoan jellyfish species in the North Sea. The more southerly species, *C. lamarckii* and *Ch. hysoscella* could expand to the northern parts of the North Sea and possibly into the Baltic Sea. The adaptable species *A. aurita* presumably will benefit from warmer temperatures, having longer strobilation periods (present study), faster growth due to higher feeding rates (Hansson, 1997; Widmer, 2005) and higher reproduction rates of medusae (Ishii & Takagi, 2003). The cold water *C. capillata* might be the only North Sea scyphozoan that could suffer from warmer temperatures; however, *C. capillata* ephyra production also occurred at the warm winter temperature of 10°C (Fig. 1). In general, I assume that the abundances of scyphozoan jellyfish in the North and Baltic seas will increase in future years if the water temperatures continue to increase as predicted (Belkin, 2009). This assessment agrees with the opinion of other authors suggesting an increase of gelatinous predators (scyphomedusae, hydromedusae, siphonophores and ctenophores) in the North Sea due to climate changes, including increasing temperatures, the reduction of the ocean pH and probably, the increasing Atlantic inflow into the North Sea (Attrill et al., 2007; Boersma et al., 2007; Doyle et al., 2008; Lilley et al., 2009; Licandro et al., 2010). The results of this study and previous studies clearly show the linkage between physical environmental factors and ephyra production and thus forecasts of the abundance and distribution of scyphomedusae might be possible by circulation models in future (Johnson et al., 2001, 2005; Barz et al., 2006). More knowledge on the locations of polyp habitats and on the reproduction cycles of the benthic polyp stage in the field is necessary for successful monitoring and understanding the population dynamics in scyphozoan jellyfish.

Acknowledgments I thank PD Dr. Gerhard Jarms for his support, for imparting his knowledge on polyp culturing and for providing the research facilities. I am grateful to the Biologische Anstalt Helgoland (Alfred Wegener Institute) for providing the guest laboratory and I am thankful for the assistance of students and technicians of the Biocenter Grindel. I thank Drs. Ilka Sötje, Jennifer Purcell, and the reviewers for their helpful comments on the manuscript, and Dr. Caroline Stolter for her help in the statistics. This study is part of a Ph. D. thesis conducted at the Biocenter Grindel and Zoological Museum in Hamburg supported by EUROGEL (EUROpean GELatinous Zooplankton, European Commission Contract no. EVK-CT-2002-00074).

References

Attrill, M. J., J. Wright & M. Edwards, 2007. Climate-related increases in jellyfish frequency suggest a more gelatinous future for the North Sea. Limnology and Oceanography 52: 480–485.

Barz, K. & H.-J. Hirche, 2007. Abundance, distribution and prey composition of scyphomedusae in the southern North Sea. Marine Biology 151: 1021–1033.

Barz, K., H.-H. Hinrichsen & H.-J. Hirche, 2006. Scyphozoa in the Bornholm Basin (central Baltic Sea) – the role of advection. Journal of Marine Systems 60: 167–176.

Bayha, K. M., 2005. The molecular systematics and population genetics of four coastal ctenophores and scyphozoan jellyfish of the United States Atlantic and Gulf of Mexico. Dissertation, University of Delaware: 283 pp.

Belkin, I. M., 2009. Rapid warming of large marine ecosystems. Progress in Oceanography 81: 207–213.

Boero, F., J. Boullion, C. Gravilli, M. P. Miglietta, T. Parsons & S. Piraino, 2008. Gelatinous plankton: irregularities rule the world (sometimes). Marine Ecology Progress Series 356: 299–310.

Boersma, M., A. M. Malzahn, W. Greve & J. Javidpour, 2007. The first occurrence of the ctenophore *Mnemiopsis leidyi* in the North Sea. Helgoland Marine Research 61: 153–155.

Brewer, R. H. & J. S. Feingold, 1991. The effect of temperature on the benthic stages of *Cyanea* (Cnidaria: Scyphozoa), and their seasonal distribution in the Niantic River estuary, Connecticut. Journal of Experimental Marine Biology and Ecology 152: 49–60.

CIESM, 2001. Gelatinous zooplankton outbreaks: theory and practice. CIESM Workshop Series 14: 112 pp.

Dawson, M. N., 2003. Macro-morphological variation among cryptic species of the moon jellyfish, *Aurelia* (Cnidaria: Scyphozoa). Marine Biology 143: 369–379.

Dawson, M. N., 2005. *Cyanea capillata* is not a cosmopolitan jellyfish: morphological and molecular evidence for *C. annaskala* and *C. rosea* (Scyphozoa: Semaeostomeae: Cyaneidae) in south-eastern Australia. Invertebrate Systematics 19: 361–370.

Di Camillo, C. G., F. Betti, M. Bo, M. Martinelli, S. Puce & G. Bavestrello, 2010. Contribution to the understanding of seasonal cycle of *Aurelia aurita* (Cnidaria: Scyphozoa) scyphopolyps in the northern Adriatic Sea. Journal of the Marine Biological Association of the United Kingdom 90: 1105–1110.

Doyle, T. K., J. D. R. Houghton, S. M. Buckley, G. C. Hays & J. Davenport, 2007. The broad-scale distribution of five jellyfish species across a temperate coastal environment. Hydrobiologia 579: 29–39.

Doyle, T. K., H. De Haas, D. Cotton, B. Dorschel, V. Cummins, J. D. R. Houghton, J. Davenport & G. C. Hays, 2008. Widespread occurrence of the jellyfish *Pelagia noctiluca* in Irish coastal and shelf waters. Journal of Plankton Research 30: 963–968.

Gröndahl, F., 1988. A comparative ecological study on the scyphozoans *Aurelia aurita*, *Cyanea capillata* and *Cyanea lamarckii* in the Gullmar Fjord, western Sweden. Marine Biology 97: 541–550.

Hansson, L. J., 1997. Effect of temperature on growth rate of *Aurelia aurita* (Cnidaria, Scyphozoa) from Gullmarsfjorden, Sweden. Marine Ecology Progress Series 161: 145–153.

Hartlaub, C., 1894. Die Coelenteraten Helgolands. Wissenschaftliche Meeresuntersuchungen Helgoland Neue Folge 1: 161–206.

Hay, S. J., J. R. G. Hislop & A. M. Shanks, 1990. North Sea scyphomedusae; summer distribution, estimated biomass and significance particularly for 0-Group gadoid fish. Netherlands Journal of Sea Research 25: 113–130.

HELCOM, 2007. Climate change in the Baltic Sea Area – HELCOM thematic assessment in 2007. Baltic Sea Environment Proceedings 111: 1–49.

Hernroth, L. & F. Gröndahl, 1985. On the biology of *Aurelia aurita* (L.): 2. Major factors regulating the occurrence of ephyrae and young medusae in the Gullmar Fjord, western Sweden. Bulletin of Marine Science 37: 567–576.

Holst, S. & G. Jarms, 2007. Substrate choice and settlement preferences of planula larvae of five Scyphozoa (Cnidaria) from German Bight, North Sea. Marine Biology 151: 863–871.

Holst, S. & G. Jarms, 2010. Effects of low salinity on settlement and strobilation of Scyphozoa (Cnidaria): is the lion's mane *Cyanea capillata* (L.) able to reproduce in the brackish Baltic Sea? Hydrobiologia 645: 53–68.

Holst, S., I. Sötje, H. Tieman & G. Jarms, 2007. Life cycle of the rhizostome jellyfish *Rhizostoma octopus* (L.) (Scyphozoa, Rhizostomeae), with studies on cnidocysts and statoliths. Marine Biology 151: 1695–1710.

IPCC, 2007. Summary for policymakers. In Solomon, S., D. Qin, M. Manning, Z. Chen, M. Marquis, K. B. Averyt, M. Tignor & H. L. Miller (eds), Climate Change 2007: The Physical Science Basis. Contribution of Working Group I to the Fourth Asessment Report of the Intergovernmental Panel on Climate Change. Cambridge University Press, Cambridge.

Ishii, H. & K. Katsukoshi, 2010. Seasonal and vertical distribution of *Aurelia aurita* polyps on a pylon in the innermost part of Tokyo Bay. Journal of Oceanography 66: 329–336.

Ishii, H. & A. Takagi, 2003. Development time of planula larvae on the oral arms of the scyphomedusa *Aurelia aurita*. Journal of Plankton Research 25: 1447–1450.

Jarms, G., 2010. The early life history of Scyphozoa with emphasis on Coronatae a review with a list of described life cycles. Verhandlungen des Naturwissenschaftlichen Vereins Hamburg 45: 17–31.

Johnson, D. R., H. M. Perry & W. D. Burke, 2001. Developing jellyfish strategy hypotheses using circulation models. Hydrobiologia 451: 213–221.

Johnson, D. R., H. M. Perry & W. M. Graham, 2005. Using nowcast model currents to explore transport of non-indigenous jellyfish into the Gulf of Mexico. Marine Ecology Progress Series 305: 139–146.

Kändler, R., 1961. Über das Vorkommen von Fischbrut, Decapodenlarven und Medusen in der Kieler Förde. Kieler Meeresforschung 17: 48–65.

Korringa, P., 1953. The shell of *Ostrea edulis* as a habitat. Archives Neerlandaises de Zoologie 10: 32–59.

Künne, C., 1952. Untersuchungen über das Großplankton in der Deutschen Bucht und im Nordsylter Wattenmeer. Helgoländer Wissenschaftliche Meeresuntersuchungen 4: 1–54.

Licandro, P., D. V. P. Conway, M. N. Daly Yahia, M. L. Fernandez De Puelles, S. Gasparini, J. H. Hecq, P. Tranter & R. R. Kirby, 2010. A blooming jellyfish in the northeast Atlantic and Mediterranean. Biology Letters 6: 688–691.

Lilley, M. K. S., J. D. R. Houghton & G. C. Hays, 2009. Distribution, extent of inter-annual variability and diet of the bloom-forming jellyfish *Rhizostoma* in European waters. Journal of the Marine Biological Association of the United Kingdom 89: 39–48.

Liu, W.-C., W.-T. Lo, J. E. Purcell & H.-H. Chang, 2009. Effects of temperature and light intensity on asexual reproduction of the scyphozoan, *Aurelia aurita* (L.) in Taiwan. Hydrobiologia 616: 247–258.

Lucas, C. H., 1996. Population dynamics of *Aurelia aurita* (Scyphozoa) from an isolated brackish lake, with particular reference to sexual reproduction. Journal of Plankton Research 18: 987–1007.

Lucas, C. H., A. G. Hirst & J. A. Williams, 1997. Plankton dynamics and *Aurelia aurita* production in two contrasting ecosystems: comparison and consequences. Estuarine, Coastal and Shelf Science 45: 209–219.

Lynam, C. P., S. J. Hay & A. S. Brierley, 2005. Jellyfish abundance and climatic variation: contrasting responses in oceanographically distinct regions of the North Sea, and possible implications for fisheries. Journal of the Marine Biological Association of the United Kingdom 85: 435–450.

Lynam, C. P., M. J. Attrill & M. D. Skogen, 2010. Climatic and oceanic influences on the abundance of gelatinous zooplankton in the North Sea. Journal of the Marine Biological Association of the United Kingdom 90: 1153–1159.

Merck, T., 1990. Untersuchungen zur ökologischen Nische von *Chrysaora hysoscella*. Jahresbericht der Biologischen Anstalt Helgoland 1989: 53–54.

Mieszkowska, N., R. Leaper, P. Moore, M. A. Kendall, M. T. Burrows, D. Lear, E. Poloczanska, K. Hiscock, P. S. Moschella, R. C. Thompson, R. J. Herbert, D. Lafolley, J. Baxter, A. J. Southward & S. J. Hawkins, 2006. Marine biodiversity and climate change. Assessing and predicting the influence of climate change using intertidal rocky shore biota. Final report for United Kingdom funders. Marine Biological Association Occasional Publications 20: 1–53.

Miyake, H., M. Terazaki & Y. Kakinuma, 2002. On the polyps of the common jellyfish *Aurelia aurita* in Kagoshima Bay. Journal of Oceanography 58: 451–459.

Molinero, J. C., M. Casini & E. Buecher, 2008. The influence of the Atlantic and regional climate variability on the long-term changes in gelatinous carnivore populations in the northwestern Mediterranean. Limnology and Oceanography 53: 1456–1467.

Möller, H., 1980a. A summer survey of large zooplankton, particularly scyphomedusae, in North Sea and Baltic. Meeresforschung 28: 61–68.

Möller, H., 1980b. Population dynamics of *Aurelia aurita* medusae in Kiel Bight, Germany (FRG). Marine Biology 60: 123–128.

Palmén, E., 1954. Seasonal occurrence of ephyrae and subsequent instars of *Aurelia aurita* (L.) in the shallow waters of Tvärminne, S. Finland. Archivum Societatis Zoologicae Botanicae Fennicae Vanamo 8: 122–138.

Pierce, J., 2009. Prediction, location, collection and transport of jellyfish (Cnidaria) and their polyps. Zoo Biology 28: 163–176.

Purcell, J. E., 2005. Climate effects on formation of jellyfish and ctenophore blooms: a review. Journal of the Marine Biological Association of the United Kingdom 85: 461–476.

Purcell, J. E., 2007. Environmental effects on asexual reproduction rates of the scyphozoan *Aurelia labiata*. Marine Ecology Progress Series 348: 183–196.

Purcell, J. E., 2009. Extension of methods for jellyfish and ctenophore trophic ecology to large-scale research. Hydrobiologia 616: 23–50.

Purcell, J. E., 2011. Jellyfish and ctenophore blooms coincide with human proliferations and environmental perturbations. Annual Review of Marine Science. doi:10.1146/annurev-marine-120709-142751.

Purcell, J. E., J. R. White, D. A. Nemazie & D. A. Wright, 1999. Temperature, salinity and food effects on asexual reproduction and abundance of the scyphozoan *Chrysaora quinquecirrha*. Marine Ecology Progress Series 180: 486–487.

Purcell, J. E., S.-I. Uye & W.-T. Lo, 2007. Anthropogenic causes of jellyfish blooms and their direct consequences for humans: a review. Marine Ecology Progress Series 350: 153–174.

Purcell, J. E., R. A. Hoover & N. T. Schwarck, 2009. Interannual variation of strobilation by the scyphozoan *Aurelia labiata* in relation to polyp density, temperature, salinity, and light conditions in situ. Marine Ecology Progress Series 375: 139–149.

Purcell, J. E., D. Atienza, V. Fuentes, A. Olariaga, U. Tilves, C. Colahan, J.-M. Gili, 2012. Temperature effects on asexual reproduction rates of scyphozoan polyps from the northwest Mediterranean Sea. Hydrobiologia, this issue. doi:10.1007/s10750-012-1047-7.

Rasmussen, E., 1973. Systematic and ecology of the Isefjord marine fauna (Denmark). Ophelia 11: 1–507.

Richardson, A. J., A. Bakun, G. C. Hays & M. J. Gibbons, 2009. The jellyfish joyride: causes, consequences and management responses to a more gelatinous future. Trends in Ecology & Evolution 24: 312–322.

Russell, F. S., 1970. The medusae of the British Isles II. Pelagic Scyphozoa with a supplement to the first volume on hydromedusae. Cambridge University Press, Cambridge.

Sabatés, A., F. Pagès, D. Atienza, V. Fuentes, J. E. Purcell & J.-M. Gili, 2010. Planktonic cnidarian distribution and feeding of *Pelagia noctiluca* in the NW Mediterranean Sea. Hydrobiologia 645: 153–165.

Sötje, I. & G. Jarms, 2009. Derivation of the reduced life cycle of *Thecoscyphus zibrowii* Werner, 1984 (Cnidaria, Scyphozoa). Marine Biology 156: 2331–2341.

Thiel, M. E., 1938. Scyphomedusae. In Bronn, H. G. (ed.), Dr. H. G. Bronns Klassen und Ordnungen des Tierreichs. Akademische Verlagsgesellschaft, Leipzig.

Thiel, H., 1962. Untersuchungen über die Strobilisation von *Aurelia aurita* LAM. an einer Population der Kieler Förde. Kieler Meeresforschungen 18: 198–230.

Tronolone, V. B., A. C. Morandini & A. E. Bigotto, 2002. On the occurrence of scyphozoan ephyrae (Cnidaria, Scyphozoa, Semaestomeae and Rhizostomeae) in the southeastern Brazilian coast. Biota Neotropica 2: 1–18.

Verwey, J., 1942. Die Periodizität im Auftreten und die aktiven und passiven Bewegungen der Quallen. Archives Neerlandaises de Zoologie 6: 363–468.

Widmer, C. L., 2005. Effects of temperature on growth of north-east Pacific moon jellyfish ephyrae, *Aurelia labiata* (Cnidaria: Scyphozoa). Journal of the Marine Biological Association of the United Kingdom 85: 569–573.

Willcox, S., N. Moltschaniwskyj & C. Crawford, 2007. Asexual reproduction in scyphistomae of *Aurelia* sp.: effects of temperature and salinity in an experimental study. Journal of Experimental Marine Biology and Ecology 353: 107–114.

Wiltshire, K. H. & B. F. J. Manly, 2004. The warming trend at Helgoland Roads, North Sea: Phytoplankton response. Helgoland Marine Research 58: 269–273.

Wiltshire, K. H., A. Kraberg, I. Bartsch, M. Boersma, H.-D. Franke, J. Freud, C. Gebühr, G. Gerdts, K. Stockmann & A. Wichels, 2010. Helgoland Roads, North Sea: 45 years of change. Estuaries and Coasts 33: 295–310.

You, K., C. Ma, H. Gao, G. Li, M. Zhang, B. Wang & R. Wei, 2008. The effects of temperature decrease on the scyphistomae strobilation of Jellyfish, *Rhopilema esculentum* Kishinouye. Journal of the World Aquaculture Society 39: 706–711.

Hydrobiologia (2012) 690:141–155
DOI 10.1007/s10750-012-1036-x

JELLYFISH BLOOMS

Ecological aspects of early life stages of *Cotylorhiza tuberculata* (Scyphozoa: Rhizostomae) affecting its pelagic population success

Diana Astorga · Javier Ruiz · Laura Prieto

Published online: 9 March 2012
© Springer Science+Business Media B.V. 2012

Abstract *Cotylorhiza tuberculata* is a common symbiotic scyphozoan in the Mediterranean Sea. The medusae occur in extremely high abundances in enclosed coastal areas in the Mediterranean Sea. Previous laboratory experiments identified thermal control on its early life stages as the driver of medusa blooms. In the present study, new ecological aspects were tested in laboratory experiments that support the pelagic population success of this zooxanthellate jellyfish. We hypothesized that planulae larvae would have no settlement preference among substrates and that temperature would affect ephyra development, ingestion rates and daily ration. The polyp budding rate and the onset of symbiosis with zooxanthellae also were investigated. Transmission electron microscopy revealed that zooxanthella infection occurred by the polyp stage. Our results showing no substrate selectivity by planulae and high polyp budding rates in high temperatures suggest increased benthic polyp populations, which would lead to higher medusa abundances. Rates of transition from ephyrae to medusae and the feeding of early medusa stages also increased with temperature. Continuing changes in coastal ecosystems such as future climate warming and marine construction may lead to increased populations of jellyfish to the detriment of fish globally.

Keywords Jellyfish · Mediterranean sea · Planulae settlement · Zooxanthellae · Feeding · Growth · Reproduction

Introduction

The worldwide proliferation of marine jellyfish has become a crucial ecological and social issue in recent decades. Most jellyfish compete with fish for food resources and are potential predators of fish eggs and larvae (Möller, 1980). Some gelatinous species appear to be responsible for abrupt changes in the species abundance and composition of zooplankton, ichthyoplankton and/or fish (Vinogradov & Shushkina, 1992; Pérez-Ruzafa et al., 2002; Richardson et al., 2009). Mass occurrences of jellyfish are numerous (Hamner & Dawson, 2009) and increasingly interfere with economic and recreational activities. Jellyfish have been reported to clog fishing nets, spoil commercial catches, cause serious damage to aquaculture, clog the cooling systems of coastal power plants, and sting or even kill tourist swimmers (Arai, 1997; Mills, 2001; Uye & Ueta, 2004; Hay, 2006; Purcell et al., 2007).

Guest editors: J. E. Purcell, H. Mianzan & J. R. Frost / Jellyfish Blooms: Interactions with Humans and Fisheries

D. Astorga · J. Ruiz · L. Prieto (✉)
Instituto de Ciencias Marinas de Andalucía (CSIC),
República Saharaui 2, 11519 Puerto Real, Cádiz, Spain
e-mail: laura.prieto@icman.csic.es

Concern about gelatinous outbreaks has resulted in extensive recent scientific interest (Mills, 2001; Shiganova et al., 2001; Purcell, 2005; Purcell et al., 2007; Pitt & Purcell, 2009; Richardson et al., 2009). Several factors have been proposed to explain their occurrence including eutrophication (Arai, 2001), an increase in hard substrates for polyp attachment (Parsons & Lalli, 2002; Holst & Jarms, 2007), exotic translocations (Purcell et al., 2001), over-fishing (Pauly et al., 2002) and climate change (Purcell, 2005). The underlying causes of blooms are difficult to determine because the processes involved are not mutually exclusive and the conclusions may depend on the focus of the study (i.e. global or local-scale; Gibbons & Richardson, 2009).

Cotylorhiza tuberculata (Macri, 1778) is a common symbiotic rhizostome scyphozoan from the Mediterranean Sea. The medusae reach very high abundances in enclosed areas such as Vlyho Bay in the Ionian Island of Lefkada-Greece (Kikinger, 1992) and the Mar Menor coastal lagoon in the western Mediterranean Sea where annual blooms have been observed since 1995 (Pérez-Ruzafa et al., 2002). Kikinger (1992) described the life history of the population of *C. tuberculata* from Lefkada Island. Prieto et al. (2010) parameterized the life cycle of *C. tuberculata* from the Mar Menor within the context of global warming and highlighted thermal control as the mechanism driving medusa blooms; low winter temperatures, which reduced polyp survival, and abrupt warming, which triggered strobilation in springtime, determined the abundance of medusae in summer. Thus, milder winters and hotter summers, as predicted by future climate scenarios, may increase blooms of this jellyfish (Prieto et al., 2010).

The life cycle of *C. tuberculata*, as in most scyphozoans, includes a benthic asexual phase and a sexually dimorphic pelagic phase. The free-swimming planulae liberated after internal fertilization is a relatively fast and resistant larval stage (Prieto et al., 2010) that ends when it reaches a suitable surface for attachment and develops into a polyp. The natural aggregating tendency of settling planulae (Kikinger, 1992) and the asexual reproduction by lateral budding of the resulting polyps can lead to formation of colonies with hundreds of individuals (Kikinger, 1992; Prieto et al., 2010). Polyps produce a single bud at a time and do not reproduce asexually by podocyst formation. Ephyrae originate from polyps after environmental changes trigger monodisc strobilation. Budding and strobilation processes do not occur simultaneously in this species (Astorga et al., unpublished data) and the rate of re-strobilation is minimal, resulting in only one ephyra per polyp per year (Prieto et al., 2010).

One hypothesis proposed for increasing jellyfish outbreaks is increased artificial hard substrates for polyp attachment in coastal areas (Parsons & Lalli, 2002). The assessment of the settlement preferences of planulae may help us determine if this explanation applies to *C. tuberculata* in the Mar Menor. The onset of blooms in the lagoon came after a shift in benthic vegetation with an increase in *Cymodocea nodosa* (Ucria) Ascherson during the 1980s (Pérez-Ruzafa et al., 2002) after the enlargement of the El Estacio channel (Pérez-Ruzafa & Marcos, 1992). If planulae have higher settlement and survival rates on live seagrass blades, then the rise of jellyfish blooms in the lagoon may be related to an increase in these natural attachment sites.

The presence of symbiotic dinoflagellates could also be required for polyp strobilation in symbiotic scyphomedusae (Table 1). The absence of strobilation in aposymbiotic polyps (Kikinger, 1992) suggests that zooxanthellae have a crucial role in the transition between the benthic and pelagic phases in *C. tuberculata*. Given that one polyp results in one ephyra, the proliferation capacity of this species depends directly on the strobilation success of the polyp population. Therefore, the onset of zooxanthellae infection could be of great importance in determining the success of the pelagic population of this species.

Temperature regulation was found to be the physical mechanism controlling polyp survival and strobilation in *C. tuberculata* (Prieto et al., 2010); however, effects on ephyra development and the consequences for medusa population success were not investigated. Ephyra, metaephyra and small medusa correspond to the sequence of stages during growth of the pelagic phase of scyphozoans. These stages are distinguishable by the development of the central disc with respect to the total body diameter, the degree of maturation of the oral system and shaping up of the umbrella (Kikinger, 1992; Prieto et al., 2010; Straehler-Pohl & Jarms, 2010). The influence of temperature on growth and ingestion during these three early medusa stages is unknown.

In this study, we hypothesized that *C. tuberculata* planulae larvae would have no settlement preference among substrates and that temperature would affect ephyra development, ingestion rates and daily ration.

Table 1 Strobilation requirements of some symbiotic rhizostome scyphomedusae

Species	Strobilation type	Zooxanthellae	Temperature	Preconditioning	Special inducers	*n*
Cassiopea andromeda	Monodiscous[a]	Not essential* (aposymbiotic planulae, symbiotic/aposymbiotic polyps)[a–c] Morphogenic effect: lower temperatures for strobilation[c]	Increase from 20 to 24°C[a] Increase from 18 to 20–30°C[c]		Accumulation of polyp factor, facilitated by zooxanthella metabolite, enables strobilation in aposymbiotic polyps[c]. Iodine[d]	>1[c]
Catostylus mosaicus	Monodiscous and polydiscous[e]	Not essential (may be absent in the whole life cycle)[e,f]	Temperature variation does not initiate strobilation[e]	Polyps need to be hanging in an inverted position[e] Food abundance[e]	Strobilation not attributed to variation in photoperiod or salinity[e]	1[e]
Cotylorhiza tuberculata	Monodiscous[g,h]	Indispensable (aposymbiotic polyps do not strobilate)[i]	Increase from 20 to 24°C[i] Increase from 17.5 to 20[j]	Zooxantellae infection[i] Food availability[i]	Potassium iodide[j]	>1[j]
Mastigias papua	Monodiscous[k]	Indispensable* (absent in eggs and planulae, aposymbiotic polyps obtained in laboratory)[k]	Increase from 20 to 25, 28–29°C[k] 20°C critical[l]	Precooling: 1 month at 20°C[l]		>1[k]
Phyllorhiza punctata	Monodiscous[m]	Not essential (symbiotic and aposymbiotic medusae)[n]	Increase from 16 to 24°C[o]		Special interaction between salinity and temperature[o]	>1[o]

Strobilation type, role of zooxanthellae, temperature change, preconditioning factors, special inducers, and number of strobilations (*n*) in life cycle are detailed

*Fast multiplication of algae related to beginning of strobilation (colour of strobilae)

[a] Hofmann et al. (1978); [b] Ludwig (1969); [c] Rahat & Adar (1980); [d] Pierce (2005); [e] Pitt (2000); [f] Pitt et al. (2005); [g] Clauss (1890); [h] Clauss (1893); [i] Kikinger (1992); [j] Prieto et al. (2010); [k] Sugiura (1964); [l] Sugiura (1965); [m] Hofmann & Crow (2002); [n] Galil et al. (2009); [o] Rippingale & Kelly (1995)

In addition, the polyp population increase by budding was investigated, and transmission electron microscopy on planulae and polyps was used to identify the life phase at which zooxanthellae infection occurred. All these ecological aspects provided insights into the factors controlling the pelagic population success of this zooxanthellate jellyfish.

Materials and methods

Benthic stage

Cotylorhiza tuberculata planula spatial and substrate settlement preferences, the suitability of *Cymodocea nodosa* as polyp attachment surface, and the budding capacity of polyps were analysed. We tested the null hypothesis that *C. tuberculata* planulae had no preference among substrates. First, planula larvae were removed from gravid female medusae collected in the Mar Menor coastal lagoon in late September 2006 (Experiment 1) and October 2010 (Experiment 2). Planulae were gathered in a container filled with unfiltered seawater from their natural habitat (temperature: 20 and 22°C, salinity: 47 and 39, for Experiments 1 and 2, respectively). Replicates of 150 ml of mixed planula-rich seawater were allocated to different cylindrical glass flasks of 6.5-cm diameter with a 5-cm water column (approximately 120 and 170 planulae per flask in Experiments 1 and 2, respectively). In addition to the container surfaces, a glass slide of 7.5 × 2.5 cm was placed diagonally to enable inverted settlement in six replicates to test for spatial preferences for planula settlement (Experiment 1).

A glass slide (7.5 × 2.5 cm), a small stone (~1.8 × 1.3 × 0.5 cm), and one piece each of brick (~1.7 × 1.4 × 0.5 cm), wood (~0.5 × 6 cm) and shell (half of empty shell of *Pholas dactylus* Linnaeus: 1 × 2 cm) were offered to planulae as hard substrates to determine their substrate preferences (Experiment 2). The glass slide and the wood stick were placed diagonally, and other substrates were placed on the bottom in each of three flasks. The inside of the shell faced down and enabled free movement of planulae on all surfaces. When no planulae remained in the experimental flasks (polyp numbers corresponding to ~40% of the introduced planulae; Prieto et al., 2010), the different substrates were transferred individually to new containers, and the number of polyps attached on all exposed surfaces counted with the aid of a stereomicroscope. The polyp abundance was standardized by area of exposed substrate surface.

The ability of polyps to settle on *Cymodocea nodosa* leaves was tested in Experiment 3. Two freshly collected plants that each included a horizontal rhizome and four ramets were placed in a glass aquarium filled with 3.2 l of filtered seawater (base: 14 × 17 cm, water column height: 13 cm, salinity: 38) with two Petri dishes (diameter: 9 cm, height: 1.3 cm) each holding one rhizomes and roots on the bottom and keeping the plants in a natural upward position. The plants had a total of 22 leaves (mean leaf dimensions: 12 × 0.3 cm). 379 detached polyps and 69 swimming buds were introduced in the aquarium right below the water surface. Four days later, the number of polyps attached to *C. nodosa* leaves and other available substrates were counted visually to avoid polyp detachment by manipulation. Polyp abundance per unit of available substrate surface was calculated.

Asexual reproduction of polyps by budding was evaluated in the laboratory. The polyp culture was maintained at a constant temperature of 17.5°C, salinity 38, with a photoperiod of 12:12, which ensured asexual reproduction only by budding (Prieto et al., 2010). An IBERCER F-4 incubator provided a light intensity of 360 μmol quanta m^{-2} s^{-1} by means of four Philips master TL-D 18 W/840 fluorescence lamps. Polyps were kept without aeration in 6.5-cm-diameter cylindrical glass flasks with 150 ml of 45-μm-filtered seawater that contained a glass slide leaning diagonally bottom-up. Rotifers (~400 l^{-1}) were provided as prey once per week. One hour after feeding, the rearing medium was exchanged with new, aerated water. Four replicates, each with 22 polyps, were monitored with the aid of a stereomicroscope at intervals of 2–4 days for 2 weeks to determine the daily budding rate of polyps (Experiment 4).

Onset of zooxanthella infection

To determine the onset of zooxanthella infection during ontogeny, planulae were carefully extracted from the oral arm grooves and the brood-carrying filaments of gravid females. Additional samples of planulae naturally liberated in the medusa collection container were taken for comparison (control group). The three different planulae sets obtained (oral arms, brood-carrying filaments, and control planulae) were treated separately for their study by means of transmission electron microscopy (TEM).

Freshly collected planulae were carefully washed with sterile filtered seawater and transported to the laboratory in sterile seawater flasks. Before providing any food items, the planulae and resulting polyps were transferred by 1.5-ml Eppendorf pipettes and fixed in 2.5% glutaraldehyde in 0.1-M sodium cacodylate for 1 h at ambient temperature. Following three 10-min rinses in 0.1-M sodium cacodylate, samples were post-fixed for 1 h in 1% osmium tetroxide in 0.1-M sodium cacodylate and rinsed three more times in 0.1-M sodium cacodylate. After a sequential dehydration of 15 min in 30 and 50% ethanol, samples were left overnight in 70% ethanol. Dehydration was completed through 90% and 100% (×3) ethanol. After being transferred to propylene oxide, samples were gradually embedded in Spurr's epoxy resin. After 48-h-polymerisation at 55°C, thin sections of the resulting capsules were cut by an ultramicrotome, mounted on copper grids, stained with lead acetate, and viewed on a JEOL transmission electron microscope. Five planulae and three polyps per set were analysed.

Growth of early medusa stages

Laboratory experiments were conducted to establish the growth of ephyrae over the range of temperatures typical of the Mar Menor (Pérez-Ruzafa et al., 2002), with 20°C as the temperature of ephyra liberation (Prieto et al., 2010). First, early medusa growth was studied at a constant temperature of 20°C, salinity 38, and photoperiod of 12:12 (normal culture conditions). For 1 month after the day of liberation from the strobila, groups of 10 ephyrae were introduced into cylindrical glass flasks with 150 ml of 45-μm-filtered seawater without aeration and fed daily with newly hatched *Artemia* nauplii at ~220 prey l^{-1}. The ephyrae were transferred daily to new containers with aerated water and fresh prey. Once the metaephyra stage was achieved, the number of animals per flask was reduced to three, and the prey items were changed to Selco-enriched *Artemia* nauplii (220 prey l^{-1}). To establish the allometric relationships in early medusa stages, 100 ephyrae, 20 metaephyrae and 10 medusae were removed from their rearing containers and measured with the aid of a stereomicroscope. In addition, 23 of these specimens (11 ephyrae, six metaephyrae and six medusae) were individually put onto pre-combusted, pre-weighed glass-fibre filters, dried at 60°C to constant weight, weighed, then ash-free dried at 450°C for 1 day and re-weighed. The correlations of diameter, dry weight (DW), and ash-free dry weight (AFDW) with age were determined.

Temperature effects on ephyra growth rates also were tested (Experiment 5). We tested the null hypothesis that ephyra development was unaffected by different temperatures. Fifteen ephyrae individually identified by micro-photograph were randomly assigned by groups of five specimens to growth temperatures of 20, 25 and 30°C. Each individual was maintained in a 3.3-cm diameter cylindrical glass flask with 40 ml of 45-μm-filtered seawater (salinity: 38, water column height: 4.3 cm) without aeration. Photoperiod was kept at 12:12, and ephyrae were fed daily with *Artemia* at ~220 nauplii l^{-1}. Every day, before adding fresh prey, the rearing medium was exchanged with new, aerated water. Once per week for 4 weeks, every ephyra was removed, allowed to relax, photographed under the stereomicroscope and returned to the rearing medium. The mean diameter between opposite lappet tips (usually eight measurements per ephyra) was measured from the photographs. At the end of the experiment, ephyrae were put individually in pre-combusted and pre-weighed glass-fibre filters, and their DW and AFDW measured as above.

Ingestion by early medusa stages

The ingestion rate and daily ration of early medusa stages were studied at 20, 25 and 30°C (Experiment 6). We tested the null hypothesis that ingestion rates of *Artemia* nauplii by ephyrae were unaffected by different temperatures. Cylindrical glass flasks (3.3-cm diameter) filled with 40 ml of 45-μm-filtered seawater (salinity: 38, water column height: 4.3 cm) were used as incubation containers providing recently hatched *Artemia* nauplii at a concentration of 400 prey l^{-1} without aeration. Twenty-seven ephyrae (mean: 3.6 ± 0.4 mm) previously unfed for 24 h were randomly placed in three temperature treatments (three per flask, three flasks per treatment). In order to compare between ephyrae and medusae in the same experiment, six unfed medusae (mean: 7.1 ± 0.4 mm) were individually placed in the experimental flasks and randomly distributed among three replicates in each of the lowest and intermediate temperature treatments. An additional container without predators served as a treatment prey control. After 6 h of incubation, the remaining nauplii in each glass container were counted under a stereomicroscope. The respective ingestion rate and daily ration of ephyrae and small medusae were determined according to the equations in Båmstedt et al. (1999, 2001, respectively). Specimens were measured from digital images before the experiment to ensure size homogeneity.

Statistical analyses

Statistical analyses of data were performed using SPSS Statistical Software. Assumptions of analysis of variance (ANOVA) were tested on datasets before statistical testing. If data failed normality and equality of variances and homogeneity could not be met by transformations, then non-parametric Kruskal–Wallis analysis of variance was applied. When looking for between-subjects and within-subjects effects as in the case of subjects measured over time, repeated ANOVA tests were applied once sphericity and homogeneity of dependent variables were confirmed. If significant differences were found between treatments, then multiple comparisons were made using the Tukey test or its non-parametric homologue.

Results

Benthic stage

Cotylorhiza tuberculata planulae had attached and developed into polyps on the glass slides and surfaces of the glass container in Experiment 1 after 20 days (Fig. 1a; Table 2). Greater relative densities of polyps on the bottom of the flasks clearly indicated their spatial settlement preference ($\chi^2_3 = 19.815, p < 0.01$). Planulae attached to the underside of the water–air interface in relative densities similar to those on the glass slides (upper and underside combined), but in significantly higher relative densities than on the sides of the glass containers (Fig. 1a). Settlement preferences on the lower and upper surfaces of the glass slides were similar ($F_{1,10} = 1.429, p > 0.05$).

Cotylorhiza tuberculata planulae also attached to all substrates provided in Experiment 2: glass slide, brick, wood, stone, shell and the surfaces of the glass container (Fig. 1b; Table 2). Therefore, the null

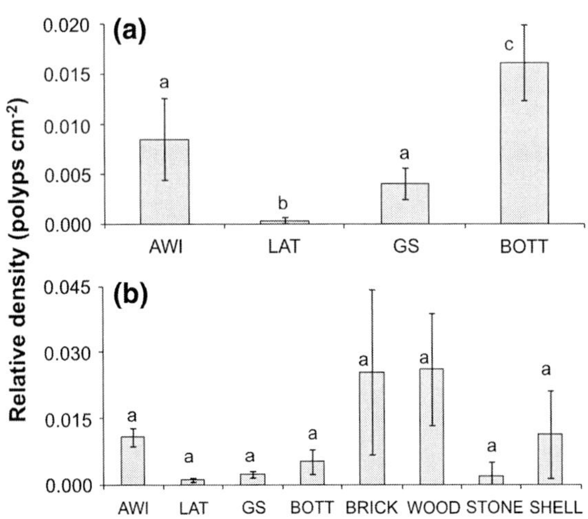

Fig. 1 Settlement preferences of *C. tuberculata* (Macri, 1778) planulae in natural seawater. Mean density of polyps per substrate (*bars* represent standard deviation) for: **a** spatial preferences (Experiment 1): $n = 6$ replicates with 50 polyps each, and **b** substrate preferences (Experiment 2): $n = 3$ replicates with 70 polyps each. *AWI* air–water interface; *LAT* sides of the glass container; *GS* glass slides; *BOTT* bottom of the glass container. Different *letters* indicate significant differences at $\alpha = 0.05$

hypothesis tested was accepted as the relative densities did not differ significantly among substrates (Fig. 1b) in spite of the apparent preferences for wood, brick, shell or the water–air interface versus the other surfaces ($\chi^2_7 = 14.021, p > 0.05$). No settlement preferences were detected between organic and inorganic substrates or natural and artificial substrates (Fig. 1b). As was observed in Experiment 1, polyp settlement on the underside of glass slides was similar to that of the upper side ($F_{1,4} = 0.571$, $p > 0.05$), and no polyps settled upside-down on the shells internal surface. Polyps were observed after settlement at densities of up to 4 cm^{-2}, but densities were reduced to 1.3 cm^{-2} within a month.

Experiment 3 showed that *Cymodocea nodosa* seagrasses were a suitable surface for polyp attachment. Four days after introduction to the *C. nodosa* aquarium, polyps got attached along the upper and underside of the leaves at a density of 0.54 polyps cm^{-2}. They also settled on the few exposed areas of the rhizomes at a density of 3.33 polyps cm^{-2}. The *C. nodosa* plants sheltered an overall polyp density of 0.79 cm^{-2}. Polyp densities on the other substrates available were glass bottom (Petri dishes, 0.34 polyps cm^{-2} and aquarium, 0.13 polyps cm^{-2}) and sides of the aquarium (0.04 polyps cm^{-2}).

Polyps maintained at 17.5°C and 12:12 photoperiod reproduced exponentially by budding during the first 13 days (Experiment 4, Fig. 2) at a rate of 3.9 ± 1.27% per day (polyp number = 22.17 $e^{0.04 \text{ day}}$, $R^2 = 0.94$).

Onset of zooxanthellae infection

The infection with zooxanthellae occurred early in the *C. tuberculata* life cycle (Fig. 3). Zooxanthellae were not observed directly on planulae ($n = 15$); however, TEM revealed the presence of algae inside the polyps from all sets of planulae ($n = 9$): (1) naturally liberated by gravid females and (2) extracted in aseptic conditions from medusa brood-carrying filaments and oral arm grooves.

Growth of early medusa stages

The diameter of *C. tuberculata* ephyrae ranged between 2.1 and 2.8 mm on the day of liberation from the strobila ($n = 60$, 2.46 ± 0.34). When maintained at a constant temperature of 20°C, the diameter of

Table 2 *Cotylorhiza tuberculata* planula settlement preferences

	Studied variable	AWI	LAT	GS	BOTT	Brick	Wood	Stone	Shell	Statistics
Exp. 1	Polyp number									
	Mean	13.67	1.83	7.50	29.83					
	SD	6.77	2.23	3.94	14.99					
	Total	82	11	45	179					
	Available surface (cm^2)	33.18	102.10	37.50	33.183					
	Attachment (%)									
	Mean	28.12	3.61	14.94	53.33					$F_{3,20} = 29.205$**
	SD	13.59	3.15	5.88	12.51					
Exp. 2	Polyp number									
	Mean	26.67	6.0	5.67	9.67	6.67	11.33	1.33	2.00	
	SD	21.38	2.00	3.21	6.81	3.05	11.93	2.31	2.00	
	Total	80	18	17	29	20	34	4	6	
	Available surface (cm^2)	33.18	102.10	37.50	25.89	5.48	5.44	6.25	1.88	
	Attachment (%)									
	Mean	35.91	10.57	8.58	13.42	14.00	14.21	1.17	2.14	$F_{7,16} = 9.516$**
	SD	6.88	4.78	2.77	7.31	10.25	6.89	2.03	1.87	

AWI air–water interface; *LAT* sides; *GS* glass slide; *BOTT* bottom, ** $p < 0.01$

Total polyp number per substrate, available surface per substrate and mean settlement percentages (standard deviation, SD) after 20 days

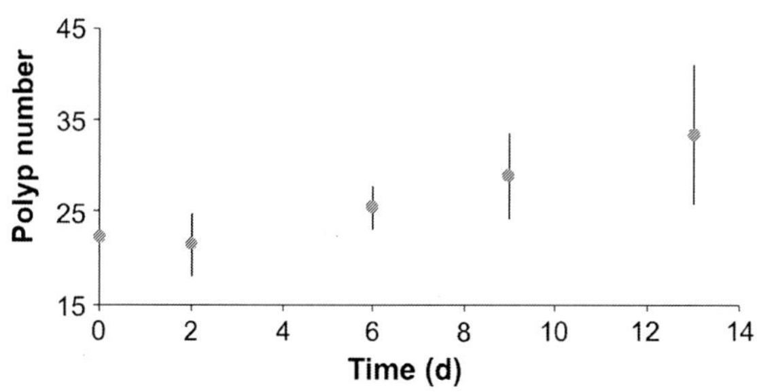

Fig. 2 *Cotylorhiza tuberculata* polyp budding rate at a constant temperature of 17.5°C, salinity of 38 and photoperiod of 12:12. Mean increase of polyp number by budding in 13 days (Experiment 4): $n = 4$, *bars* represent standard deviation

early medusae (Φ) increased 0.08 mm per day ($n = 130$, $\Phi = 0.081$ day $+ 2.049$, $R^2 = 0.90$, Fig. 4). Linear correlations were found between individual size, DW and AFDW. An increase of 1 mm in diameter represented ~213 μg DW and ~90 μg AFDW ($n = 23$; $DW = 212.850\Phi - 496.750$, $R^2_{DW} = 0.82$; $AFDW = 89.677\Phi - 173.560$, $R^2_{AFDW} = 0.67$).

Different rearing temperatures had no effect on ephyra growth in diameter increase; however, ephyrae grown at 20, 25 and 30°C differed in the time to attain

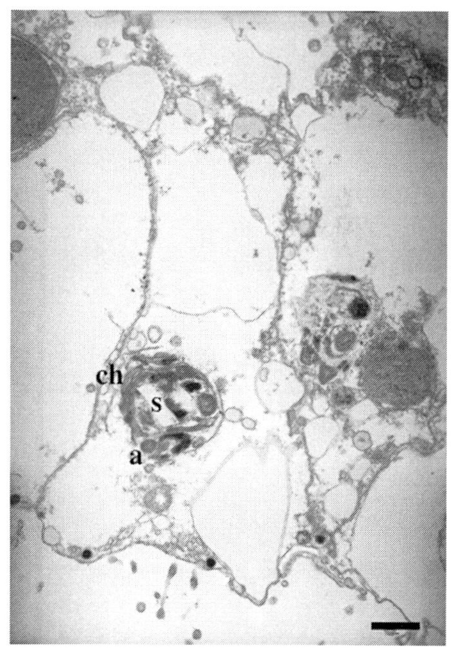

Fig. 3 Transmission electron microscope photograph of a zooxanthella in an a priori aposymbiotic *Cotylorhiza tuberculata* polyp (scale bar: 5 μm). *a* accumulation body; *ch* chloroplast; *s* starch body

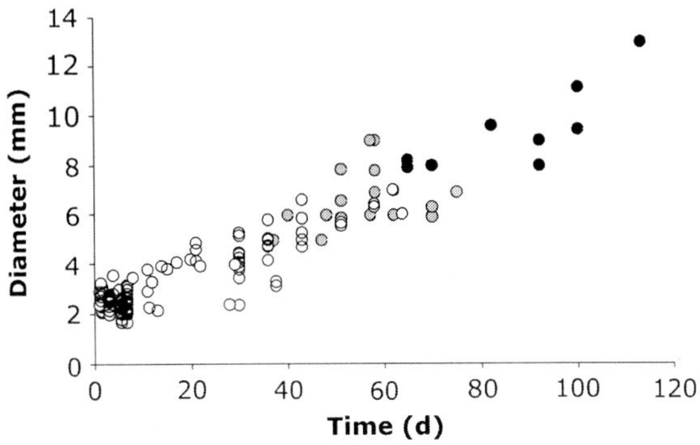

Fig. 4 Allometric relationships in early medusa stages of *C. tuberculata* incubated at 20°C. Each *dot* represents an individual ($n = 130$). *Open, grey* and *black circles* represent ephyrae, metaephyrae and small medusae (up to 113 days), respectively

the medusa stage (Experiment 5: temperature: $F_{2,12} = 3.075$, $p > 0.05$; week: $F_{3,36} = 15.629$, $p < 0.01$; interaction: $F_{6,36} = 7.863$, $p < 0.01$, Fig. 5). By day 21, none of the ephyrae reared at 20°C had reached the metaephyra stage, but 80% of those incubated at 25°C and 100% at 30°C had already developed into medusae. Thus, the null hypothesis tested was rejected as ephyra development was affected by temperature.

Ingestion by early medusa stages

The ingestion rate and daily ration of early medusa stages depended on the incubation temperature, rejecting the null hypothesis tested (Experiment 6; Table 3). Ephyrae maintained at 20°C ingested a similar amount of prey daily as those incubated at 30°C, but significantly less than those at 25°C. Ephyra daily ration varied among treatments with the highest DW digestion of prey at 25°C.

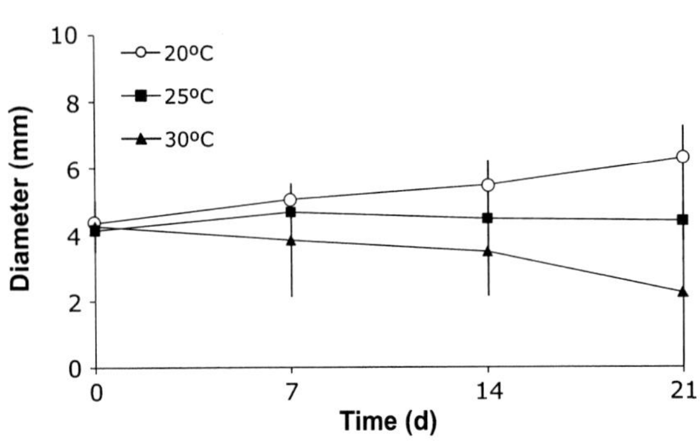

Fig. 5 *Cotylorhiza tuberculata* ephyra growth at three temperatures (20, 25, 30°C; Experiment 5). *Bars* represent standard deviations ($n = 5$)

Differences between the ingestion rates of ephyrae and medusae were highly significant, with medusae consuming more prey than ephyrae in all incubation temperatures (Table 3). Medusae as well as ephyrae ingested more prey at 25°C than at 20°C. The small medusae in Experiment 6 were twice as wide as the ephyrae (2.0 ± 0.3) and consumed 2–3-times more *Artemia* nauplii daily (2.8 ± 1.2 medusa^{-1}); however, the respective daily rations were determined by the incubation temperature and were independent of stage. The AFDW-specific percentage ingested was higher at 25°C than at 20°C, but similar for ephyrae and medusae (Table 3).

Discussion

Benthic stage

The settlement preferences of planulae confirmed that these ciliated larvae represent a highly versatile stage of *C. tuberculata* development (Prieto et al., 2010). Planulae attached to all surfaces available, with a clear spatial preference for the bottom but no preference between organic or inorganic substrates and/or natural or artificial substrates.

Settlement of planulae at the water–air interface also has been observed in other scyphozoan species like *Lychnorhiza lucerna* Haeckel (Schiriati et al., 2008) and *Aurelia aurita* Linnaeus (Holst & Jarms, 2007), and probably is an artefact related to the motionless water within laboratory containers. Although Kroiher & Berking (1999) suggested that planula settlement on the water surface is normal in natural conditions, evidence is missing for *C. tuberculata* given that few polyps have been observed in situ (only three polyps). A stable air–water interface is unlikely in almost all natural environments.

Planula settlement on the water–air interface could reflect the preference of scyphozan polyps to attach to the underside of surfaces (Pitt, 2000; Holst & Jarms, 2007). Although polyps of *Cyanea* sp. discriminated among different textures and preferred rough substrates (Brewer, 1989), settlement did not differ significantly for the upper and lower surfaces of diverse substrates, including shell, glass and seagrass leaves, for *C. tuberculata* planulae. Indeed, planulae were not particularly selective, even though artificial

Table 3 Ingestion rate and daily ration of early medusa stages incubated at 20, 25 and 30°C

Variable	Temperature			Statistics	
	20°C	25°C	30°C	Factor	$F_{1,28}$
Diameter (mm)					
Ephyra					
Mean	3.50	3.68	3.72	Temp	0.576^{NS}
SD	0.33	0.45	0.46	Stage	354.732**
Medusa					
Mean	7.40	6.95		Temp × stage	2.667^{NS}
SD	0.35	0.26			
AFDW (μg)					
Ephyra					
Mean	140.18	156.29	160.41	Temp	0.577^{NS}
SD	29.64	40.27	41.45	Stage	354.537**
Medusa					
Mean	489.58	449.92		Temp × stage	2.667^{NS}
SD	30.95	23.18			
Ingestion (prey day^{-1})					
Ephyra					
Mean	121.67	392.00	230.67	Temp	82.491**
SD	24.29	55.43	50.60	Stage	192.492**
Medusa					
Mean	419.00	783.67		Temp × stage	3.608^{NS}
SD	83.47	83.58			
Daily ration (AFDW%)					
Ephyra					
Mean	286.48	855.79	471.85	Temp	29.503**
SD	75.26	276.60	101.61	Stage	3.100^{NS}
Medusa					
Mean	272.36	556.41		Temp × stage	2.356^{NS}
SD	38.33	34.16			

Mean and standard deviation (SD) of specimen diameter, AFDW, number of prey items ingested per day and daily ration

temp temperature

** $p < 0.01$; ^{NS} $p > 0.05$

surfaces are often preferred by scyphopolyps (Pitt, 2000; Holst & Jarms, 2007; Hoover & Purcell, 2009).

The lack of substrate preference suggests that the increase in *Cymodocea nodosa* was unlikely to be a factor responsible for the onset of blooms in the Mar Menor, because similar surface areas of submerged macrophytes were present in the lagoon before *C. nodosa* dominance (Pérez-Ruzafa et al., 1991). There is evidence that other macrophyte species such as *Zostera marina* Linnaeus, *Zostera noltii* Horneman (Pérez-Ruzafa et al., 1987; Calvín et al., 1999) and the green alga *Caulerpa prolifera* (Forsskål) Lamouroux (Pérez-Ruzafa et al. 1991; Calvín et al., 1999) may be suitable for colonization by *C. tuberculata* polyps. Two polyps on *Zostera* sp. were observed by Kikinger in 1981 (Kikinger, 1992) and 2010 (personal communication) in Vlyhho Bay, and a third polyp was found attached to a *C. prolifera* leaf in Mar Menor (personal observation). This is in contrast with the scyphozoan *Catostylus mosaicus* Quoy & Gaimard, which avoided seagrasses when offered glass, shell, wood or sandstone (Pitt, 2000).

Nevertheless, the suitability of *C. nodosa* as attachment surface for polyps is important, given that without this seagrass, sandy and muddy sediments, which predominate in the Mar Menor, do not allow successful settlement (Holst & Jarms, 2007). Rocky substrates in the lagoon are limited to the islands and El Estacio channel, and compact red clay sediments with *Pholas dactylus* shells and *C. prolifera* are exclusively located on the central eastern shore (Pérez-Ruzafa et al., 1987, 2008). The increase in total available artificial surface resulting from the growing anthropogenic activities in the littoral zone is difficult to assess (24 km of the 58-km shoreline is affected) (Pagès, 2001). Retaining walls, sport harbours, bypasses, artificial channels, jetties, docks and boats around the Mar Menor shore greatly increase the suitable settling surface for planula and polyp attachments. Given the absence of substrate selectivity and the high asexual reproduction by budding compared to other scyphozoans (reviewed in Purcell et al., 2012), this extension of available surface in the lagoon potentially could have increased the benthic population of *C. tuberculata*, which combined with the appropriate environmental conditions (Prieto et al., 2010) may have contributed to increase of medusa blooms (Parsons & Lalli, 2002; Holst & Jarms, 2007). Worldwide substrate additions by human modification of shorelines are considered to favour the sessile stages of scyphozoans (Purcell, 2012), and the Mediterranean coastline is a very anthropogenic modified area (Halpern et al., 2008). Among all the Mediterranean sub-basins, *C. tuberculata* has been reported in the Catalan Sea (Fuentes et al., 2010), Ligurian Sea (Carli et al., 1991), Strait of Sicily (Daly Yahia et al., 2003), Adriatic Sea (Kogovsek et al., 2010), Ioanian Sea (Kikinger, 1992), Aegean Sea (Gülsahin & Tarkan, 2011) and in the Levantine Basin (Lakkis, 1991). This ubiquity of *C. tuberculata* combined with the plasticity to attach to any type of substrate allows us to extrapolate the implications of the present study to the whole Mediterranean. Indeed, *C. tuberculata* is the most common rhizostomae in the Mediterranean Sea (Kikinger, 1992).

Onset of zooxanthellae infection

Among the six species of rhizostomae jellyfish occurring in the Mediterranean Sea, only three of them are symbiotic: *Cassiopeia andromeda* Eschscholtz, *C. tuberculata* and *Phyllorhiza punctata* von Lendelfeld. Of these three species, the symbiotic zooxanthellae are not always essential for the phase transition between the benthic and the pelagic stages (Table 1). *Cassiopeia andromeda* aposymbiotic polyps can strobilate and aposymbiotic ephyrae can be obtained (Rahat & Adar, 1980). The same occurred with *Phyllorhiza punctata* (Galil et al., 2009), a Pacific jellyfish observed in several regions in the Mediterranean recently (Cevik et al., 2011). The crucial role of zooxanthellae in medusa formation in *C. tuberculata* was discovered by Kikinger (1992) when he observed no strobilation in hundreds of laboratory aposymbiotic polyps during a 2-year period. We believe that planulae are infected by *Symbiodinium* sp. while still remaining within the mother medusa, because the polyps we obtained in aseptic conditions had zooxanthellae that could only have been transmitted previously.

Zooxanthellae generally are absent in the eggs and planulae in most symbiotic scyphozoans (Table 1) and must be acquired from the environment during the scyphistoma stage (Arai, 1997; Thornhill et al., 2006). In contrast, the coronate *Linuche unguiculata* Swartz releases eggs in mucus strands containing zooxanthellae that infect embryos and planulae during the 24 h after fertilization (Montgomery & Kremer, 1995). Algal infection at the planula stage is also known for other symbiotic marine invertebrates, such as the octocoral *Xenia macrospiculata* Gohar (Achituv et al., 1992) and the scleractinian coral *Fungia scutaria* Lamarck (Schwarz et al., 1999).

Infection of *C. tuberculata* planulae from the brood-carrying filaments was possible given the high zooxanthella content of the surrounding mucus (Kikinger, 1992). By contrast, planulae from the oral arm canals were expected not to harbour any zooxanthellae (Kikinger, 1992); however, algal infection also could have occurred during the embryonic development, given the abundance of zooxanthellae in the ovarial mesoglea (Kikinger, 1992). The likelihood of zooxanthella infection at the planula stage of *C. tuberculata* would decrease the possibility of aposymbiotic polyps in nature and increase the strobilation success of the population in the appropriate environmental conditions (Prieto et al., 2010). Nevertheless, further studies should be conducted to find zooxanthellae in the embryos or planulae, as found in *Linuche unguiculata* (Montgomery & Kremer, 1995).

Growth of early medusa stages

Field observations of a natural population of *C. tuberculata* from the Ionian Island of Lefkada estimated over 8–10 weeks for the newly liberated ephyrae to reach 3-cm diameter at temperatures above 24°C (Kikinger, 1992). Live *C. tuberculata* ephyrae measured since production at a constant temperature of 20°C would need 2–3-times longer to attain that size. Development of *Rhizostoma octopus* Macri ephyrae in laboratory cultures also was very slow compared to natural growth (Holst et al., 2007). They thought the differences were due to measurements made on living versus preserved specimens and/or to disparities between natural and laboratory conditions. In fact, the youngest ephyra stages in our study (2.1–2.8 mm in diameter) were larger than the preserved samples previously described for Vlyho Bay by Kikinger (1.5–2 mm). Feeding of ephyrae in small culture flasks often relies on *Artemia* sp. nauplii and omits any natural prey (Issel, 1922; Pérez-Ruzafa et al., 2002); the highest daily rations of *Aurelia aurita* ephyrae occurred only when mixed zooplankton was available (Båmstedt, 1990), which would promote rapid growth. A constant rearing temperature in the laboratory also may have contributed to low ephyra growth rates. The beneficial consequences of natural temperature fluctuations for ephyra growth are clearly shown for *Cyanea capillata* Linnaeus in Gullmar Fjord (Gröndahl & Hernroth, 1987) and *A. aurita* in Tapong Bay (Lo & Chen, 2008).

Growth of scyphomedusae is generally enhanced at higher temperatures when food is not a limiting factor (Lucas, 2001; Widmer, 2005); however, the growth rates of early medusa stages of *C. tuberculata* were similar at 20, 25 and 30°C. Nevertheless, the rate of transition between ephyra, metaephyra and small medusa was strongly controlled by temperature. The medusa stage was rapidly attained at the highest temperature, resulting in small, completely developed medusae. Similar results were obtained for *Aurelia aurita* juveniles; equal-sized individuals with no obvious behavioural differences were classified as ephyrae or as medusae depending on their rearing temperature (13 or 21°C, respectively; Nawroth et al., 2010). This phenotypic plasticity in ephyra development could be beneficial for scyphozoan species that encounter large temperature fluctuations as those induced by climate and ocean circulation changes (Nawroth et al., 2010). Mediterranean Sea water temperature already has increased 0.67–0.89°C from 1982 to 2006 (Belkin, 2009). Predicted warming scenarios for the end of the twenty-first century are between 1.8 and 6°C (best estimates of temperature change at 2090–2099 relative to 1980–1999; IPCC, 2007). For *C. tuberculata*, higher temperatures would enable faster medusa development, with the potential to accelerate sexual maturity and spawning.

Ingestion by early medusa stages

The ingestion rate and daily ration of *C. tuberculata* ephyrae depended on the incubation temperature as previously shown for the semaestome *Aurelia aurita* (Båmstedt et al., 1999, 2001) and the hydrozoan *Moerisia lyonsi* Boulenger (Ma & Purcell, 2005). *C. tuberculata* ephyrae consumed between 290 and 850% of their DW daily at suitable growth temperatures (20–25°C), which, as observed also for *Aurelia aurita* (Båmstedt et al., 1999), implies that these ephyrae may be of significant relevance in decreasing dense patches of zooplankton prey in their natural environment. Although the size reached by ephyrae after 21 days in culture at 30°C was similar to their sizes at 20 and 25°C, high mortality occurred at 30°C that could not be attributed to prey limitation, given that prey concentrations corresponded to saturated feeding conditions in similar experiments with *Aurelia aurita* ephyrae (Båmstedt et al., 1999, 2001) and never decreased below 100 Artemia sp. nauplii l^{-1}. Therefore, we conclude that 30°C is not appropriate for the successful growth of *C. tuberculata* ephyrae. It is unlikely that they experience 30°C in the Mediterranean Sea during spring when strobilation occurs and ephyrae are present (Prieto et al., 2010) because the temperature range within the Mar Menor, much more confined and land thermal affected, is 16–25°C.

The ingestion rates and daily rations of small *C. tuberculata* medusae also depended on their sizes. Larger specimens consumed more prey than ephyrae, but the DW-specific daily ration of prey they ingested was relatively lower, suggesting an increasing importance of zooxanthellae for the nutrition and development of the pelagic stage. In fact, despite the large number of ingested prey predicted for *C. tuberculata* medusae of 35-cm diameter, zooplankton feeding may not be adequate to support medusa abundances of up to 0.9 individuals m^{-3} in the Mar Menor (Mas, 1999), <1 individuals m^{-3} in the Gulf of Tunis (Daly Yahia et al., 2003) and up to 4,000 individuals in the small Bay of Vlyho (Kikinger, 1992). Also, in the Aegean Sea was reported *C. tuberculata* in large densities as 4–5 individuals m^{-2} in the Güllük Bay, 3 individuals m^{-2} in the Gökova Bay and 2 individuals m^{-2} in the coast of Milas (Gülsahin & Tarkan, 2011). This suggests that zooxanthellae must have a substantial contribution to the nutrition of this species (Pagès, 2001). Further studies are necessary to clarify the importance of zooxanthellae symbiosis during the pelagic stage of *C. tuberculata*.

Planktivorous gelatinous species are known to be important consumers in both low and high productivity marine ecosystems (Mills, 1995). *C. tuberculata* effectively exert a strong top-down control on the food web by selective grazing on large diatoms, ciliates, veliger larvae and copepods (Pérez-Ruzafa et al., 2002). The diet overlap between zooplanktivorous jellyfish and pelagic fish (Purcell & Sturdevant, 2001, Hiromi et al., 2005; Brodeur et al., 2008), combined with predictions of increased jellyfish populations (Lynam et al., 2005; Purcell, 2005; Attrill et al., 2007; Richardson et al., 2009), suggests potential changes in future pelagic communities (e.g. Vinogradov & Shushkina, 1992; Richardson et al., 2009) that may have detrimental consequences for fisheries and economies worldwide.

In summary, part of the success of the *C. tuberculata* benthic phase may be due to the lack of preference among substrates for planulae settlement and the high rate of asexual reproduction by budding at mild winter temperatures (17.5°C). These aspects combined with the proliferation of artificial substrates and the recovery of seagrass beds may increase both the availability of suitable surfaces for the development of polyps and the viable benthic population, leading to a rise of medusa abundances. Moreover, the proximity of zooxanthellae in mother medusae facilitates infection early in the developmental of *C. tuberculata*. Hence, zooxanthellae infection is unlikely to constitute a limiting factor for the proliferation of this species. Finally, warmer temperatures accelerated the transition from ephyrae to medusae, which occurred at a smaller size, and increased their food ingestion. The high feeding rates measured in early medusa stages at 25°C suggest the potential for changes in the pelagic communities of coastal anthropogenically altered ecosystems, especially considering predicted warming scenarios (IPCC, 2007) that may benefit these jellyfish.

Acknowledgments The authors thank A. Moreno, F. Rey and J. Pampín for their support on medusa sampling; Dr. A. García and O. Aliseda for assistance with electronic microscopy; Dr. A. Medina and A. Santos for their support and advice regarding microscopic preparations; and the reviewers and editor for helpful suggestions. This work was financially supported by projects PERSEUS (FP7-287600), JELLY-PHYS (CTM 2011-22856), Junta de Andalucía (P07-RNM-02976), and Observatorio del Estrecho. A research grant to L. Prieto from Ramon y Cajal Programme of Spanish MIC, and an I3P-CSIC (partly funded by the European Social Fund, ESF) pre-doctoral fellowship to D. Astorga, are also acknowledged.

References

Achituv, Y., Y. Benayahu & J. Hanania, 1992. Planulae brooding and acquisition of zooxanthellae in *Xenia macrospiculata* (Cnidaria: Octocorallia). Helgoländer Meeresuntersuchungen 46: 301–310.

Arai, M. N., 1997. A Functional Biology of Scyphozoa. Chapman and Hall, London.

Arai, M. N., 2001. Pelagic coelenterates and eutrophication: a review. Hydrobiologia 451: 69–87.

Attrill, M. J., J. Wright & M. Edwards, 2007. Climate-related increases in jellyfish frequency suggest a more gelatinous future for the North Sea. Limnology and Oceanography 52: 480–485.

Båmstedt, U., 1990. Trophodynamics of the scyphomedusae *Aurelia aurita*. Predation rate in relation to abundance, size and type of prey organism. Journal of Plankton Research 12: 215–229.

Båmstedt, U., J. Lane & M. B. Martinussen, 1999. Bioenergetics of ephyrae larvae of the scyphozoan jellyfish *Aurelia aurita* in relation to temperature and salinity. Marine Biology 135: 89–98.

Båmstedt, U., B. Wild & M. B. Martinussen, 2001. Significance of food type for growth of ephyrae *Aurelia aurita* (Scyphozoa). Marine Biology 139: 641–650.

Belkin, I. M., 2009. Rapid warming of Large Marine Ecosystems. Progress in Oceanography 81: 207–213.

Brewer, R. H., 1989. The annual pattern of feeding, growth, and sexual reproduction in *Cyanea* (Cnidaria: Scyphozoa) in the Niantic River Estuary, Connecticut. Biological Bulletin 176: 272–281.

Brodeur, R. D., C. L. Suchman, D. C. Reese, T. W. Miller & E. A. Daly, 2008. Spatial overlap and trophic interactions between pelagic fish and large jellyfish in the northern California Current. Marine Biology 154: 649–659.

Calvín, J. C., I. Franco Navarro, A. Marín, A. M. Martínez Inglés, A. Belmonte Ríos, J. M. Ruiz Fernández, A. Belando Franco, M. Vicente Albaladejo & J. M. Rocamora Tomás, 1999. El Litoral Sumergido de la Región de Murcia. Cartografía Bionómica y Valores Ambientales. Consejería de Medio Ambiente, Agricultura y Agua, Región de Murcia.

Carli, A., V. Pane, T. Valente & S. Cotta, 1991. Lipid and protein content of jellyfish from the Ligurian Sea. First results. In Jellyfish Blooms in the Mediterranean. Proceedings of the II Workshop on Jellyfish in the Mediterranean Sea. MAP Technical Report Series 47 UNEP, Athens: 236–240.

Cevik, C., O. B. Derici, F. Cevik & L. Cavas, 2011. First record of *Phyllorhiza punctata* von Lendenfeld, 1884 (Scyphozoa: Rhizostomeae: Mastigiidae) from Turkey. Aquatic Invasions 6: S27–S28.

Claus, C., 1890. Über die Entwicklung des Scyphostoma von *Cotylorhiza*, *Aurelia*, und *Chrysaora*, sowie ueber die systematische Stellung der Scyphomedusen. I Arbeiten aus den Zoologischen Instituten der Universität Wien 9: 85–128.

Claus, C., 1893. Über die Entwicklung des Scyphostoma von *Cotylorhiza*, *Aurelia*, und *Chrysaora*, sowie ueber die systematische Stellung der Scyphomedusen. II Arbeiten aus den Zoologischen Instituten der Universität Wien 10: 1–70.

Daly Yahia, M. N., J. Goy & O. Daly Yahia-Kéfi, 2003. Distribution et écologie des Méduses (Cnidaria) du golfe de Tunis (Méditerranée sud occidentale). Oceanologica Acta 26: 645–655.

Fuentes, V. L., D. L. Angel, K. M. Bayha, D. Atienza, D. Edelist, C. Bordehore, J. M. Gili & J. E. Purcell, 2010. Blooms of the invasive ctenophore, *Mnemiopsis leidyi*, span the Mediterranean Sea in 2009. Hydrobiologia 645: 23–37.

Galil, B. S., L. Shoval & M. Goren, 2009. *Phyllorhiza punctata* von Lendenfeld, 1884 (Scyphozoa: Rhizostomeae: Mastigiidae) reappeared off the Mediterranean coast of Israel. Aquatic Invasions 4: 481–483.

Gibbons, M. J. & A. J. Richardson, 2009. Patterns of jellyfish abundance in the North Atlantic. Hydrobiologia 616: 51–65.

Gröndahl, F. & L. Hernroth, 1987. Release and growth of *Cyanea capillata* (L.) ephyrae in the Gullmar Fjord, western Sweden. Journal of Experimental Marine Biology and Ecology 106: 91–101.

Gülsahin, N. & A. N. Tarkan, 2011. A familiar organism in Mugla Region *Cotylorhiza tuberculata* (Macri, 1778). In Turan, C. & B. Öztürk (eds), First National Workshop on Jellyfish and Other Gelatinous Species in Turkish Marine Waters. Turkish Marine Research Foundation 35, Istanbul: 53–57.

Halpern, B. S., S. Walbridge, K. A. Selkoe, C. V. Kappel, F. Micheli, C. D'Agrosa, J. F. Bruno, K. S. Casey, C. Ebert, H. E. Fox, R. Fujita, D. Heinemann, H. S. Lenihan, E. M. P. Madin, M. T. Perry, E. R. Selig, M. Spalding, R. Steneck & R. Watson, 2008. A global map of human impact on marine ecosystems. Science 319: 948–952.

Hamner, W. M. & M. N. Dawson, 2009. A review and synthesis on the systematics and evolution of jellyfish blooms: advantageous aggregations and adaptive assemblages. Hydrobiologia 616: 161–191.

Hay, S., 2006. Marine ecology: gelatinous bells may ring change in marine ecosystems. Current Biology 16: 679–682.

Hiromi, J., T. Kasuya & H. Ishii, 2005. Impacts of massive occurrence of jellyfish on pelagic ecosystem. Bulletin of Plankton Society of Japan 52: 82–90.

Hofmann, D. K. & G. Crow, 2002. Induction of larval metamorphosis in the tropical scyphozoan *Mastigias papua*: striking similarity with upside down-jellyfish *Cassiopea* spp. (with notes on related species). Vie et Milieu 52: 141–147.

Hofmann, D. K., R. Neuman & K. Henne, 1978. Strobilation, budding and initiation of scyphistoma morphogenesis in the rhizostome *Cassiopea andromeda* (Cnidaria: Scyphozoa). Marine Biology 47: 161–176.

Holst, S. & G. Jarms, 2007. Substrate choice and settlement preferences of planula larvae of 453 five Scyphozoa (Cnidaria) from German Bight, North Sea. Marine Biology 151: 863–871.

Holst, S., I. Sötje, H. Tiemann & G. Jarms, 2007. Life cycle of the rhizostome jellyfish *Rhizostoma octopus* (L.) (Scyphozoa, Rhizostomeae), with studies on cnidocysts and statoliths. Marine Biology 151: 1695–1710.

Hoover, R. A. & J. E. Purcell, 2009. Substrate preferences of scyphozoan *Aurelia labiata* polyps among common dock-building materials. Hydrobiologia 616: 259–267.

IPCC, 2007. Climate Change 2007: Synthesis Report. In Core Writing Team, Pachauri, R. K. & Reisinger, A. (eds), Contribution of Working Groups I, II and III to the Fourth Assessment Report of the Intergovernmental Panel on Climate Change. IPCC, Geneva.

Issel, R., 1922. Nuove indagini sul plankton nelle aquae di Rovigno. Memorie R. Comitato Talassografico Italiano 102: 1–36.

Kikinger, R., 1992. *Cotylorhiza tuberculata* (Cnidaria: Scyphozoa)- life history of a stationary population. PSZNI Marine Ecology 13: 333–362.

Kogovsek, T., B. Bogunovic & A. Malej, 2010. Recurrence of bloom-forming scyphomedusae: wavelet analysis of a 200-year time series. Hydrobiologia 645: 81–96.

Kroiher, M. & S. Berking, 1999. On natural metamorphosis inducers of the cnidarian *Hydractinia echinata* (Hydrozoa) and *Aurelia aurita* (Scyphozoa). Helgoland Marine Research 53: 118–121.

Lakkis, S., 1991. Aggregations of the scyphomedusa *Rhizostoma pulmo* in the Lebanese coastal waters during the summer of 1986. In Jellyfish Blooms in the Mediterranean. Proceedings of the II Workshop on Jellyfish in the Mediterranean Sea. MAP Technical Report Series 47. UNEP, Athens: 119–127.

Lo, W. T. & I. L. Chen, 2008. Population succession and feeding of scyphomedusae *Aurelia aurita* in a eutrophic tropical lagoon in Taiwan. Estuarine, Coastal and Shelf Science 76: 227–238.

Lucas, C. H., 2001. Reproduction and life history strategies of the common jellyfish, *Aurelia aurita*, in relation to its ambient environment. Hydrobiologia 451: 55–68.

Ludwig, F. D., 1969. Die Zooxanthellen bei *Cassiopea andromeda*, Eschscholtz 1829 (Polyp Stadium) und Ihre Bedeutung für die Strobilation. Zoologische Jahrbbucher-Abteilung für Anatomie und Ontogenie der Tiere 86: 238–277.

Lynam, C. P., S. J. Hay & A. S. Brierley, 2005. Jellyfish abundance and climatic variation: contrasting responses in oceanographically distinct regions of the North Sea, and possible implications for fisheries. Journal of the Marine

Biological Association of the United Kingdom 85: 435–450.

Ma, X. & J. E. Purcell, 2005. Temperature, salinity, and prey effects on polyp versus medusa bud production by the invasive hydrozoan *Moerisia lyonsi*. Marine Biology 147: 225–234.

Mas, J., 1999. Estudio de la dinámica de poblaciones de las medusas *Cotylohiza tuberculata*, *Rhizostoma pulmo* y *Aurelia aurita* en el Mar Menor y su problemática asociada. Instituto Español de Oceanografía, La Coruña, Internal Report.

Mills, C. E., 1995. Medusae, siphonophores, and ctenophores as planktivorous predators in changing global ecosystems. ICES Journal of Marine Science 52: 575–581.

Mills, C. E., 2001. Jellyfish blooms: are populations increasing globally in response to changing ocean conditions? Hydrobiologia 451: 55–68.

Möller, H., 1980. Population dynamics of *Aurelia aurita* medusae in the Kiel Bight, Germany (FRG). Marine Biology 60: 123–128.

Montgomery, M. K. & P. M. Kremer, 1995. Transmission of symbiotic dinoflagellates through the sexual cycle of the host scyphozoan *Linuche unguiculata*. Marine Biology 124: 147–155.

Nawroth, J. C., K. E. Feitl, S. P. Colin, J. H. Costello & J. O. Dabiri, 2010. Phenotypic plasticity in juvenile jellyfish medusae facilitates effective animal-fluid interaction. Biology Letters 6: 389–393.

Pagès, F., 2001. Past and present anthropogenic factors promoting the invasion, colonization and dominance by jellyfish of a Spanish coastal lagoon. In CIESM Gelatinous Zooplankton Outbreaks: Theory and Practice 14. CIESM Workshop Series, Monaco: 69–71.

Parsons, T. R. & C. M. Lalli, 2002. Jellyfish population explosions: revisiting a hypothesis of possible causes. La Mer 40: 111–121.

Pauly, D., V. Christensen, S. Guénette, T. J. Pitcher, U. R. Sumaila, C. J. Walters, R. Watson & D. Zeller, 2002. Towards sustainability in world fisheries. Nature 418: 695–986.

Pérez-Ruzafa, A. & C. Marcos, 1992. Colonization rates and dispersal as essential parameters in the confinement theory to explain the structure and horizontal zonation of lagoon benthic assemblages. Rapports et Proces-Verbaux des Reunions Commission Internationale pour l'Exploration Scientifique de la Mer Mediterranee 33: 100.

Pérez-Ruzafa, A., C. Marcos, M. Pérez-Ruzafa & J. D. Ros, 1987. Evolución de las características ambientales y los poblamientos del Mar Menor (Murcia, SE de España). Anales de Biología 12(Biología Ambiental 3): 53–65.

Pérez-Ruzafa, A., C. Marcos-Diego & J. D. Ros, 1991. Environmental and biological changes related to recent human activities in the Mar Menor (SE of Spain). Marine Pollution Bulletin 23: 747–751.

Pérez-Ruzafa, A., J. Gilabert, J. M. Gutiérrez, A. I. Fernández, C. Marcos & S. Sabah, 2002. Evidence of a planktonic food web response to changes in nutrient input dynamics in the Mar Menor coastal lagoon, Spain. Hydrobiologia 475–476: 359–369.

Pérez-Ruzafa, A., M. I. Hegazi, I. M. Pérez-Ruzafa & C. Marcos, 2008. Differences in spatial and seasonal patterns of macrophyte assemblages between a coastal lagoon and the open sea. Marine Environmental Research 65: 291–314.

Pierce, J., 2005. A system for mass culture of upside-down jellyfish *Cassiopea* spp. as a potential food item for medusivores in captivity. International Zoo Yearbook 39: 62–69.

Pitt, K. A., 2000. Life history and settlement preferences of the edible jellyfish *Catostylus mosaicus* (Scyphozoa: Rhizostomeae). Marine Biology 136: 269–279.

Pitt, K. A. & J. E. Purcell (eds), 2009. Jellyfish blooms: causes, consequences, and recent advances, Hydrobiologia, Vol. 616. Springer, Dordrecht.

Pitt, K. A., K. Koop & D. Rissik, 2005. Contrasting contributions to inorganic nutrient recycling by the co-occurring jellyfishes, *Catostylus mosaicus* and *Phyllorhiza punctata* (Scyphozoa, Rhizostomeae). Journal of Experimental Marine Biology and Ecology 315: 71–86.

Prieto, L., D. Astorga, G. Navarro & J. Ruiz, 2010. Environmental control of phase transition and polyp survival of a massive-outbreaker jellyfish. PLoS ONE 5: e13793.

Purcell, J. E., 2005. Climate effects on formation of jellyfish and ctenophore blooms: a review. Journal of the Marine Biological Association of the United Kingdom 85: 461–476.

Purcell, J. E., 2012. Jellyfish and ctenophore blooms coincide with human proliferations and environmental perturbations. Annual Review of Marine Science 4: 209–235.

Purcell, J. E. & M. V. Sturdevant, 2001. Prey selection and dietary overlap among zooplanktivorous jellyfish and juvenile fishes in Prince William Sound, Alaska. Marine Ecology Progress Series 210: 67–83.

Purcell, J. E., T. A. Shiganova, M. B. Decker & E. D. Houde, 2001. The ctenophore *Mnemiopsis* in native and exotic habitats: US estuaries versus the Black Sea basin. Hydrobiologia 451: 145–176.

Purcell, J. E., S. I. Uye & W. T. Lo, 2007. Anthropogenic causes of jellyfish blooms and direct consequences for humans: a review. Marine Ecology Progress Series 350: 153–174.

Purcell, J. E., D. Atienza, V. Fuentes, A. Olariaga, U. Tilves, C. Colahan, J.-M. Gili, 2012. Temperature effects on asexual reproduction rates of Scyphozoan species from the northwest Mediterranean Sea. Hydrobiologia. doi:10.1007/s10750-012-1047-7.

Rahat, M. & O. Adar, 1980. Effect of symbiotic zooxanthellae and temperature on budding and strobilation in *Cassiopea andromeda* (Eschscholtz). Biological Bulletin 159: 394–401.

Richardson, A. J., A. Bakun, G. C. Hays & M. J. Gibbons, 2009. The jellyfish joyride: causes, consequences and management to a more gelatinous future. Trends in Ecology & Evolution 24(6): 312–322.

Rippingale, R. J. & S. J. Kelly, 1995. Reproduction and survival of *Phyllorhiza punctata* (Cnidaria: Rhizostomeae) in a seasonally fluctuating salinity regime in Western Australia. Marine Freshwater Research 46: 1145–1151.

Schiriati, A., M. Kawahara, S. Uye & H. W. Mianzan, 2008. Life cycle of the jellyfish *Lychnorhiza lucerna* (Scyphozoa: Rhizostomeae). Marine Biology 156: 1–12.

Schwarz, J. A., D. A. Krupp & V. M. Weis, 1999. Late larval development and onset of symbiosis in the scleractinian coral *Fungia scutaria*. Biological Bulletin 196: 70–79.

Shiganova, T. A., Z. A. Mirzoyan & E. A. Studenikina, 2001. Population development of the invader ctenophore

Mnemiopsis leidyi, in the Black Sea and in other seas of the Mediterranean basin. Marine Biology 139: 431–445.
Straehler-Pohl, I. & G. Jarms, 2010. Identification key for young ephyrae: a first step for early detection of jellyfish blooms. Hydrobiologia 645: 3–21.
Sugiura, Y., 1964. On the life-history of rhizostome medusae. II. Indispensability of zooxanthellae for strobilation in *Mastigias papua*. Embryologia 8: 223–233.
Sugiura, Y., 1965. On the life–history of rhizostomae medusae. III On the effects of temperature on the strobilation of *Mastigias papua*. Biological Bulletin 28: 493–496.
Thornhill, D. J., M. W. Daniel & T. C. LaJeunesse, 2006. Natural infections of aposymbiotic *Cassiopea xamachana* scyphistomae from environmental pools of *Symbiodinium*. Journal of Experimental Marine Biology and Ecology 338: 50–56.
Uye, S. & Y. Ueta, 2004. Recent increase of jellyfish populations and their nuisance to fisheries in the Inland Sea of Japan. Bulletin of the Japanese Society of Fisheries and Oceanography 68: 9–19.
Vinogradov, M. E. & E. A. Shushkina, 1992. Temporal changes in community structure in the open Black Sea. Oceanology 32: 485–491.
Widmer, C. L., 2005. Effects of temperature on growth of north-east Pacific moon jellyfish ephyrae, *Aurelia labiata* (Cnidaria: Scyphozoa). Journal of the Marine Biological Association of the United Kingdom 85: 569–573.

Hydrobiologia (2012) 690:157–167
DOI 10.1007/s10750-012-1045-9

JELLYFISH BLOOMS

The potential role of podocysts in perpetuation of the common jellyfish *Aurelia aurita* s.l. (Cnidaria: Scyphozoa) in anthropogenically perturbed coastal waters

Htun Thein · Hideki Ikeda · Shin-ichi Uye

Published online: 16 March 2012
© Springer Science+Business Media B.V. 2012

Abstract Common moon jellyfish, *Aurelia* spp. bloom seasonally in eutrophic or polluted coastal waters around the world. We hypothesized that podocysts, a part of asexual reproduction of the benthic polyps, were important in perpetuating populations of *Aurelia aurita* s.l. in anthropogenically perturbed waters. We examined the effects of temperature, salinity, dissolved oxygen concentrations, and food on the encystment and excystment of *A. aurita* podocysts. Podocysts were formed only by unfed or poorly-fed polyps ($\leq$4.8 μg C polyp^{-1} day^{-1}), indicating that starvation was the primary cause of encystment, while increased temperatures accelerated podocyst production rate. Encystment was never induced by changed salinity (15–32) or dissolved oxygen concentration (1–5 mg O_2 l^{-1}). Excystment occurred only when podocysts were returned to 19°C from 28°C and to oxic waters from hypoxic (0.2–1.0 mg O_2 l^{-1}). The podocysts were capable of surviving for up to 3.2 year. Histology revealed that newly-formed podocysts contained rich organic reserves (e.g., carbohydrate, protein, and lipid) that were gradually consumed while encysted. Podocysts may contribute minimally to increasing *A. aurita* polyp abundance, but they can insure maintenance of the population in adverse environmental conditions and in predator attacks. Podocysts may also enable the population to extend to areas where polyp survival is marginal.

Keywords Temperature · Salinity · Predation · Starvation · Dissolved oxygen · Dormant stage

Electronic supplementary material The online version of this article (doi:10.1007/s10750-012-1045-9) contains supplementary material, which is available to authorized users.

Guest editors: J. E. Purcell, H. Mianzan & J. R. Frost / Jellyfish Blooms: Interactions with Humans and Fisheries

H. Thein · H. Ikeda · S. Uye (✉)
Graduate School of Biosphere Science,
Hiroshima University, 4-4 Kagamiyama 1 Chome,
Higashi-Hiroshima 739-8528, Japan
e-mail: suye@hiroshima-u.ac.jp

Introduction

In world's temperate coastal waters, *Aurelia aurita* (Linnaeus) sensu lato (see Dawson & Martin, 2001; Dawson, 2003; Ki et al., 2008 for sibling species of this genus) is the most common scyphozoan. This species has frequently caused problematic blooms particularly in waters with significant anthropogenic influence (e.g., Kuwabara et al., 1969; Graham, 2001; Uye et al., 2003; Uye & Ueta, 2004; Lee et al., 2006; Purcell et al., 2007; Dong et al., 2010). The life cycle of *A. aurita* is comprised of an alternation of planktonic sexual medusa phase and benthic asexual polyp phase; the polyps play a key role in determining the medusa population size. Previous studies (e.g., Chapman, 1968; Coyne, 1973; Arai, 1997; Miyake et al., 2002; Han & Uye, 2010) have demonstrated that *A. aurita* polyps have three methods of asexual reproduction (budding, fission, and podocyst formation) in addition to strobilation, which transforms a polyp to a strobila that produces several tiny jellyfish (ephyrae). The role of podocysts in jellyfish population dynamics has seldom been investigated (Arai, 2009).

According to Chapman (1968), the dome-shaped podocysts of *A. aurita* were first described more than a century ago by Hyde (1894), but not until 1907 did Hérouard (1907) suggest that the podocysts of *Chrysaora* sp. might be an important part of the scyphozoan life cycle. Preliminary studies on cytology, formation, and excystment of podocysts were conducted by Hérouard (1911, 1912a, b) and Tchéou-Tai-Chuin (1930). Later, Chapman (1968) and Blanquet (1972) revealed that the podocysts of *A. aurita* and *Chrysaora quinquecirrha* (Desor) were covered with a chitin–protein complex cuticle tanned by phenolic substances and stored carbohydrate–protein–lipid reserves. The seasonal temperature increase and decrease were responsible for podocyst encystment and excystment, respectively, of *Cyanea* sp. in Connecticut, USA (Brewer & Feingold, 1991). It was suggested that podocyst formation was induced by unfavorable environmental factors to enable their survival of inclement periods (Cargo & Schultz, 1966, 1967; Brewer & Feingold, 1991; Arai, 2009). The podocysts later undergo excystment when environmental conditions are favorable to form active polyps.

Because of increased jellyfish blooms world wide during recent decades, the importance of examining the reproduction of benthic polyps is widely recognized (Purcell et al., 1999; Ishii & Watanabe, 2003; Purcell, 2007; Willcox et al., 2007; Liu et al., 2009; Purcell et al., 2009, submitted). Among bloom-forming jellyfish species, *A. aurita* can inhabit most heavily eutrophic or polluted waters, such as Tokyo Bay (Omori et al., 1995; Ishii, 2001; Ishii & Tanaka, 2001) and Honjo Area of Lake Nakaumi (Han et al., 2009), Japan, where severe benthic deoxygenation (<1 mg O_2 l^{-1}) prevails during summer (Unoki & Kishino, 1977; Yamamuro et al., 2000; Ishii et al., 2008). Although polyps can be substantially tolerant of hypoxia (Condon et al., 2001), when severe hypoxia is prolonged, it can be lethal to polyps (Ishii et al., 2008). In contrast, podocysts are both physical robustness and capable of dormancy. Hence, we hypothesize that podocysts may enable *A. aurita* to endure unfavorable periods to maintain the population in waters where polyp survival is sometimes marginal. To test this hypothesis, we first examined various environmental factors to induce polyps to produce podocysts. Second, we studied which environmental conditions caused podocyst excystment into polyps. Finally, we investigated the longevity of podocysts along with histological examination of podocysts of various ages.

Materials and methods

Podocyst formation

Polyps of *A. aurita* were grown from planulae from a single mature female medusa caught during August 2007 from the Inland Sea of Japan, where water temperature was 28°C and salinity was 32. The polyps then were maintained as a stock-culture at 22°C in filtered seawater of salinity 32. On October 9, 2007, experimental polyps were removed gently from the wall of stock-culture containers using a thin metal blade and placed individually in wells of 6-well polystyrene culture plates, each containing 10 ml of filtered seawater of salinity 32. The plates were kept at 22°C for 1 week in darkness to insure attachment to the well bottom until the start of podocyst-formation experiments. Effects of four environmental parameters were examined: temperature, salinity, dissolved oxygen concentration, and food supply. During the experiment, newly-budded polyps were excised with forceps (Han & Uye, 2010), and strobilated polyps were monitored. Throughout the experiments, polyps were maintained in dark incubators (Nihon-ika Co.) except during feeding (10 min $week^{-1}$), water change (5 min 3 day^{-1}), and observation (1–10 min day^{-1}). The numbers of podocysts produced in different treatments of each parameter were tested by one-way analysis of variance (ANOVA). If the differences among treatments were significant overall, differences between means were tested using Tukey pair-wise comparisons.

To test the effect of temperature on podocyst formation, one culture plate with six polyps was placed at each of six temperatures (i.e., 5, 11, 18, 22, 26, and 28 ± 0.3°C) in well-aerated (≥5.0 mg O_2 l^{-1}), 32 salinity seawater (Table 1). Because podocysts were produced only by poorly-fed polyps (i.e., 1.7–3.3 μg C $polyp^{-1}$ day^{-1}; Han & Uye, 2010), the

Table 1 Experimental conditions to examine the effects of temperature, salinity, dissolved oxygen (DO) concentration, and food supply on podocyst formation of *Aurelia aurita* s.l.

Environmental factor tested	Experimental conditions				
	Temperature (°C)	Food supply (μg C $polyp^{-1}$ day^{-1})	Salinity	DO (mg O_2 l^{-1})	Duration (weeks)
Temperature	5, 11, 18, 22, 26, 28	2.4	32	>5.0	8
Salinity	22	2.4	5, 10, 15, 20, 25,30, 32	>5.0	8
DO	22	0	32	0.2, 0.5, 1.0, 5.0	4
Food supply	22	0, 2.4, 4.8, 12.1,16.9	32	>5.0	8

There are six polyps in each condition

polyps were fed with seven newly hatched *Artemia* sp. nauplii once per week (i.e., 2.4 μg C $polyp^{-1}$ day^{-1}) by placing a pipette containing nauplii near each polyp's tentacles.

The effect of salinity on podocyst formation was tested by placing each plate with six polyps in a container of well-aerated (≥5.0 mg O_2 l^{-1}) seawater having different salinities (i.e., 5, 10, 15, 20, 25, 30, and 32; Table 1). The temperature was 22°C and the food regime was 2.4 μg C $polyp^{-1}$ day^{-1}.

To test the effect of dissolved oxygen concentration on podocyst formation, the polyps were exposed to seawater of four concentrations (i.e., 0.2, 0.5, 1.0, and 5.0 mg O_2 l^{-1}), which were established by bubbling N_2 gas in 8-l, air-tight glass bottles for 4 weeks (Table 1). The oxygen concentration was monitored twice daily by an oxygen meter (LDO™ HQ10, Hach Co.) and lowered to the original levels if a significant increase was detected. The range of the variation was usually ≤20% around each target level. The polyps were starved throughout the experiment at 22°C.

The effect of food supply was tested by feeding polyps daily at five food levels (i.e., 0, 1, 2, 5, and 7 nauplii $polyp^{-1}$ day^{-1}, corresponding to a carbon supply of 0, 2.4, 4.8, 12.1, and 16.9 μg C $polyp^{-1}$ day^{-1}, respectively). Polyps were kept in well-aerated (≥5 mg O_2 l^{-1}), 32 salinity seawater at 22°C (Table 1).

Podocyst excystment

To test the effect of temperature on excystment, we obtained podocysts of two different ages, i.e., young (≤1 month old) and old (17–20 months old). To induce formation of podocysts, polyps were cultured in polystyrene dishes (92 × 92 × 18 mm) at either 28 or 18°C, fed with excess *Artemia* nauplii for 2 weeks, and then starved for 1 month. Then, the parent polyps were removed from the dishes and the podocysts were kept in 32 salinity seawater at their respective temperatures before the experiment. Both young and old podocysts initially at 28°C were immediately transferred to lower temperatures (i.e., 19 and 11°C) and those at 18°C were transferred to higher temperatures (i.e., 22 and 28°C) to examine their excystment during 12 weeks. As controls, the podocysts were maintained at their initial temperatures throughout the experiment. In addition, the podocysts that were initially at 28°C were exposed to lower temperatures in stages, each lasting 3 weeks (i.e., 25, 22, 19, and 13°C), which resembled the seasonal seawater temperature decline. In each treatment, 34–43 young podocysts and 76–96 old podocysts were used.

The effect of salinity on excystment was tested only on young podocysts (37–42 podocysts $treatment^{-1}$), which were produced at 22°C at two salinities (32 and 20). The podocysts that initially were at 32 salinity were immediately transferred to 25 and 15 salinities; those at 20 salinity were transferred to 25 and 32 salinities. Then, their excystment was monitored for 12 weeks at 22°C. Controls consisted of constant salinity at 32 throughout the experiment.

The effect of dissolved oxygen concentration was tested on young podocysts, which were produced in well-aerated (≥5.0 mg O_2 l^{-1}), 32 salinity seawater at 22°C and then transferred to 0.2, 0.5, and 1.0 mg O_2 l^{-1} for 2 weeks and returned to the aerated condition to examine their excystment for 4 weeks. Oxygen concentrations were monitored as in the podocyst formation experiment; the variation was ≤20% around each target concentration. As the control, the podocysts were kept at the initial aerated conditions throughout the experiment. The podocyst numbers were 52–63 $treatment^{-1}$.

Podocyst viability

To examine the ability of unexcysted podocysts to transform into polyps, following the excystment experiments, the chitinous covering of about one-half of the unexcysted podocysts (number: 5–30) was artificially removed with a dissecting needle (Chapman, 1968). In addition, the cuticle was removed artificially at irregular intervals (ca. 3–12 month) from podocysts (number: 5–20) that were produced at 18, 22, and 28°C in well-aerated ($\geq$5.0 mg O_2 l^{-1}), 32 salinity seawater, and kept under their respective conditions for various periods (up to 3.7 year) and still appeared to be viable.

Histological studies of podocysts

Histochemical examinations were made using 1-month-old podocysts. They were fixed in a solution of 4% formaldehyde buffered with 30 mM HEPES (pH 7.4) for 1.5–2 h at room temperature. Fixed specimens were dehydrated using an ethanol series (50, 70, 90, and 100%) and embedded in LR-White resin (London Resin Company) polymerized at 60°C. Sections (1-μm thick) were cut on an ultramicrotome (Leica EM UC6rt) using a glass knife and stained using Mayer's Haematoxylin/Eosin technique for general morphology, the periodic acid schiff (PAS) technique for demonstration of carbohydrates, the acid solochrome cyanine technique for basic proteins and nucleic acids, the alcian blue technique for acidic and neutral mucopolysaccharides, and the methyl green-pyronin (MGP) technique for DNA and RNA. To identify lipids, podocysts were fixed in 2.5% glutaraldehyde and 2% formaldehyde buffered with 30 mM HEPES and post-fixed in 1% OsO_4 in the buffer for 1 h. Fixed specimens were dehydrated with the ethanol series and propylene oxide and then embedded in Quetol 651 resin (Nissin EN, Japan) at 60°C. Sliced sections were stained by the Sudan black B techniques (McGee-Russell & Smale, 1963). As controls, the sections were soaked in 2% potassium hydrate–ethanol solution to remove resin and lipids (Wada et al., 1993) and were stained using the same methods. To validate the different histological examinations, at least five podocysts were used in each staining method.

We also examined the changes in the internal structural of podocysts with age. When newly-produced podocysts were detached from the parent polyps at 22°C and 32 salinity, their locations and birth dates were recorded. Podocysts of known ages that looked viable, judging from the color of the inner cell mass, were removed from the substrate using a fine dissecting needle. They then were fixed in a solution of 2.5% glutaraldehyde and 2% formaldehyde buffered with 0.1 M phosphate containing 0.4 M NaCl (pH: 7.4) and stored overnight at room temperature. After the fixation, a small hole was opened with the tip of a needle on the side of the podocyst to insure penetration of reagents. The fixed podocysts were rinsed with the buffer three times at 15-min intervals using a shaker. These specimens then were post-fixed in a solution of 1% OsO_4 made up in 0.1 M PO_4 buffer at 0°C for 1 h and dehydrated through a graded series of ethanol. Finally, the podocysts were embedded in Quetol 65 resin at 60°C and stored overnight. Sections were mounted on glass slides, stained with 0.5% toluidine blue, and examined with the aid of an Olympus BX5 light microscope.

Results

Podocyst formation

The podocysts of *A. aurita* were opaque, whitish or light-brown in color, and dome-shaped, with diameter and height varying widely from 200 to 700 μm and from 60 to 110 μm, respectively. Two types of podocysts formed, even in a single polyp; some formed at the base of the polyp (Fig. 1A) and some at the end of an extended stolon (Fig. 1B). The first type accounted for 85% of the 320 podocysts observed. Because there were no obvious differences in their shapes and sizes, both types of podocysts were used together in the experiments.

Temperature strongly affected podocyst formation. All six polyps at 5°C strobilated and no podocysts were formed. At 11°C, four polyps strobilated without producing podocysts and two polyps did not strobilate, but produced podocysts with an overall average production of only 0.3 podocysts polyp^{-1}. At temperatures ranging from 11 to 28°C, the numbers of podocysts produced differed significantly among temperatures ($F_{3,20} = 59.933$, $P < 0.001$). In general, the higher the temperature, the earlier and the more the podocysts formed. The highest production was at 28°C

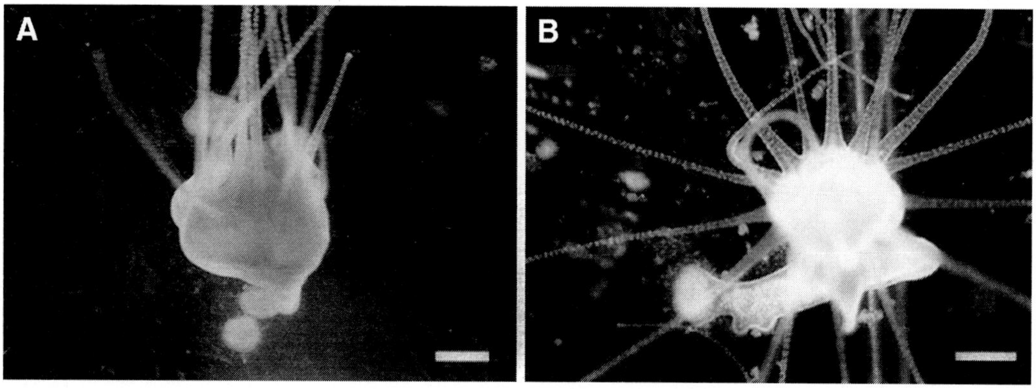

Fig. 1 Formation of *Aurelia aurita* s.l. podocysts at the base of polyp stalk (**A**) and at the end of extended stolon (**B**). Scale bars 500 μm

(mean: 5.0 podocysts polyp^{-1}). After podocyst formation commenced, their production was relatively constant over the experimental period at a given temperature, yielding a linear relationship between cumulative numbers of podocysts and duration (Fig. 2A).

Most of the podocysts were produced at high salinity. At salinity 5, all polyps died within 2 weeks. At salinity 10, polyps appeared to be less active, have less-extended tentacles than polyps in other treatments and they produced no podocysts. At salinity 15, polyps produced an average of 2.3 podocysts. Maximum production (mean: 3.3 podocysts polyp^{-1}) was attained at salinity 32. The effect of salinity on encystment was not significant at the 0.05 level for the salinity range of 15 to 32 ($F_{4,25} = 2.591$, $P = 0.061$; Fig. 2B).

Only the lowest dissolved oxygen concentration (0.2 mg O_2 l^{-1}) affected the polyps, which did not produce any podocysts and died within 1 week. At 0.5 mg O_2 l^{-1}, the polyps looked healthy and produced an average of 0.8 podocysts polyp^{-1}. There was no significant effect of dissolved oxygen concentration on podocyst formation at oxygen concentrations ranging from 0.5 to 5.0 mg O_2 l^{-1} ($F_{2,15} = 0.227$, $P = 0.799$).

Food supply had a significant effect on podocyst production ($F_{4,25} = 147.586$, $P < 0.001$). No podocysts were formed by well-fed polyps at food regimes of 12.1 and 16.9 μg C polyp^{-1} day^{-1}. The mean numbers of podocysts formed increased with decreasing food supply, from 2.0 podocysts in 4.8 μg C polyp^{-1} day^{-1} to 5.2 podocysts in the unfed treatment (Fig. 2C).

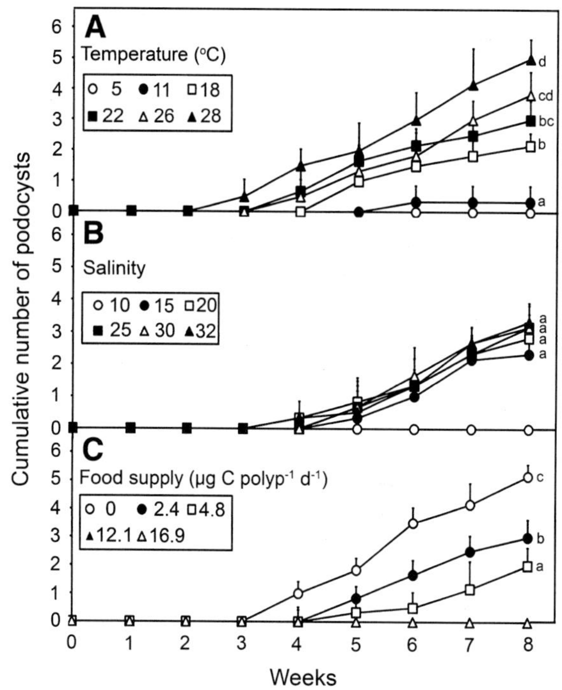

Fig. 2 Cumulative numbers of *Aurelia aurita* s.l. podocysts produced during 8 week at different temperatures (**A**), salinities (**B**), and food supplies (**C**). Vertical bars denote standard deviations. Means with different letters are significantly different

Podocyst excystment

Before the transfer from 28 to 19°C, the 1-month-old podocysts were filled with whitish-colored cell mass (Fig. 3A). Approximately 1–2 weeks before excystment, the cell mass began to shrink to form a round shape and moved to the upper part of the podocyst near an emergence hole. Excystment was accomplished by

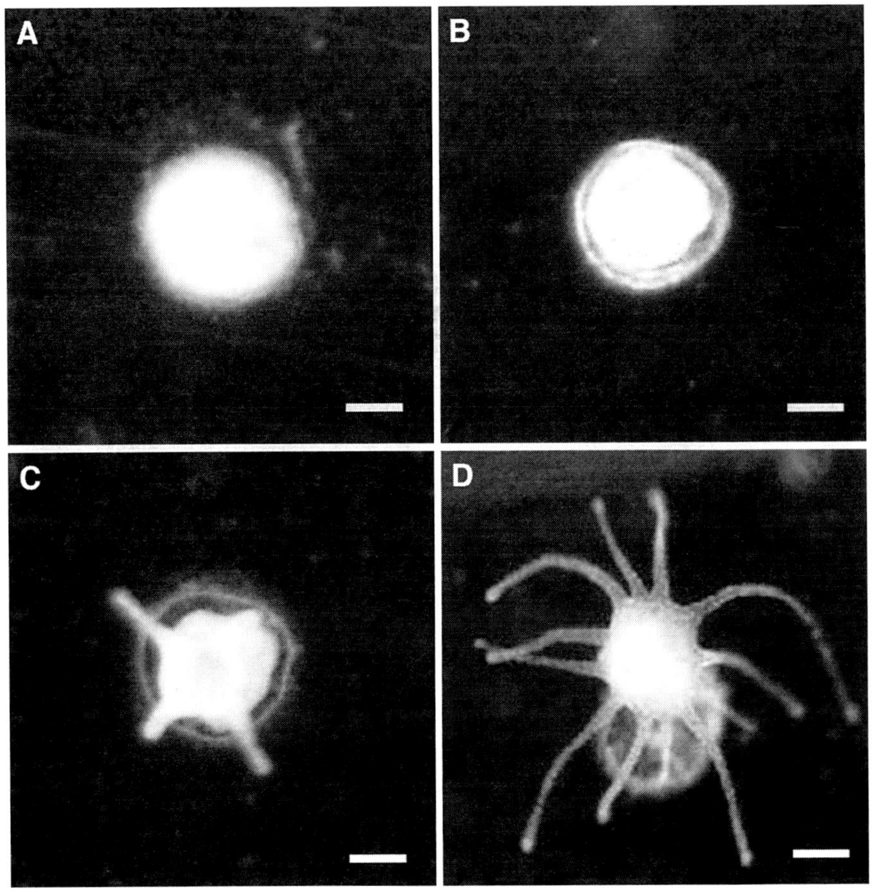

Fig. 3 Top views of the excystment process of 1-month-old *Aurelia aurita* s.l. podocysts when they were transferred directly from 28 to 19°C. **A** A podocyst before the transfer is filled with a whitish-colored cell mass. **B** An excysting podocyst 29 days after the transfer, with the inner cell mass protruding through the emergent hole. **C** A 4-tentacled polyp 3 days after excystment. **D** A 12-tentacled polyp 2 week after excystment. Scale bars 100 μm

the cell mass extruding through the hole (Fig. 3B). Three to 5 days after excystment, the cell mass developed into a polyp with four rudimentary tentacles (Fig. 3C) and simultaneously began feeding. After an additional 4–10 days, polyps had 12–16 tentacles (Fig. 3D).

Decrease in temperature resulted in podocyst excystment. Excystment of young podocysts (≤1 month old) produced at 28°C occurred only after transfer to 19°C, where 7% excysted within 4 weeks and a total of 39% within 8 weeks; thereafter, no further excystment occurred (Fig. 4). In contrast, only 3% of the old podocysts (17–20 months old) excysted with the same treatment (Fig. 4). Neither the young nor the old podocysts produced at 18°C excysted when transferred to warmer temperatures (22 and 28°C).

Results from the gradual decrease in temperature starting at 28°C indicated that 6% of young podocysts excysted after they were transferred to 25°C and then to 22°C. The percentage of excystment increased to 15% after they were subsequently transferred to 19°C, but no further excystment occurred despite additional cooling to 13°C (Fig. 5). A similar gradual temperature decrease did not result in excystment of old podocysts.

Changes in salinity did not cause excystment. No excystment occurred during any salinity treatment.

When young podocysts were returned to well-aerated conditions from the low-oxygen conditions (i.e., 0.2, 0.5, and 1.0 mg O_2 l^{-1}), 25–31% excysted within 1 week. No excystment occurred for podocysts that were kept in aerated conditions throughout the experiment.

Podocyst viability

Even after the excystment experiments, the color and density of cell mass from the unexcysted podocysts looked the same as those before the experiments. When their cuticle cover was removed, the exposed cell mass was induced to transform from a planula-like form into a polyp. Such artificial transformations took

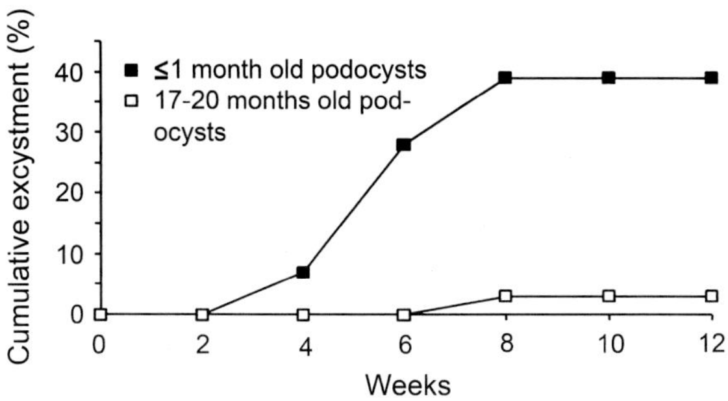

Fig. 4 Cumulative excystment of young (≤1-month-old) and old (17–20-months-old) *Aurelia aurita* s.l. podocysts when they were transferred directly from 28 to 19°C

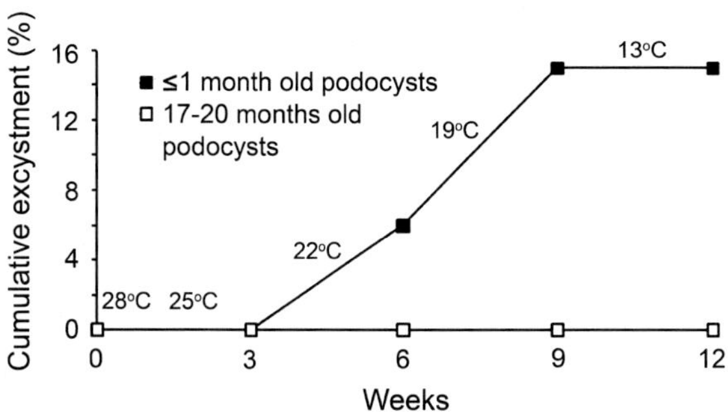

Fig. 5 Cumulative excystment of young (≤1-month-old) and old (17–20-months-old) *Aurelia aurita* s.l. podocysts when they were cooled from 28 to 13°C in stages, each lasting 3 weeks (i.e., 25, 22, 19, and 13°C)

place in 80–100% of the podocysts examined, demonstrating that most of them could survive during adverse conditions that included a low salinity of 15 for at least 12 weeks and hypoxia of 0.2 mg O_2 l^{-1} for at least 2 weeks.

When podocysts were kept for periods longer than a year, the cell mass became darker in color and markedly shrank in some podocysts (Fig. 6). Some of them acquired bacterial and/or fungal infections, which resulted in them becoming empty shells (Fig. 6). Our sporadic examinations of their viability by artificial shell removal revealed that the maximum longevity in the laboratory was 3.2, 3.2, and 2.4 years at 18, 22, and 28°C, respectively. All podocysts died within 3.7 years.

Histology of podocysts

One-month-old podocysts contained a mass of dormant cells of undefined shape in a dome-shaped capsule (Appendix 1; Supplementary material). Each cell had a very small nucleus (ca. 0.5-μm diameter) and many granules of various sizes. Cnidoblasts occurred in the cell mass.

The histochemical methods demonstrated that 1-month-old podocysts contained carbohydrates and proteins, because the cell masses were extensively stained purple with the PAS method and red with acid solochrome cyanin (Appendix 2; Supplementary material). Traces of mucopolysaccharides, which were stained blue by alcian blue, were contained specifically within the cnidoblasts. Although nuclei containing DNA were clearly positive with methyl green, the reaction of pyronin to RNA was very weak. Pyronin-stained cnidoblasts were notable, but this reaction was probably false. Small granules stained black with Sudan black corresponded to lipids. Particles from the control treatment showed diminished staining of lipids. These lipid granules were distinct from those of carbohydrates and proteins.

The internal structure of podocysts changed with age. In newly formed podocysts, the cell mass filled the entire internal space (Appendix 3; Supplementary material). The upper dome cuticle (2–4 μm-thick) was relatively smooth, but the bottom cuticle was thicker and often irregular in thickness. One month later, the cell mass shrank slightly to leave small gaps. The gaps gradually became larger with the age and the innermost layer of the cuticle became denser and thicker. Twelve-month-old podocysts looked mostly hollow, with only approximately 50% of the inner cell mass remaining.

Discussion

Among various modes of asexual reproduction of *A. aurita* polyps, budding and strobilation are the major means to increase population abundance. The rate of budding accelerates with increases of temperature or food supply (Purcell et al., 1999; Ishii & Watanabe, 2003; Purcell, 2007; Willcox et al., 2007; Liu et al., 2009; Han & Uye, 2010). Strobilation, by which a single polyp produces several ephyrae, is induced primarily by lowering temperature below a threshold (ca. 15°C, Kakinuma, 1962; Arai, 1997; Lucus, 2001; Ishii & Watanabe, 2003). On the other hand, the podocyst production rate is much lower (maximum: 0.63 podocysts $polyp^{-1}$ $week^{-1}$ at 28°C, Fig. 1) as compared to budding (maximum: 8.1 buds

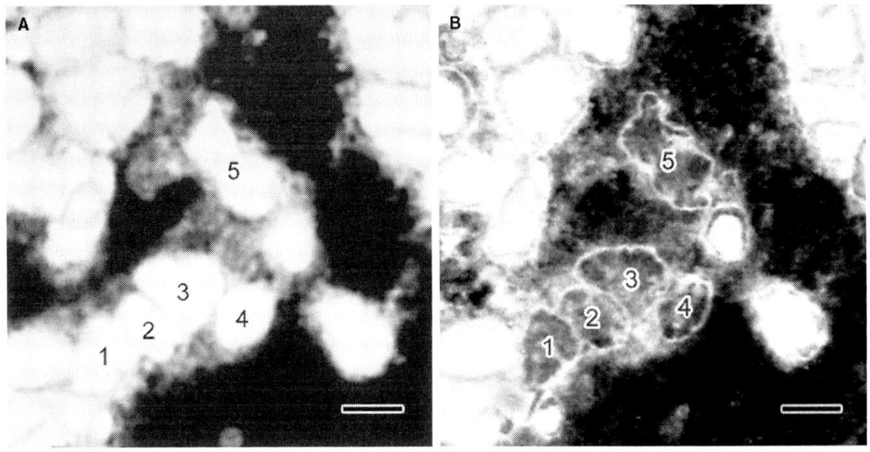

Fig. 6 Photograph of 6-month-old (**A**) and 25-month-old (**B**) *Aurelia aurita* s.l podocysts kept at 22°C. All the 6-month-old podocysts looked intact, but 5 podocysts (with numerals) were degraded by the age of 25-months. Scale bars 300 μm

week^{-1} at 28°C, Han & Uye, 2010), suggesting that podocysts contribute minimally to the immediate increase of the polyp abundance.

The results of our study supported our hypothesis that podocysts were important in perpetuating populations of *A. aurita* s.l. in anthropogenically perturbed waters. Both food supply and temperature were significant environmental factors to induce encystment of *A. aurita* polyps. Our results agreed with those of Han & Uye (2010), in that podocysts were never formed by well-fed polyps but only by unfed and poorly-fed ones, suggesting that starvation was the primary cue to induce podocyst formation in *A. aurita*. Below some food threshold, which may exist between 4.8 and 12.1 μg C polyp^{-1} day^{-1} (Fig. 2), *A. aurita* polyps may shift the allocation of nutrition from bud formation to podocyst production. Podocysts were produced by starved polyps even at 11°C, and their production rate accelerated with increasing temperature to 28°C.

Several previous studies demonstrated the importance of temperature change (either increase or decrease), rather than food supply, on podocyst production. In *Rhopilema esculentum* (Kishinouye), podocyst production increased with temperature increase from 15 to 30°C (Lu et al., 1997). In *Cyanea capillata* (Linnaeus) polyps from both the Chesapeake Bay and Niantic River estuary, Connecticut, podocyst formation occurred when the water temperature underwent seasonal warming (Cargo, 1974; Brewer & Feingold, 1991). In contrast, podocyst formation of *C. quinquecirrha* from the Chesapeake Bay increased when water temperatures cooled (Cargo & Schultz, 1967). For both of those species, podocyst formation coincided with the dormant period of the polyps.

Both cooling and re-oxygenation were effective environmental factors to induce excystment of *A. aurita* podocysts. Interestingly, sudden cooling from 28 to 19°C was more effective than gradual cooling. Temperature also was important for excystment of *C. capillata* podocysts from the Chesapeake Bay, where it occurred upon seasonal cooling (Cargo, 1974). In contrast, excystment of *C. quinquecirrha* podocysts from the Chesapeake Bay occurred when the water temperature rose (Cargo & Schultz, 1967; Cargo & Rabenold, 1980). For both those species, podocyst excystment coincided with the active period of the polyps.

Neither salinity nor dissolved oxygen concentration affected podocyst production in *A. aurita* within the ranges of salinity (15–32) and dissolved oxygen concentration (0.5–5.0 mg O_2 l^{-1}) usually encountered by polyps in coastal waters. Salinity also was not a significant factor for excystment; however, the effect of oxygen was striking, with all excystments occurring within 1 week after returning podocysts to well-aerated seawater from hypoxic (i.e., $\leq$1.0 mg O_2 l^{-1}). All the above results support the idea that podocysts enable survival during stressful environmental conditions.

Podocysts also may provide protection from predators. Previous studies reported that some nudibranch species consumed polyps of *A. aurita*, *C. quinquecirrha*, and *Cyanea* spp., but they did not consume the podocysts (Thiel, 1962; Cargo & Schultz, 1967; Hérnroth & Gröndahl, 1985; Gröndahl, 1988; Gröndahl

& Hernroth, 1987). We found that a nudibranch, *Hermissenda crassicornis* (Eschscholtz), a gastropod, *Calliostoma unicum* (Dunker), and a shrimp, *Rhynchocinetes uritai* (Kubo), consumed >300 *A. aurita* polyps day^{-1}, but they did not consume any podocysts (Uye, unpubl.).

The general morphology, internal structure, and major chemical contents of organic reserves from *A. aurita* podocysts in our study were basically the same as those reported by Chapman (1968). Our study showed that staining of RNA from the organic-matter-rich dormant cells was very weak, indicating that gene expression activity was low during encystment. Thus, basal metabolism of the dormant cells might be low (Black, 1981), enabling the podocyst to live longer. Our histological study revealed that approximately half of the initial reserves were consumed during 1 year of dormancy. Black (1981) reported a similar amount of reduction for *C. quinquecirrha* podocysts, in which half of the DNA, one-third of the proteins, and one-fifth of the lipids contained in newly-formed podocysts were lost during 1 year. In our study, the maximum longevity of *A. aurita* podocysts was 3.2 years, similar to the duration of dormancy reported by Hérouard (1911). Because the success of excystment from the 17–20-month-old podocysts was very low (3%), actual excystment of older podocysts may diminish over time due to the decrease in their reserves.

In temperate coastal waters, where the water temperature ranges from ca. 5 to 30°C over the year, the occurrence of *A. aurita* medusae shows a remarkable seasonal pattern as schematically depicted in Fig. 7 (Yasuda, 1983; Omori et al., 1995; Uye & Shimauchi, 2005). Typically, they appear in the plankton as ephyrae during late winter and early spring, grow to a bell diameter of ca. 20 cm, and attain sexual maturity in early summer, when the population biomass reaches its annual peak. After mid-summer, the medusae become senescent and disappear rapidly from the plankton. Sexually-produced planula larvae attach to substrates and metamorphose to polyps, entering the benthic phase of the life cycle, which may be perennial. The polyps increase their numbers by asexual reproduction (i.e., budding) at high rates when food supply and water temperature are both high. As water temperature decreases, they reproduce at lower rates, and when it is ≤15°C, they transform to strobilae, which later give birth medusa population for the next season. This study adds new ecological significance of podocysts in the seasonal life cycle of *A. aurita*. When polyps are exposed to low food conditions, they reduce or stop budding and instead produce podocysts. Summer is the major season for podocyst production because most were produced at the highest temperatures. In fall, induced by exposure to cooling, these podocysts can excyst to polyps, which maintain the benthic population by budding until they strobilate. Unexcysted podocysts remain dormant until further stimulated.

The physio-ecological abilities of podocysts enabling them to survive stressful environmental conditions may be an important reason that *A. aurita* populations thrive and medusae bloom in heavily

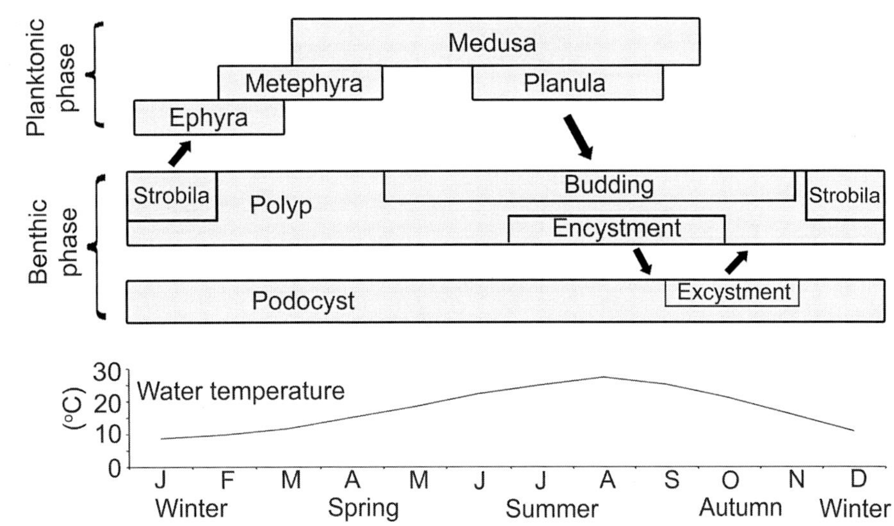

Fig. 7 Schematic representation of the typical seasonal life cycle of *Aurelia aurita* s.l. in temperate East Asian coastal waters with special emphasis on the role of podocysts. Seasonal water temperature is the average for the last 30 years in Hiroshima Bay, the Inland Sea of Japan (Hiroshima Prefectural Technology Research Institute, http://www2.ocn.ne.jp/~hfes/current.html)

eutrophicated or polluted coastal waters world wide. Where polyp survival is sometimes marginal due to adverse conditions, such as low salinity or hypoxia, their physical robustness and dormancy may maintain the population. These endurance potentials may allow podocysts to act as temporal population outposts in anthropogenically perturbed coastal areas. Once these podocysts are established in such areas, future blooms of *A. aurita* medusae could occur.

Acknowledgments We thank Masato Kawahara and Chang-Hoon Han for their help with the experiments. English of the early manuscript was edited by the Sea Pen Scientific Writing, LLC. This work was partially supported by a research grant from the Agriculture, Forestry and Fisheries Research Council (POMAL-STOPJELLY Project).

References

Arai, M. N., 1997. A Functional Biology of Scyphozoa. Chapman & Hall, New York: 316 pp.

Arai, M. N., 2009. The potential importance of podocysts to the formation of scyphozoan blooms: a review. Hydrobiologia 616: 241–246.

Black, R. E., 1981. Metabolism and ultrastructure of dormant podocysts of *Chrysaora quinquecirrha* (Scyphozoa). Journal of Experimental Zoology 218: 175–182.

Blanquet, R. S., 1972. Structure and chemical aspects of the podocyst cuticle of the scyphozoan medusa, *Chrysaora quinquecirrha*. Biological Bulletin 142: 1–10.

Brewer, R. H. & J. S. Feingold, 1991. The effect of temperature on the benthic stages of *Cyanea* (Cnidaria: Scyphozoa), and their seasonal distribution in the Niantic River estuary, Connecticut. Journal of Experimental Marine Biology and Ecology 152: 49–60.

Cargo, D. G., 1974. Comments on the laboratory culture of Scyphozoa. In Smith, W. L. & M. H. Chanley (eds), Culture of Marine Invertebrate Animals. Plenum Publishing Corporation, New York: 145–154.

Cargo, D. G. & G. E. Rabenold, 1980. Observations on the asexual reproductive activities of the sessile stages of the sea nettle *Chrysaora quinquecirrha* (scyphozoan). Estuaries 3: 20–27.

Cargo, D. G. & L. P. Schultz, 1966. Notes on the biology of sea nettle, *Chrysaora quinquecirrha*, in Chesapeake Bay. Chesapeake Science 7: 95–100.

Cargo, D. G. & L. P. Schultz, 1967. Further observations on the biology of the sea nettle and jellyfishes in Chesapeake Bay. Chesapeake Science 8: 209–220.

Chapman, D. M., 1968. Structure, histochemistry and formation of the podocyst and cuticle of *Aurelia aurita*. Journal of the Marine Biological Association of the United Kingdom 48: 187–208.

Condon, R. H., M. B. Decker & J. E. Purcell, 2001. Effects of low dissolved oxygen on survival and asexual reproduction of scyphozoan polyps (*Chrysaora quinquerirrha*). Hydrobiologia 451: 89–95.

Coyne, J. A., 1973. An investigation of the dynamics of population growth and control in scyphistomae of the scyphozoan *Aurelia aurita*. Chesapeake Science 14: 55–58.

Dawson, M. N., 2003. Macro-morphological variation among cryptic species of the moon jellyfish, *Aurelia* (Cnidaria: Scyphozoa). Marine Biology 143: 369–379.

Dawson, M. N. & L. E. Martin, 2001. Geographic variation and ecological adaptation in *Aurelia* (Scyphozoa, Semaeostomeae): some implications from molecular phylogenetics. Hydrobiologia 451: 259–273.

Dong, Z., D. Liu & K. K. John, 2010. Jellyfish blooms in China: dominant species, causes and consequences. Marine Pollution Bulletin 60: 954–963.

Graham, W. M., 2001. Numerical increase and distribution shifts of *Chrysaora quinquecirrha* (Desor) and *Aurelia aurita* (Linne) (Cnidaria: Scyphozoa) in northern Gulf of Mexico. Hydrobiologia 451: 97–111.

Gröndahl, F., 1988. A comparative ecological study on the scyphozoans *Aurelia aurita*, *Cyanea capillata* and *C. lamarckii* in the Gullmar Fjord, western Sweden, 1982–1986. Marine Biology 97: 541–550.

Gröndahl, F. & L. Hernroth, 1987. Release and growth of *Cyanea capillata* (L.) ephyrae in the Gullmar Fjord, western Sweden. Journal of Experimental Marine Biology and Ecology 106: 91–101.

Han, C. H. & S. Uye, 2010. Combined effects of food supply and temperature on asexual reproduction and somatic growth of polyps of the common jellyfish *Aurelia aurita* s.l. Plankton Benthos Research 5: 98–105.

Han, C. H., M. Kawahara & S. Uye, 2009. Seasonal variations in the trophic relationship between scyphomedusa *Aurelia aurita* s.l. and mesozooplankton in a eutrophic brackish-water lake. Japan. Plankton and Benthos Research 4: 14–22.

Hérnroth, L. & F. Gröndahl, 1985. On the biology of *Aurelia aurita* (L.): 3. Predation by *Coryphella verrucosa* (Gastropoda, Opisthobranchia), a major factor regulating the development of *Aurelia* populations in the Gullmar Fjord, western Sweden. Ophelia 24: 37–45.

Hérouard, E., 1907. Existence de statoblasts chez le scyphistome. Comptes Rendus Hebdomadaires des Seances de l Academie des Sciences, Paris 145: 601–603 (from Chapman, D. M., 1968).

Hérouard, E., 1911. Sur la progenèse parthénogénésique à longue échéance de *Chrysaora*. Comptes Rendus Hebdomadaires des Seances de l Academie des Sciences, Paris 153: 1094–1095 (from Chapman, D. M., 1968).

Hérouard, E., 1912a. Histoire du kyste pédieux de *Chrysaora* et sa signification. Archives of Zoological Experimental Genetics 10: 11–25 (from Chapman, D. M., 1968).

Hérouard, E., 1912b. Préparations concernant la formation du kyste pédieux de *Chrysaora*. Bulletin of the Zoological Society of France 37: 44–45 (from Chapman, D. M., 1968).

Hyde, I. H., 1894. Entwicklungsgeschichte einiger Scyphomedusen. Zeitschrift Fur Wissenschaftliche Zoologie 58: 531–565 (from Chapman, D. M., 1968).

Ishii, H., 2001. The influence of environmental changes upon the coastal plankton ecosystems, with special reference to mass occurrence of jellyfish. Bulletin of the Plankton

Society of Japan 48: 55–61 (in Japanese with English abstract).

Ishii, H. & F. Tanaka, 2001. Food and feeding of *Aurelia aurita* in Tokyo Bay with an analysis of stomach contents and a measurement of digestion time. Hydrobiologia 451: 311–320.

Ishii, H. & T. Watanabe, 2003. Experimental study of growth and asexual reproduction in *Aurelia aurita* polyps. Sessile Organisms 20: 69–73.

Ishii, H., T. Ohba & T. Kobayashi, 2008. Effects of low dissolved oxygen on planula settlement, polyp growth and asexual reproduction of *Aurelia aurita*. Plankton Benthos Research 3: 107–113.

Kakinuma, Y., 1962. On some factors for the differentiations of *Cladonema uchidai* and of *Aurelia aurita*. Bulletin Marine Biological Station of Asamushi 11: 81–85.

Ki, J. S., D. S. Hwang, K. Shin, W. D. Yoon, D. Lim, Y. S. Kang, Y. Lee & J. S. Lee, 2008. Recent moon jelly (*Aurelia* sp.) blooms. I. Korean coastal waters suggest global expansion: examples inferred from mitochondrial COI and nuclear ITS-5.85 rDNA sequences. ICES Journal of Marine Science 65: 443–452.

Kuwabara, R., S. Sato & N. Noguchi, 1969. Ecological studies on the medusa, *Aurelia aurita*. 1. Distribution of *Aurelia* patches in the north-eastern region of Tokyo Bay in summer of 1966 and 1967. Bulletin of the Japanese Society of Fisheries Oceanography 35: 156–162 (In Japanese with English abstract).

Lee, J. H., H. W. Choi, J. Chae, D. S. Kim & S. B. Lee, 2006. Performance analysis of intake screens in power plants on mass impingement of marine organisms. Ocean Polar Research 28: 385–393.

Liu, W. C., W. T. Lo, J. E. Purcell & H. H. Chang, 2009. Effects of temperature and light intensity on asexual reproduction of the scyphozoan *Aurelia aurita* (L.) in Taiwan. Hydrobiologia 616: 247–258.

Lu, N., S. Jiang & J. Chen, 1997. Effect of temperature, salinity and light on the podocyst generation of *Rhopilema esculenta* Kishnouye. Fisheries Science 16: 3–8 (Chinese with English abstract).

Lucus, H. L., 2001. Reproduction and life history strategies of the common jellyfish, *Aurelia aurita*, in relation to its ambient environment. Hydrobiologia 451: 229–246.

McGee-Russell, S. M. & N. B. Smale, 1963. On colouring epon-embedded tissue sections with Sudan black B or Nile blue A for light microscopy. Journal of Microscopical Science 104: 109–115.

Miyake, Y., M. Terazaki & Y. Kakinuma, 2002. On the polyps of the common jellyfish *Aurelia aurita* in Kagoshima Bay. Journal of Oceanography 58: 451–459.

Omori, M., H. Ishii & A. Fujinaga, 1995. Life history strategy of *Aurelia aurita* (Cnidaria, Scyphomedusae) and its impact on the zooplankton community of Tokyo bay. ICES Journal of Marine Science 52: 597–603.

Purcell, J. E., 2007. Environmental factors on asexual reproduction rates of the scyphozoan *Aurelia labiata*. Marine Ecology Progress Series 348: 183–196.

Purcell, J. E., D. Atienza, V. Fuentes, A. Olariaga, U. Tilves, C. Colahan & J.-M. Gili, submitted. Temperature effects on asexual reproduction rates of scyphozoan polyps from the NW Mediterranean Sea. Hydrobiologia. doi:10.1007/s10750-012-1047-7.

Purcell, J. E., J. R. White, D. A. Nemazie & D. A. Wright, 1999. Temperature, salinity and food effects on asexual reproduction and abundance of the scyphozoan *Chrysaora quinquecirrha*. Marine Ecology and Progress Series 180: 187–196.

Purcell, J. E., S. Uye & W. T. Lo, 2007. Anthropogenic causes of jellyfish blooms and their direct consequence for humans: a review. Marine Ecology Progress Series 350: 153–174.

Purcell, J. E., R. A. Hoover & N. T. Schwarck, 2009. Interannual variation of strobilation by the scyphozoan *Aurelia labiata* in relation to polyp density, temperature, salinity, and light conditions in situ. Marine Ecology Progress Series 375: 139–149.

Tchéou-Tai-Chuin, 1930. Le cycle evolutif du scyphistome de *Chrysaora*. Travel Station Biology, Roscoff 8: 1–179 (from Chapman, D. M., 1968).

Thiel, H., 1962. Untersuchungen über die Strobilisation von *Aurelia aurita* Lam. an einer Population der Kieler Förde. Kieler Meeresforschungen 18: 198–230 (from Brewer, R. H. & J. S. Feingold, 1991).

Unoki, S. & M. Kishino, 1977. Average ocean condition and water exchange in Tokyo Bay, Vol. 1. Technical Reports of Physical Oceanography Laboratory, Institute of Physics and Chemistry Research: 1–89 (in Japanese)

Uye, S. & T. Shimauchi, 2005. Population biomass, feeding, respiration and growth rates, ad carbon budget of the scyphomedusae *Aurelia aurita* in the Inland Sea of Japan. Journal of Plankton Research 27: 237–248.

Uye, S. & Y. Ueta, 2004. Recent increase of jellyfish populations and their nuisance to fisheries in the Inland Sea of Japan. Bulletin of the Japanese Society of Fisheries Oceanography 68: 9–19 (In Japanese with English abstract).

Uye, S., N. Fujii & H. Takeoka, 2003. Unusual aggregations of the scyphomedusa *Aurelia aurita* in coastal waters along western Shikoku, Japan. Plankton Biology and Ecology 50: 17–21.

Wada, T., H. Miyake, H. Suzuki, K. Suzuki & Y. Takeoka, 1993. An improved method for light microscopy using extraction techniques of Spurr's resin sections. Japanese Journal of Crop Science 62: 128–129 (in Japanese).

Willcox, S., N. A. Moltschaniwskyj & C. Crawford, 2007. Asexual reproduction in scyphistoma of *Aurelia* sp.: effects of temperature and salinity in an experimental study. Journal of Experimental Marine Biology and Ecology 353: 107–114.

Yamamuro, M., J. Hiratuska & Y. Ishitobi, 2000. Seasonal change in filter-feeding bivalve *Musculista senhousia* population of a eutrophic estuarine lagoon. Journal of Marine Systems 26: 117–126.

Yasuda, T., 1983. Studies on the common jellyfish *Aurelia aurita* (Linne). Japan Fisheries Resource Conservation Association, Tokyo: 189 pp. (in Japanese with English abstract).

Hydrobiologia (2012) 690:169–180
DOI 10.1007/s10750-012-1047-7

Temperature effects on asexual reproduction rates of scyphozoan species from the northwest Mediterranean Sea

Jennifer E. Purcell · Dacha Atienza · Verónica Fuentes · Alejandro Olariaga · Uxue Tilves · Chandler Colahan · Josep-María Gili

Published online: 15 March 2012
© Springer Science+Business Media B.V. 2012

Abstract In recent decades, many areas worldwide have experienced mass occurrences of jellyfish. To determine how temperature may affect jellyfish populations in the northwest (NW) Mediterranean Sea, we maintained polyps of three scyphozoan species, *Aurelia aurita, Rhizostoma pulmo,* and *Cotylorhiza tuberculata* in the laboratory at three temperatures (14, 21, 28°C) to test effects on survival and production of new polyps and ephyrae. Temperature significantly affected survival of all species, with longest survival of *A. aurita* and *R. pulmo* at 14°C and of *C. tuberculata* at 21°C. More polyps were budded by all species at temperatures above 14°C. *A. aurita* produced the most buds polyp^{-1} (43.5) and *R. pulmo* the fewest (8.8). Strobilation occurred only at 14°C for *A. aurita* and at 21°C for *C. tuberculata*. For *R. pulmo*, fewer polyps strobilated and strobilated later at 14°C. These patterns of survival and asexual reproduction were seasonally appropriate for each species in the NW Mediterranean, where *A. aurita* medusae occur earliest (~April–May) in cool waters, followed by *R. pulmo* during May–June, and then by *C. tuberculata* in mid-summer. Comparisons among scyphozoan species suggested that many may be restricted by low temperatures, and that global warming may benefit temperate species, but not tropical or boreal species.

Keywords Global warming · Temperature · Jellyfish · Zooplankton · Bloom · Climate

Introduction

Outbreaks of gelatinous zooplankton species can have detrimental effects on human enterprise and such outbreaks seem to have increased in some areas, perhaps due to human-caused deterioration of the environment (reviewed in Purcell et al., 2007). Among the factors that may favor jellyfish blooms is warming of the global ocean. Recent correlations of gelatinous predator populations with climate variables make a compelling case for the importance of climate in determining their population sizes (reviewed in Purcell, 2005, 2012; Purcell et al., 2007; see also Brodeur et al., 2008; Kogovšek et al., 2010; Licandro et al., 2010). For the temperate scyphozoan species studied, including *Chrysaora quinquecirrha* (Desor 1848) and *Pelagia noctiluca* (Forskål 1795), large populations occurred in warm conditions (Goy et al., 1989; Cargo

Guest editors: J. E. Purcell, H. Mianzan & J. R. Frost / Jellyfish Blooms: Interactions with Humans and Fisheries

J. E. Purcell (✉) · C. Colahan
Shannon Point Marine Center, Western Washington University, 1900 Shannon Point Road, Anacortes, WA 98221, USA
e-mail: purcelj3@wwu.edu

D. Atienza · V. Fuentes · A. Olariaga · U. Tilves · J.-M. Gili
Institut de Ciencies del Mar, CSIC, P. Marítim de la Barceloneta, 37-49 08003 Barcelona, Spain

& King, 1990; Kogovšek et al., 2010; Licandro et al., 2010). Additional evidence that jellyfish may flourish in warm conditions comes from laboratory experiments showing that more buds and new medusae (ephyrae) are produced from the benthic polyps at increasingly higher temperatures in *C. quinquecirrha* and *Aurelia* spp. (Purcell et al., 1999; Purcell, 2007; Liu et al., 2009, 2010).

Earlier studies on scyphozoan polyps sought to determine the stimuli for strobilation (reviews Spangenberg, 1968; Arai, 1997; Lucas, 2001). Changes in several factors, including temperature, food, and light, seemed to trigger strobilation in *Aurelia aurita* Linnaeus 1758 polyps (e.g., Ishii & Watanabe, 2003). Many scyphozoan polyps strobilate in early spring as temperature, light, and prey increase (e.g., *C. quinquecirrha* and *Aurelia* spp.; Purcell et al., 1999, 2009).

The Western Mediterranean Sea has populations of several species of scyphomedusae, most abundantly, *P. noctiluca*, *Rhizostoma pulmo* (Macri 1778), *Cotylorhiza tuberculata* (Macri 1778), and *A. aurita* (in Mariottini & Pane, 2010). There is concern that these species, and the problems they cause, may be increasing (Licandro et al., 2010; Mariottini & Pane, 2010).

Moon jellyfish in the genus *Aurelia* have a global distribution and are responsible for problem blooms in many areas (reviewed in Purcell et al., 2007). *Aurelia aurita* is the native species found throughout the Mediterranean Sea, with high populations in some areas (Papathanassiou et al., 1987; reviewed in Mariottini & Pane, 2010). *Rhizostoma pulmo* was shown to be the most abundant Mediterranean coastal species in recent years, showing marked interannual fluctuations in population densities. Large numbers of *R. pulmo* medusae have been reported in the Northern and Southern Adriatic, the Ionian, and the Eastern and Western Mediterranean seas (reviewed in Mariottini & Pane, 2010). Recent proliferations of *R. pulmo* also have occurred in the Mar Menor, southeastern Spain (Pagés, 2001). *Cotylorhiza tuberculata* is found throughout the Mediterranean Sea and outbreaks have been observed in numerous coastal areas and in the Mar Menor, where it alternates seasonally with *R. pulmo* (reviewed in Mariottini & Pane, 2010; Prieto et al., 2010). In this study, we quantify budding and ephyra production rates of three scyphozan species, *A. aurita, R. pulmo,* and *C. tuberculata*, in response to different temperatures. The null hypotheses tested were that survival, the numbers of buds and ephyrae produced, and the timing and durations of strobilation did not differ among the species or temperatures.

Materials and methods

Polyps were grown from sexually produced planulae released by mature jellyfish reared in the laboratory at the Marine Science Institute in Barcelona, Spain. Ambient 5-μm-filtered sea water was delivered continuously to all aquaria. Cultures of *R. pulmo*, *C. tuberculata*, and *A. aurita* medusae produced new polyps each year. Polyp cultures were maintained at ambient temperatures, which varied from 14 to 25°C seasonally, and salinities (38–39) in 12-h light:12-h dark with fluorescent lighting.

On 11 February 2009, one polyp was placed in each well of six-well polycarbonate culture plates with 10 ml of 5-μm-filtered ambient seawater and allowed to attach. Eighteen polyps of each species were placed in each of three temperatures (14, 21, and 28°C) maintained by three water baths. These temperatures were chosen to test responses over the range of local conditions, with 14 and 28°C representing the typical winter and summer temperatures, respectively. Temperatures measured continuously in the water baths by MicroLog temperature loggers showed little variation within each bath (14.3 ± 0.3, 21.0 ± 0.5, 28.0 ± 0.5°C; means ± standard errors). Light was provided by fluorescent fixtures on a 12 h on–off cycle in the constant temperature (16°C) room housing the water baths. After polyps had 1 week to reattach and acclimate to the experimental temperatures, beginning on 19 February 2009 newly hatched *Artemia salina* nauplii were fed in excess every other day. After the polyps fed for 1 h, the wells were cleaned with swabs; seawater and uneaten food were discarded and replaced with filtered seawater of the same temperature. This feeding protocol was intended to provide saturating prey briefly, resulting in equal feeding in all treatments by minimizing enhanced feeding at warmer temperatures (Ma & Purcell, 2005). The numbers of buds and ephyrae were counted for each polyp at each cleaning. After enumeration, ephyrae and buds that had detached from the initial polyps were removed. The experiment was terminated after 108 days (12 June 2009). Substantial mortality occurred between 11 and 24 February 2009. Therefore, 24 February was considered the first day of the experiment and the numbers of polyps tested adjusted accordingly (Table 1).

We tested the null hypothesis that the percentages of polyps surviving did not differ by temperature or species by χ^2. Other data first were tested for normality and equality of variances. We chose not to test the data for two factors (species and temperature) by two-way analysis of variance (ANOVA) because interactions could not be tested due to missing data (100% mortality, no strobilation). Therefore, we used *t* tests or one-way ANOVAs to test the null hypotheses that days of survival and asexual reproduction variables did not differ among species or among temperatures. No transformation corrected normality or variances for some data, which then were analyzed by non-parametric analogues (Mann–Whitney Rank Sum Test and Kruskal–Wallis one-way ANOVA on ranks, respectively); pair-wise comparisons were by Tukey Tests or Dunn's method (non-parametric).

Results

Survival

Considerable mortality occurred in two species soon after the experiment was set up. At 28°C, eight *A. aurita* polyps died 4 days after feeding began and excessive evaporation in the wells killed the remaining *A. aurita* 2 weeks later. Four *R. pulmo* polyps at 21°C and eleven at 28°C died before feeding began. To remove that early mortality from the analysis, we set 24 February 2009 as day 1 of the experiment and eliminated *A. aurita* at 28°C. Polyps surviving <70 days also were removed from all analyses, except for the numbers of days survived. Thus, the numbers of polyps of each species in the analyses were less than the original 18, especially at 28°C (Table 1). Additional mortality occurred after 108 days, especially for *C. tuberculata*; therefore, we set 108 day as the end of the experiment.

The percentages of polyps surviving at least 70 days differed significantly among species-temperature treatments and the null hypothesis was rejected (χ^2; $P < 0.001$; Table 1). The three species survived best at different temperatures. Most *A. aurita* and *R. pulmo* polyps survived at 14°C (94.4 and 100%, respectively). Their survival was lower at 21°C (75 and 30%, respectively), and high mortality occurred at 28°C before the evaporation accident. Percent survival for 70 days was consistent across temperatures for *C. tuberculata* (~43–56%), but only four and three *C. tuberculata* polyps survived to 108 days at 14 and 28°C, respectively.

The number of days that polyps survived differed significantly across temperatures for each species (rows) and across species at 14 and at 21°C and the null hypothesis was rejected for those temperatures (columns; Table 1). The length of survival showed similar patterns to percentage survival. *A. aurita* and *R. pulmo* polyps survived for the duration of the experiment at 14°C. Polyps of *C. tuberculata* survived longest at 21°C. Survival duration at 14°C differed among species, with *A. aurita* and *R. pulmo* surviving longer (>100 days) than *C. tuberculata* (79; Table 1). At 21°C, *R. pulmo* had shorter survival (51 days) than the other species (>80 days). Days of survival (~50–64) did not differ significantly among species at 28°C.

Bud production

Polyps began producing buds within a week after attaching in the experimental wells, except for numerous unattached *R. pulmo* polyps. The number of buds produced per polyp differed significantly across temperatures for each species and the null hypothesis was rejected (Table 1). The most buds were produced by *A. aurita* polyps, with more produced at 21 than at 14°C (43.5 vs. 14.0, respectively). The fewest buds of all species were produced by *R. pulmo*; no polyps budded at 14°C where many polyps remained unattached throughout the experiment, 3–7 buds were produced at 21°C, but one of the two surviving polyps at 28°C produced 31 buds. The few *R. pulmo* polyps surviving at 21 and 28°C were attached to the well surfaces. Polyps of *C. tuberculata* produced many more buds (22 buds polyp^{-1}) at the higher temperatures than at 14°C (1 bud polyp^{-1}).

Bud production differed significantly among species at 14 and at 21°C and the null hypothesis was rejected (Table 1). At 14°C, *A. aurita* produced more buds polyp^{-1} (14) than did *R. pulmo* (0) and *C. tuberculata* (1). At 21°C, all species differed significantly, ranging from 2.5 buds polyp^{-1} for *R. pulmo* to 43.5 for *A. aurita*. *C. tuberculata* polyps

Table 1 *Aurelia aurita, Rhizostoma pulmo,* and *Cotylorhiza tuberculata* asexual reproduction and statistical results

Species	Temperature			Test statistic	*P* value
	14°C	21°C	28°C		
No. of polyps strobilating/surviving ≥70 days/on day 1					
A. aurita	10/17/18	0/12/16	0/0/10	$\chi^2_4 = 72.99$	$P < 0.001$
R. pulmo	2/16/16	4/3/10	2/2/5	–	
C. tuberculata	0/9/16	4/10/18	0/7/16		
Mean no. days survival (SE)					
A. aurita	108.1 (2.9)^a a	89.6 (7.2)^a b	NA	$U_{25} = 199.5$	$P = 0.011$
R. pulmo	111.0 (0)^a a	51.2 (10.6)^b b	59.4 (21.8) b	$H_2 = 16.98$	$P < 0.001$
C. tuberculata	78.8 (9.0)^b a	84.8 (6.4)^a a	49.6 (11.0) b	$H_2 = 6.77$	$P = 0.034$
Test statistic	$H_2 = 21.75$	$H_2 = 9.92$	$t_{19} = 0.42$		
P value	$p < 0.001$	$p = 0.007$	$p = 0.676$		
Mean no. of buds polyp^{-1} (SE)					
A. aurita	14.0 (1.3)^a a	43.5 (2.6)^a b	NA	$t_{25} = -11.3$	$P < 0.001$
R. pulmo	0 (0)^b a	2.5 (1.0)^b b	8.8 (5.6) b	$H_2 = 20.96$	$P < 0.001$
C. tuberculata	1.0 (0.4)^b a	16.4 (1.9)^c b	22.0 (5.6) b	$H_2 = 14.99$	$P < 0.001$
Test statistic	$H_2 = 34.77$	$F_{2,23} = 84.73$	$t_9 = -1.65$		
P value	$P < 0.001$	$P < 0.001$	$P = 0.133$		
Mean no. days until first ephyrae (SE)					
A. aurita	18.0 (1.9)	NA	NA		
R. pulmo	85.5 a	15.0 (2.4) b	9.5 b	$F_{2,7} = 83.21$	$P < 0.001$
C. tuberculata	NA	83.5 (3.4)	NA		
Test statistic	$t_{12} = -12.47$	$U = 16.00$			
P value	$P < 0.001$	$P = 0.029$			
Mean duration of strobilation (SE)					
A. aurita	18.2 (1.3)	NA	NA		
R. pulmo	40.5 a	12.2 (1.6) b	13.5 b	$F_{2,7} = 22.43$	T $P = 0.003$
C. tuberculata	NA	6.8 (1.2)	NA		
Test statistic	$t_{12} = -6.40$	$t_6 = 2.82$			
P value	$P < 0.001$	$P = 0.030$			
Mean no. ephyrae polyp^{-1} (SE)					
A. aurita	13.8 (0.9)	NA	NA		
R. pulmo	13.5 a	6.8 (0.8) b	8.5 ab	$F_{2,7} = 11.11$	$P = 0.014$
C. tuberculata	NA	1.0 (0)	NA		
Test statistic	$t_{12} = 0.11$	$U = 0.00$			
P value	$P = 0.914$	$P = 0.029$			
Mean buds + ephyrae polyp^{-1} (SE)					
A. aurita^a	23.1 (1.4)^a a	43.5 (2.6)^a b	NA	$t_{25} = -7.46$	$P < 0.001$
R. pulmo^b	1.8 (1.2)^b a	9.2 (3.1)^b b	14.8 (5.5) b	$H_2 = 13.75$	$P = 0.001$
C. tuberculata^{bc}	1.0 (0.4)^b a	16.8 (1.9)^b b	22.0 (5.6) b	$H_2 = 14.99$	$P < 0.001$
Test statistic	$H_2 = 32.17$	$F_{2,23} = 52.20$	$t_9 = -0.90$		
P value	$P < 0.001$	$P < 0.001$	$P = 0.390$		

Numbers of polyps at the beginning of the experiment, surviving, and producing ephyrae (strobilating) at three temperatures during a 108-day experiment. The percentages of polyps surviving at least 70 days of the starting number were tested by a χ^2 test. One-way ANOVAs or *t* tests tested for differences among species and temperatures; significant pair-wise differences among temperatures for each species (*rows*) are indicated by different lower-case letters (a, b) and among species for each temperature (*columns*) by different superscript letters (a, b, c). Statistical tests for 28°C had low power because of small sample sizes and results should be interpreted cautiously. *NA* not applicable produced 22 buds polyp^{-1} at 28°C. In addition to buds, seven polyps of *C. tuberculata* at 21°C each produced one free-swimming ball of tissue before the experiment began. Because similar tissue balls did not appear during the experiment, in other temperatures, or in the other species, they were not considered further. Other methods of asexual production, such as from cysts, were not observed.

Strobilation

Strobilation did not occur in many species-temperature treatments; therefore, statistical tests were possible in few cases (Table 1). Strobilation by *A. aurita* occurred only at 14°C in 70.6% of the polyps. *R. pulmo* was the only species that strobilated at all temperatures, but small sample sizes make our conclusions preliminary; 12.5% of the surviving polyps at 14°C and all surviving polyps at 21 and 28°C strobilated. Polyps (33%) of *C. tuberculata* strobilated only at 21°C.

On average at 14°C, *A. aurita* ephyrae appeared after 18 days and ephyra production continued for 18 days, yielding 13.8 ephyrae polyp^{-1} (Table 1). Polyps of *R. pulmo* at 14°C were slow to start (85.5 days) and complete (40.5 days) ephyra production, but much shorter times were required at the higher temperatures ($\leq$15 days for initiation and duration). More *R. pulmo* ephyrae were produced at 14°C (13.5 ephyrae polyp^{-1}) than at 21°C (6.8 ephyrae polyp^{-1}) in the first strobilation. This species also differed from the others in that two polyps strobilated twice. The average number of ephyrae was greater for the second (13 ephyrae polyp^{-1}) than for the first strobilation (8 ephyrae polyp^{-1}). Polyps of *C. tuberculata* at 21°C began to produce a single ephyra after 84 days and released it 7 days later. These results showed that the numbers of ephyrae produced and the times for initiation and duration of production differed by temperature for the tested species and the null hypotheses were rejected.

Similarly high numbers of ephyrae ($\sim$13.5 ephyrae polyp^{-1}) were produced by *A. aurita* and *R. pulmo* at 14°C (Table 1). The averages for *R. pulmo* were 13.5, 20.0, and 15.0 ephyrae polyp^{-1} at 14, 21, and 28°C, respectively, and ephyrae were produced much more quickly at 21 and 28°C than at 14°C (10–16 vs. 85 days). Therefore, the higher temperatures may be better than 14°C for *R. pulmo* ephyra production overall.

Total production

Total asexual production (buds plus ephyrae per polyp) differed significantly across temperatures for each species and the null hypothesis was rejected (Table 1). The most production of the three species was by *A. aurita* polyps at 21°C (44 buds) and at 14°C (14 buds and 14 ephyrae; Fig. 1). Total production was greatest at 28°C for *R. pulmo* (9 buds and 10 ephyrae) and *C. tuberculata* (22 buds), although differences between numbers at 21 and 28°C were not significant. Production of buds relative to ephyrae increased with temperature in *R. pulmo* from 0% at 14°C to 51% at 28°C (Fig. 1). Production of ephyrae relative to buds was low in *C. tuberculata* ($\sim$6% only at 21°C).

Total production differed significantly among species at 14 and at 21°C and the null hypothesis was rejected for those temperatures (Table 1). At 14°C, production of *A. aurita* and *R. pulmo* were higher than of *C. tuberculata*, which had very low production. At 21°C, production of *A. aurita* (buds only) differed significantly from the other species. Differences in production at 28°C were not significant among species.

Discussion

Surface water temperatures in northeastern Spain, where *Aurelia aurita, Rhizostoma pulmo*, and *Cotylorhiza tuberculata* occur, range from about 12°C in winter to 27°C in summer. Thus, the polyps were tested over the range of temperatures in their natural environment. The period of this experiment, from early February to early June, coincided with the probable times of strobilation of these species in the Northwest (NW) Mediterranean Sea; therefore, they were likely to have experienced appropriate seasonal temperature and salinity changes from the ambient seawater prior to the experiment.

Survival

The ability to determine temperature effects on survival of the three species was compromised by accidental mortality, especially in *Aurelia aurita* and *Rhizostoma pulmo* at 28°C. In a similar experiment, all *A. aurita* polyps from the Mediterranean Sea survived at 14, 21, and 28°C (Pascual et al., pers. comm.);

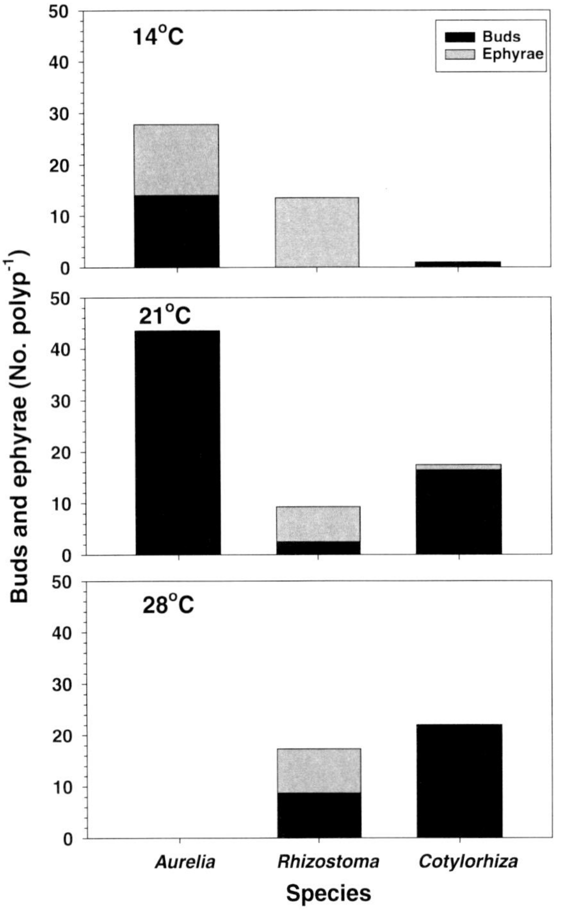

Fig. 1 Numbers of buds and ephyrae produced per polyp *Aurelia aurita, Rhizostoma pulmo*, and *Cotylorhiza tuberculata* (numbers of polyps in each treatment and statistical analyses are in Table 1) in a 108-day experiment at three temperatures (14, 21, and 28°C)

therefore, we conclude that mortality of *A. aurita* at 28°C was due to experimental error in our study and not to a temperature effect. For the species we tested, slightly fewer polyps survived at least 70 days, and survival durations were shorter at 28°C than at the cooler temperatures. Polyps of *A. aurita* from the Baltic Sea had slightly lower survival at 28°C (94%) than at 14 or 21°C (100%), and polyps from the Red Sea had the lowest survival (50%) at 14°C (Pascual et al., pers. comm.). Tropical *A. aurita* from Taiwan had lower survival at 30°C (64%) than at 20 (85%) or 25°C (94%) in a similar experiment (Liu et al., 2009). Thus, populations from different regions are affected differently by temperature extremes.

Among species, both *A. aurita* and *R. pulmo* had the highest and longest survival at 14°C, while *C. tuberculata* had slightly higher and longer survival at 21°C. Survival of *C. tuberculata* polyps from the Mar Menor was 100% at 16°C, but all polyps at lower temperatures (2, 4, and 9°C) died in ≤32 days (Prieto et al., 2010). Polyps of *Rhopilema nomadica* Galil, Spanier & Ferguson, 1990 from Mediterranean waters off Israel (16–28°C) survived at temperatures between 11 and 26°C, but did not feed below 13°C (Lotan et al., 1994). Similar results were reported for *Rhopilema esculentum* Kishinouye 1891 polyps (Chen et al., 1985). Survival of *Aurelia labiata* Chamisso and Eysenhardt 1821 polyps was very high (83–100%) in all experimental temperatures (7, 10, 15, 20°C), but they fed poorly at 5°C in the northeastern Pacific where ambient temperatures range from ~7–14°C (Purcell, 2007). *Pelagia noctiluca*, which lacks a benthic polyp, has not established a resident population in the Northern Adriatic, in part because of low winter temperatures that are frequently below 10°C (Malej & Malej, 2004). Thus, the distribution of scyphozoan polyps and blooms of the jellyfish may be limited by low winter temperatures. By contrast, the polyps of scyphozoans appear tolerant of temperatures higher than normal ambient temperatures.

Budding

Temperature significantly affected budding of new polyps by the three species tested here ($p < 0.001$), with more buds produced at 21 or 28°C than at 14°C. In some polyps tested previously, the numbers of buds produced were inversely related to temperature and strobilation (Table 2). In those species, there appears to be an energetic or metamorphic trade-off between polyp production, which predominated in cool temperatures, and strobilation, which increased as temperatures rose. By contrast, bud production continued to increase with temperature when strobilation was absent or reduced in high temperatures, as for *A. aurita* polyps from the NW Mediterranean and Japan and for *C. tuberculata* polyps. Production of both polyps and ephyrae by other polyps (i.e., *A. aurita* in Taiwan, *A. labiata* in the USA, *Mastigias* sp. in Palau) decreased at unusually high temperatures.

Comparisons among species (Table 2) show that *C. quinquecirrha* and *R. pulmo* produced very few buds (≤0.01 and 0.04 polyp^{-1} d^{-1}, respectively). Several species produced maxima of 0.2–0.3 buds polyp^{-1} d^{-1} (*C. tuberculata, C. andromeda*). *A. aurita*

Table 2 Effects of temperature on asexual reproduction of scyphozoans

Species	Habitat	Env. T (°C)	Exp. T (°C)	Buds (No. d^{-1})	Ephyrae (No. $strob^{-1}$)	Reference
Aurelia aurita	NW Med	12–28	14	0.13	13.8	This study
			21	0.40	NA	
A. aurita	NW Med	12–28	14	0.24	16.8	Pascual et al. (pers. comm.)
			21	0.35	NA	
			28	0.57	NA	
A. aurita	Baltic Sea	7–14	14	0.14	17.5	Pascual et al. (pers. comm.)
			21	0.33	9.7	
			28	0.52	5.1	
A. aurita	North Sea	7–14	10	ND	5.0	Holst (2012)
A. aurita	Red Sea	24–28	14	0.11	3.1	Pascual et al. (pers. comm.)
			21	0.29	3.7	
			28	0.39	7.4	
A. aurita	S Taiwan	20–30	20	0.11	1.0	Liu et al. (2009)
			25	0.04	3.2	
			30	0.03	0.7	
A. aurita	E Japan	5–30	18	0.09–0.48	NA	Han & Uye (2010)
			22	0.15–0.82	NA	
			26	0.15–1.15	NA	
			28	0.12–1.13	NA	
A. labiata	Washington USA	7–14	7	0.08	21.9	Purcell (2007)
			10	0.09	25.3	
			15	0.06	35.2	
			20	0.04	29.6	
Chrysaora quinquecirrha	Chesapeake Bay USA	4–25	5–25	0.005–0.01	1–13	Purcell et al. (1999)
Chrysaora hysoscella	North Sea	7–14	5	ND	NA	Holst (2012)
			10	ND	1.9	
			15	ND	1.6	
Cyanea capillata	North Sea	7–14	5	ND	2.4	Holst (2012)
			10	ND	3.5	
			15		NA	
Cyanea lamarckii	North Sea	7–14	5	ND	2.2	Holst (2012)
			10	ND	2.7	
			15	ND	4.8	
Cassiopea andromeda	Israel	17–27	17–18	0.24	0.3	Rahat & Adar (1980)
			20–22	0.04	3.4	
			25–26	0.01	2.8	
			28–30	0.006	3.1	
Catostylus mosaicus (Quoy and Gaimard, 1824)	SE Australia	10–30	21	ND	1–5	Pitt (2000)
Cotylorhiza tuberculata	NW Med	12–28	14	0.01	0	This study
			21	0.16	40%	
			28	0.23	0	

Table 2 continued

Species	Habitat	Env. T (°C)	Exp. T (°C)	Buds (No. d^{-1})	Ephyrae (No. $strob^{-1}$)	Reference
C. tuberculata	Mar Menor SE Spain	10–31	18	ND	~3%	Prieto et al. (2010)
			19	ND	~5%	
			20–21	ND	~10%	
			23	ND	~25%	
C. tuberculata	Mar Menor SE Spain	10–31	17.5	0.04	ND	Astorga et al. (2012)
Rhizostoma pulmo	NW Med	12–28	14	0	13.5	This study
			21	0.04	6.8	
			28	0.03	8.5	
Nemopilema nomurai Kishinouye 1922	Sea of Japan	10–27	23	ND	3–7	Kawahara et al. (2006)
Mastigias sp.	Goby Lake Palau	30–32	28.7	ND	4%	Dawson et al. (2001)
			31.5	ND	12%	
			33.3	ND	5%	
			34.4	ND	4%	
Rhopilema nomadica	SE Med	16.5–28	18–22	ND	~5	Lotan et al. (1994)
			24–26	ND	~3	

Data are numbers of buds $polyp^{-1}$ d^{-1} or ephyrae $polyp^{-1}$ $strobilation^{-1}$. *Cotylorhiza tuberculata* and *Mastigias* sp. produce only 1 ephyra per strobilation; therefore those data are in percentages of polyps that strobilated. *Env.* environmental; *T* temperature; *Exp.* experimental; *NW Med* northwestern Mediterranean; *S* south; *E* east; *strob* strobilation; *NA* not applicable, no strobilation occurred; *ND* no data

from the NW Mediterranean, Baltic, and Red seas produced as many as 0.4–0.6 buds $polyp^{-1}$ d^{-1}. The amount of asexual reproduction also depends on the amount of food available, which confounds comparisons among species in different experiments.

Strobilation

Of the scyphozoan species tested to date (Table 2), most polyps strobilated over broad temperature ranges. Polyps from habitats with marked seasons, *A. aurita*, *A. labiata*, *C. quinquecirrha*, *Chrysaora hysoscella* (Linneaus, 1767), *Cyanea capillata* (Linnaeus, 1758), and *Cyanea lamarkii* Péron & Lesueur, 1810 strobilated at temperatures ≤15°C, which correspond to late-winter to early spring temperatures in their habitats. Unlike the other species, polyps of boreal *C. capillata* strobilated at cold temperatures (5 vs. 10°C) and did not strobilate at 15°C (Holst, 2012). Polyps of *R. pulmo* from the NW Mediterranean strobilated at 14–28°C, but did not attach or produce buds at 14°C, which appears to be cooler than optimal. Polyps of *A. aurita* from the NW Mediterranean were different from those tested from other locations because they strobilated only at 14°C (Pascual et al., pers. comm.; this study). Polyps of *A. aurita* from other habitats strobilated at all temperatures tested (Baltic and Red seas at 14–28°C; Pascual et al. (pers. comm.), and Taiwan at 20–30°C; Liu et al. (2009)). In these studies on *A. aurita*, experimental temperatures were the same as or higher than the culture temperature and 33–100% of the polyps strobilated. By contrast, experimental temperatures were the same as or lower than culture temperature in Holst (2012) and fewer (5–60%) *A. aurita* polyps from the North Sea strobilated after a 0–10°C temperature decrease and did not strobilate again when temperatures were raised in the spring to autumn. Thus, the different results may have been due to differences in the experimental protocols or to the different population characteristics.

In the boreal and temperate species tested, polyps produced more ephyrae in higher temperatures (Table 2). Differences in strobilation of *A. aurita* polyps from Germany at 14–15°C (100% of polyps from the Baltic Sea, Kiel strobilated vs. 5–6.5% from the North Sea, Helgoland) and in the numbers ephyrae produced (17.5 vs. 5.0 ephyrae polyp^{-1} strobilation^{-1}) could have been due to different feeding regimes (three times per week in Pascual et al., pers. comm. vs. once per 7–10 days in Holst, 2012). Ephyra production was increasingly greater in higher temperatures (to 23°C) in *Cotylorhiza tuberculata* polyps from the Mar Menor Lagoon (southeastern Spain) where temperatures reach 31°C, in contrast to polyps from northeastern Spain, which strobilated at 21°C, but not at 28°C, which may be too high (Table 2). Four species from warmer waters, *A. aurita* (Taiwan), *Rhopilema nomadica, Cassiopea andromeda* (Forskål, 1775), and *Mastigias* sp. had highest ephyra production at intermediate temperatures (Table 2). The range of temperatures tested in our experiments was very broad and was intended to distinguish differences among species and not to determine optimal temperatures.

The numbers of ephyrae produced differed substantially among the various species and populations tested (Table 2). *Aurelia* spp. from cool habitats (Baltic and North seas and Washington, USA) produced the greatest numbers of ephyrae (18–35 polyp^{-1} strobilation^{-1}); however, *A. aurita* from tropical climates (Taiwan and the Red Sea) produced only ~3–7 ephyrae polyp^{-1} strobilation^{-1}. Similarly, other species from tropical climates produced few ephyrae (*R. nomadica, C. andromeda*) (~14 polyp^{-1}). The NW Mediterranean Sea, where the polyps in our study originated, has temperatures that range from cool to very warm (12–28°C), and the polyps produced intermediate numbers of ephyrae (~13–17) or had moderate percentages of the polyps strobilating (40%).

High temperatures consistently shortened the times until strobilation of the several scyphozoan species tested (Fig. 2), which would result in medusae appearing earlier seasonally. *Cotylorhiza tuberculata* is the only species that has been tested twice. The reason for the difference between the mean time to strobilate at 21°C (83.5 days) in our experiment and 22 days in Prieto et al. (2010) is not clear. The feeding regime differed (daily before and on alternate days during our experiment versus fed once weekly before and unfed during the experiment in Prieto et al., 2010), which seems unlikely to explain the difference. Although the date of the experiment was not given in Prieto et al. (2010), it is possible that their polyps were stimulated to strobilate closer to their natural time than occurred in our experiment.

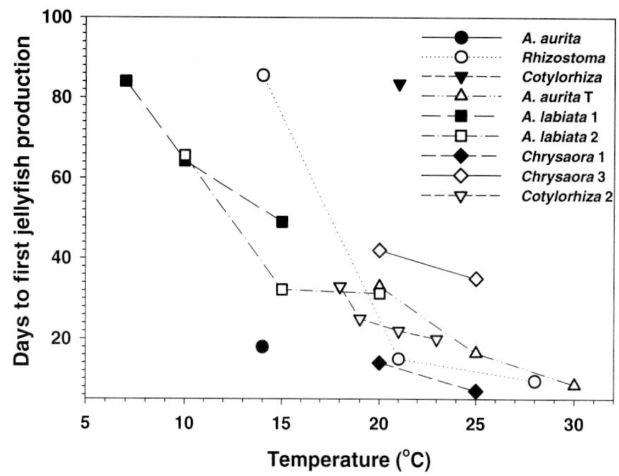

Fig. 2 Effect of temperature on the numbers of days until new jellyfish were produced by polyps of *Aurelia aurita, Rhizostoma pulmo*, and *Cotylorhiza tuberculata* (this study), as compared with *A. aurita* from Taiwan (Lui et al., 2009), *Aurelia labiata* (Exp 1 7–15°C, salinity 20; Exp 2 10–20°C salinity 27; Purcell, 2007), *Cotylorhiza* 2 (Prieto et al., 2010), *Chrysaora quinquecirrha* (Exp 1 and 3; Purcell et al., 1999). Figure updated from Purcell (2007)

Our study showed that *A. aurita*, *R. pulmo*, and *C. tuberculata* from the NW Mediterranean have different temperatures at which strobilation occurs and is greatest. Comparison of these results with the natural arrivals of these three species along the NW Mediterranean coast shows clear coherence of our experimental results and in situ conditions (Fig. 3). Natural populations of *A. aurita* medusae occur mostly during May and June, and strobilation would start a few months earlier when temperatures were about 13–14°C. Natural populations of *R. pulmo* medusae appear during June and July, and strobilation would occur at about 15–16°C (Fig. 3). Finally, *C. tuberculata* medusae appear in situ in August and September, and strobilation would occur at temperatures between 20 and 22°C, which corresponds with the results of our experiments.

Jellyfish and global climate change

Scyphozoan species appear to be able to thrive over wide ranges of temperature and salinity conditions (Purcell et al., 1999; Licandro et al., 2010). Scyphozoans and hydrozoans have especially versatile reproductive strategies (Lucas, 2001; Boero et al., 2002; Fuentes et al., 2011). This versatility may predispose them to exploit changing conditions that may prove detrimental to other taxa. We are only

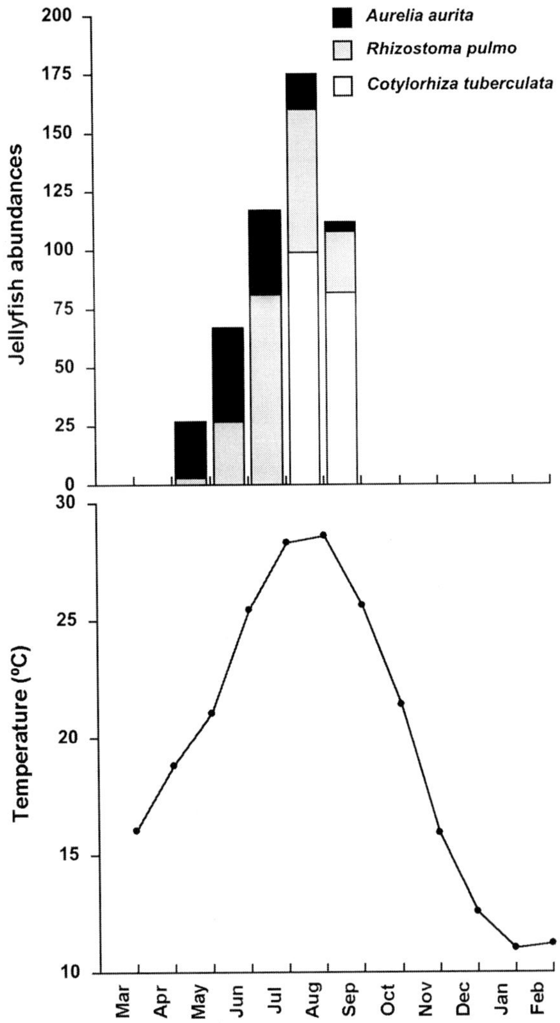

Fig. 3 Temporal patterns of natural populations of *Aurelia aurita* (*black*), *Rhizostoma pulmo* (*gray*), and *Cotylorhiza tuberculata* (*white*) and surface sea water temperatures in the NW Mediterranean Sea in 2008. Scyphozoan data were collected only in May through September

beginning to understand the causes of jellyfish and ctenophore blooms.

Climate models project that global warming due to human influences will be 0.1–0.2°C per decade, and that ocean surface temperatures will rise nearly everywhere (Alley, 2007). Experimental data show that, in most cases, higher temperatures increase asexual production of jellyfish. Long-term variations in jellyfish populations have been correlated with variations in temperature related to climatic cycles (reviewed in Purcell, 2005, 2012; Purcell et al., 2007). Most of the analyses to date show high medusa abundances in warm years: *Aurelia aurita* and *Cyanea* spp. in the Irish Sea (1994–2009, Lynam et al., 2011), and in the North Sea west of Denmark (1971–1986, Lynam et al., 2010). Regional differences in climate effects on medusa abundances are due to incursions of cool oceanic water that enhance *Aurelia aurita* and *Cyanea lamarckii* populations in the northern part of the North Sea, which is opposite to association of those species with warm conditions in the southern part (1971–1986, Lynam et al., 2010). High abundances of *Pelagia noctiluca, Cotylorhiza tuberculata,* and *Rhizostoma pulmo* in 2003 coincided with exceptionally high temperatures and low river flow in the Northern Adriatic Sea (Kogovšek et al., 2010).

In the NW Mediterranean, a warming trend has been reported during the last decades, with an increase by 1.1°C (4×10^{-2}°C year^{-1}) at surface waters since 1970 (Salat & Pascual, 2002). In particular, an increasing trend was identified after 1994, which was confirmed by predominant positive temperature anomalies, with the greatest increase for Catalan Sea in the spring (Sabatés et al., 2006). In a nearby area of the NW Mediterranean, a clear change in climate factors showed a noticeable decrease in precipitation after the 1980s accompanied by higher temperatures (Molinero et al., 2005, 2008).

Therefore, the observed warming in the NW Mediterranean may cause jellyfish to appear earlier and persist longer, perhaps even surviving through the winter. The greatest warming has occurred in spring when polyps strobilate. Jellyfish distributions also may broaden or shift as a result of temperature changes. In summary, ocean warming may increase the population sizes of gelatinous species, lengthen their seasons, increase their distributions, and lead to increasing problems for tourism, fisheries, and other industries in the region.

Acknowledgments We thank the Catalan Water Agency (ACA) for funding the Medusa Project, Miriam Gentile for maintaining the jellyfish cultures, and reviewers for their helpful suggestions.

References

Alley, R. et al., 2007. Climate change 2007: the physical science basis, summary for policymakers: contribution of Working Group I to the fourth assessment report of the Intergovernmental Panel on Climate Change (IPCC). http://www.IPCC.ch. Accessed on 2 Apr 2007.

Arai, M. N., 1997. A Functional Biology of Scyphozoa. Chapman & Hall, London.
Astorga, D., J. Ruiz & L. Prieto, 2012. Ecological aspects of early life stages of *Cotylorhiza tuberculata* (Cnidaria: Scyphozoa) affecting its pelagic population success. Hydrobiologia. doi:10.1007/s10750-012-1036-x.
Boero, F., J. Bouillon, S. Piraino & V. Schmid, 2002. Asexual reproduction in the Hydrozoa (Cnidaria). In Hughes, R. N. (ed.), Reproductive Biology of Invertebrates XI: Progress in Asexual Reproduction. Oxford and IBH Publishing Company, New Delhi: 141–158.
Brodeur, R. D., M. B. Decker, L. Ciannelli, J. E. Purcell, N. A. Bond, P. J. Stabeno, G. L. Hunt Jr & E. Acuna, 2008. Rise and fall of jellyfish in the Bering Sea in relation to climate regime shifts. Progress in Oceanography 77: 103–111.
Cargo, D. G. & D. R. King, 1990. Forecasting the abundance of the sea nettle, *Chrysaora quinquecirrha*, in the Chesapeake Bay. Estuaries 13: 486–491.
Chen, J. K., B. W. Ding & C. Y. Liu, 1985. Effect of nutritional conditions on the strobilation of edible medusa, *Rhopilema esculenta* Kishinouye. Shuichan Xuebao (Journal of Fisheries China) 9: 321–329.
Dawson, M. N., L. E. Martin & L. K. Penland, 2001. Jellyfish swarms, tourists, and the Christ-child. Hydrobiologia 451: 131–144.
Fuentes, V., I. Straehler-Pohl, D. Atienza, I. Franco, U. Tilves, M. Gentile, M. Acevedo, A. Olariaga & J. -M. Gili, 2011. Life cycle of the jellyfish *Rhizostoma pulmo* (Scyphozoa: Rhizostomeae) and its distribution, seasonality and interannual variability along the Catalan coast and the Mar Menor (Spain, NW Mediterranean). Marine Biology 158: 2247–2266.
Goy, J., P. Morand & M. Etienne, 1989. Long-term fluctuations of *Pelagia noctiluca* (Cnidaria, Scyphomedusa) in the western Mediterranean Sea. Prediction by climatic variables. Deep-Sea Research 36: 269–279.
Han, C.-H. & S.-I. Uye, 2010. Combined effects of food supply and temperature on asexual reproduction and somatic growth of polyps of the common jellyfish *Aurelia aurita* s.l. Plankton and Benthos Research 5: 98–105.
Holst, S., 2012. Effects of climate warming on strobilation and ephyra production of North Sea scyphozoan jellyfish. Hydrobiologia. doi:10.1007/s10750-012-1043-y.
Ishii, H. & T. Watanabe, 2003. Experimental study of growth and asexual reproduction in *Aurelia aurita* polyps. Sessile Organisms 20: 69–73.
Kawahara, M., S.-i. Uye, K. Ohtsu & H. Iizumi, 2006. Unusual population explosion of the giant jellyfish *Nemopilema nomurai* (Scyphozoa: Rhizostomeae) in East Asian waters. Marine Ecology Progress Series 307: 161–173.
Kogovšek, T., B. Bogunović & A. Malej, 2010. Recurrence of bloom-forming scyphomedusae: wavelet analysis of a 200-year time series. Hydrobiologia 645: 81–96.
Licandro, P., D. V. P. Conway, M. N. Daly Yahia, M. L. Fernandez de Puelles, S. Gasparini, J. H. Hecq, P. Tranter, et al., 2010. A blooming jellyfish in the northeast Atlantic and Mediterranean. Biology Letters 6: 688–691.
Liu, W.-C., W.-T. Lo, J. E. Purcell & H.-H. Chang, 2009. Effects of temperature and light intensity on asexual reproduction of the Scyphozoan, *Aurelia aurita* (L.) in Taiwan. Hydrobiologia 616: 247–258.
Lotan, A., M. Fine & R. Ben-Hillel, 1994. Synchronization of the life cycle and dispersal pattern of the tropical invader scyphomedusan *Rhopilema nomadica* is temperature dependent. Marine Ecology Progress Series 109: 59–65.
Lucas, C. H., 2001. Reproduction and life history strategies of the common jellyfish, *Aurelia aurita*, in relation to its ambient environment. Hydrobiologia 451: 229–246.
Lynam, C. P., M. J. Attrill & M. D. Skogen, 2010. Climatic and oceanic influences on the abundance of gelatinous zooplankton in the North Sea. Journal of the Marine Biological Association of the United Kingdom 90: 1153–1159.
Lynam, C. P., M. K. S. Lilley, T. Bastian, T. K. Doyle, S. E. Beggs & G. C. Hays, 2011. Have jellyfish in the Irish sea benefited from climate change and overfishing? Global Change Biology 17: 767–782.
Ma, X. & J. E. Purcell, 2005. Temperature, salinity and prey effects on polyp versus medusa bud production of the invasive hydrozoan, *Moerisia lyonsi*. Marine Biology 147: 225–234.
Malej, A. & A. Malej Jr, 2004. Invasion of the jellyfish *Pelagia noctiluca* in the Northern Adriatic: a non-success story. In Dumont, H., et al. (eds), Aquatic Invasions in the Black, Caspian, and Mediterranean Seas. Kluwer Academic Publishers, Dordrecht: 273–285.
Mariottini, G. L. & L. Pane, 2010. Mediterranean jellyfish venoms: a review on Scyphomedusae. Marine Drugs 8: 1122–1152.
Molinero, J. C., F. Ibanez, P. Nival, E. Buecher & S. Souissi, 2005. North Atlantic climate and northwestern Mediterranean plankton variability. Limnology and Oceanography 50: 1213–1220.
Molinero, J. C., M. Casini & E. Buecher, 2008. The influence of the Atlantic and regional climate variability on the long-term changes in gelatinous carnivore populations in the northwestern Mediterranean. Limnology and Oceanography 53: 1456–1467.
Pagés, F., 2001. Past and present anthropogenic factors promoting the invasion, colonization and dominance by jellyfish of a Spanish coastal lagoon. In Briand, F. (ed.), Gelatinous Zooplankton Outbreaks: Theory and Practice. CIESM Workshop Series 14, Monaco: 69–71.
Papathanassiou, E., P. Panayotidis & K. Anagnostaki, 1987. Notes on the biology and ecology of the jellyfish *Aurelia aurita* Lam. in Elefsis Bay (Saronikos Gulf, Greece). PSZNI Marine Ecology 8: 49–58.
Pitt, K. A., 2000. Life history and settlement preferences of the edible jellyfish *Catostylus mosaicus* (Scyphozoa: Rhizostomeae). Marine Biology 136: 269–280.
Prieto, L., D. Astorga, G. Navarro & J. Ruiz, 2010. Environmental control of phase transition and polyp survival of a massive-outbreaker jellyfish. PLoS ONE 5(11): e13793. doi:10.1371/journal.pone.0013793.
Purcell, J. E., 2005. Climate effects on jellyfish populations. Journal of the Marine Biological Association United Kingdom 85: 461–476.
Purcell, J. E., 2007. Environmental effects on asexual reproduction rates of the Scyphozoan, *Aurelia labiata*. Marine Ecology Progress Series 348: 183–196.

Purcell, J. E., 2012. Jellyfish and ctenophore blooms coincide with human proliferations and environmental perturbations. Annual Review of Marine Science 4: 209–235.

Purcell, J. E., J. R. White, D. A. Nemazie & D. A. Wright, 1999. Temperature, salinity and food effects on asexual reproduction and abundance of the Scyphozoan *Chrysaora quinquecirrha*. Marine Ecology Progress Series 180: 187–196.

Purcell, J. E., S.-I. Uye & W.-T. Lo, 2007. Anthropogenic causes of jellyfish blooms and direct consequences for humans: a review. Marine Ecology Progress Series 350: 153–174.

Purcell, J. E., R. A. Hoover & N. T. Schwarck, 2009. Interannual variation of strobilation by the scyphozoan *Aurelia labiata* in relation to polyp density, temperature, salinity, and light conditions in situ. Marine Ecology Progress Series 375: 139–149.

Rahat, M. & O. Adar, 1980. Effect of symbiotic zooxanthellae and temperature on budding and strobilation of *Cassiopeia andromeda* (Eschscholz). Biological Bulletin 159: 394–401.

Sabatés, A., P. Martín, J. Lloret & V. Raya, 2006. Sea warming and fish distribution: the case of the small pelagic fish, *Sardinella aurita*, in the western Mediterranean. Global Change Biology 12: 2209–2219.

Salat, J. & J. Pascual, 2002. The oceanographic and meteorological station at L'Estartit (NW Mediterranean). Tracking long-term hydrological change in the Mediterranean sea. CIESM Workshop Series 16: 29–32.

Spangenberg, D. B., 1968. Recent studies of strobilation in jellyfish. Oceanography and Marine Biology Annual Reviews 6: 231–247.

Hydrobiologia (2012) 690:181–187
DOI 10.1007/s10750-012-1056-6

JELLYFISH BLOOMS

Predator-induced vertical behavior of a ctenophore

Josefin Titelman · Lars Johan Hansson · Trygve Nilsen · Sean P. Colin · John H. Costello

Published online: 15 March 2012
© Springer Science+Business Media B.V. 2012

Abstract Although many studies have focused on *Mnemiopsis leidyi* predation, little is known about the role of this ctenophore as prey when abundant in native and invaded pelagic systems. We examined the response of the ctenophore *M. leidyi* to the predatory ctenophore *Beroe ovata* in an experiment in which the two species could potentially sense each other while being physically separated. On average, *M. leidyi* responded to the predator's presence by increasing variability in swimming speeds and by lowering their vertical distribution. Such behavior may help explain field records of vertical migration, as well as stratified and near-bottom distributions of *M. leidyi*.

Keywords *Beroe* spp. · *Mnemiopsis leidyi* · Ctenophore · Behavior · Vertical distribution · Predator–prey

Guest editors: J. E. Purcell, H. Mianzan & J. R. Frost / Jellyfish Blooms: Interactions with Humans and Fisheries

J. Titelman (✉)
Department of Biology, University of Oslo, PO Box 1066, Blindern, 0316 Oslo, Norway
e-mail: e.r.j.titelman@bio.uio.no

L. J. Hansson
Department of Marine Ecology-Göteborg, University of Gothenburg, Box 461, 405 30 Göteborg, Sweden

T. Nilsen
Department of Mathematics, University of Bergen, PO Box 7800, 5020 Bergen, Norway

S. P. Colin
Environmental Sciences and Marine Biology, Roger Williams University, Bristol, RI 02809, USA

J. H. Costello
Biology Department, Providence College, 549 River Ave, Providence, RI 02918-0001, USA

Introduction

The ctenophore *Mnemiopsis leidyi* (A. Agassiz, 1865) persists in high numbers during the summer to winter, both in its native range along the American Atlantic coasts (Costello et al., 2006; 2012) and in invaded habitats like the Black and Caspian Seas (Vinogradov et al., 2005), the North Sea (Riisgård et al., 2007), the Baltic Sea (Javidpour et al., 2009), and the Mediterranean Sea (Fuentes et al., 2010). The recent invasions of northern European waters have stimulated heightened interest in the role of *M. leidyi* as a competitor and predator of crustacean zooplankton, fish eggs, and larvae (e.g., Colin et al., 2010; Jaspers et al., 2011). An understanding of its ecology also requires quantification of its role as prey, but such studies are sparse (e.g., Oviatt & Kremer, 1977; Purcell & Cowan, 1995; Kreps et al., 1997; Hosia et al., 2011; Hosia & Titelman, 2011).

Although *M. leidyi* remains among the most frequently studied gelatinous plankton, the sensory and behavioral ecology involved in its distributions and its interactions with prey and predators remains poorly understood (Purcell & Cowan, 1995). Many predators exploit lobate ctenophores (Oviatt & Kremer, 1977; Condon & Steinberg, 2008; Hosia & Titelman, 2011). Despite its various post-encounter escape behaviors (reviewed in Titelman et al., 2007), *M. leidyi* is vulnerable to predation or partial predation from gelatinous predators (e.g., Purcell & Cowan, 1995; Kreps et al., 1997; Hosia et al., 2011). In particular, ctenophores in the genus *Beroe* feed on many ctenophores (reviewed in Purcell, 1991). Population control on various ctenophore species have been implicated from field studies in the northwestern Atlantic Ocean (*Beroe ovata* Bruguière, 1789; in Swanberg, 1974; Purcell et al., 2001), the Black Sea (*B. ovata*; in Stone, 2005; Vinogradov et al., 2005), Norwegian coastal waters (*Beroe cucumis* Fabricius, 1780; in Falkenhaug, 1996), and the North Sea (*Beroe gracilis* Künne, 1939; in Greve & Reiners, 1988).

Upon encountering a predator such as *Beroe* spp, a prey ctenophore stands little chance of survival (Swanberg, 1974; Harbison et al., 1978; Falkenhaug, 1996; Hosia et al., 2011). The chemical presence of prey ctenophores triggers search behavior of *Beroe* spp. and engulfment occurs almost instantaneously upon encounter (Swanberg, 1974; Falkenhaug & Stabell, 1996; Hosia et al., 2011). The ability to remotely detect predators could enhance survival probability. Jellyfish exude various dissolved chemicals into the environment (Hansson & Norrman, 1995; Riemann et al., 2006; Titelman et al., 2006; Pitt et al., 2009) that could potentially be used as cues. Some ctenophores possess chemoreceptors (Horridge, 1965; Kass-Simon & Hufnagel, 1992; Aronova & Alekseeva, 2002, 2003). However, documented escape behaviors from predators by *M. leidyi* are generally elicited after direct contact. Such escape behaviors include altering swimming direction and speed (Kreps et al., 1997), as well as tearing away and losing tissue when caught by predators (Purcell & Cowan, 1995; Kreps et al., 1997; Hosia & Titelman, 2011). Responses, such as crumpling, to remote fluid disturbances also exist (Moss et al., 2004). In contrast, escape strategies such as migration and vertical habitat shelters in response to perceived risk are virtually unexplored for ctenophores (e.g., Esser et al., 2004), despite being widespread amongst zooplankton (e.g., reviews in Ohman, 1988; Hays, 2003), including scyphozoan jellyfish (Albert, 2011). Chemical cues from jellyfish can induce vertical behavior in crustacean zooplankton (McKelvey & Forward, 1995; Cohen & Forward, 2003). In our experimental study, we test the hypothesis that *M. leidyi* may remotely perceive risk from predatory ctenophores and adjust their vertical position accordingly.

Methods

Mnemiopsis leidyi and *B. ovata* were collected from Eel Pond, Woods Hole, Massachusetts, USA on the same day as the experiment in August 2008. Experiments were conducted at the Marine Biological Laboratory in natural sea water (22°C, 32‰) that was collected at the same time as the specimens. In our experimental setup, *M. leidyi* and *B. ovata* ctenophores potentially could sense each other remotely, while being spatially separated. The experiment consisted of two treatments (predator: with *B. ovata*, and control: without *B. ovata*), each with three replicates. The order of the experimental trials was randomized (control, predator, predator, control, predator, control) and trials were conducted immediately after one another.

The setup consisted of a 5-l glass aquarium (25 × 25 × 8 cm, length × height × width) with a holder for the predator at the top of the aquarium (Fig. 1). The holder was a funnel made from a PET bottle with the bottom cut off and was centered at the top of the aquarium with the neck (2.2-cm diameter) facing downward about 9 cm below the surface. The submerged part of the holder created a 150-ml isolation chamber where *B. ovata* could be placed. Water could exchange freely between the holder and the rest of the tank, but *B. ovata* could not escape from the holder. The setup was lit from the side with cold white light. The experiment was video recorded in 2D at 30 frames s^{-1} with a SONY HDV camera (HVR-Z7U) equipped with a Carl Zeiss 1.6/4.4–52.8 lens.

In each of the six experimental trials, the tank was first filled with seawater and then 10 *M. leidyi* (total length 31.9 ± 9.3 mm, mean ± SD) were added. The water level was then adjusted to a set mark (0.9 cm from the top). *M. leidyi* were allowed to acclimatize for 10 min. Each *B. ovata* was rinsed in seawater to avoid addition of already released chemicals, and then gently poured with a glass beaker into the holder together with a small amount of filtered seawater (total volume 150 ml). In the control treatment, 150 ml of water was poured into the holder. The introduction of

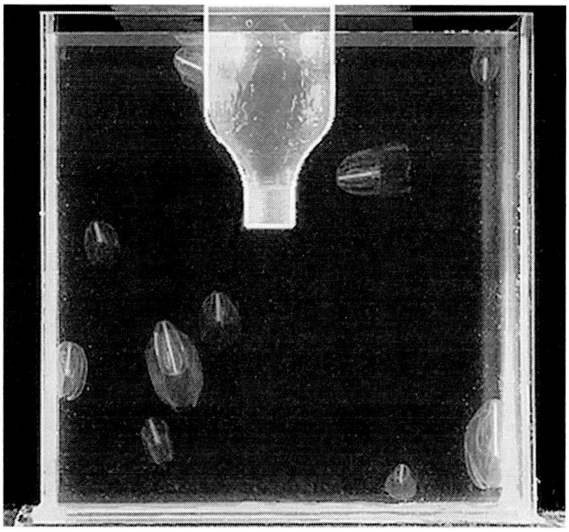

Fig. 1 Experimental tank with funnel containing one *Beroe ovata* predator at *top center* and 10 *Mnemiopsis leidyi* ctenophores in the water. The *outlines* of the holder and ctenophore guts have been enhanced for clarity

B. ovata or the water was considered to be the start of the experiment. The setup was then left undisturbed and video recorded for about 30 min. Each treatment was replicated 3 times using new water and animals. The aquarium was rinsed with hot water and then with natural seawater between trials. To avoid contamination by *B. ovata* chemicals, we used different bottles and transfer jars for the predator and control treatments. The volume of *B. ovata* was measured after each trial (41 ± 18 ml, mean ± SD), and *M. leidyi* total body lengths were measured from the videos. There were no differences in mean *M. leidyi* length between trials (ANOVA, $F_{5,54} = 0.426$, $P = 0.829$). The flow patterns in the tank were not quantified; however, we assume that the water circulation caused by 10 *M. leidyi* in the tank by far exceeded that of one *B. ovata* (Colin et al., 2010).

The position of each *M. leidyi* in the tank at 1-s intervals was determined manually from the video recordings using Image J (Rasband, 2008), with the aboral apex of the ctenophore as the tracking point. The x, y coordinates were smoothed by a running average of 3 steps prior to calculations. To compare positions and motility parameters between the predator and the control treatments, we first calculated each parameter for every individual in all replicates. All statistical analyses, including those for test assumptions, were done by SPSS (14.0), R (R Development Core Team, 2008), or SigmaPlot (10.0 or 11.0). The position of each ctenophore in the tank during the ~30 min of experimentation can be considered independent of their initial position because *M. leidyi* could easily swim across the tank (personal observations) (e.g., Kreps et al., 1997). There were no significant differences in initial distribution between tanks at t_0 (Kruskal–Wallis, $\chi^2 = 9.590$, df = 5, $P = 0.088$). All individual *M. leidyi* were tracked over time (1,780 s after stimulus introduction).

Results

The motility tracks appeared to differ between the two treatments. In the controls, *M. leidyi* used both the upper and lower parts of the tank and generally wandered over much of the available space during the observation period (Fig. 2). In contrast, in the predator treatments, many individuals displayed seemingly more convoluted tracks with a smaller vertical component than those observed in the controls and longer residence times at the lower part of the tank. The variability in apparent behavior was analyzed by considering the data of vertical movement as two panels of time series (one for each treatment) (Fig. 2). The vertical motions of the individuals ("ups and downs") were cyclic but without a fixed frequency (Fig. 2). We therefore modeled the correlations in locations for the individual *M. leidyi* as an autoregressive process of order 2 (AR, $P = 2$). This was done by the function *gls* from the package *nlme* (Pinheiro et al., 2008) of *R* (R Development Core Team, 2008) to the data in Fig. 2. We used time as a covariate and included tank as a factor variable to account for any possible differences between the six trials. We then compared the fits from a homogenous model, in which we forced the variances within the two treatments to be equal, with a heterogeneous model, in which we allowed unequal variance for the two treatments (i.e., H_0: $\sigma_{\text{predator}} = \sigma_{\text{control}}$ vs. $\sigma_{\text{predator}} \neq \sigma_{\text{control}}$ are equal). These models differed significantly from one another (L ratio = 12.26, df = 10, 11, $P = 0.0005$), indicating a significant effect of treatment on individual variability in vertical position (Fig. 2). The better heterogeneous model yielded $\Phi_1 = 0.767$ and $\Phi_2 = -0.107$.

In the second set of analyses, we examined how position (x, y) changed as a function of time (t_i) by using the average tank values ($n = 3$ per treatment) for each time step. To examine the dynamics of these

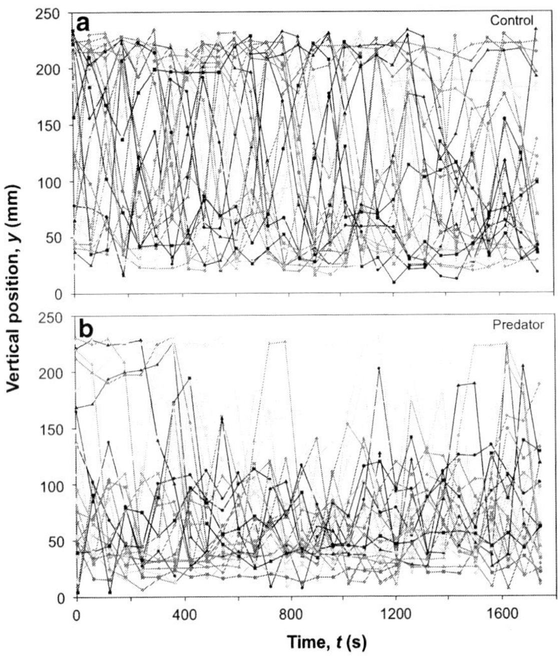

Fig. 2 Vertical positions of *Mnemiopsis leidyi* ctenophores in the water column as a function of time for all individuals in the controls (**a**) and in *Beroe ovata* predator treatments (**b**). *Line color* indicates individual *M. leidyi*, and *symbol type* represents the three replicate tanks. Data are only shown for every 60 s for clarity

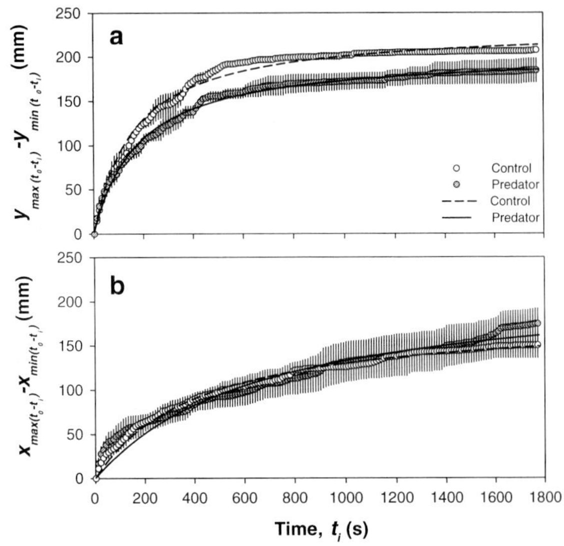

Fig. 3 Spans of vertical, y (**a**) and horizontal, x (**b**) distances covered by *Mnemiopsis leidyi* ctenophores as a function of time in the predator and the control treatments. Data points are mean values from the replicates ± SE ($n = 3$). Every 10th data point is shown for clarity. **a–b** Curve fits $\left(f(t) = \frac{at}{b+t}\right)$ yielded the following coefficients ± SE for **a**: $a_{\text{control}} = 232.7 \pm 0.32$, $b_{\text{control}} = 159.8 \pm 1.15$, $R^2 = 0.95$, and $a_{\text{predator}} = 203.9 \pm 0.51$, $b_{\text{predator}} = 178.2 \pm 2.20$, $R^2 = 0.87$; for **b**: $a_{\text{control}} = 183.7 \pm 0.99$, $b_{\text{control}} = 415.0 \pm 7.03$, $R^2 = 0.82$, and $a_{\text{predator}} = 222.3 \pm 2.51$, $b_{\text{predator}} = 674.2 \pm 18.89$, $R^2 = 0.66$. $P < 0.0001$ for all coefficients and curve fits

parameters over time, we plotted the difference between the maximum and minimum values observed from t_0 to t_i. Patterns were analyzed by fitting the hyperbolic function $f(t) = \frac{at}{b+t}$ to the data. This confirmed that *M. leidyi* in the predator treatment were located lower in the tank than in the controls (Fig. 3). The variability in the mean vertical position of animals in the predator treatment generally decreased with time (cf Fig. 2). The dependency of the fitted coefficients a and b on treatment was tested by comparing mixed effects models with and without the factor treatment and including a random tank effect. This analysis was done using the *nlme* package (Pinheiro et al., 2008) in *R* (R Development Core Team, 2008). Treatment had a significant effect on a and b (L ratio = 14.12, $P = 0.0009$); *M. leidyi* in the predator treatment took longer to explore the vertical range of the aquarium ($b_{\text{control}} = 160$ s vs. $b_{\text{predator}} = 178$ s) and, on average, they used less of the vertical range than did *M. leidyi* in the control treatment ($a_{\text{control}} = 233$ mm, $a_{\text{predator}} = 204$ mm; Fig. 3a). As expected, there was no significant effect of treatment on horizontal placement (Fig. 3b).

In both treatments, *M. leidyi* alternated between slow and faster swimming. When at the bottom of the tank, the ctenophores either rested with their lobe tips at the bottom or moved upwards intermitted with sinking at regular intervals. Similarly, animals were often stationary at the surface for some time before descending. Although plots of mean speed over time suggested few differences, there was higher variation in speed in the predator treatment than in the controls (data not shown, but see Fig. 4).

We tested for differences in vertical position (y) and speed (v) parameters (mean, median, max–min, and variance) in the time-integrated data (Fig. 4). The analyses conducted for the individual variances of vertical position and speed become relevant when animals alter their behavior with time (i.e., all variables here) or when differences in mean or median values are expected to be small because the control treatment is expected to be uniformly distributed

Fig. 4 Individual median (**a**), mean (**b**), maximum–minimum (**c**), and variance (**d**) in vertical (*y*) position of *Mnemiopsis leidyi* ctenophores, and median (**e**), mean (**f**), maximum–minimum (**g**), and variance (**h**) of their speed (*v*) as a function of treatment. Bars represent mean ± SE of all 30 *M. leidyi* in each treatment. Data for each *M. leidyi* were integrated over 1,775 s

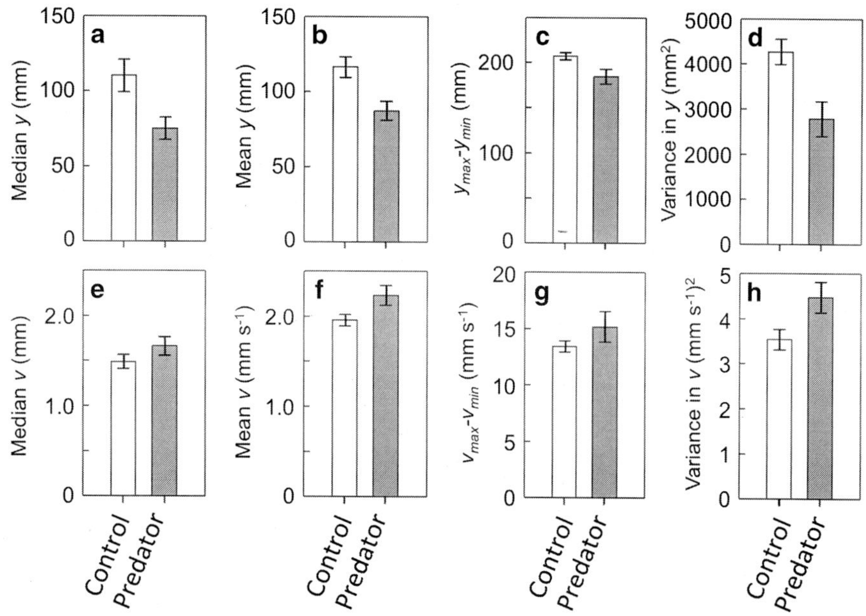

across the entire measured range (i.e., *x*, *y*). To test if ctenophore behavior differed in the two treatments, we fitted linear models to each of the dependent variables using the *nlme* package (Pinheiro et al., 2008) of *R* (R Development Core Team, 2008). We included size as a covariate, added a random component to account for any possible tank effects, and allowed for unequal variance. The test without and with the factor treatment included (i.e., models with 5 and 6 degrees of freedom, respectively) were compared by likelihood tests. We found a significant treatment effect on several vertical distribution and speed parameters. Median *y* (L ratio $= 5.045$, $P = 0.025$) and mean *y* (L ratio $= 4.263$, $P = 0.039$) differed between treatments (Fig. 4a, b). Treatment also affected $y_{max} - y_{min}$ (L ratio $= 5.906$, $P = 0.015$), but not the variance in *y* (L ratio $= 2.240$, $P = 0.135$) (Fig. 4c, d). Although *M. leidyi* in the predator treatment explored a large part of the tank (Fig. 4c), they spent much less time in the top section than did the *M. leidyi* in the controls (Fig. 4a, b). Treatment had a significant effect on individual variance in speed (L ratio $= 4.959$, $P = 0.026$), with *M. leidyi* in the predator treatment being more variable. In contrast, treatment effects on other median *v* (L ratio $= 1.419$, $P = 0.234$), mean *v* (L ratio $= 3.078$, $P = 0.079$), and $v_{max} - v_{min}$ (L ratio $= 1.473$, $P = 0.225$) were not significant (Fig. 4e, h). We found no effects of individual size of *M. leidyi*.

Discussion

The statistical analyses assume that *M. leidyi* in the same tank behaved independently of one another, or in other words that the experimental signal was caused by the treatment itself and not by a dominant *M. leidyi*. Although opportunities for physical interactions occurred in a tank of this size, dominant group behavior in ctenophores has not been documented in the literature.

We demonstrate a suite of behavioral responses of *M. leidyi* to the presence of their predator, *B. ovata*. Our results suggest that lobate ctenophores may actively use remote signals and alter their behavior to avoid risky habitats. In contrast to previously documented escape behaviors of *M. leidyi,* which occur post-encounter, vertical positioning may enhance survival by limiting predator encounters. Such avoidance behaviors are common among smaller pelagic crustaceans (Titelman & Fiksen, 2004) and have been suggested for the ctenophore *Pleurobrachia pileus* (Esser et al., 2004). *M. leidyi* populations may be dense both close to the bottom and the surface (Miller, 1974; Costello & Mianzan, 2003). Vertically heterogeneous distributions in nature may also be attributed to both passive downward mixing and active surface avoidance during periods of heavy wind mixing, because high turbulence supposedly interferes with maintenance of the feeding position (Miller,

1974; Purcell et al., 2001; Mianzan et al., 2010). Also, *M. leidyi* tolerates hypoxia well (Thuesen et al., 2005) and may utilize poorly oxygenated deep water layers for spatial refuge (Decker et al., 2004). Nevertheless, our experiment suggests that predation risk may be involved in governing behavior of *M. leidyi*.

Probably, chemical cues from the predator *B. ovata* triggered a response from *M. leidyi*. The alternative explanation that fluid disturbance caused by the predator elicited a response seems unlikely because *M. leidyi* themselves created considerable fluid motion (Colin et al., 2010) in both treatments. Regardless of the nature of the cue involved, our results indicate that *M. leidyi* may actively adjust their position in the water column in response to remote cues and perceived risk from predators. *M. leidyi* responded to the predator presence by altering directional movement, reducing their vertical range, changing their motility patterns, and increasing the variability in swimming speed. Given that *M. leidyi* responds behaviorally to at least one of its major predators, *B. ovata*, and reacts by adjusting its swimming behavior and position in an experimental water column, it seems likely that vertically distinct distribution patterns of lobate ctenophores in the field may also be influenced by risk-sensitive behaviors.

Acknowledgments We thank J. E. Skjæraasen, P. Tiselius, E. Selander, T. Andersen, and reviewers for constructive comments. This is a contribution to Baltic Zooplankton Cascades (BAZOOCA) program funded by Baltic Organizations Network for funding Science, European Economic Interest Grouping (BONUS EEIG). We received financial support from Swedish Research Council for Environment, Agricultural Sciences and Spatial Planning (FORMAS), Sweden (grant numbers 2007-1105 to LJH and JT; 2006-1054 and 2008-1586 to JT), the National Science Foundation, USA (grant numbers OCE-0350834 and OCE-0623508 to JHC; OCE-0351398 and OCE-0623534 to SPC.), Stiftelsen Konung Carl XVI Gustafs 50-årsfond för vetenskap, teknik och miljö (to JT), and Birgit och Birger Wåhlströms minnesfond (to JT).

References

Albert, D. J., 2011. What's on the mind of a jellyfish? A review of behavioural observations on *Aurelia* sp. jellyfish. Neuroscience and Biobehavioral Reviews 35: 474–482.

Aronova, M. Z. & T. M. Alekseeva, 2002. Second messengers in the locomotor–sensory system of first multicellulars. The IP3-containing compartments in chemoreceptor cells of the comb jelly *Beroe cucumis* from data of cytochemical study. Journal of Evolutionary Biochemistry and Physiology 38: 318–329.

Aronova, M. Z. & T. M. Alekseeva, 2003. Development of chemoreceptor cells in oral epithelium of adult jelly-fish *Beroe cucumis*. Journal of Evolutionary Biochemistry and Physiology 39: 714–723.

Cohen, J. H. & R. B. Forward Jr., 2003. Ctenophore kairomones and modified aminosugar disaccharides alter the shadow response in a larval crab. Journal of Plankton Research 25: 203–213.

Colin, S. P., J. H. Costello, L. J. Hansson, J. Titelman & J. O. Dabiri, 2010. Stealth predation and the predatory success of the invasive ctenophore *Mnemiopsis leidyi*. Proceedings of the National Academy of Sciences of the United States of America 107: 17223–17227.

Condon, R. H. & D. K. Steinberg, 2008. Development, biological regulation, and fate of ctenophore blooms in the York River estuary, Chesapeake Bay. Marine Ecology Progress Series 369: 153–168.

Costello, J. H., K. M. Bayha, H. W. Mianzan, T. A. Shiganova & J. E. Purcell, 2012. Transitions of *Mnemiopsis leidyi* (Ctenophora: Lobata) 3 from a native to an exotic species: a review. Hydrobiologia. doi:10.1007/s10750-012-1037-9.

Costello, J. H. & H. W. Mianzan, 2003. Sampling field distributions of *Mnemiopsis leidyi* (Ctenophora, Lobata): planktonic or benthic methods? Journal of Plankton Research 25: 455–459.

Costello, J. H., B. K. Sullivan, D. J. Gifford, D. Van Keuren & L. J. Sullivan, 2006. Seasonal refugia, shoreward thermal amplification, and metapopulation dynamics of the ctenophore *Mnemiopsis leidyi* in Narragansett Bay, Rhode Island. Limnology and Oceanography 51: 1819–1831.

Decker, M. B., D. L. Breitburg & J. E. Purcell, 2004. Effects of low dissolved oxygen on zooplankton predation by the ctenophore, *Mnemiopsis leidyi*. Marine Ecology Progress Series 280: 163–172.

Esser, M., W. Greve & M. Boersma, 2004. Effects of temperature and the presence of benthic predators on the vertical distribution of the ctenophore *Pleurobrachia pileus*. Marine Biology 145: 595–601.

Falkenhaug, T., 1996. Distributional and seasonal patterns of ctenophores in Malangen, northern Norway. Marine Ecology Progress Series 140: 59–70.

Falkenhaug, T. & O. B. Stabell, 1996. Chemical ecology of predator–prey interactions in ctenophores. Marine and Freshwater Behavior and Physiology 27: 249–260.

Fuentes, V. L., D. L. Angel, K. M. Bayha, D. Atienza, D. Edelist, C. Bordehore, J. M. Gili & J. E. Purcell, 2010. Blooms of the invasive ctenophore, *Mnemiopsis leidyi*, span the Mediterranean Sea in 2009. Hydrobiologia 645: 23–37.

Greve, W. & F. Reiners, 1988. Plankton time–space dynamics in German Bight—a systems-approach. Oecologia 77: 487–496.

Hansson, L. J. & B. Norrman, 1995. Release of dissolved organic carbon (DOC) by the scyphozoan jellyfish *Aurelia aurita* and its potential influence on the production of planktic bacteria. Marine Biology 121: 527–532.

Harbison, G. R., L. P. Madin & N. R. Swanberg, 1978. On the natural history and distribution of oceanic ctenophores. Deep Sea Research 25: 233–256.

Hays, G. C., 2003. A review of the adaptive significance and ecosystem consequences of zooplankton diel vertical migrations. Hydrobiologia 503: 163–170.

Horridge, G. A., 1965. Non-motile sensory cilia and neuromuscular junctions in a ctenophore independent effector organ. Proceedings of the Royal Society B 162: 333–350.

Hosia, A. & J. Titelman, 2011. Intraguild predation between the native North Sea jellyfish *Cyanea capillata* and the invasive ctenophore *Mnemiopsis leidyi*. Journal of Plankton Research 33: 535–540.

Hosia, A., J. Titelman, L. J. Hansson & M. Haraldsson, 2011. Interactions between native and alien ctenophores: *Beroe gracilis* and *Mnemiopsis leidyi* in Gullmarsfjorden. Marine Ecology Progress Series 422: 129–138.

Jaspers, C., J. Titelman, L. J. Hansson, M. Haraldsson & C. Ditlefsen, 2011. The invasive ctenophore *Mnemiopsis leidyi* poses no direct threat to Baltic cod eggs and larvae. Limnology and Oceanography 56: 431–439.

Javidpour, J., J. Molinero, J. Peschutter & U. Sommer, 2009. Seasonal changes and population dynamics of the ctenophore *Mnemiopsis leidyi* after its first year of invasion in the Kiel Fjord, Western Baltic Sea. Biological Invasions 11: 873–882.

Kass-Simon, G. & L. A. Hufnagel, 1992. Suspected chemoreceptors in coelenterates and ctenophores. Microscopy Research and Technique 22: 265–284.

Kreps, T. A., J. E. Purcell & K. B. Heidelberg, 1997. Escape of the ctenophore *Mnemiopsis leidyi* from the scyphomedusa predator *Chrysaora quinquecirrha*. Marine Biology 128: 441–446.

McKelvey, L. M. & R. B. Forward Jr, 1995. Activation of brine shrimp nauplii, photoresponses involved in diel vertical migration by chemical cues from visual and non-visual planktivores. Journal of Plankton Research 17: 2191–2206.

Mianzan, H. W., P. M. Martos, J. H. Costello & R. A. Guerro, 2010. Avoidance of hydrodynamically mixed environments by *Mnemiopsis leidyi* (Ctenophora: Lobata) in open-sea populations from Patagonia, Argentina. Hydrobiologia 645: 113–124.

Miller, R. J., 1974. Distribution and biomass of an estuarine ctenophore population, *Mnemiopsis leidyi* (A. Agassiz). Chesapeake Science 15: 1–8.

Moss, A. G., B. Wells & L. Muellner, 2004. A mechanosensory system that controls feeding in adult *Mnemiopsis*. Hydrobiologia 530–531: 145–153.

Ohman, M. D., 1988. Behavioral responses of zooplankton to predation. Bulletin of Marine Science 43: 530–550.

Oviatt, C. A. & P. M. Kremer, 1977. Predation on the ctenophore, *Mnemiopsis leidyi*, by butterfish, *Periplus triacanthus*, in Narragansett Bay, Rhode Island. Chesapeake Science 18: 236–240.

Pinheiro, J., S. DebRoy, D. Sarkar & The R Core team, 2008. nlme: linear and nonlinear mixed effects models. R package version 3.1-89.

Pitt, K. A., D. T. Welsh & R. H. Condon, 2009. Influence of jellyfish blooms on carbon, nitrogen and phosphorous cycling and plankton production. Hydrobiologia 616: 133–149.

Purcell, J. E., 1991. A review of cnidarians and ctenophores feeding on competitors in the plankton. Hydrobiologia 216–217: 335–342.

Purcell, J. E. & J. H. Cowan, 1995. Predation by the scyphomedusan *Chrysaora quinquecirrha* on *Mnemiopsis leidyi* ctenophores. Marine Ecology Progress Series 129: 63–70.

Purcell, J. E., T. A. Shiganova, M. B. Decker & E. D. Houde, 2001. The ctenophore *Mnemiopsis* in native and exotic habitats: U.S. estuaries versus the Black Sea basin. Hydrobiologia 451: 145–176.

R Development Core Team, 2008. R: a language and environment for statistical computing. R Foundation for Statistical Computing, Vienna. ISBN 3-900051-07-0 [available on internet at http://www.R-project.org].

Rasband, W. S., 2008. Image J. U. S. National Institutes of Health, Bethesda, MD [available on internet at http://imagej.nih.gov/ij/, 1997–2011]. Accessed 23.08.2008.

Riemann, L., J. Titelman & U. Båmstedt, 2006. Links between jellyfish and microbes in a jellyfish dominated fjord. Marine Ecology Progress Series 325: 29–42.

Riisgård, H. U., L. Bøttiger, C. V. Madsen & J. E. Purcell, 2007. Invasive ctenophore *Mnemiopsis leidyi* in Limfjorden (Denmark) in late summer 2007—assessment of abundance and predation effects. Aquatic Invasions 2: 395–401.

Stone, R., 2005. Attack of the killer jellies. Science 309: 1805–1806.

Swanberg, N., 1974. The feeding behavior of *Beroe ovata*. Marine Biology 24: 69–76.

Thuesen, E. V., L. D. Rutherford & P. L. Brommer, 2005. The role of aerobic metabolism and intragel oxygen in hypoxia tolerance of three ctenophores: *Pleurobrachia bachei*, *Bolinopsis infundibulum* and *Mnemiopsis leidyi*. Journal of Marine Biological Association of the United Kingdom 85: 627–633.

Titelman, J. & Ø. Fiksen, 2004. Ontogenetic vertical distribution patterns in small copepods: field observations and model predictions. Marine Ecology Progress Series 284: 49–63.

Titelman, J., L. Gandon, A. Goarant & T. Nilsen, 2007. Intraguild predatory interactions between the jellyfish *Cyanea capillata* and *Aurelia aurita*. Marine Biology 152: 745–756.

Titelman, J., L. Riemann, T. A. Sørnes, T. Nilsen, P. Griekspoor & U. Båmstedt, 2006. Turnover of dead jellyfish: stimulation and retardation of microbial activity. Marine Ecology Progress Series 325: 43–58.

Vinogradov, M. E., E. A. Shushkina & T. A. Lukasheva, 2005. Population dynamics of the ctenophores *Mnemiopsis leidyi* and *Beroe ovata* as a predator–prey system in the near-shore communities of the Black Sea. Oceanology 45(Suppl 1): S161–S167.

Hydrobiologia (2012) 690:189–197
DOI 10.1007/s10750-012-1038-8

JELLYFISH BLOOMS

Predation potential of the jellyfish *Drymonema larsoni* Bayha & Dawson (Scyphozoa: Drymonematidae) on the moon jellyfish *Aurelia* sp. in the northern Gulf of Mexico

Keith M. Bayha · William M. Graham · John E. Higgins III · Heather A. Fletcher

Published online: 15 March 2012
© Springer Science+Business Media B.V. 2012

Abstract The jellyfish *Drymonema larsoni* bloomed in the northern Gulf of Mexico in the Fall of 2000 and fed voraciously on the moon jellyfish *Aurelia* sp., especially where they were concentrated in frontal convergence. We evaluated the predation potential of *D. larsoni* on *Aurelia* sp. medusa using laboratory and field data. Our data set represents the most complete study to date on the new scyphozoan family Drymonematidae and indicates that *D. larsoni* may be one of the most effective predators on other jellyfish recorded to date. On average, each *D. larsoni* medusa contained 2.7 *Aurelia* sp. prey, but as many as 34. In addition, 94% of moon jellyfish unassociated with *D. larsoni* showed scarring from previous contact with *D. larsoni* tentacles. Digestion times for *D. larsoni* feeding on individual *Aurelia* sp. ranged from 2 to 3 h and averaged 2.7 h. Potential clearance rates for predation on *Aurelia* sp. were extremely high (320–1043.5 m^3 d^{-1}) and indicate that *D. larsoni* is potentially an important predator on *Aurelia* sp. blooms where the species co-occur. When the two species co-occur in numbers, predation by *D. larsoni* medusae could reduce moon jellyfish blooms, possibly alleviating predation pressure on lower trophic levels utilized by *Aurelia* sp., such as copepods and the early life history stages of ecologically and economically important fish and invertebrate species.

Keywords Scyphomedusae · Gelatinous zooplankton · Intraguild predation · Digestion · Feeding · Gut contents

Guest editors: J. E. Purcell, H. Mianzan & J. R. Frost / Jellyfish Blooms: Interactions with Humans and Fisheries

K. M. Bayha (✉) · J. E. Higgins III · H. A. Fletcher
Dauphin Island Sea Lab, 101 Bienville Blvd.,
Dauphin Island, AL 36528, USA
e-mail: kbayha@disl.org

W. M. Graham
University of Southern Mississippi, 1020 Balch Blvd.,
Stennis Space Center, MS 39529, USA

J. E. Higgins III
University of South Alabama, Mobile, AL 36688, USA

Introduction

Drymonema larsoni Bayha & Dawson is a recently described species of the new scyphozoan family Drymonematidae (Cnidaria, Discomedusa) found in the Caribbean Sea, Gulf of Mexico, U.S. Atlantic coast, and Bermuda (Bayha & Dawson, 2010). Originally identified as its congener, *Drymonema dalmatinum* Haeckel, this jellyfish was long thought to be a member of the same family as the lion's mane jellyfish, *Cyanea capillata* Linnaeus (Family Cyaneidae) because of their superficial morphological similarities (Mayer, 1910; Kramp, 1961). However, *Drymonema* species were shown to be morphologically and molecularly distinct from medusae in all other scyphozoan families, warranting the classification of a new family (Bayha & Dawson, 2010; Bayha et al., 2010).

The extreme rarity of *Drymonema* spp. throughout most of the world likely contributed to this historical oversight and has hampered efforts to effectively study their ecology. In the Mediterranean, *D. dalmatinum* may appear in cycles of nearly 30 years (Stiasny, 1940) and *Drymonema gorgo* Müller is observed only sporadically along the South American coast (Mianzan, 1989; Morandini, personal communication). Consequently, ecological studies on *Drymonema* spp. were nonexistent until the latter half of the twentieth century, when large populations were recognized in the Caribbean Sea (Larson, 1987; Williams et al., 2001) and the northern Gulf of Mexico (Williams et al., 2001), leading to the first ecological study of a *Drymonema* species (Larson, 1987). Larson (1987) found extremely large *D. larsoni* medusae that fed largely or entirely on the moon jellyfish *Aurelia* sp., something reported anecdotally from in the Mediterranean for *D. dalmatinum* by Stiasny (1940). Larson (1987) observed *Drymonema* sp. jellyfish capturing multiple *Aurelia* sp. individuals simultaneously in their large oral arms, with extracellular digestion of the moon jellyfish occurring in the oral arm region.

Interestingly, *Aurelia* spp. tends to be a preferred prey item among scyphozoan jellyfish that feed on other scyphozoans. Both *Aurelia aurita* Linnaeus (Möller, 1980) and *Chrysaora hysoscella* Linnaeus (Lebour, 1923) feed on the ephyrae of *Aurelia aurita*. In the Pacific, the jellyfish *Phacellophora camtschatica* Brandt feeds heavily on *Aurelia* sp. and exhibits behaviors that increase the chances of encountering the drifting moon jellyfish (Strand & Hamner, 1988). *Cyanea capillata* preys on *A. aurita* in northern European waters where they co-occur (Båmstedt et al., 1994; Hansson, 1997; Martinussen & Båmstedt, 1999). Lastly, species of *Drymonema* feed on *Aurelia* sp. in the Mediterranean (Stiasny, 1940) and western Atlantic (Larson, 1987; Williams et al., 2001; Bayha & Dawson 2010).

During the bloom of *D. larsoni* medusae in the northern Gulf of Mexico in the fall of 2000, we found large numbers feeding on *Aurelia* sp. medusae concentrated along frontal convergence zones. During this period, we collected data on jellyfish abundance, inter-specific contact rates, digestion time, and clearance rates in order to assess predation potential and possible impact of *D. larsoni* on *Aurelia* sp. populations in the northern Gulf of Mexico. These data represent only the second and most complete ecological study of the family Drymonematidae as of date. Over the past few decades, moon jellyfish blooms have become increasingly recognized as problematic in coastal zones where, among others things, they detrimentally affect fishing activities and aquaculture and clog power plant intakes (reviewed in Purcell et al., 2007). The goal of this study was to examine the predation potential of *D. larsoni* on *Aurelia* spp. to determine if it is larger than that found for other *Aurelia* predators, as suggested by the medusa's large size and perceived fishing volume. This represents a first step toward understanding if *D. larsoni* predation may help limit destructive moon jellyfish blooms where the two species co-occur.

Materials and methods

Collection

Jellyfish, both *D. larsoni* and *Aurelia* sp., were collected in late October, 2000 from the northern Gulf of Mexico, either on the north side of Dauphin Island, Alabama (U.S.A.) near the Mississippi Sound or in regions of frontal convergence south of Dauphin Island in the open Gulf of Mexico. Collections were made boat-side using a 1.3-m wide dip net with 0.5-cm mesh. Individual *D. larsoni* were transferred to a small plastic bucket containing a plastic bag filled with seawater. The plastic bags were subsequently tied and placed in a large barrel of ambient water for temperature control and transport of the jellyfish to the laboratory. Four to seven *Aurelia* sp. medusae also were placed together in plastic bags containing seawater and kept in large coolers for transport. All collected jellyfish were kept separately by species in special 180-l kreisels (flow-through aquaria for jellyfish) at the Dauphin Island Sea Lab (Dauphin Island, Alabama).

"Gut" content analysis

The term "gut" as applied to *D. larsoni* medusae was defined as the central subumbrellar pit, including gonadal and oral arm tissue radiating distally from the bell. Processing of the gut contents occurred both in the lab and in the field. In the field, collected animals were placed exumbrellar-side down inside the collection net and the oral arm-gonad mass was closely examined. Because *D. larsoni* feeds voraciously on *Aurelia* sp., we searched for *Aurelia* sp. remains within the oral arm and gonad regions. Any jellyfish prey remains were removed from each animal, identified and counted. The *D. larsoni* then were carefully released from the net and bagged for transport to the lab. Some jellyfish were taken in seawater-filled plastic bags to the laboratory where gut contents were processed similarly. Each *D. larsoni* medusa was placed in a large plastic bin and the contents of each bag were examined for *Aurelia* sp. remains, which were identified and counted. A total of 23 *D. larsoni* medusae were examined for gut contents.

In situ abundance estimates

Aurelia sp. medusa abundances were estimated using in situ videographic surveys. The "jellycam", an underwater video sled equipped with a flow meter (Graham et al., 2003a, b), was towed along a 5.46 km transect south of Dauphin Island in a region of frontal convergence. Videotapes were audited for flow volume, species, and numbers of medusae passing through the 1 m by 1 m field-of-view of the camera. The density of *Aurelia* sp. medusae was calculated for the entire transect and for a smaller concentrated patch within the transect.

Drymonema larsoni medusa abundances were estimated from shipboard observations made continuously from atop the mast, bow, stern, starboard, and port areas of the ship while it steered in and out of the frontal convergence zone over a 2-h period. The distance estimated that medusae could be sighted on each side of the ship was approximately 50 m.

In situ observations were made at collection locations of *D. larsoni* medusa swimming and feeding behaviors. Because of their large size (up to 70 cm bell diameter), great care was taken not to disturb their natural behavior. All underwater observations were recorded by underwater digital video camera. Because *D. larsoni* medusae are slow swimmers, SCUBA divers were able to record each jellyfish for extended periods of time. Careful observations were made of the region surrounding the oral arms, gonads, and tentacles, including captures of *Aurelia* sp. medusae as much as 30 m from the bell margin.

Scarring on *Aurelia* sp. medusae

Fifty *Aurelia* sp. medusae were collected from the boat within the frontal convergence zone in the northern Gulf of Mexico. Initial observations suggested the possibility of scarring from contact with *D. larsoni* tentacles. Therefore, each individual was inspected for evidence of scarring on the exumbrella. Evidence of scarring was determined by visual observation and manual examination for changes in surface integrity (i.e., the presence of a groove) suggesting scarring from contact with *D. larsoni*.

To verify that the field observations of scarring of *Aurelia* sp. medusae were caused by contact with *D. larsoni*, individual unscarred *Aurelia* sp. were placed exumbrella-side-up in a plastic pan and contacted with a *D. larsoni* tentacle. Individual tentacles were removed from live *D. larsoni* medusae and placed across the exumbrella of the *Aurelia* sp. medusae. The tentacle was removed at 15-min intervals over a 2-h period to determine if any surface damage or scarring was evident. These tests were conducted similarly onboard ship and in the lab. This experiment was repeated three times using different *Aurelia* sp. and *D. larsoni* medusae in each trial.

Digestion times

Each *D. larsoni* medusa was kept separately in a 180-l flow-through kreisel. Seven to ten *Aurelia* sp. medusae (average bell diameter = 20 cm) were kept in each holding kreisel. Two *D. larsoni* medusae (bell diameters = 21.0 and 25.5 cm) were used for a total of five digestion experiments, with water temperatures and salinities ranging from 27.0 to 28.5°C and from 30.8 to 31.4, respectively.

One *Aurelia* sp. medusa was used as a prey item for each digestion experiment. Bell diameters of both *Aurelia* sp. and *D. larsoni* were measured and prey wet weights were taken using an OHAUS Scout II digital scale. Excess mucus on the *Aurelia* sp. was removed by patting it dry with a paper towel before each weighing. One *Aurelia* sp. medusa was placed directly in the oral arm-gonad mass beneath the bell of each *D. larsoni*. Prey was removed and the wet weight measured every 0.5 h until no prey was found within the oral arm-gonad-tentacle mass (~100% digestion).

Results

Observations in situ

Individual *D. larsoni* medusae found in the northern Gulf of Mexico ranged in size from 20 to 70 cm diameter (Fig. 1). Typical bell pulsations were slow and resulted in little forward progress, which may have been caused by drag created by the extensive oral arm-gonad-tentacle mass below the bell margin. In addition, tentacle length was as much as 25–30 m, which would cause additional drag, but also provide an extensive fished volume for capture of *Aurelia* sp. medusae.

Drymonema larsoni medusae were typically observed in a fishing posture (Fig. 2) and were oriented in numerous variations of this posture (exumbrella-side-up, exumbrella-side-down, or at an angle to ambient flow within the water column). Tentacles were sometimes extended upward above and away from the bell, extended below the bell margin, or extended horizontally from the bell. Newly caught and partially digested *Aurelia* sp. medusae were observed within the oral arm-gonadal mass and trapped in distal tentacles. Solitary medusae typically had extensive tentacle nets cast while groups of medusae along frontal convergence zones had tentacles retracted with numerous *Aurelia* sp. prey within the oral arm-gonadal mass.

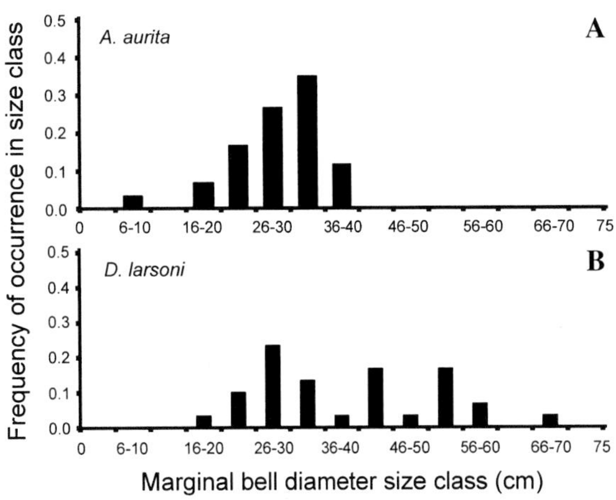

Fig. 1 Marginal bell diameter size class frequency distributions for *Aurelia* sp. medusae (**A**) and *Drymonema larsoni* medusae (**B**) from the northern Gulf of Mexico in October 2000

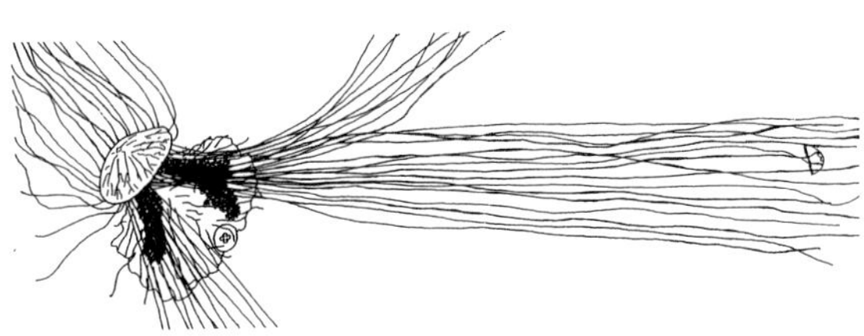

Fig. 2 Drawing of a *Drymonema larsoni* medusa in a typical fishing posture. The medusa bell is at the upper left of the animal, while the oral arm mass (*white*), gonads (*black*), and tentacles extend from the subumbrellar side of the medusa bell. The oral arm mass is considerably larger than the bell and extends far from it. *Aurelia* sp. medusae can be seen ensnared in the oral arms and in the tentacles. Although the drawing is not to scale, the maximum tentacle length reached 25–30 m

Medusa densities

Aurelia sp. densities along the 5.46-km-long transect averaged 0.023 medusae m^{-3}, with one concentrated patch covering a distance of 1.65 km reaching 0.075 medusae m^{-3}. *Aurelia* sp. bell diameters ranged from 6 to 40 cm (Fig. 1). *D. larsoni* were observed either as solitary individuals or in patches of ~10 medusae per 10 m in areas of frontal convergence.

Scarring

Aurelia sp. medusae, unassociated with *D. larsoni* at the time of capture, were inspected for evidence of contact with *D. larsoni*, based on the presence of scarring (Fig. 3). Scarring was evident on 94% of the

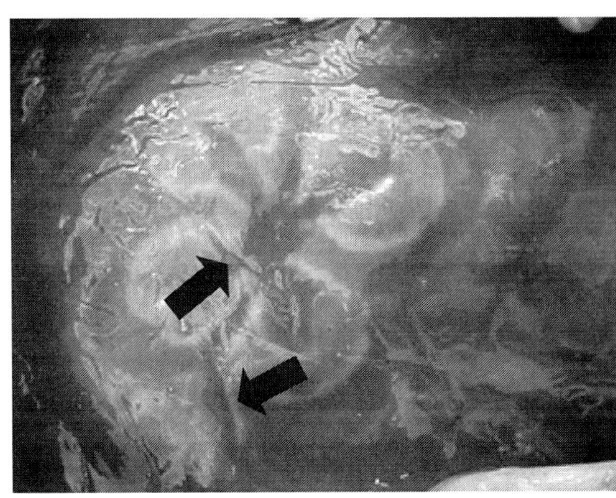

Fig. 3 Scarring along the bell of *Aurelia* sp. medusae caused by contact with *Drymonema larsoni* tentacles. *Arrows* point to scar grooves. Approximate marginal bell diameter of the *Aurelia* sp. was 20 cm

50 *Aurelia* sp. medusae examined. In the laboratory, a contact period of 0.5–1 h was necessary for scarring to be visible. With increased contact time, the scar groove was both deepened and widened. Damage caused in the laboratory was consistent with scarring observed in the field.

Gut contents

Aurelia sp. constituted 98% of all prey items found within the gut and oral arm-gonad mass of *D. larsoni*. Only two animals of the 23 examined contained remnants of other medusae, including the sea nettle *Chrysaora* sp. and the cannonball jellyfish *Stomolophus meleagris* L. Agassiz.

Drymonema larsoni medusae sampled in the northern Gulf of Mexico were voracious predators, containing 0–34 *Aurelia* sp. medusae (average = 4.1; 2.7 if the outlying 34 prey item sample is ignored) in the oral arm-gonad mass at any time. Although it was expected that larger medusae would contain more *Aurelia* sp. than smaller ones, the correlation was not a significant ($r^2 = 0.1304$, $p = 0.129$) (Fig. 4).

Digestion experiments

Digestion of one *Aurelia* sp. by one *D. larsoni* was complete in 2–3 (Fig. 5) with an average of 2.7 h for five trials. In each case, both the digestion rate and wet weight digested each 0.5 h decreased as the remaining prey biomass decreased.

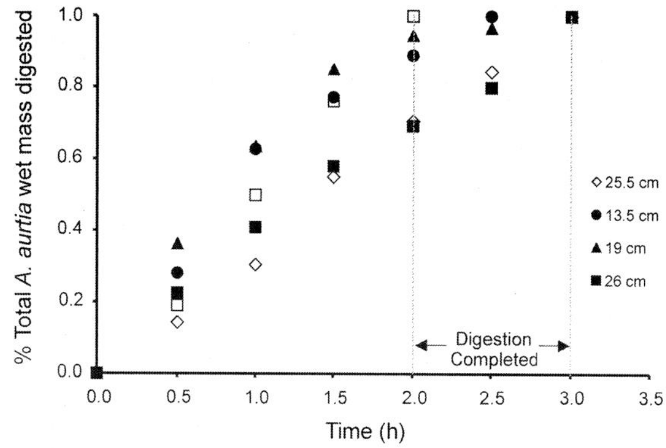

Fig. 5 Digestion of single *Aurelia* sp. medusae by *Drymonema larsoni* given as the percent prey wet weight digested each 0.5 h. *Symbols* represent individual feeding experiments and correspond to different prey *Aurelia* sp. marginal bell diameters. *Open symbols* are for a *D. larsoni* medusa 21.0 cm diameter of; *solid symbols* are for a *D. larsoni* 25.5 cm diameter. All prey items were digested within 2–3 h

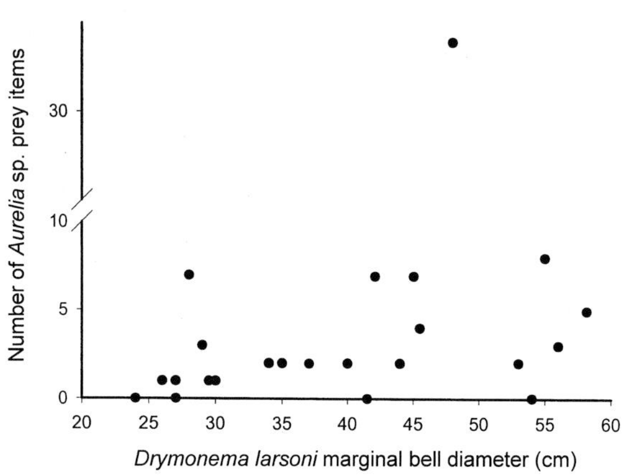

Fig. 4 Numbers of *Aurelia* sp. medusae in the oral arms of *Drymonema larsoni* of different marginal bell diameters. A linear regression fit to the data indicated there was no significant relationship between *D. larsoni* diameter and the number of *Aurelia* sp. medusae ($r^2 = 0.1304$, $p = 0.129$)

The larger of the two *D. larsoni* (diameter = 25.5 cm) consumed three *Aurelia* sp. medusae, with bell diameters of 13.5, 19, and 26 cm. The smallest *Aurelia* sp. was digested in 2.5 h, and the two larger medusae were completely digested in 3 h. The smaller *D. larsoni* (diameter = 21.0 cm) consumed two *Aurelia* sp. with diameters of 17.2 and 25.5 cm. Again, the smaller medusa required 2 h for complete digestion and the larger required 3 h. During digestion, *Aurelia* sp. mesoglea was liquefied from the outer bell toward the center, leaving the gonadal tissue until the final stages of digestion. Digestion occurred in the oral arms, proceeding from the outer bell margin toward the gonads and mouth, as was observed for *Cyanea capillata* medusae feeding on *Aurelia aurita* by Hansson (1997).

Estimation of clearance potential

The potential clearance rate for *D. larsoni* feeding on *Aurelia* sp. medusae (*F*) was calculated using the following equation:

$$F = (H/A) \times D$$

where *H* was the average number of *Aurelia* sp. contained in each *D. larsoni* medusa, *A* was the ambient concentration of *Aurelia* sp. (medusae m^{-3}), and D was the average digestion time (d^{-1}) of one *Aurelia* sp. Based on this equation (assuming the conservative average of 2.7 *Aurelia* sp. per *D. larsoni*), our estimates ranged from 320 m^3 d^{-1} in the concentrated patch of *Aurelia* sp. (0.075 medusae m^{-3}) to 1044 m^3 d^{-1} for the average density of *Aurelia* sp. in the convergence zone (0.023 medusae m^{-3}). If we assumed the greatest number of prey items per predator (34), clearance would be as high as 13,140 m^3 d^{-1}.

Discussion

Our data indicate that *Drymonema larsoni* medusae feed voraciously on *Aurelia* sp. medusae and likely can exert significant predation pressure on their populations, especially where moon jellyfish are aggregated. Our findings of extremely high contact rates between *D. larsoni* and *Aurelia* sp., large numbers of *Aurelia* sp. per predator, and high clearance potentials all suggest that *D. larsoni* predation could severely curtail large blooms of *Aurelia* sp. and alleviate the predation pressure those medusae place on zooplankton populations. In addition, our data further support the view that *D. larsoni* medusae, and possibly the Drymonematidae in general, have adaptations for predation on other gelatinous zooplankton that help make it one of the most effective jellyfish predators recorded.

Digestion rates

Estimates of digestion time are vital to many studies seeking to estimate predation potential and our digestion experiments indicate some of the fastest digestion rates to date for a scyphozoan feeding on *Aurelia* spp. In the five experiments run, all digestion was complete in 2–3 h, with an average of 2.7 h, quicker than the 9.5 h (Plotnikova, 1961) or 38 h (Hansson, 1997) calculated for *Cyanea capillata* medusae feeding on *Aurelia aurita*. While some have explained such differences to be due to lower temperatures resulting in slower digestions rates (Hansson, 1997; Martinussen & Båmstedt, 1999), Martinussen & Bämstedt (2001) indicated that temperature explained very little of the variation in digestion rate, which may be the result of a variety of factors such as predator and prey-size and -concentration. Larson (1987) reported a digestion time of 6–8 h for *D. larsoni* medusae feeding on *Aurelia* sp. based on a single-digestion experiment. It is unclear why our values for digestion rate exceed those of Larson (1987), because those were based on a similarly sized *D. larsoni* (25 cm) feeding on a smaller *Aurelia* sp. (10 cm) than in our study (Fig. 5). Digestion times would be expected to shorten when a predator fed on a smaller animal (e.g., Hansson, 1997). Because our small sample size (five) precludes us from examining the many factors that affect digestion rate in gelatinous predators (Plotnikova, 1961; Hansson, 1997; Martinussen & Båmstedt, 1999, 2001), it is clear that additional studies examining a wider variety of predator and prey sizes are necessary to examine this variation in *D. larsoni*. Nevertheless, our study represents the most robust estimates of digestion rate in *D. larsoni* as of date.

Contact rate and gut contents

The large size, extensive lengths and numbers of the tentacles, and size of the oral arm mass (Fig. 2) all likely result in each *D. larsoni* medusa covering much of the water column and contacting a large proportion of the available moon jellyfish. Of the *Aurelia* sp. we collected unassociated with *D. larsoni* medusae, nearly all of them (94%) showed bell scarring indicating extended contact with *D. larsoni* tentacles. Because laboratory experiments showed that 0.5–1 h of contact was necessary for scarring to become apparent, nearly all free-swimming *Aurelia* sp. medusae probably had been captured by *D. larsoni* at some time and able to escape. Although we have no data on the ability of moon jellyfish to escape from *D. larsoni* tentacles, substantial escape success has been recorded for other gelatinous animals fed on by scyphozoan predators, including *Aurelia aurita* escaping from *Phacellophora camtschatica* (Strand & Hamner, 1988) and *Cyanea capillata* (Martinussen & Båmstedt, 1999), as well as the ctenophore *Mnemiopsis leidyi* L. Agassiz escaping from various scyphozoan predators (Kreps et al., 1997; Hosia & Titelman, 2010). The high-contact rate and the large volume of the water column encompassed by *D. larsoni*'s fishing tentacles suggest that, even if they escape, *Aurelia* sp. medusae are likely to be recaptured and eventually eaten.

The large proportion of *D. larsoni* medusae with ingested *Aurelia* sp., as well as the numerous prey items ingested for each medusa, suggest that *D. larsoni* is a significant predator on *Aurelia* sp. Of the 23 *D. larsoni* collected in the field, 87% contained at least one moon jellyfish and each medusa contained on average 4.1 *Aurelia* sp. prey items (Fig. 4). A single individual contained an extraordinary number of prey items (34) and when this animal is ignored, the average becomes 2.7. Comparatively, Hansson (1997) found an average of 1.2 *Aurelia aurita* per *Cyanea capillata*, less than half of the value recorded for *D. larsoni*.

In addition, we did not find a correlation between predator-size and -number for *D. larsoni* (Fig. 4), which may indicate a higher prey capacity for *D. larsoni* as compared to other *Aurelia* spp. predators. Studies done on *Phacellophora camtschatica* and *Cyanea capillata* medusae showed a positive correlation between predator-size and -number (Strand & Hamner, 1988; Hansson, 1997), indicating that consumption may have been limited by predator size. Because the oral arm elaboration is so extensive in *D. larsoni* medusae (Larson, 1987; Bayha & Dawson, 2010; Fig. 2), even at small sizes, they have been able to ingest more prey than they typically encounter. As a result, *D. larsoni* may rarely reach predatory capacity and the amount of prey ingested may be more a function of the number available than the predator capacity.

Clearance potential and purported ecological impacts

Our laboratory and field-derived calculations of clearance potential are considerable and indicate that *D. larsoni* medusae likely have the potential to significantly modulate moon jellyfish blooms, which could have concomitant effects on lower trophic levels such as those fed on by *Aurelia* sp. Based on our experimentally derived estimate of digestion time (2.7 h per prey medusa) and a conservative average of 2.7 prey items per predator, a *D. larsoni* medusa can feed on an average of 24.9 moon jellyfish per day, although this number could be as high as 313.8 jellyfish. A potential clearance rate of 320–1043.5 $m^3\ d^{-1}$ is markedly higher than the 2.37 $m^3\ h^{-1}$ (56.88 $m^3\ d^{-1}$) determined for *C. capillata* feeding in the laboratory under saturated *Aurelia aurita* concentrations (Titelman et al., 2007). Our high estimate of clearance potential indicates that *D. larsoni* could consume a very large number of *Aurelia* sp., which would then release predation pressure on the zooplankton foods of *Aurelia* sp., possibly benefitting young *D. larsoni* that still feed primarily on zooplankton. Therefore, although *D. larsoni*'s large size makes the medusae a nuisance to humans by causing fishing gear damage, stinging bathers, etc., it is possible that they may ameliorate potential competition for zooplankton of blooming *Aurelia* sp. populations with zooplanktivorous.

The impact of *D. larsoni* is likely to be greatest in frontal convergence zones in the northern Gulf of Mexico, in which converging currents entrain and concentrate moon jellyfish, an important fact since the northern Gulf of Mexico is an open shelf system. Other studies on the potential impact of gelatinous predation on other gelatinous organisms have been performed in enclosed systems that tend to concentrate jellyfish populations (Feigenbaum & Kelly, 1984; Strand & Hamner, 1988; Brewer, 1989; Purcell, 1991; Hansson, 1997).

Post-2000 *D. larsoni* medusa dynamics

Since the bloom in 2000, *D. larsoni* medusae have been sighted sporadically in the northern Gulf of Mexico (K.M.B, personal observation), but not in the numbers observed in 2000. Anecdotal evidence suggests this may be due to lower *Aurelia* sp. densities after 2000. Bayha & Dawson (2010) reasoned that, since morphological evidence indicates that small medusae feed on zooplankton while larger ones (>100 mm bell diameter) feed almost entirely on *Aurelia* sp. medusae, effective *D. larsoni* populations may be dependent upon population dynamics of both zooplankton and *Aurelia* sp. and may require a close match among abundances of all three. Given the high-zooplankton productivity of the northern Gulf of Mexico, it is unlikely that zooplankton populations would be limiting to small *D. larsoni*, leaving *Aurelia* sp. as the important player. Starting in 2001, *Aurelia* sp. populations decreased significantly relative to 2000, rebounding in 2008 to levels similar to 2000 (Robinson, unpublished data). Observations indicated noticeably larger numbers of *D. larsoni* medusae in 2008 as compared to previous years (Miller, personal communication; Bayha & Dawson, 2010). Therefore, *D. larsoni* abundances in the Gulf of Mexico may be mostly dependent upon large *Aurelia* sp. abundances. Although this is based on anecdotal evidence, this compelling correlation can be tested with ongoing surveys.

Evolutionary adaptation in the Drymonematidae

That the predation potential of *Drymonema larsoni* medusae on *Aurelia* sp. exceeds that of other moon jellyfish predators is likely due to morphological adaptations in *D. larsoni* that enhance its ability as a medusivore. In the description of *D. larsoni* and family Drymonematidae, Bayha & Dawson (2010) pointed to numerous morphological adaptations for feeding on medusae, such as the large number of tentacles situated along most of the oral surface of the bell that would allow for a large fishing volume, as well as the massive elaboration of the oral arms that would aid in catching and digesting multiple moon jellyfish at a time (Larson, 1987; Bayha et al., 2010). Our data reinforce these contentions, as large tentacle coverage is indicated by nearly 100% contact rate with *Aurelia* sp. That the oral arm structure is even more elaborated in *D. larsoni* than in other moon jellyfish predators is supported by *D. larsoni* ingesting and digesting a larger number of prey items then either *C. capillata* (Hansson, 1997) or *Phacellophora camtschatica* (Strand & Hamner, 1988). Furthermore, *D. larsoni* did not appear to reach predatory capacity, as evidenced by a lack of relationship between prey number and predator width, while both *C. capillata* (Hansson, 1997) and *P. camtschatica* (Strand & Hamner, 1988) did. Therefore, *D. larsoni* may be even better adapted to feeding on *Aurelia* than all previously studied moon jellyfish predators.

Acknowledgments The authors would like to acknowledge the following for help with field and/or lab work on this project: M. Dardeau, A. Gunter, K. Kirsch, L. Linn, D. Martin, and J. Martin. Feedback from H. Mianzan, J. Purcell and 2 anonymous reviewers greatly improved the manuscript. This work was funded by National Science Foundation Grant OCE-9733441 to WMG.

References

Båmstedt, U., M. B. Martinussen & S. Matsakis, 1994. Trophodynamics of the two scyphozoan jellyfishes, *Aurelia aurita* and *Cyanea capillata*, in western Norway. ICES Journal of Marine Science 51: 369–382.

Bayha, K. M. & M. N. Dawson, 2010. New family of allomorphic jellyfishes, Drymonematidae (Scyphozoa, Discomedusae), emphasizes evolution in the functional morphology and trophic ecology of gelatinous zooplankton. Biological Bulletin 219: 249–267.

Bayha, K. M., M. N. Dawson, A. G. Collins, M. S. Barbeitos & S. H. D. Haddock, 2010. Evolutionary relationships among scyphozoan jellyfish families based on complete taxon sampling and phylogenetic analyses of 18S and 28S ribosomal DNA. Integrative and Comparative Biology 50: 436–455.

Brewer, R. H., 1989. The annual pattern of feeding, growth, and sexual reproduction in *Cyanea* (Cnidaria: Scyphozoa) in the Niantic River estuary, Connecticut. Biological Bulletin 176: 272–281.

Feigenbaum, D. & M. Kelly, 1984. Changes in the lower Chesapeake Bay food chain in presence of the sea nettle *Chrysaora quinquecirrha* (Scyphomedusa). Marine Ecology Progress Series 19: 39–47.

Graham, W. M., D. L. Martin & J. C. Martin, 2003a. In situ quantification and analysis of large jellyfish using a novel video profiler. Marine Ecology Progress Series 254: 129–140.

Graham, W. M., D. L. Martin, D. Felder, V. Asper & H. Perry, 2003b. Ecological and economic implications of a tropical jellyfish invader in the Gulf of Mexico. Biological Invasions 5: 53–69.

Hansson, L. J., 1997. Capture and digestion of the scyphozoan jellyfish *Aurelia aurita* by *Cyanea capillata* and prey response to predator contact. Journal of Plankton Research 19: 195–208.

Hosia, A. & J. Titelman, 2010. Intraguild predation between the native North Sea jellyfish *Cyanea capillata* and the invasive ctenophore *Mnemiopsis leidyi*. Journal of Plankton Research 33: 535–540.

Kramp, P. L., 1961. Synopsis of the medusae of the world. Journal of the Marine Biological Association of the United Kingdom 40: 7–382.

Kreps, T. A., J. E. Purcell & K. B. Heidelberg, 1997. Escape of the ctenophore *Mnemiopsis leidyi* from the scyphomedusa predator *Chrysaora quinquecirrha*. Marine Biology 128: 441–446.

Larson, R. J., 1987. First report of the little-known scyphomedusa *Drymonema dalmatinum* in the Caribbean Sea, with notes on its biology. Bulletin of Marine Science 40: 437–441.

Lebour, M. V., 1923. The food of plankton organisms. II. Journal of the Marine Biological Association of the United Kingdom 13: 70–92.

Martinussen, M. B. & U. Båmstedt, 1999. Nutritional ecology of gelatinous planktonic predators. Digestion rate in relation to type and amount of prey. Journal of Experimental Marine Biology and Ecology 232: 61–84.

Martinussen, M. B. & U. Bämstedt, 2001. Digestion rate in relation to temperature of two gelatinous planktonic predators. Sarsia 86: 21–36.

Mayer, A. G., 1910. Medusae of the world, III: the Scyphomedusae. Carnegie Institute, Washington.

Mianzan, H. W., 1989. Sistemática y zoogeografía de Scyphomedusae en aguas neríticas argentinas. Investigaciones Marinas CICIMAR 4: 15–34.

Möller, H., 1980. Scyphomedusae as predators and food competitors of larval fish. Kieler Meeresforsch 28: 90–100.

Plotnikova, E. D., 1961. On the diet of medusae in the littoral of eastern Murman. In Kamshilov, M. (ed.), Hydrological and

Biological Features of the Shore Waters of Murman. Akademia Nauk SSSR, Murmansk: 153–166.

Purcell, J. E., 1991. Predation by *Aequorea victoria* on other species of potentially competing pelagic hydrozoans. Marine Ecology Progress Series 72: 255–260.

Purcell, J. E., S. Uye & W. Lo, 2007. Anthropogenic causes of jellyfish blooms and their direct consequences for humans: a review. Marine Ecology Progress Series 350: 153–174.

Stiasny, G., 1940. Über Drymonema dalmatina Haeckel. Zoologische Jahrbücher 66: 437–462.

Strand, S. W. & W. M. Hamner, 1988. Predatory behavior of *Phacellophora camtschatica* and size-selective predation upon *Aurelia aurita* (Scyphozoa: Cnidaria) in Saanich Inlet, British Columbia. Marine Biology 99: 409–414.

Titelman, J., L. Gandon, A. Goarant & T. Nilsen, 2007. Intraguild predatory interactions between the jellyfish *Cyanea capillata* and *Aurelia aurita*. Marine Biology 152: 745–756.

Williams, E. H., L. Bunkley-Williams, C. G. Lilyestrom, R. J. Larson, N. A. Engstrom, E. A. R. Ortiz-Corps & J. H. Timber, 2001. A population explosion of the rare tropical/subtropical purple sea mane, *Drymonema dalmatinum*, around Puerto Rico in the summer and fall of 1999. Caribbean Journal of Science 37: 127–129.

Hydrobiologia (2012) 690:199–213
DOI 10.1007/s10750-012-1044-x

Nudibranch predation and dietary preference for the polyps of *Aurelia labiata* (Cnidaria: Scyphozoa)

Richard A. Hoover · Ruth Armour · Ian Dow · Jennifer E. Purcell

Published online: 5 April 2012
© Springer Science+Business Media B.V. 2012

Abstract There is concern that jellyfish blooms may be increasing worldwide. Some factors controlling population size, such as temperature and food, often have been studied; however, the importance of predators is poorly known. Aeolid nudibranchs feed on cnidarians, but their predation on the benthic polyps of scyphozoan rarely has been documented. To understand the potential of nudibranchs to consume polyps, we tested several predation preference hypotheses with the generalist feeding nudibranch, *Hermissenda crassicornis*, and polyps of the common moon jellyfish, *Aurelia labiata*. Of the six prey species tested during feeding experiments, *A. labiata* polyps and the tunicate *Distaplia occidentalis* were significantly preferred. Nudibranch size, diurnal cycle, and ingestive conditioning did not significantly influence prey choice. Nudibranchs showed significant positive chemotaxis toward living polyps, hydroids, and tunicates, but not to sea anemones. Nudibranch chemotaxis was significantly more positive to polar extract of *A. labiata* than of *D. occidentalis*. Consumption of polyps was correlated with nudibranch size, with mean consumption by large nudibranchs (>0.92 g) of about 31 polyps h^{-1}. Three other nudibranch species also ate *A. labiata* polyps. Our results emphasize the potential importance of predation for controlling jellyfish benthic polyp populations and consequent jellyfish blooms.

Keywords Jellyfish · Predator · Selection · Hydroid · Tunicate · Sea anemone · Chemotaxis

Guest editors: J. E. Purcell, H. Mianzan & J. R. Frost / Jellyfish Blooms: Interactions with Humans and Fisheries

R. A. Hoover (✉) · J. E. Purcell
Shannon Point Marine Center, Western Washington University, 1900 Shannon Point Rd, Anacortes, WA 98221, USA
e-mail: rhoovz@gmail.com

J. E. Purcell
e-mail: purcelj3@wwu.edu

R. Armour
California Polytechnic University, Pomona, CA, USA

I. Dow
University of South Florida, Tampa, FL, USA

Introduction

There is concern that jellyfish blooms have increased in recent decades and, consequently, have had increased effects on ecosystem dynamics and human enterprises. Large blooms may reduce zooplankton biomass to such an extent that they can alter entire trophic webs (Mills, 1995; Brodeur et al., 2002; Purcell, 2003; Ruzicka et al., 2007; Pauly et al., 2009). They degrade fisheries by consuming ichthyoplankton and potentially competing with fish for food resources (Purcell & Arai, 2001; Purcell & Sturdevant, 2001), and they also impede the fishing industry by clogging fishing nets (Uye & Ueta, 2004; Purcell et al., 2007).

Jellyfish blooms also clog water-intake screens of coastal power and desalination plants and reduce tourism revenues by increased stinging at beaches (UNEP, 1991; Purcell et al., 2007; Mariottini & Pane, 2010). Common moon jellyfish in the cosmopolitan genus *Aurelia* are key problem species around the world.

To understand the causes of large blooms of medusae, increased attention is being paid to the importance of the benthic asexual stage (polyp) of scyphozoan jellyfish. The asexual strobilation of the polyps is directly responsible for producing new jellyfish. Studies on *Aurelia* spp. show that polyp population dynamics are affected by several factors, including environmental and climatic conditions (Lucas, 2001; Purcell, 2005; Purcell et al., 2009, 2012; Holst, 2012; Thein et al., 2012), substrate preference and availability (Lucas et al., 1997; Miyake et al., 2002; Holst & Jarms, 2007; Willcox et al., 2008; Hoover & Purcell, 2009), food availability (Buss, 1990; Gong, 2002), and predation (Hernroth & Gröndahl, 1985a, b; Gröndahl, 1988; Keen, 1991). Predation, the least studied of the above factors, is the topic of our study.

Nudibranchs are a group of shell-less marine gastropods, commonly called sea slugs. Many species from the nudibranch suborders Dendronotacea and Aeolidacea utilize benthic cnidarians as food sources. Those in the Suborder Dendronotacea commonly have generalist feeding habits (McDonald & Nybakken, 1996); species in the Suborder Aeolidacea harvest the unfired nematocysts of their prey and incorporate them into their own tissues for defense (Cargo & Burnett, 1982).

Although very few studies exist, some show important effects of nudibranch predation on *Aurelia* spp. polyp populations. In Gullmar Fjord, Sweden, the nudibranch *Coryphella verrucosa* (Sars, 1829) ingested the polyps of *Aurelia aurita* (Linnaeus, 1758) at rates up to 200 polyps d^{-1} on settling plates; this predation was believed to be responsible for a drastic decline in polyp abundance (Hernroth & Gröndahl, 1985a). Further, Gröndahl (1988) believed that inter-annual variation in ephyra (and thus medusa) abundance was controlled by that predation. A 2-year study by Keen (1991) showed that the effect of predation by the nudibranch *Hermissenda crassicornis* (Eschscholtz, 1831) on *Aurelia* sp. polyps was highly dependent on the density and extent of polyp colonies. Keen (1991) found that large patches of polyps could be broken up into smaller patches by predation and that small patches (<100 cm^2) frequently were consumed within a month. Keen (1991) also observed that the numbers of experimental sites in situ that lost all polyps during monthly census intervals were positively correlated with the numbers of large (>4 cm in length) nudibranchs present.

Hermissenda crassicornis is a common aeolid nudibranch species found in a wide variety of habitats (e.g., rocky intertidal, mud flats, and boat docks) along the Pacific coasts of North America, from Alaska to Mexico, and of Asia (Morris et al., 1980; Behrens, 1991; Thein et al., 2012). Mating animals and egg masses of *H. crassicornis* occur all year in the Puget Sound (Morris et al., 1980). *H. crassicornis* is a generalist, preying on many cnidarians, tunicates, bryozoans, sponges, annelids, and other gastropods, including con-specifics (McDonald & Nybakken, 1996). Because of its availability and generalized diet, *H. crassicornis* was chosen for experiments on nudibranch predation on *A. labiata* (Chamisso & Eysenhardt, 1821) polyps.

It is not known if the polyps of *Aurelia* spp. are a preferred food of *H. crassicornis.* The necessity of harvesting fresh nematocysts to maintain their defenses (Cargo & Burnett, 1982) may influence their dietary preferences, but the use of scyphozoan polyps as food has rarely been studied. Avila et al. (1998) compared *H. crassicornis* growth and survival on three cnidarian diets, the hydroid *Tubularia crocea* (Agassiz, 1862) or either of the sea anemones *Haliplanella luciae* (Verrill, 1870) and *Metridium senile* (Linnaeus, 1761), but did not distinguish preference. In a second study, Avila (1998) used the cnidarians above, as well as the hydroid *Pennaria* sp., the tunicate *Ciona intestinalis* (Linnaeus, 1767), and the mussel *Mytilus edulis* (Linnaeus, 1758), concluding that ingestive conditioning influenced chemotactic preferences of *H. crassicornis*, but ingestive preferences were not tested.

Some nudibranchs exhibit ingestive conditioning based on dietary history (Hall et al., 1984; Avila, 1998). Early ingestive conditioning may occur when larval *H. crassicornis* complete metamorphosis on several species of hydroids that it later consumes (Harrigan & Alkon, 1978); however, adult *H. crassicornis* also lives well on a diet of tunicates (Harrigan & Alkon, 1978), which would provide more energy-efficient foraging.

In one study, *H. crassicornis* displayed chemotaxis toward the hydroid *Pennaria* sp., which it had never fed on, but not toward the conditioned diet of the sea anemone *M. senile* (Avila, 1998).

Several nudibranch species use chemotaxis in choosing prey items (Willows, 1978; Todd, 1981; Seavy & Muller-Parker, 2002). *H. crassicornis* exhibits chemotaxis to several species of hydroids in a simple Y-maze, and choices via chemotaxis have been observed (Tyndale et al., 1994; Avila, 1998). Because *H. crassicornis* is a generalist feeder, many known prey items have not been tested for chemotaxis by the nudibranch. Whether any nudibranch species shows chemotaxis to scyphozoan polyps of any species is unknown. Furthermore, to our knowledge, chemotaxis has never been compared to ingestive preference in nudibranchs.

Factors affecting the polyp stage could greatly affect medusa abundances of species globally. Our study examined how the predatory behavior of the nudibranchs could affect scyphozoan polyp populations by testing the following null hypotheses: H_{01} *H. crassicornis* preference does not differ between paired food choices; H_{02} *H. crassicornis* shows no preferences among six food choices; H_{03} nudibranch size does not affect food preference; H_{04} feeding preferences do not differ between daytime and nighttime; H_{05} ingestive conditioning does not affect food preferences of *H. crassicornis*; H_{06} *H. crassicornis* shows no taxis to living food choices or to seawater controls; H_{07} *H. crassicornis* shows no taxis to polar or non-polar prey extracts or to control seawater blanks; H_{08} *H. crassicornis* shows no chemotactic preference between the polar extracts of *A. labiata* polyps and *D. occidentalis*; H_{09}, H_{10}, and H_{11} nudibranch size, polyp size, and polyp density do not affect the total number of *A. labiata* polyps consumed; and H_{12} nudibranch size does not affect the polyp size class consumed. We tested six additional nudibranch species to determine if consumption of *A. labiata* polyps was common.

Methods

Collection and maintenance of organisms

The food organisms used in ingestion and chemotaxis experiments were chosen based on field observations near a large colony of *A. labiata* polyps and known foods of nudibranchs (McDonald & Nybakken, 1996). Animals were collected from various sites surrounding Shannon Point Marine Center (SPMC) in Anacortes, Washington (48°30′N, 122°41′W). The collection sites were intertidal or slightly subtidal and chosen only for the abundance of the desired organisms found there.

Experimental organisms were maintained at SPMC in sea tables with a constant supply of flow-through ambient seawater. Water temperatures averaged 12.4 ± 1.3°C and salinities averaged 30.3 ± 1.1 ppt during experimentation. Organisms to be used as food were separated by species into 33 × 25.4 × 15 cm flow-through plastic mesh baskets. All food organisms were offered newly hatched *Artemia* sp. nauplii once per week. *H. crassicornis* nudibranchs were kept individually in 25 × 17 × 13 cm flow-through plastic mesh baskets to prevent cannibalism.

Maintenance diets of the nudibranchs differed among experiments. For the 30-min, 2-choice experiments, nudibranchs were maintained on mixed diets of the hydroid *Obelia geniculata* (Linnaeus, 1758), the tunicates *Distaplia occidentalis* Bancroft, 1899 and *Corella willmeriana* Herdman, 1898, and *A. labiata* polyps. The test species were used in the diet regimen, but the nudibranchs to be tested were not fed their test prey species. For the more robust 6-choice, 24-h ingestive preference testing, individual *H. crassicornis* were maintained on the predetermined diets described below to test for possible effects of ingestive conditioning. For single-choice chemotaxis experiments, nudibranchs were not fed between collection and testing 3 days later. For 2-choice chemotaxis experiments, the nudibranchs were maintained on a diet of *O. geniculata* hydroids. All organisms were used within 3 weeks of collection and determined to be in healthy condition.

Ingestive preference: 30-min, 2-choice experiments

We first tested the prey choices of *H. crassicornis* in short experiments with pairs of four prey species: *O. geniculata* (hydroid), *D. occidentalis* (colonial tunicate), *Epiactis prolifera* Verrill, 1869 (sea anemone), and polyps of *A. labiata* (jellyfish). Nudibranchs were unfed for 24 h before testing in order to standardize conditions for each nudibranch (Hall et al., 1984;

Avila et al., 1998). The testing arenas were 21-cm-diameter × 10-cm-deep glass bowls filled with seawater filtered to 50 μm. To maintain ambient water temperature, the bowls were placed nearly immersed in a flow-through sea table. Two prey specimens were placed on opposite sides of the testing arena 15 min before each test began to allow their scents to disperse in the bowl.

Each unfed nudibranch was transferred in a large spoon from its holding pen to the test arena immediately before the 30-min testing period. One nudibranch was placed in the middle of the bottom of each bowl, equidistant and oriented away from both prey items. The prey that was fed upon first and the duration of time each nudibranch spent eating each prey item was recorded. Due to the limited number of nudibranchs available, subsequent tests were conducted at 48-h intervals with the same nudibranchs using different prey choices. The order of the test treatments was determined by availability and freshness of prey items collected. In total, 26 nudibranchs were tested for the polyp-tunicate choice, and 16 nudibranchs each for the polyp-hydroid and polyp-anemone choices. The data were analyzed using paired *t* tests. The null hypothesis (H_{01}) was that *H. crassicornis* feeding shows no differences between paired food choices.

Ingestive preference: 24-h, 6-choice experiments

To assess the food preferences of *H. crassicornis* when offered a wider selection of foods over a longer period, 6-choice ingestive preference experiments ($n = 18$) were run for 24 h per trial. Nudibranchs were placed in the center of a circular flow-through arena (diameter = 40 cm, height = 15 cm) containing six food choices that included the polyp, hydroid, and tunicate species used in the 2-choice tests, plus the sea anemone *Anthopleura elegantissima* (Brandt, 1835), the bryozoan *Bugula* sp., and the sponge *Halichondria bowerbanki* Burton, 1930. The arena apparatus was adapted from Seavy & Muller-Parker (2002). To eliminate any bias from currents or gradients present in the sea table, the arena was surrounded by a circular seawater delivery hose that introduced water into the arena from all directions through small holes at 10-cm intervals around the circumference. Water flow within the arena was tested prior to each trial by adding food coloring to the food choice locations and observing mixing patterns. Food choice locations appeared to receive similar flows and the water in the arena was well mixed.

To record the ingestive preferences of the nudibranchs, a Sony DCR-TRV900 digital video camera was mounted 180 cm above the sea table. The 24-h experiments were recorded using the time-lapse video function, which recorded for 2 s at 30-s intervals. Tests were conducted in natural light, which averaged 14.3 h of daylight and 9.7 h of darkness during experimentation. The low light exposure setting was used to ensure adequate exposure during darkness. An 80-min Sony miniDV cassette was used at LP speed (120 min) to record each trial.

Experimental food organisms were chosen based on records of known *H. crassicornis* foods (McDonald & Nybakken, 1996) and food resources near the nudibranchs when collected. To standardize the amounts of different foods presented, all samples were gently scooped in a small spoon from the holding baskets and placed it into seawater-filled plastic weighing boats on a Mettler Toledo digital balance tared to include the weight of the seawater and weighing boat. The samples were then adjusted to as similar wet weights as possible without damaging the organisms. The food choices were in 30-ml glass Petri dishes that were placed at equidistant marks 5 cm from the perimeter of the arena. To further reduce potential bias, the position of each food choice was chosen from a random number table for each new trial and the nudibranch was always placed with its head oriented to the north. After each trial, the arena was removed and nudibranch waste and slime trails were cleaned from the arena using hot freshwater and an abrasive pad. Each nudibranch was unfed for 24 h before testing (Avila et al., 1998) and its wet weight (g) measured as above.

The data determined from the recordings were the times spent ingesting each food in daylight and in darkness, and the total time spent ingesting each food. The data were analyzed using log likelihood *G* tests to rank preferences, paired *t* tests to test day/night patterns in feeding, a Wilcoxon signed-rank test for day/night data that were not normally distributed, and a type II regression to examine the effect of nudibranch size on food choice. Three null hypotheses were tested: (H_{02}) *H. crassicornis* shows no preferences among 6 food choices, (H_{03}) nudibranch size does not affect food preference, and (H_{04}) feeding preferences do not differ between daytime and nighttime.

Ingestive conditioning

To ensure the reliability of the preference testing, we tested the potential for ingestive conditioning to bias food preference. The nudibranchs were divided into six groups of three individuals, and each group was conditioned to one of the experimental foods (the polyp, hydroid, tunicate, sea anemone, bryozoan, or sponge species above) for 1 week. The conditioning was limited to 1 week because nudibranch mortality increased greatly with prolonged exposure to a single food. The food preferences of the nudibranchs were then tested for positive preference for the conditioned food and top preference (most-preferred) for the conditioned food. To examine if ingestive conditioning affected preference, the multidimensional non-parametric statistical program PRIMER was used to map nudibranch preference. The null hypothesis (H_{05}) was that ingestive conditioning does not affect food preferences of *H. crassicornis*.

Food preference analysis

Microsoft Excel™ was used to calculate the maximum likelihood ratio (G) for food preference tests. The G statistic is used to evaluate goodness of fit much the same way as Chi-square (χ^2) values and is used specifically in preference testing whenever any observed outcome is more than twice the expected outcome (Williams, 1976). Zar (1996) calculates G for an individual treatment as:

$$G_i = f_o[\ln(f_o/f_e)]$$

and G for all treatments is calculated as:

$$G_T = 2\sum f_o[\ln(f_o/f_e)]$$

where f_o is the number of observed outcomes and f_e is the number of expected outcomes for each treatment in the preference test. Expected outcomes were calculated by dividing the total time spent feeding by six. For further exploration of the degree and direction of preference (i.e., preference for vs. preference against), the normal standard deviates (d) for each G_i were calculated by adjusting the calculated standard residuals (e) by variance (v) and comparing the result to a z statistic:

$$e = (f_o - f_e)/\sqrt{(f_e)}$$

$$v = \left[1 - \left((f_o + f_e)/\left(2\sum f_o\right)\right)\right] \times \left[1 - \left(\left(\sum f_o\right)/\left(2\sum f_o\right)\right)\right]$$

$$d = e/\sqrt{(v)}$$

If the normal standard deviate (d) has a greater absolute value than 1.96 ($p = 0.05$), then a positive value for d represents preference for a treatment and a negative value represents preference against. The greater the absolute value of d, the greater the degree of preference.

Chemotaxis by *Hermissenda crassicornis*: experimental Y-maze

To test the chemotactic preferences of *H. crassicornis*, we tested chemotaxis to living prey items and to the polar and non-polar extracts of selected prey species. Because the classic Y-maze design has been found to hinder natural behavior due to constriction of movement (Zimmer-Faust et al., 1996), preference testing was conducted using a modified Y-maze designed by Seavy & Muller-Parker (2002). The modified Y-maze consisted of a circular, clear, 30-cm-diameter plexiglass arena connected to two flow-through catch chambers. During experimentation, ambient seawater from the flow-through seawater system constantly filled a 30-l tank that was connected by valved 1.27-cm-diameter surgical tubing to two flow-through holding boxes that contained the prey items. The holding boxes had opaque sides to prevent visual response and contained baffles with screened holes near the bottom to ensure that seawater flowed directly over the prey items. The seawater then flowed from the holding boxes into the Y-maze arena catch chambers via valved 0.95-cm-diameter surgical tubing. The seawater drained from the arena through two screened outlets at the rear of the arena.

To ensure proper Y-conformation flow with minimal mixing in the arena, dye tests using food coloring were performed before each treatment. Dyes of different colors were placed in each respective holding box and allowed to flow through the catch chambers into the arena. The valves attached to the catch chambers were adjusted to equalize flow and a center line and small circle marking the point of convergence were used to calibrate the flow to be identical for each treatment. The small circle was also used to ensure consistent positioning of the nudibranch for each trial. Once Y-conformation flow was established, the entire apparatus was flushed with seawater for 10 min before the start of a trial.

The entire apparatus was disassembled and cleaned rigorously with hot freshwater and brushes after each trial to remove nudibranch mucus trails and chemical residue to eliminate those potential biases. After thorough rinsing and reassembly, the Y-flow was recalibrated. In addition, the order of prey species presentation and which basket held prey were randomized for each set of trials and for each nudibranch. The time between trials was approximately 1 h. To minimize handling effects, nudibranchs were transferred to and from the experimental chamber in new 60 × 15 mm plastic Petri dishes filled with seawater. Each nudibranch was unfed for 3 days before testing to ensure a rapid response (Seavy & Muller-Parker, 2002).

Chemotaxis: living prey items

Hermissenda crassicornis nudibranchs first were tested for chemotaxis to whole, living prey items using single-choice (prey vs. control) experiments. The prey items tested were *A. elegantissima* sea anemones ($n = 9$), *A. labiata* polyps ($n = 9$), *D. occidentalis* tunicates ($n = 9$), and *O. geniculata* hydroids ($n = 8$). With the exception of one nudibranch that was not tested on *O. geniculata*, each was tested for chemotaxis toward each prey item in a randomized series. The chemotaxis testing procedure was as follows: With the Y-maze calibrated and flowing, a prey item was placed in one of the holding boxes while the other remained empty (control). The scent of the prey item was then allowed to effuse into the arena for 1 min. A trial was initiated by carefully positioning a nudibranch in a Petri dish with its head toward the back of the arena on the circle marking the point of flow convergence. Chemotaxis (choice) was defined as entry of the nuibranch's head (including rhinophores) into one of the catch chambers. If no choice was made by 1 h, the trial was deemed "no choice" and removed from analysis. Results were analyzed using a χ^2 contingency table. The null hypothesis (H_{06}) was that *H. crassicornis* shows no taxis to living food choices or to seawater controls.

Chemotaxis: polar and non-polar extracts

Based on results of the living prey tests, polar and non-polar extracts were made from *A. labiata* polyps and *D. occidentalis* tunicates. For extraction of the polar and non-polar compounds from the test species, 6-ml samples were first frozen at −70°C in a So-Low Ultra Low Freezer, then soaked in a 2:1 dichloromethane:methanol solvent solution for approximately 5 min inside an explosion-proof Isotemp Fisher Scientific refrigerator. Remaining solids were removed using a Buckner funnel, and the eluate was placed in a separatory funnel with a small amount of reverse osmosis (RO) water until the polar and non-polar layers separated. To ensure complete separation, samples were centrifuged using a Jouan Inc. Br4i centrifuge for 1 min at 1,000 rpm. A glass Pasteur pipette was used to transfer the methanol layers to separate containers. Complete separation of the two layers was verified by centrifuging again for 2 min at 1,000 rpm. The polar and non-polar extracts were then recovered by evaporating the solvents using a Büchi Rotovapor R-114 in a 40°C water bath. To further remove impurities, methanol was added to the non-polar extracts and the samples were again centrifuged for 2 min at 1,000 rpm. The solutions were then filtered using a 0.2-μm 20-ml syringe and dried using a Savant SpeedVac Plus concentrator. To standardize volume for testing, the polar and non-polar extracts were diluted in RO water and methanol, respectively.

Preliminary single-choice tests (polar or non-polar extract vs. seawater control) were run to determine if the nudibranchs exhibited chemotaxis to the extracts. Then, 2-choice tests between the polar extracts of *A. labiata* polyps and *D. occidentalis* were performed to determine chemotactic preference. Testing procedures were as follows for single-choice tests: 160 μl of extract was randomly injected in one of the holding boxes while the other was left empty (control). For 2-choice tests, 160 μl of the two extracts were injected simultaneously into the two holding boxes, which were alternated for each new trial. The trial then was immediately initiated by carefully positioning a nudibranch in a Petri dish with its head facing the back of the arena on the circle marking the point of flow convergence. Chemotaxis (choice) was defined as the entry of the nudibranch's head (including rhinophores) into one of the catch chambers. If no choice was made within 5 min, the trial was deemed "no choice" and removed from analysis. Results were analyzed using χ^2 contingency tables. The null hypotheses were (H_{07}) that *H. crassicornis* shows no taxis to polar or non-polar prey extracts or to control seawater blanks and (H_{08}) that *H. crassicornis* shows no chemotactic preference between the polar extracts of *A. labiata* polyps and *D. occidentalis*.

Feeding potential

Hermissenda crassicornis was used to estimate the feeding potential of nudibranchs on *A. labiata* polyps. Thirty-five 1-h feeding experiments were conducted during which individual nudibranchs were allowed to feed on a known number of polyps. Brown algal blades with attached *A. labiata* polyps were harvested and used for feeding potential experiments. Polyp colony densities were determined by cutting the algal blades into known areas and counting the polyps in each area. To ensure that feeding was not limited by food availability, each of the 35 algal blades contained more than 100 polyps, which exceeded nudibranch consumption during preliminary feeding tests. Polyp size was determined by measuring diameters with a caliper tool. Polyps were categorized into "small" (<3 mm diameter) or "large" (>3 mm diameter) size classes based on an apparent discontinuity in size. The algal blades with polyps then were attached perpendicular to the water flow with dissection pins to the bottom of individual 25 × 17 × 13 cm plastic flow-through experimental cages in a sea table with flowing seawater.

To avoid possible ingestive conditioning to other prey, the nudibranchs were supplied with only *A. labiata* polyps as a food source for 1 week. Each nudibranch was then unfed for 24 h before testing to standardize feeding conditions (Avila et al., 1998). The wet weight (g) of each nudibranch was measured as described above immediately before testing. A separation of large and small nudibranchs was set as the median value of 0.912 g for ease of analysis.

After weighing, each nudibranch was placed in an experimental cage with polyps as detailed above. Feeding was allowed for 1 h beginning with first contact with the polyps. After 1 h, the nudibranch was removed and the polyps were recounted and remeasured to determine the numbers of small, large, and total polyps consumed. A multiple stepwise major axis regression was used to determine the relative importance of nudibranch sizes, polyp sizes, and polyp colony densities to the total numbers of polyps consumed. The null hypotheses (H_{09}, H_{10}, H_{11}) were that nudibranch size, polyp size, and polyp density do not affect the total number of *A. labiata* polyps consumed. The null hypothesis (H_{12}) was that nudibranch size does not affect the polyp size class consumed.

Predation on *A. labiata* polyps by other nudibranchs

During monthly surveys of a large colony of *A. labiata* polyps from January 2004 to April 2006 (Hoover & Purcell, 2009; Purcell et al., 2009), six additional nudibranch species found near or on the colony were collected and tested for predation on the polyps. The nudibranchs were unfed for 24 h and then each was placed individually inside an experimental arena with a known number of polyps, as in the feeding potential experiments. The polyps were recounted after 24 h and examined for evidence of predation.

Results

Ingestive preference: 30-min, 2-choice experiments

Two-choice tests indicated that the polyps of *A. labiata* were the preferred prey of *Hermissenda crassicornis* nudibranchs among the 4 choices offered in 30-min trials. The polyps were the first prey chosen in 66% of *Distaplia occidentalis* versus polyp trials, 81% of *Epiactis prolifera* versus polyp trials, and 75% of *Obelia geniculata* versus polyp trials. The nudibranchs also spent significantly more time feeding on polyps than any other prey item ($t_{25} = -2.83$, $P = 0.009$ vs. *D. occidentalis*; $t_{15} = -3.81$, $P = 0.002$ vs. *E. prolifera*, and $t_{15} = -3.34$, $P = 0.004$ vs. *O. geniculata*; Fig. 1). The null hypothesis H_{01}, that no preferences existed between prey pairs, was rejected.

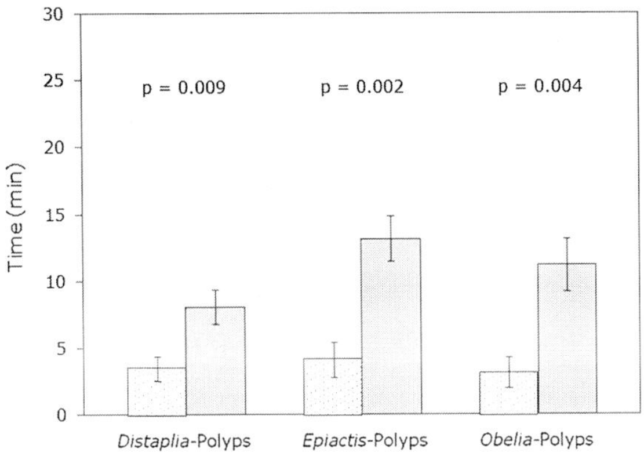

Fig. 1 Feeding times of *Hermissenda crassicornis* nudibranchs during 30-min, 2-choice ingestive preference experiments testing *Distaplia occidentalis* (colonial tunicate), *Epiactis prolifera* (sea anemone), and *Obelia geniculata* (hydroid) against polyps of *A. labiata* (jellyfish). Data are mean ± standard error for 26, 16, and 16 trials, respectively

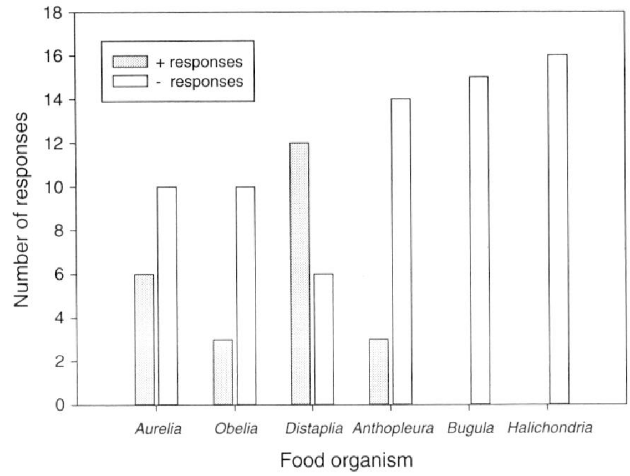

Fig. 2 Numbers of food preference responses by prey type for 17 *Hermissenda crassicornis* nudibranchs. Prey species were *A. labiata* polyps, *Obelia geniculata* hydroids, *Distaplia occidentalis* colonial tunicates, *Anthopleura elegantissima* sea anemones, *Bugula* sp. bryozoans, and *Halichondria bowerbanki* sponges. *Gray bars* indicate preference toward (+) and *white bars* indicate selection against (−) the prey items

Ingestive preference: 24-h, 6-choice experiments

We tested the null hypothesis that no significant differences existed in the preferences of *H. crassicornis* nudibranchs among six food choices. Sixteen of 18 nudibranchs explored the arena before making a choice and all nudibranchs actively traversed the arena during the 24-h trial. One nudibranch did not eat and that trial was omitted from analysis. The remaining 17 nudibranchs averaged 9.0 ± 7.2 h d^{-1} feeding. *D. occidentalis* tunicates, *O. geniculata* hydroids, and *A. labiata* polyps all were consumed by more than half of the nudibranchs tested (Table 1).

The log likelihood G_T values were greater than the critical χ^2 of 11.07, indicating significant preferences in food choice in all 17 trials. Calculation of the normal standard deviates of the G_i values for each food type permitted determination of the degree and direction (+/−) of each nudibranch's preference. *D. occidentalis* was the only food choice for which more than half of the 17 nudibranchs (58.8%) showed a positive preference (Fig. 2). The polyps of *A. labiata* were the second-most-preferred food, with 35.3% of nudibranchs showing a positive preference (Fig. 2); however, only three nudibranchs chose polyps as their most-preferred prey (Table 1), and 58.8% showed selection against the polyps (Fig. 2). All nudibranchs that exhibited a positive preference for the polyps had a very strong preference toward them (Table 2). The nudibranchs showed 17.6% positive preference for *O. geniculata* hydroids and *Anthopleura elegantissima* sea anemones, but there were no positive responses to *Bugula sp.* bryozoans or *Halichondria bowerbanki* sponges (Fig. 2). H_{02}, that no preferences existed among six prey species, was rejected.

To determine if food preference was influenced by nudibranch size, regressions were run between the wet weights (g) of the *H. crassicornis* nudibranchs and the

Table 1 Indices of *Hermissenda crassicornis* nudibranch predation on six test food organisms

Food organism	Nudibranchs consuming (%)	Feeding time (%)	Top preference (% of nudibranchs)
A. labiata polyps	52.9	18.4 ± 35.8	17.6
Obelia geniculata hydroids	58.8	6.7 ± 26.1	11.8
Distaplia occidentalis tunicates	64.7	56.2 ± 42.9	52.9
Anthopleura elegantissima sea anemones	35.3	18.3 ± 38.1	17.6
Bugula sp. bryozoans	17.6	0.2 ± 1.4	0
Halichondria bowerbanki sponges	0	0	0

Table 2 Results of log likelihood *G* preference tests for all *Hermissenda crassicornis* nudibranchs that showed positive preference for the polyps of *A. labiata*

Conditioning prey type	Wet weight (g)	Feeding time on polyps (%)	Standard deviate	*P* value
Distaplia occidentalis tunicates	2.23	22.4	2.997	0.003
A. labiata polyps	1.53	83.9	18.382	<0.001
A. labiata polyps	1.48	16.4	2.449	0.014
Anthopleura elegantissima anemone	0.892	85.7	22.553	<0.001
Halichondria bowerbanki sponges	0.373	34.5	5.498	<0.001
Obelia geniculata hydroids	0.285	99.1	133.386	<0.001

Standard deviates were used to measure the direction (+ or −) and strength of preference; large numbers indicate stronger preferences. Wet weights are included to display the range of nudibranch sizes showing positive preference to polyps. The variety of conditioning prey types illustrates the lack of conditioning effects

G_i preference value standard deviates for each food. Nudibranch size did not significantly affect preferences of any food choice (Table 3); H_{03} was not rejected. All sizes of nudibranchs, from the smallest (0.285 g) to the largest (2.230 g) tested, consumed *A. labiata* polyps.

Comparison of *H. crassicornis* feeding during daylight and nighttime showed that the nudibranchs spent the same amount of time eating in daylight (29.6 ± 13.0% of total time feeding) as at nighttime (33.1 ± 10.5% of total time feeding). No significant difference was found between daytime and nighttime preferences ($t_{16} = -0.112$, $P = 0.913$ for *A. labiata* polyps and $T = 429$, $N = 102$, $P = 0.172$ for all foods, Fig. 3). H_{04} was not rejected.

The potential for ingestive conditioning to affect the ingestive preferences of the *H. crassicornis* nudibranchs was tested. Although nudibranchs eating the three most-consumed foods (tunicates, polyps, and hydroids) appeared to show predation responses based on ingestive conditioning, only 6 of 17 (35.3%) of all nudibranchs exhibited a positive preference toward their conditioned food source. All three nudibranchs conditioned to *D. occidentalis* tunicates preferred it to all other food choices. Two of three nudibranchs conditioned to *A. labiata* polyps showed a positive preference for polyps, but polyps were the top preference for only one of them. One of three nudibranchs conditioned to *O. geniculata* hydroids showed a positive preference for the hydroid, but it was not its top preference (Fig. 4). The multidimensional non-parametric statistical program PRIMER analyzed similarity by proximity of like treatments on a two-dimensional plane. The absence of significant clusters with a goodness-of-fit stress level of 0.07 indicated that ingestive conditioning did not affect preference and H_{05} was not rejected.

Table 3 Probabilities that the size of the nudibranch *Hermissenda crassicornis* ($n = 17$) affected preferences for six food choices

Food organism	r^2	$F_{1,16}$	P
A. labiata polyps	0.175	3.393	0.084
Obelia geniculata hydroids	0.018	0.299	0.592
Distaplia occidentalis tunicates	0.023	0.381	0.546
Anthopleura elegantissima sea anemones	0.002	0.024	0.879
Bugula sp. bryozoans	<0.001	0.005	0.947
Halichondria bowerbanki sponges	0.033	0.544	0.471

Probabilities were determined with regressions between the wet weights (g) and the G_i preference value standard deviates for each food

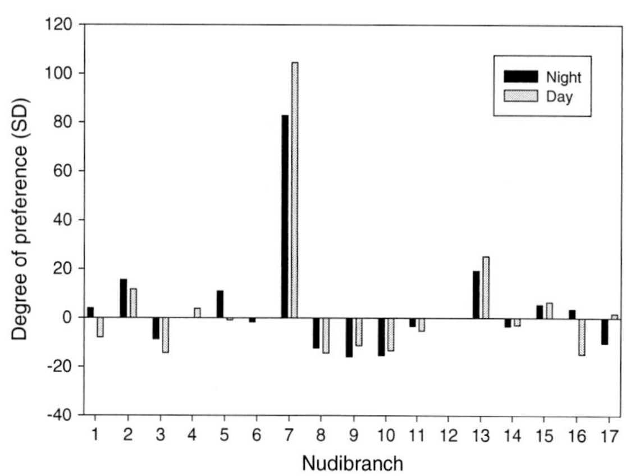

Fig. 3 Day versus night preferences of *Hermissenda crassicornis* nudibranchs for the polyps of *A. labiata*. The calculated standard deviates (SD) of *G* statistics for daytime and nighttime preferences are compared for 17 nudibranchs

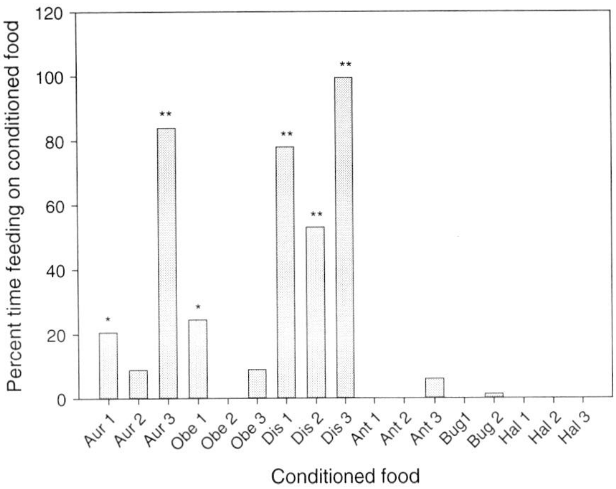

Fig. 4 Percent of total time individual *Hermissenda crassicornis* nudibranchs spent feeding on conditioned foods. Aur, *A. labiata* polyps; Obe, *Obelia geniculata* hydroids; Dis, *Distaplia occidentalis* tunicates; Ant, *Anthopleura elegantissima* sea anemones; Bug, *Bugula* sp. bryozoans; Hal, *Halichondria bowerbanki* sponges. Two or three nudibranchs were tested for each prey species. *Positive preference; **top preference

Chemotaxis: whole, living prey

Chemotaxis of *H. crassicornis* nudibranchs to living prey was observed during single-choice tests between prey items and seawater. Significant responses were observed for *A. labiata* polyps ($\chi^2_7 = 8$, $P = 0.005$), *D. occidentalis* tunicates ($\chi^2_8 = 9$, $P = 0.003$), and *O. geniculata* hydroids ($\chi^2_6 = 7$, $P = 0.008$). The null hypothesis H_{06}, that no chemotaxis to living prey occurred, was rejected. By contrast, no responses were observed in tests using *A. elegantissima* sea anemones ($n = 8$). The seawater blank was not chosen during any trial. One nudibranch from *A. labiata* trials and one nudibranch from *O. geniculata* trials did not make a choice and these trials were excluded from analysis. Nudibranchs responded more quickly to *A. labiata* and *D. occidentalis* (within 5–10 min of prey introduction) than to *O. geniculata* (20–30 min).

Chemotaxis: polar and non-polar extracts

Preliminary single-choice tests using *H. crassicornis* showed that the nudibranch exhibited chemotaxis to the polar extracts of *A. labiata* polyps and *D. occidentalis* tunicates. Immediate responses to the polar extracts occurred in all trials ($n = 4$ for each extract) and H_{07} was rejected; however, no responses were observed toward the non-polar extracts ($n = 4$ for each extract). Therefore, the polar extracts of *A. labiata* and *D. occidentalis* were chosen for use in 2-choice chemotaxis experiments.

Two-choice tests showed a significant chemotactic preference of *H. crassicornis* for the polar extract of *A. labiata* polyps over that of *D. occidentalis* tunicates ($\chi^2_{10} = 7.36$, $P = 0.007$; Fig. 5). H_{08} was rejected. One nudibranch that had just finished laying an egg mass before the trial did not make a choice and was excluded from analysis.

Feeding potential

Feeding rates of *H. crassicornis* nudibranchs on *A. labiata* polyps were determined to assess their effect on polyp populations. Feeding data were +1 $\log_{10}$ transformed to include zero values in the analysis. The numbers of polyps consumed increased with the wet weights of the nudibranchs. The percentages of large (>3 mm) and small (<3 mm) polyps consumed both increased proportionally with the wet weight of the nudibranchs ($r^2 = 0.156$, $F_{1,34} = 6.08$, $P = 0.019$ and $r^2 = 0.124$, $F_{1,34} = 4.68$, $P = 0.038$, respectively; Fig. 6) and H_{09} was rejected. There was no significant difference between the slopes of the two regression lines ($F = 2.33$, $P = 0.132$), therefore H_{10} was not rejected. Results of the multiple stepwise major axis regression indicated that only nudibranch size significantly affected the total number of polyps consumed ($r^2 = 0.341$, $F_{3,34} = 5.334$, $P < 0.001$); therefore, polyp colony density (10–65 polyps cm^{-2};

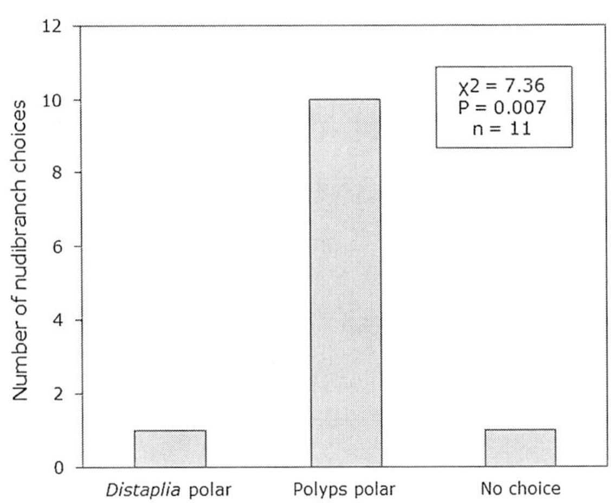

Fig. 5 Chemotactic choices made by *Hermissenda crassicornis* between the polar extracts of *Distaplia occidentalis* colonial tunicates and *A. labiata* polyps

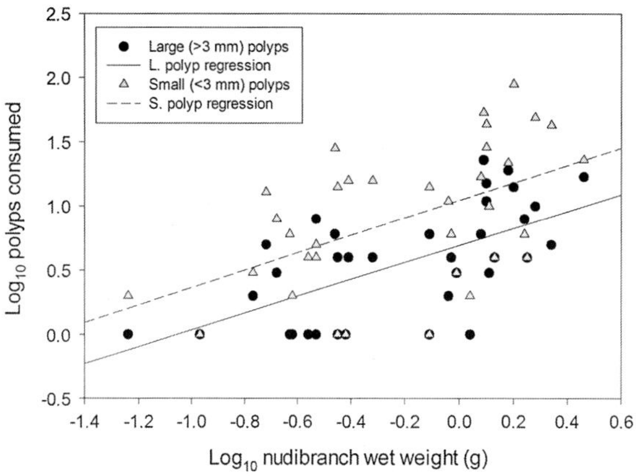

Fig. 6 Effect of the size of *Hermissenda crassicornis* nudibranchs on the numbers of large and small *A. labiata* polyps consumed. Data were +1 Log_{10} transformed to account for zeros in the data

$P = 0.621$) and polyp size class ($P = 0.498$) were removed from the regression, and H_{11}, H_{12} were not rejected. The greatest consumption was 102 polyps in 1 h by a nudibranch weighing 1.576 g. The mean rate ± standard error of consumption for large nudibranchs weighing more than 0.912 g (median) was 31.3 ± 28.9 polyps h^{-1} ($n = 17$), while the mean rate of consumption for nudibranchs weighing less than 0.912 g was 8.6 ± 9.7 polyps h^{-1} ($n = 18$). The mean rate of consumption for all nudibranchs was 19.6 ± 23.9 polyps h^{-1} ($n = 35$). The total number of polyps (p) that would be consumed in an hour by a *H. crassicornis* nudibranch could be approximated from its wet weight (ww) based on the following equation:

$$\log_{10}(p) = 1.337 + (1.347 * \log_{10}(\text{ww})).$$

Predation on *A. labiata* polyps by other nudibranchs

Six species of nudibranchs, in addition to *H. crassicornis*, were tested for predation on *A. labiata* polyps in the laboratory. Four of the species consumed polyps (Table 4), but five arminid nudibranchs (*Janolus fuscus* O'Donohue, 1924) collected in March, April, and September and two nudibranchs in each of the aeolid species, *Dirona albolineata* MacFarland, 1905 (March and August) and *Dirona aurantia* Hurst, 1966 (March), did not. Nudibranchs were observed near or on the *A. labiata* colony during the spring, summer, and autumn when the colony was most actively adding individuals through budding and possible planula recruitment (Purcell et al., 2009). No nudibranchs were observed during the winter. Few other nudibranch species are known to be predators of scyphozoan polyps (Table 4).

Discussion

Nudibranch preference for the polyps of *A. labiata*

Our results clearly demonstrate feeding preference by *Hermissenda crassicornis* nudibranchs for *A. labiata* polyps among several prey taxa in live prey choice and chemotaxis experiments. Thus, null hypotheses H_{01}, H_{02}, H_{06}, H_{07}, and H_{08} concerning choices were rejected. Preferences were unaffected by nudibranch size, day versus night, or ingestive conditioning; therefore, H_{03}, H_{04}, and H_{05} were not rejected.

The ingestive preferences of the nudibranch on *A. labiata* polyps and other foods differed somewhat in 2-choice and 6-choice tests. During the 2-choice tests, the nudibranchs greatly preferred *A. labiata* polyps over the three other food choices, including the colonial tunicate *Distaplia occidentalis*. The polyps were chosen first in ≥66% of the trials and were fed on longest (≥70% of total feeding time). In contrast, in the 6-choice tests the nudibranchs preferred *D. occidentalis* to all other food choices (58.8% of nudibranchs showed positive preference), and *A. labiata* polyps were the second-most-preferred food (~50% of the nudibranchs tested consumed polyps, but only 35.3% showed a positive preference for polyps). Apparently, the availability of additional food items and the longer foraging time during 6-choice testing influenced preferences for these two organisms. Alternatively, the preferences observed during 6-choice testing may have been related to the costs versus benefits of foraging on the dense tissues of the tunicate versus the more diffuse tissues of the other organisms. *H. crassicornis* has been shown to survive and grow well on a diet exclusively of tunicates (Harrigan & Alkon, 1978).

Results of chemotactic preference testing indicated that the *H. crassicornis* actively uses chemotaxis in determining prey choice and that the polar compounds extracted from *A. labiata* polyps were significantly preferred to those from *D. occidentalis* tunicates. Of interest in these experiments is the complete and

Table 4 Known nudibranch predators of scyphozoan polyps

Nudibranch species	Months observed	Number	Scyphozoan species	Reference
Dendronotacea				
Dendronotus dalli (Bergh, 1879)	Aug	1	*A. labiata*	This study
Dendronotus rufus (O'Donoghue, 1921)	Sep, Oct	3	*A. labiata*	Kozloff (1983) This study
Aeolidacea				
Flabellina fusca (Bergh, 1894)	Mar	12	*A. labiata*	This study
Hermissenda crassicornis	Mar, Aug, Sep	5	*A. labiata*	This study
Cratena pilata (Gould, 1870)			*Chrysaora quinquecirrha* (Desor, 1848)	Schultz & Cargo (1971)
Austraeolis catina (Marcus, 1962)			*Cassiopea* sp.	Clark & Goetzfried (1978)
Flabellina verrucosa (Sars, 1829) (as *Coryphella verrucosa*)	Oct–Nov		*A. aurita*	Hernroth & Gröndahl (1985a, b)
Dondice parguerensis (Brandon & Cutress, 1985)	Nov–Feb		*Cassiopea xamachana* (Forsskal), *C. frondosa*	Brandon & Cutress (1985)

definitive nature of the choices made. These results suggest that chemotaxis plays a strong role in nudibranch foraging behavior.

Also of interest in the chemotaxis results was the complete lack of response to the anemone, *Anthopleura elegantissima*, and to the non-polar extracts of *A. labiata* and *D. occidentalis.* Our results for polar and non-polar extracts are consistent with previous studies showing that many of the compounds that elicit chemoreception in marine organisms are polar (Croll, 1983; Purcell & Anderson, 1995). This may be due to their water-soluble nature, or the fact that polar compounds are generally metabolites such as proteins, carbohydrates, and amino acids (Croll, 1983; Christie, 1993). Studies have confirmed amino acids as feeding stimulants in other marine gastropods such as *Aplysia* sp., and they are thought to be stimulants for the snail, *Nassarius obsoletus* (Say, 1822) (reviewed in Croll, 1983). In contrast, many non-polar compounds are composed of lipids that are not water soluble (Christie, 1993).

The combined results of the prey choice experiments suggest that *A. labiata* polyps, along with *D. occidentalis* tunicates, are top prey choices of *H. crassicornis.* As they are both highly preferred, it is likely that the nudibranch preys upon the two food sources to fulfill different biological needs. While the tissues of the tunicate may be consumed to provide simple nourishment, the polyps are consumed, in part, to harvest nematocysts. Because aeolid nudibranchs utilize harvested nematocysts of their cnidarian prey in their own defense mechanisms, it is necessary for the nudibranchs to periodically replenish nematocysts to the cnidosacs (Martin, 2003).

The results of our prey choice experiments suggest that among cnidarian prey, *H. crassicornis* nudibranchs prefer feeding on small, soft-bodied (athecate), colonial cnidarians, such as scyphozoan polyps. Ingestive preference experiments showed that *A. labiata* polyps were preferred twice more than any other cnidarian food choice. In chemotaxis experiments, no chemotaxis was exhibited toward *A. elegantissima* anemones. Despite its generalist feeding habits on organisms from different phyla, *H. crassicornis* has distinct favorites within phyla. Other species of aeolid nudibranchs, such as *Aeolidia papillosa* (Linnaeus, 1761), specialize on one class of cnidarian prey; for example, *A. papillosa* eats sea anemones, such as *A. elegantissima* (Waters, 1973; Edmunds et al., 1974; Hall et al., 1984; Seavy & Muller-Parker, 2002). In addition, *H. crassicornis* fed exclusively on the soft tissue of hydroid polyps, but ignored the harder stalks when offered a choice (Avila et al., 1998). Small, soft-bodied, colonial cnidarians would provide the most cost-effective foraging for nudibranchs that include cnidarians in their diets.

Two sources of potential error were unavoidable in our nudibranch food preference experiments. First, ingestive conditioning may have occurred prior to capture of the nudibranchs. We tested for ingestive conditioning, which was negligible for diets assigned after capture; however, the feeding habits of the nudibranchs pre-capture were unknown. The potential of larval stage conditioning also cannot be discounted. It would be preferable to start with laboratory-reared naive subjects. Second, although all food organisms used in testing appeared healthy, their health and attractiveness to the nudibranchs may have been compromised during <3 weeks in the seawater table. Unfortunately, analysis of how the duration of captivity affected preference was not possible.

Nudibranch feeding potential on the polyps of *A. labiata*

We assessed the potential for *H. crassicornis* nudibranchs to affect the population dynamics of *A. labiata* polyps. Data from laboratory feeding trials support previous findings that *H. crassicornis* can be a significant predator of *Aurelia* spp. polyps (Keen, 1991). Only one of the four null hypotheses tested (H_{09}) was rejected; the number of polyps consumed increased with nudibranch size, but were unaffected by polyp size or density. Large *H. crassicornis* (>0.912 g wet weight, $n = 17$) consumed a mean of 31.3 polyps h^{-1}, with a maximum of 102 polyps h^{-1}. Because *H. crassicornis* consumed prey for an average of 9 ± 7.2 h d^{-1}, the maximum feeding potential of the nudibranchs appears to be substantially greater than previously measured (120 polyps d^{-1}) for a ~5.7 g nudibranch (Keen, 1991).

Polyp colony density and size may play significant roles in limiting predation pressure. In laboratory experiments, Keen (1991) found significantly greater effects of *H. crassicornis* predation on polyp mortality at low densities (2 polyps cm^{-2}) than at high densities (10 polyps cm^{-2}). Those results would give consumption rates by a ~2.4 g nudibranch of 49.6 polyps d^{-1} at low polyp density but only 5.0 polyps d^{-1} at high polyp density. By contrast, our results indicated that polyp densities (10–65 polyps cm^{-2}) did not affect nudibranch consumption rates; however, we suspect that the small sizes of the polyp clusters on algal blades used in our experiment allowed nudibranchs access to the polyps without being stung. Field surveys by Keen (1991) on a large polyp colony where nudibranchs were present showed that 27% of areas with polyp patches <100 cm^2 had losses each month, but only 16% of areas with patches 100–1,000 cm^2 had losses. In our own field surveys, nudibranchs were observed only at the edges of the polyp colony or where the polyps were sparse.

The potential for nudibranch predation to control scyphozoan polyp populations may be underappreciated. Nudibranch predation may be important in controlling establishment and early growth of polyp colonies along with small, fringe populations of large colonies. During our experiments, we confirmed that four nudibranch species were predators of *A. labiata* polyps, two of which were new records of polyp predation. Because there are hundreds of aeolid and dendronotid nudibranch species that consume cnidarians, and hundreds of scyphozoan polyps and hydroid species that produce jellyfish, further studies of other nudibranch species and their feeding potentials are needed to better understand the potential of predation to control jellyfish blooms.

Acknowledgments The authors thank the director and staff of Shannon Point Marine Center for supplying the facilities, equipment, and support to make this study possible, the Charles and June Ross Foundation and the Biology Department of Western Washington University for their financial support, Dr. Brian Bingham for advice concerning experimental design, and Nathan Schwarck for his support and friendship throughout. This research was funded in part by National Science Foundation ADVANCE Fellows Award OCE-0137419 and Research Experiences for Undergraduates program NSF 09-598.

References

Avila, C., 1998. Chemotaxis in the nudibranch *Hermissenda crassicornis*: does ingestive conditioning influence its behaviour in a Y-maze? Journal of Molluscan Studies 64: 215–222.

Avila, C., E. Tyndale & A. M. Kuzirian, 1998. Feeding behaviour and growth of *Hermissenda crassicornis* (Mollusca: Nudibranchia) in the laboratory. Marine and Freshwater Behaviour and Physiology 31: 1–19.

Behrens, D. W., 1991. Pacific Coast Nudibranchs, 2nd ed. Sea Challengers Publishing, Monterey: 97 pp.

Brandon, M. & C. E. Cutress, 1985. A new *Dondice* (Opisthobranchia: Favorinidae), predator of *Cassiopea* in southwest Puerto Rico. Bulletin of Marine Science 36: 139–144.

Brodeur, R. D., H. Sugisaki & G. L. Hunt Jr., 2002. Increases in jellyfish biomass in the Bering Sea: implications for the ecosystem. Marine Ecology Progress Series 233: 89–103.

Buss, L. W., 1990. Competition within and between encrusting clonal invertebrates. Trends in Ecology & Evolution 5: 352–356.

Cargo, D. G. & J. W. Burnett, 1982. Observations on the ultrastructure and defensive behavior of the cnidosac of *Cratena pilata*. Veliger 24: 325–327.

Christie, W. W., 1993. In Christie, W. W. (ed.), Advances in Lipid Methodology – Two. Oily Press, Dundee: 195–213.

Clark, K. B. & A. Goetzfried, 1978. Zoogeographic influences on development of North Atlantic Ascoglossa and Nudibranchia, with a discussion of factors affecting egg size and number. Journal of Mollusc Studies 44: 283–294.

Croll, R. P., 1983. Gastropod chemoreception. Biological Reviews of the Cambridge Philosophical Society 58: 293–319.

Edmunds, M., G. W. Potts, R. C. Swinfen & V. L. Waters, 1974. The feeding preferences of *Aeolidia papillosa* (L.) (Mollusca: Nudibranchia). Journal of the Marine Biological Association of the United Kingdom 54: 939–947.

Gong, A. J., 2002. Allocations to clonal replication in a marine scyphozoan (*Aurelia*). Science and Engineering 62: 3516–3635.

Gröndahl, F., 1988. Interactions between polyps of *Aurelia aurita* and planktonic larvae of scyphozoans: an experimental study. Marine Ecology Progress Series 45: 87–93.

Hall, S. J., C. D. Todd & A. D. Gordon, 1984. Prey-species selection by the anemone predator *Aeolidia papillosa* (L.): the influence of ingestive conditioning and previous dietary history, and a test for switching behaviour. Journal of Experimental Marine Biology and Ecology 82: 11–33.

Harrigan, J. F. & D. L. Alkon, 1978. Larval rearing, metamorphosis, growth and reproduction of the aeolid nudibranch *Hermissenda crassicornis* (Eschscholtz, 1831). Biological Bulletin 154: 430–439.

Hernroth, L. & F. Gröndahl, 1985a. On the biology of *Aurelia aurita* (L.). 3. Predation by *Coryphella verrucosa* (Gastropoda, Opisthobranchia), a major factor regulating the development of *Aurelia* populations in the Gullmar Fjord, Western Sweden. Ophelia 24: 37–45.

Hernroth, L. & F. Gröndahl, 1985b. On the biology of *Aurelia aurita* (L.). 2. Major factors regulating the occurrence of ephyrae and young medusae in the Gullmar Fjord, Western Sweden. Bulletin of Marine Science 37: 567–576.

Holst, S., 2012. Effects of climate warming on strobilation and ephyra production of North Sea scyphozoan jellyfish. Hydrobiologia. doi:10.1007/s10750-012-1043-y.

Holst, S. & G. Jarms, 2007. Substrate choice and settlement preferences of planula larvae of five Scyphozoa (Cnidaria) from German Bight, North Sea. Marine Biology 151: 863–871.

Hoover, R. A. & J. E. Purcell, 2009. Substrate preferences of scyphozoan *Aurelia labiata* polyps among common dock-building materials. Hydrobiologia 616: 259–267.

Keen, S. L., 1991. Clonal dynamics and life history evolution in the jellyfish *Aurelia aurita*. Ph.D. thesis, University of California, Berkeley.

Kozloff, E. N., 1983. Seashore Life of the Northern Pacific Coast: An Illustrated Guide to Northern California, Oregon, Washington, and British Columbia. University of Washington Press, Seattle: 83 pp.

Lucas, C. H., 2001. Reproduction and life history strategies of the common jellyfish, *Aurelia aurita*, in relation to its ambient environment. Hydrobiologia 451: 229–246.

Lucas, C. H., A. G. Hirst & J. A. Williams, 1997. Plankton dynamics and *Aurelia aurita* production from two contrasting ecosystems: causes and consequences. Estuarine, Coastal and Shelf Science 45: 209–219.

Mariottini, G. L. & L. Pane, 2010. Mediterranean jellyfish venoms: a review on scyphomedusae. Marine Drugs 8: 1122–1152.

Martin, R., 2003. Management of nematocysts in the alimentary tract and in cnidosacs of the aeolid nudibranch gastropod *Cratena peregrina*. Marine Biology 143: 533–541.

McDonald, G. R. & J. W. Nybakken, 1996. A list of the worldwide food habits of nudibranchs. http://www.the veliger.org/nudibranch_food.html.

Mills, C. E., 1995. Medusae, siphonophores and ctenophores as planktivorous predators in changing global ecosystems. ICES Journal of Marine Science 52: 575–581.

Miyake, H., M. Terazaki & Y. Kakinuma, 2002. On the polyps of the common jellyfish *Aurelia aurita* in Kagoshima Bay. Journal of Oceanography 58: 451–459.

Morris, R. H., D. P. Abbot & E. C. Haderlie, 1980. Intertidal Invertebrates of California. Stanford University Press, Stanford, CA.

Pauly, D., W. Graham, S. Libralato, L. Morissette & M. L. D. Palomares, 2009. Jellyfish in ecosystems, online databases, and ecosystem models. Hydrobiologia 616: 67–85.

Purcell, J. E., 2003. Predation on zooplankton by large jellyfish, *Aurelia labiata, Cyanea capillata* and *Aequorea aequorea* in Prince William Sound, Alaska. Marine Ecology Progress Series 246: 137–152.

Purcell, J. E., 2005. Climate effects on formation of jellyfish and ctenophore blooms. Journal of the Marine Biology Association of the UK 85: 1–16.

Purcell, J. E. & P. A. V. Anderson, 1995. Electrical responses to water-soluble components of fish mucus recorded from the cnidocytes of a fish predator, *Physalia physalis*. Marine and Freshwater Behaviour and Physiology 26: 149–162.

Purcell, J. E. & M. N. Arai, 2001. Interactions of pelagic cnidarians and ctenophores with fish: a review. Hydrobiologia 451: 27–44.

Purcell, J. E. & M. V. Sturdevant, 2001. Prey selection and dietary overlap among zooplanktivorous jellyfish and juvenile fishes in Prince William Sound, Alaska. Marine Ecology Progress Series 210: 67–83.

Purcell, J. E., S.-I. Uye & W.-T. Lo, 2007. Anthropogenic causes of jellyfish blooms and direct consequences for humans: a review. Marine Ecology Progress Series 350: 153–174.

Purcell, J. E., R. A. Hoover & N. T. Schwarck, 2009. Interannual variation of strobilation by the scyphozoan *Aurelia labiata* in relation to polyp density, temperature, salinity. And light conditions in situ. Marine Ecology Progress Series 375: 139–149.

Purcell, J. E., D. Atienza, V. Fuentes, A. Olariaga, U. Tilves, C. Colahan & J.-M. Gili, 2012. Temperature effects on asexual reproduction rates of scyphozoan species from the northwest Mediterranean Sea. Hydrobiologia. doi:10.1007/s10750-012-1047-7.

Ruzicka, J. J., R. D. Brodeur & T. C. Wainwright, 2007. Seasonal food web models for the Oregon inner-shelf ecosystem: investigating the role of large jellyfish. California Cooperative Oceanic Fisheries Investigations Reports 48: 106–128.

Schultz, L. P. & D. G. Cargo, 1971. The sea nettle of Chesapeake Bay. Natural Resources Institute, Educational Series No. 93. University of Maryland: 8 pp.

Seavy, B. E. & G. Muller-Parker, 2002. Chemosensory and feeding responses of the nudibranch *Aeolidia papillosa* to the symbiotic sea anemone *Anthopleura elegantissima*. Invertebrate Biology 121: 115–125.

Thein, H., H. Ikeda & S. I. Uye, 2012. The potential role of podocysts in perpetuation of the common jellyfish *Aurelia aurita* s.l. (Cnidaria: Scyphozoa) in anthropogenically perturbed coastal waters. Hydrobiologia. doi:10.1007/s10750-012-1045-9.

Todd, C. D., 1981. The ecology of nudibranch molluscs. Oceanography and Marine Biology Annual Review 19: 141–234.

Tyndale, E., C. Avila & A. M. Kuziruan, 1994. Food detection and preferences of the nudibranch mollusk *Hermissenda crassicornis*: experiments in a Y-maze. Biological Bulletin 187: 274–275.

UNEP, 1991. Jellyfish blooms in the Mediterranean. In Proceedings of II Workshop on Jellyfish in the Mediterranean Sea, Mediterranean Action Plan Technical Reports Series: 47 pp.

Uye, S. & U. Ueta, 2004. Recent increase of jellyfish populations and their nuisance to fisheries in the inland Sea of Japan. Bulletin of the Japanese Society of Fisheries Oceanography 68: 9–19.

Waters, V. L., 1973. Food-preference of the nudibranch *Aeolidia papillosa*, and the effect of the defenses of the prey on predation. Veliger 15: 174–192.

Willcox, S., N. A. Moltschaniwskyj & C. M. Crawford, 2008. Population dynamics of natural colonies of *Aurelia* sp. scyphistomae in Tasmania, Australia. Marine Biology 154: 661–670.

Williams, K., 1976. The failure of Pearson's goodness of fit statistic. Statistician 25: 49.

Willows, A. O. D., 1978. Physiology of feeding in *Tritonia* I. Behavior and mechanics. Marine Behavioral Physiology 5: 115–135.

Zar, J. H., 1996. Biostatistical Analysis, 3rd ed. Prentice-Hall, Inc, Upper Saddle River.

Zimmer-Faust, R. K., P. B. O'Neill & D. W. Schar, 1996. The relationship between predator activity state and sensitivity to prey odor. Biological Bulletin 190: 82–85.

Hydrobiologia (2012) 690:215–226
DOI 10.1007/s10750-012-1040-1

Parasitism (Trematoda, Digenea) in medusae from the southwestern Atlantic Ocean: medusa hosts, parasite prevalences, and ecological implications

Luciana M. Diaz Briz · Sergio R. Martorelli · Gabriel N. Genzano · Hermes W. Mianzan

Published online: 12 April 2012
© Springer Science+Business Media B.V. 2012

Abstract Digenean are important endoparasites of fish with complex life cycles; some genera include medusae as secondary hosts. Their transmission to fish occurs when fish prey on these jelly hosts. Fish predation on jellyfish is a widespread phenomenon, even though predation by fish on jellyfish has not been determined through parasitism yet. We hypothesized that medusae with high prevalences of digeneans could be important for their transmission to fish. A total of 48,900 specimens of 50 medusa species were analyzed; 2,181 harbored digeneans. *Opechona* sp. and *Monascus filiformis* were the most frequent and abundant parasites with the widest range of hosts. Hemiuridae gen. sp. and *Bacciger* sp. were found in few specimens of some medusa species. Prevalences were unevenly distributed in the region. Three groups with high prevalence values were identified mainly related to frontal areas: Río de la Plata, Bahía Blanca, and North Patagonian tidal front. *Eucheilota ventricularis*, *Clytia hemisphaerica*, *Proboscidactyla mutabilis*, *Liriope tetraphylla*, and *Aequorea* spp. were the medusae that contributed the most as secondary hosts to *M. filiformis* and *Opechona* sp. The high prevalences found in these medusae suggest that may be a fundamental part of the life cycles of both parasites in these areas.

Keywords Digenean parasites · Secondary medusa hosts · Final fish hosts · Prevalence · Estuary and front zones · Southwestern Atlantic Ocean

Guest editors: J. E. Purcell, H. Mianzan & J. R. Frost / Jellyfish Blooms: Interactions with Humans and Fisheries

L. M. Diaz Briz (✉) · G. N. Genzano
Estación Costera Nágera, Facultad de Ciencias Exactas y Naturales, Universidad Nacional de Mar del Plata, CONICET, Funes 3350, CP 7600 Mar del Plata, Argentina
e-mail: diazbriz.luciana@gmail.com

L. M. Diaz Briz · G. N. Genzano · H. W. Mianzan
Instituto de Investigaciones Marinas y Costeras (IIMyC), Universidad Nacional de Mar del Plata, Consejo Nacional de Investigaciones Científicas y Técnicas (CONICET), Funes 3350, CP 7600 Mar del Plata, Argentina

S. R. Martorelli
Centro de Estudios Parasitológicos y Vectores (CEPAVE), CCT LA PLATA, CONICET, Calle 2 no 584 e/43 y 44, CP 1900 La Plata, Buenos Aires, Argentina

H. W. Mianzan
Instituto Nacional de Investigación y Desarrollo Pesquero (INIDEP), CONICET, Paseo Victoria Ocampo no 1, B7602HSA Mar del Plata, Argentina

Introduction

Helminths are common parasites that occur in the marine environment. Among these, the digeneans are considered an important endoparasitic group of vertebrates (mostly fish). Digeneans have complex life cycles, during which several groups of marine animals are used as intermediate hosts that harbor their larval stages to insure the transmission toward the final hosts. The transmission of these parasites often involves different predator–prey interactions between the hosts (Rohde, 1993; Marcogliese, 1995, 2004; Martorelli, 2001).

Zooplankters, including copepods, amphipods, and chaetognaths, are considered important intermediate or paratenic hosts of many digenean parasites of marine fish (Marcogliese, 1995; Daponte et al., 2008); however, some genera of these parasites, such as *Monascus* Looss, 1907 and *Opechona* Looss, 1907, include medusae and ctenophores as secondary hosts. Their transmission to fish, in which they culminate their life cycles, occurs when fish consume the jellyfish hosts (Lauckner, 1980; Girola et al., 1992; Marcogliese, 1995; Cremonte & Sardella, 1997; Martorelli & Cremonte, 1998; Martorelli, 2001).

Few complete life cycles are described for these parasites that involve medusae or ctenophores as secondary intermediate hosts, and only a limited number of fish species are known to be final hosts of these parasites (Cremonte & Sardella, 1997; Martorelli & Cremonte, 1998; Kohn et al., 2007; Averbuj & Cremonte, 2010; Diaz Briz personal observations). Most of the fish species are not documented as "jellyfish eaters".

The number of fish species that feed exclusively on jellies is small (see Arai, 2005). However, direct observations of fish stomach contents have shown that predation on jelly organisms is more common and frequent than previously thought. Many fish species with broad diets (i.e., spiny dogfish *Squalus acanthias* Linnaeus, 1758, chum salmon *Oncorhynchus keta* (Walbaum, 1792), and Atlantic mackerel *Scomber scombrus* Linnaeus, 1758) feed at times on jelly organisms (Arai, 1988, 2005; Ates, 1988, 1991; Mianzan et al., 1996; Arai et al., 2003). This behavior, named "feeding on survival food" (Mianzan et al., 2001), may occur when other prey are not available, which implies a complex, and adaptive food web.

Analysis of stomach contents provides an overall picture of the number of fish that consume jelly organisms (Mianzan et al., 1996), but the identity of the prey may be difficult to obtain. These fragile organisms are rapidly digested in the fish gut, thus identification usually refers to broad "jelly groups" (e.g., ctenophores, medusae, etc.). Even though predation by fish on jellyfish has not yet been determined through parasitism (Arai, 2005), a high prevalence of digenean parasites in medusae or ctenophores allows us to infer that jellies may have an important role in the transmission of parasites to their final fish hosts (Lauckner, 1980; Marcogliese, 2002). In particular, knowledge of species richness of medusae as hosts of digenean parasites of fish can be useful to suggest feeding interactions between medusae and fish (Mianzan et al., 1996).

Although several studies have reported on parasite–host interactions in the Northern Hemisphere (Lebour, 1916; Stunkard, 1967, 1969; Køie, 1975; Lauckner, 1980), such information from the Southern Hemisphere is scarce. In the southwestern Atlantic Ocean, the few records available only described the morphology of the metacercariae found in one ctenophore and several medusa hosts, usually collected by hand (Martorelli, 1991, 2001; Girola et al., 1992; Martorelli & Cremonte, 1998; Morandini et al., 2005). Only one digenean life cycle is known from this area (Martorelli & Cremonte, 1998). Because fish predation on jellyfish is a widespread phenomenon in the southwestern Atlantic Ocean (up to 35% of fish consume jelly prey; Mianzan et al., 1996), it is hypothesized that medusae with high prevalences of parasitism could be important for the transmission of digenean to fish.

To test this hypothesis several medusae species from an extensive geographical area of the southwestern Atlantic Ocean (26–55°S) were analyzed for first time, with the objectives to (1) expand the knowledge about different medusa species hosts of digenean parasites of fish, (2) establish the geographic distribution of the parasitized medusae and determine which zones have the highest parasite prevalences, and (3) indicate which of these parasitized species of medusae may transmit the parasites to fish.

Materials and methods

Data collection

The study area comprised the southern Brazilian, Uruguayan, and Argentine Continental Shelf (26–55°S; Fig. 1). Zooplankton samples were collected during 115 cruises carried out by the Instituto Nacional de Investigación y Desarrollo Pesquero (INIDEP) between 1987 and 2010. Hensen and Bongo plankton nets were used mostly and Calvet, Biomoc,

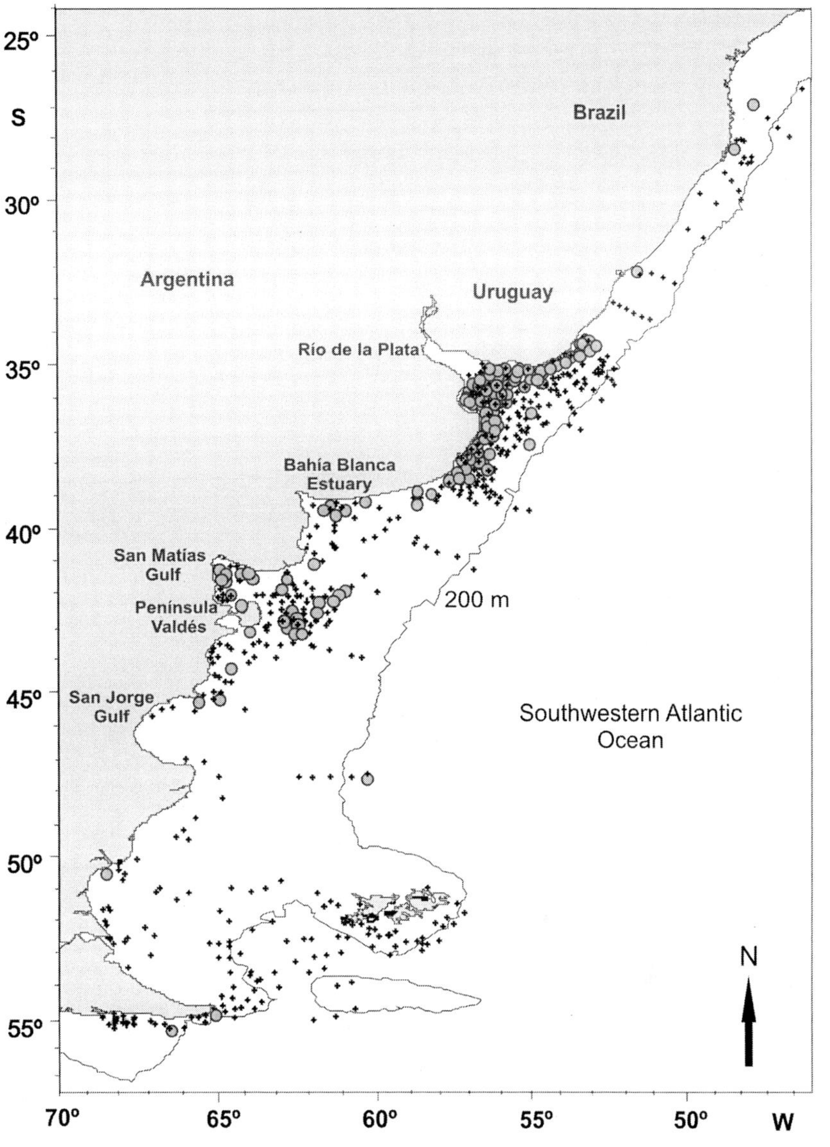

Fig. 1 Geographic distributions of all zooplankton samples collected in the southwestern Atlantic Ocean between 1987 and 2010 that contained medusae (*shaded circle* samples with medusae parasitized by digenean metacercariae; *plus* samples with medusae without parasites)

and Multired nets were used occasionally (see Wiebe & Benfield, 2003 for descriptions of nets). All samples were preserved in a 5% formalin-seawater solution. The resulting database is unique for the study area and contains all available biological and ecological information on medusae species.

A total of 3,335 zooplankton samples were analyzed. All medusa specimens were separated, identified with the aid of a stereomicroscope, and quantified. The larval stages (metacercariae) of the digenean parasites were removed with dissection needles from the mesoglea of the medusae. These parasites then were stained with Gill's Hematoxylin, dehydrated in an ethanol series, cleared in clove oil, and mounted in natural Canada balsam to be identified at the lowest taxonomical level possible. To corroborate the taxonomic identification, measurements of 10 specimens from each metacercariae group were taken by use of an ocular micrometer in the stereomicroscope. The morphological features and measurements of these metacercariae agreed with those previously reported by Martorelli & Cremonte (1998) and Martorelli (2001). Prevalence values were calculated according to Bush et al. (1997). Medusa specimens and mounted digeneans were kept in the collection of the UNMdP-INIDEP.

Data analysis

To identify areas with similar parasitic prevalences, the study area was divided into 33 one-degree grid squares. In each square, the mean prevalence of each parasite species per species of medusa host was calculated. Medusae present in just one square were excluded from the analysis (e.g., *Coryne eximia* Allman, 1859 and *Leuckartiara octona* (Fleming, 1823)). Classification methods (group average sorting of the Bray–Curtis similarity measures based on log $(X + 1)$ transformed prevalence data) were carried out using the PRIMER 5 software package (Clarke & Warwick, 2001). A one-way, non-parametric, multivariate analysis of similarity (ANOSIM) was implemented to evaluate differences among groups obtained by the cluster analysis. ANOSIM was used to test the null hypothesis that there was no difference in the composition of medusa species and their parasites between the groups. ANOSIM is an analogue of one-factor analysis of variance based on multispecies data (Chapman & Underwood, 1999), and it calculated the statistic R (Clarke & Warwick, 2001). SIMPER analysis (similarity percentages) was used to identify the medusa species that contributed most to (dis)similarities among and within groups. This analysis calculates the "average similarity" (contribution of the *i*th species to the overall dissimilarity between the groups considered) and the "internal similarity" (contribution each species makes to the average similarity within each group considered) (see Clarke & Warwick, 2001 for details).

Results

Approximately 30% of the analyzed zooplankton samples contained medusae, and of those, about 20% had parasitized medusae (Fig. 1). A total of 48,900 medusa specimens were analyzed including 50 species (Table 1). Of those, 2,181 (16 species of hydromedusae and 1 scyphomedusa) harbored the larval stage of digenean parasites in their mesoglea (Fig. 2a, b). The total prevalence was approximately 5%. Almost all the parasitized medusae species (16) were new records of secondary hosts to at least one taxon of digenean parasite (Table 1).

Four species of digenean metacercariae were found in the medusae: *Monascus filiformis* (Rudolphi, 1819)

Table 1 Taxonomic list of analyzed medusae following the classification of Marques & Collins (2004), Cartwright et al. (2008) and Collins et al. (2008)

Medusa species analyzed	Number of analyzed specimens
Phylum Cnidaria	-
Subphylum Medusozoa	
Class Hydrozoa	
Subclass Hydroidolina	
Order Anthoathecata	
Amphinema dinema (Péron & Lesueur, 1810)	9
Amphinema rugosum (Mayer, 1900)	3
Bougainvillia frondosa Mayer, 1900	1
Bougainvillia macloviana (Lesson, 1830)	130
Bougainvillia muscus Allman, 1863	23
Bougainvillia sp.[a]	361
Corymorpha gracilis (Brooks, 1882)	11
Corymorpha januarii Steenstrup, 1854	6
Coryne eximia Allman, 1859[a]	421
Dipurena reesi Vannucci, 1956	1
Euphysa aurata Forbes, 1848	342
Hybocodon spp.	40
Hydractinia spp.	13
Rathkea formosissima (Browne, 1902)	1
Tiaricodon coeruleus Browne, 1902	5
Turritopsis nutricola McCrady, 1859	11
Leuckartiara octona (Fleming, 1823)[a]	17
Proboscidactyla mutabilis (Browne, 1902)[a]	3,589
Order Leptothecata	
Aequorea spp.[a]	60
Blackfordia virginica Mayer, 1910	8,975
Clytia gracilis (Sars, 1851)	15
Clytia hemisphaerica (Linnaeus, 1767)[a]	956
Clytia lomae (Torrey, 1909)[a]	8
Clytia simplex (Browne, 1902)[a]	20
Cosmetirella davisi (Browne, 1902)[a]	39
Eucheilota ventricularis McCrady, 1859[a]	9,861
Eutonina scintillans (Bigelow, 1909)	2
Halopsis ocellata A. Agassiz, 1863[a]	6
Laodicea pulchra Browne, 1902	3
Laodicea undulata (Forbes and Goodsir, 1851)	193
Mitrocomella brownei (Kramp, 1930)[a]	229
Mitrocomella frigida (Browne, 1910)	5
Modeeria rotunda (Quoy and Gaimard, 1827)	1
Obelia spp.	4,864
Phialella falklandica Browne, 1905[a]	99

Table 1 continued

Medusa species analyzed	Number of analyzed specimens
Rhacostoma atlanticum L. Agassiz, 1850	1
Subclass Trachylina	
Order Limnomedusae	
Aglauropsis conanti Browne, 1902	1
Aglauropsis kawarii Moreira and Yamashita, 1972	10
Gossea brachymera Bigelow, 1909[a]	25
Olindias sambaquiensis Müller, 1861[b]	18
Olindiidae indet.	2
Liriope tetraphylla (Chamisso and Eysenhardt, 1821)[a]	17,152
Order Narcomedusae	
Cunina octonaria McCrady, 1859	11
Pegantha laevis H. B. Bigelow, 1909	79
Solmundella bitentaculata (Quoy and Gaimard, 1833)	9
Order Trachymedusae	
Amphogona apicata Kramp, 1957	1
Aglaura hemistoma Pèron and Lesueur, 1810	58
Rhopalonema velatum Gegenbaur, 1857	1,116
Sminthea eurygaster Gegenbaur, 1857	1
Class Scyphozoa	
Subclass Discomedusae	
Order Semaeostomeae	
Chrysaora lactea Eschscholtz, 1829[a]	96
Total number of analyzed specimens	48,900
Total number of analyzed species	50

[a] New records of secondary medusa hosts of at least one digenean parasite

[b] Secondary medusa hosts previously reported for the study area. The total numbers of specimens analyzed per medusa are given

Looss, 1907, *Opechona* sp., *Bacciger* sp. Nicoll, 1914, and Hemiuridae gen. sp. (Fig. 2c–f). *Opechona* sp. and *M. filiformis* were the most frequent parasites and occurred in 13 and 11 of the 17 parasitized species of medusae, respectively. Moreover, the total prevalences of both species were high, with *M. filiformis* found in 68.2% of the parasitized medusae and *Opechona* sp. in 39.7%. In general, the prevalence values of *Monascus filiformis* were higher than of those observed for *Opechona* sp. (Table 2). Hemiuridae gen. sp. and *Bacciger* sp. infected only 1.4 and 1.0% of the specimens, respectively, of three species of medusae (Table 2).

Most of the parasitized species of medusae only hosted *M. filiformis* and *Opechona* sp. By contrast, *Liriope tetraphylla* and *Eucheilota ventricularis* were parasitized by four taxa of metacercariae, and *Proboscidactyla mutabilis* and *Clytia simplex* were parasitized by three taxa (Table 2). Usually, a single medusa specimen harbored more than one parasite species. *L. tetraphylla*, *E. ventricularis*, *C. hemisphaerica*, *Aequorea* spp., and *P. mutabilis* medusae were often infested by both *M. filiformis* and *Opechona* sp. Three parasite species (*M. filiformis*, *Opechona* sp., and Hemiuridae gen. sp.) were found in only one *P. mutabilis* medusa. *Eucheilota ventricularis* medusa had the highest frequency of infection on the total of parasitized medusa specimens (54.6%), followed by *L. tetraphylla* (29.0%), and *C. hemisphaerica* (7.3%). The remaining species had infection frequencies less than 3.9%.

Parasitized medusae were unevenly distributed within the study area (Fig. 1). Some zones had a high concentration of medusa species and parasitized specimens. In semi-enclosed areas such as the Río de La Plata Estuary, Bahía Blanca Estuary, San Matías Gulf, and the North Patagonia tidal front (Península Valdés), the prevalences of *Opechona* sp. and *M. filiformis* were high. By comparison, *Bacciger* sp. and Hemiuridae gen. sp. had relatively low prevalences even though both usually occurred in the same zones as *Opechona* sp. and *M. filiformis* (Fig. 3a–d).

Prevalence indexes varied considerably among the different groups defined by the cluster analysis (ANOSIM, global $R = 0.598$, $P < 0.001$). Group 1 (33.4% internal similarity SIMPER analysis) cluster samples were mainly from the Río de La Plata and Bahía Blanca estuaries. Parasitized medusa hosts more important were *Eucheilota ventricularis* and *Clytia hemisphaerica* with *M. filiformis* (26.9 and 23.5%, cumulative contribution respectively, SIMPER analysis), *E. ventricularis*, and *C. hemisphaerica* with *Opechona* sp. (21.2 and 9.7%, respectively), and *Liriope tetraphylla* with *M. filiformis* (7.4%) and with *Opechona* sp. (6.1%) (Fig. 4a–c). Group 2 (29.1% average similarity) mostly corresponded to the San Matías Gulf, Peninsula Valdés and surrounding areas (tidal front of Patagonia), and the south coast of Buenos Aires. Only *P. mutabilis* medusae with *M. filiformis* and *Opechona* sp. contributed to this group (71.1 and 25.9%, respectively, SIMPER analysis)

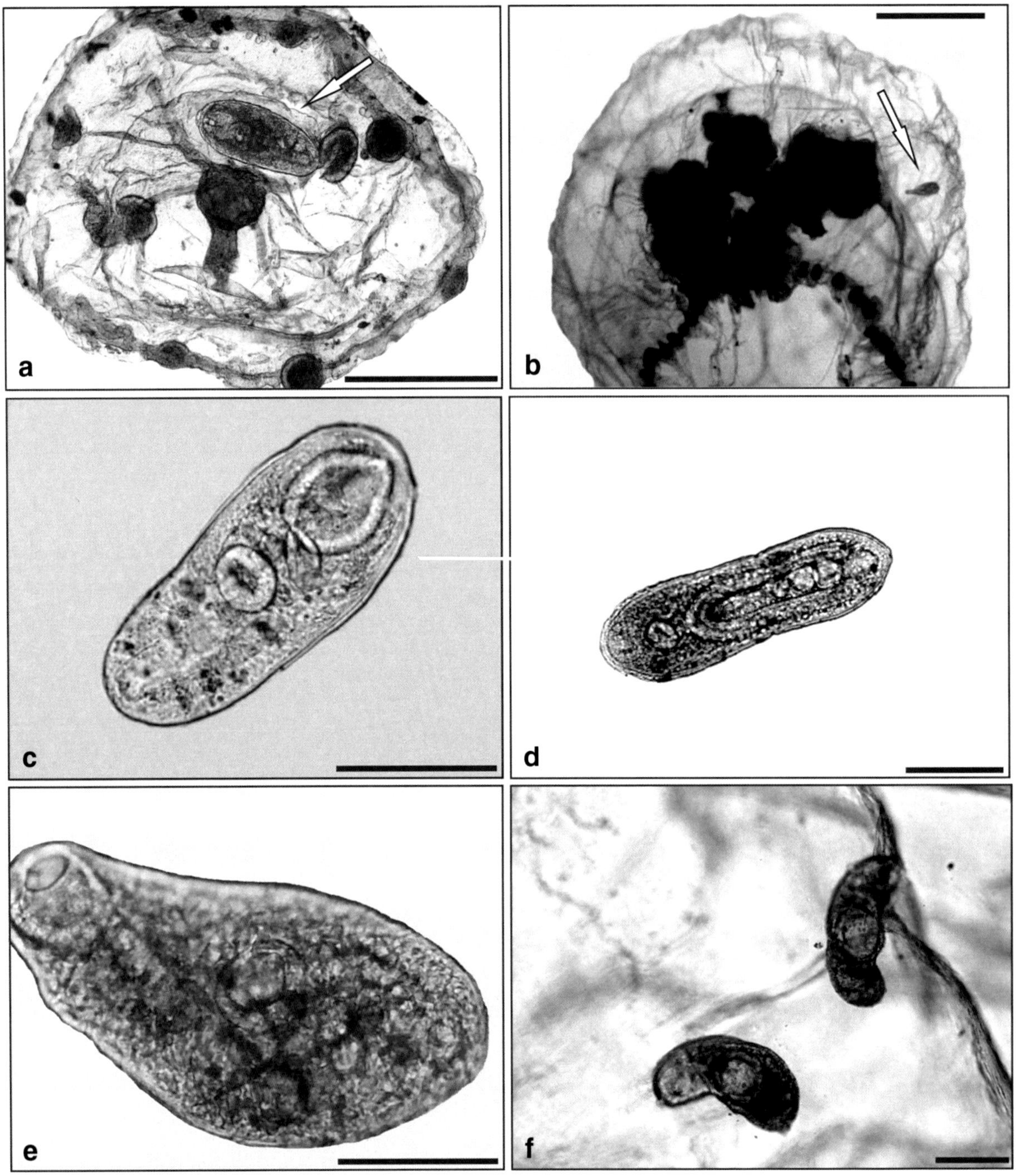

Fig. 2 Metacercariae of **a** *Monascus filiformis* in a *Eucheilota ventricularis* medusa, **b** *Opechona* sp. in a *Proboscidactyla mutabilis* medusa, and **c** *M. filiformis*, **d** *Opechona* sp., **e** *Bacciger* sp., **f** Hemiuridae gen. sp. *Scale bars* 1 mm (**a**, **b**), 200 μm (**c**, **f**), 100 μm (**d**, **e**)

(Fig. 4a–c). Group 3 (55.6% average similarity) was heterogeneous and clustered samples from distal sectors, such as the tidal front of Patagonia, Samborombóm Bay, and Laguna dos Patos (south of Brazil). Only one medusa contributed in this group, *Aequorea* spp., which was parasitized by *Opechona* sp. and *M. filiformis* (50.45 and 49.5%, respectively) (Fig. 4a–c). Thus, we rejected the null hypothesis that there was no difference in the composition of medusa species and their parasites between the groups.

Table 2 List of parasitized medusae by species for the study area (26–55°S), including the total numbers of specimens examined and parasitized per species, and total numbers of medusae parasitized by different digenean metacercariae

Medusae	Total number examined	Total number parasitized and prevalence	Total number of medusae parasitized by			
			Monascus filiformis and prevalence	*Opechona* sp. and prevalence	Hemiuridae and prevalence	*Bacciger* sp. and prevalence
Liriope tetraphylla	17,163	633 (3.7)	508 (80.2)	120 (18.9)	28 (4.4)	1 (0.1)
Eucheilota ventricularis	9,861	1,193 (12.1)	772 (64.7)	559 (46.8)	2 (0.2)	20 (1.7)
Clytia simplex	20	3 (15.0)	1 (33.3)	1 (33.3)	0	1 (33.3)
Proboscidactyla mutabilis	3,589	85 (2.4)	83 (97.6)	6 (7.0)	1 (1.2)	0
Clytia hemisphaerica	956	161 (16.8)	105 (65.2)	84 (52.2)	0	0
Aequorea spp.	60	13 (21.7)	12 (92.3)	10 (76.9)	0	0
Cosmetirella davisi	39	2 (5.1)	2 (100)	0	0	0
Clytia lomae	8	3 (37.5)	2 (66.7)	1 (33.3)	0	0
Mitrocomella brownei	229	2 (0.9)	2 (100)	0	0	0
Halopsis ocellata	6	1 (16.7)	1 (100)	0	0	0
Leuckartiara octona	17	1 (5.9)	1 (100)	0	0	0
Coryne eximia	421	5 (1.2)	0	5 (100)	0	0
Bougainvillia sp.	361	21 (5.8)	0	21 (100)	0	0
Phialella falklandica	99	1 (1.0)	0	1 (100)	0	0
Chrysaora lactea	96	53 (55.2)	0	53 (100)	0	0
Gossea brachymera	25	1 (4.0)	0	1 (100)	0	0
Olindias sambaquiensis	18	3 (16.7)	0	3 (100)	0	0
Total	32,968	2,181	1,489 (68.2)	865 (39.7)	31 (1.4)	22 (1.0)

Prevalence (%) is in parentheses

Discussion

In this study, 16 of the 17 parasitized species of medusae were reported as new records of intermediate hosts of digenean parasites of fish: *Monascus filiformis*, *Opechona* sp., *Bacciger* sp., and Hemiuridae gen. sp. (Tables 1, 2). Although previous studies pointed to the presence of these parasites in some medusae and ctenophores as well as their prevalence values for the area (see Martorelli, 1991, 1996, 2001; Girola et al., 1992; Martorelli & Cremonte, 1998; Morandini et al., 2005), our finding triples the number of medusa hosts known previously. Our results indicate that the interaction between digenean parasites and their jellyfish hosts is more common than previously thought. In addition, the parasite prevalences were higher than those previously recorded for the region.

It is well known that parasites use predator–prey relationships among their hosts to insure their transmission (Marcogliese, 1995). Their presence in a host population provides information on the host diet and predators of the host, as well as the trophic role of the host in the marine trophic web (see Marcogliese, 2004, 2005). Marcogliese (2002) proposed that a high

Fig. 3 Spatial distribution of the prevalence of each digenean parasite species on their medusa hosts in the southwestern Atlantic Ocean. **a** *Monascus filiformis*, **b** *Opechona* sp., **c** Hemiuridae gen. sp., and **d** *Bacciger* sp.

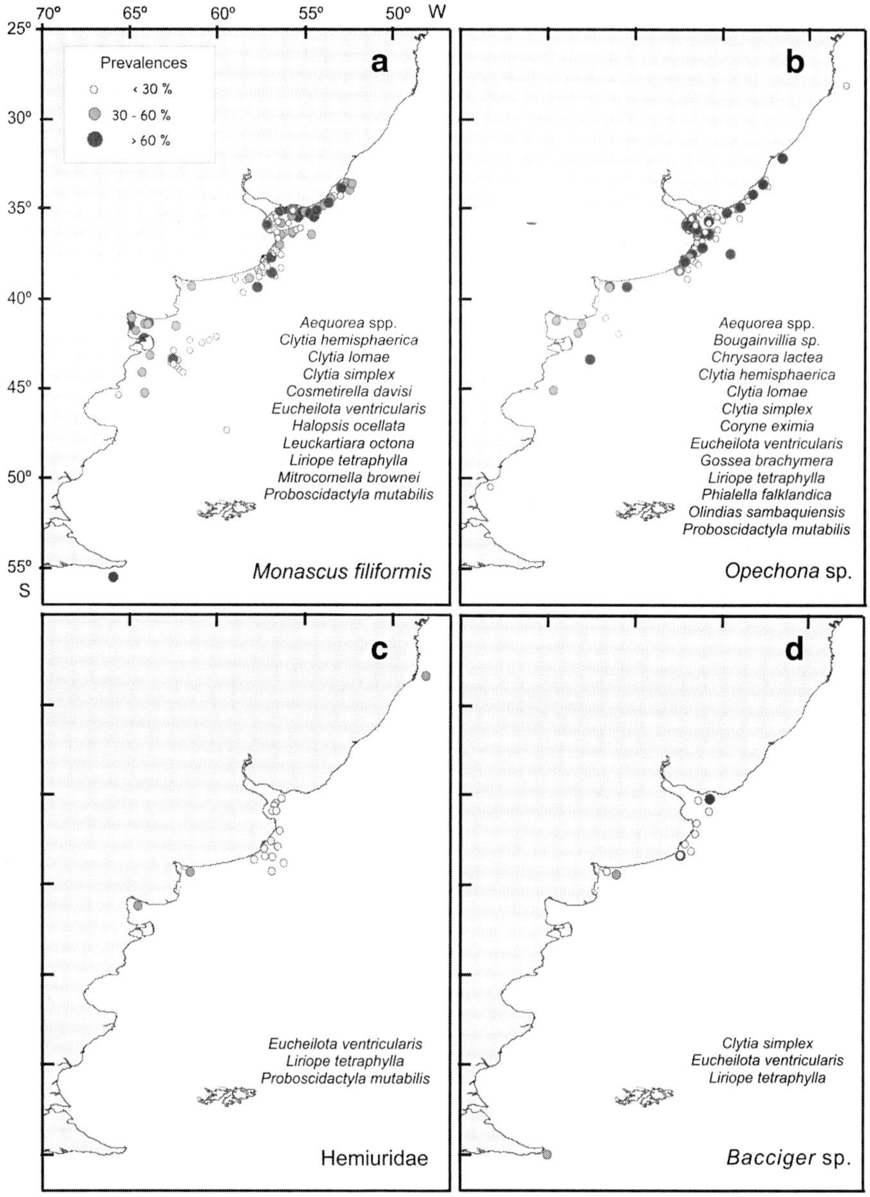

increment in number of some parasites in a host species may reflect more predation upon that species. Therefore, if one medusa species is commonly used as secondary host by a digenean parasite of fish and it has a high prevalence, then we can assume that fish predation on the jellies increases the probability of parasite transmission. Thus, it seems unlikely that the many parasitized medusae are "dead ends" for these parasites, especially considering that many fish species consume jelly organisms in the area (Mianzan et al., 1996, 2001). The high percentage of parasitized species of medusae found in this study (34% of the total analyzed) and the high parasite prevalences observed in the majority of these medusae (Table 2) suggest that many of them are important in the transmission of digenean parasites to their final fish hosts in the southwestern Atlantic Ocean, supporting our hypothesis.

Opechona sp. and *Monascus filiformis* were the most frequent and abundant parasite species, with the widest range of medusae hosts (occurring in 13 and 11 species of medusae, respectively). Adults of genus

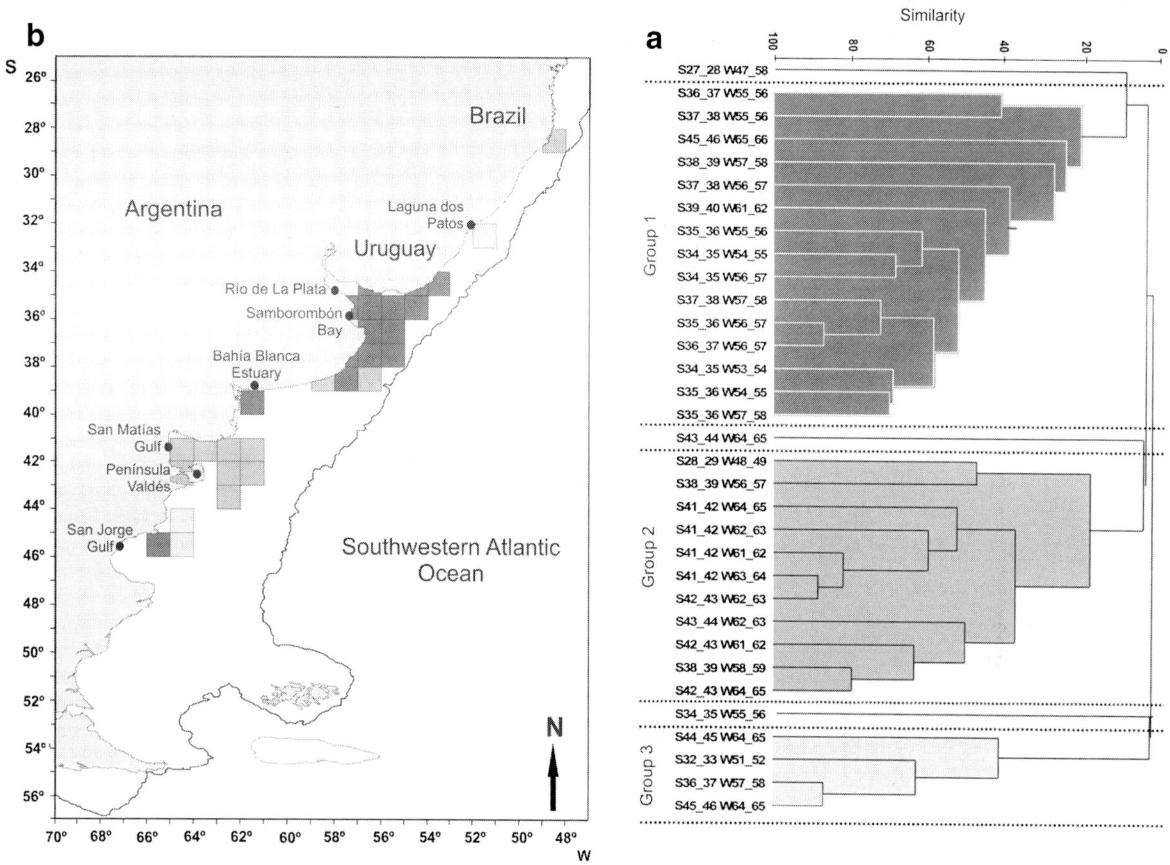

c

SIMPER analysis

Group 1				Group 2	
Eucheilota ventricularis (*Monascus filiformis*)	26.97 %	*Proboscidactyla mutabilis* (*Monascus filiformis*)	1.66 %	*Proboscidactyla mutabilis* (*Monascus filiformis*)	71.15 %
Clytia hemisphaerica (*Monascus filiformis*)	23.52 %	*Eucheilota ventricularis* (*Bacciger* sp.)	1.38 %	*Proboscidactyla mutabilis* (*Opechona* sp.)	25 .72 %
Eucheilota ventricularis (*Opechona* sp.)	21.25 %	*Cosmetirella davisi* (*Monascus filiformis*)	0.66 %	*Liriope tetraphylla* (Hemiuridae)	2.93 %
Clytia hemisphaerica (*Opechona* sp.)	9.75 %	*Liriope tetraphylla* (Hemiuridae)	0.61 %	**Group 3**	
Liriope tetraphylla (*Monascus filiformis*)	7.37 %	*Chrysaora lactea* (*Opechona* sp.)	0.50 %	*Aequorea* spp. (*Opechona* sp.)	50.45 %
Liriope tetraphylla (*Opechona* sp.)	6.13 %	*Eucheilota ventricularis* (Hemiuridae)	0.19 %	*Aequorea* spp. (*Monascus filiformis*)	49.55 %

Fig. 4 **a** Cluster groups of mean prevalence for the four parasites found across all parasitized medusae by species, **b** spatial distribution of the three groups obtained in the cluster analysis, and **c** results of the SIMPER analysis. Percentage (%) of parasitized medusae contribution for each digenean parasite is indicated for each group

Opechona and *M. filiformis* were found in eight and six fish species of the region, respectively (Travassos et al., 1965, 1967; Amato, 1982, 1983; Fernandes et al., 1985; Wallet & Kohn, 1987; Girola et al., 1992; Cremonte & Sardella, 1997; Martorelli & Cremonte, 1998; Pereira et al., 2000; Abdallah et al., 2002; Kohn et al., 2007). The fact that nearly 70% of the parasitized medusae harbored *M. filiformis* and 40% had *Opechona* sp. metacercariae (Table 2) indicates the importance of these medusae as secondary hosts for these parasites.

Bacciger sp. and Hemiuridae gen. sp. were found in only three species of medusae at low prevalences (Table 2). These metacercariae are known to utilize several marine invertebrates and vertebrates (primarily crustaceans and fish) as second intermediate hosts, in addition to medusae and ctenophores (Marcogliese, 1995; Martorelli, 2001; Daponte et al., 2006, 2008; Rocka, 2006). Several fish species act as their final hosts (Rocka, 2006; Kohn et al., 2007; Alarcos et al., 2008; Guagliardo et al., 2010). The use of a wide range of secondary hosts and the low prevalences found in this study (Table 2) suggests that parasitized medusae may act as paratenic hosts (see Marcogliese, 2005) for both parasites.

Digenean parasites were found in several species of medusae that covered a vast area but were not evenly distributed (Fig. 1). High prevalences of digeneans were concentrated in two major areas, the temperate estuarine zones of the Río de La Plata and Bahía Blanca, and the North Patagonian tidal front (Península Valdés) zone (Acha et al., 2004 and references therein) (Fig. 4b). In these zones, high prevalences (up to 100%) were observed mostly for *M. filiformis* and *Opechona* sp. (Fig. 3a, b).

Infection rates of parasites are extremely low in the pelagic realm due to its dilute nature (Marcogliese, 1995, 2002); however, certain areas like fronts may be paramount in parasites' success. In these zones, exceptionally high primary production provides adequate feeding or reproductive habitats for nektonic species, such as fish and squids, and act as retention areas for larvae of benthic species, which promotes establishment of adult beds (Acha et al., 2004). Gelatinous organisms also are very abundant in these frontal zones (Mianzan & Guerrero, 2000). Therefore, parasites may be aided by retention in frontal zones, maximizing their encounters with secondary and final hosts. The digenean parasites of this study have life cycles that usually involve a benthic mollusk as a primary intermediate host, a gelatinous zooplankter as a secondary or paratenic host, and a fish as a final definitive host. In these frontal areas, the primary hosts are the gastropods *Buccinanops monilifer* (Kiener, 1834) and *Buccinanops cochlidium* (Dillwyn, 1817) for *Opechona* sp., and the bivalve *Nucula obliqua* Lamarck, 1819 for *Monascus filiformis* (Martorelli, 1991; Martorelli & Cremonte, 1998; Averbuj & Cremonte, 2010).

Marine front areas favor the aggregation of prey organisms as well as their predators and can increase the transmission of parasites. It is unclear, however, how small cercaria finds a secondary host after leaving the first intermediate host, which is usually a benthic mollusk. Cercaria may use behavioral traits of the next host species. The infestation mechanisms (active penetration or by eating free-swimming cercariae) used by cercariae when infesting their jellyfish hosts (Stunkard, 1969; Martorelli, 1991; Martorelli & Cremonte, 1998; Morandini et al., 2005) could be enhanced by vertical migration of many medusae and ctenophores. The aggregation of some jellyfish species near the seafloor (Alvarez Colombo et al., 2003; Costello & Mianzan, 2003; Mianzan et al., 2010) may facilitate encounters between cercariae released from a benthic mollusk with medusae and between medusae and fish. The potential trophic importance of jellyfish near the seafloor is indicated by the presence of gelatinous prey in the gut contents of a variety of demersal fishes from the Argentine continental shelf (Mianzan et al., 1996). The vertical movements of medusae create trophic linkages between zones and opportunities for parasites to traverse different habitats (Marcogliese, 2002).

On the other hand, the southern Patagonian region had very few parasitized medusae and the prevalences were quite low (Fig. 3c, d). This may be related to fewer hydromedusa species being observed in high than in low latitudes (Genzano et al., 2008); however, large hydromedusae like *Aequorea* spp. that typically inhabit the pelagic zone showed high prevalences of infestation (Table 2). The fact that these large medusae are poorly sampled by traditional plankton nets makes the study of their parasitosis difficult. Also, prevalences of *Opechona* sp. in their first intermediate hosts (mollusks) are generally lower when the water temperature decreases (Averbuj & Cremonte, 2010), which could potentially limit the infection of medusae in high latitudes.

The high number of species and specimens of medusae parasitized by *M. filiformis* and *Opechona* sp. allowed us to increase knowledge about the use of medusae as secondary hosts in the southwestern Atlantic Ocean. The high prevalence values found indicate that medusae species parasitized by these digeneans may be a fundamental part of the life cycles of these parasites. It is proposed that the temperate estuarine zones and the North Patagonian tidal fronts above mentioned would provide ideal environments for *M. filiformis* and *Opechona* sp. to thrive.

Future studies about the life cycles of these digeneans, seasonality in medusae hosts, and distribution of their first intermediate and final fish hosts will allow us to achieve a better understanding of the role that gelatinous plankton have in local pelagic food webs.

Acknowledgments We are especially grateful to Ines Silvia Incorvaia (INIDEP) for providing a critical review on the early manuscript, Joanna Browne (Griffith University School of Environment) for correcting grammar and style, Carolina Rodriguez (UNMdP-CONICET) for identifying medusae, and the librarians of INIDEP for providing literature. This study was supported by the following grants to HM and GG: FONCyT PICT 2006 No 1553, the Inter-American Institute for Global Change Research (IAI) CRN 2076, which is supported by the US National Science Foundation (grant GEO-0452325), and Universidad Nacional de Mar del Plata EXA 546/11, UNMdP. This is INIDEP Contribution No 1724. L. D. B. is supported by CONICET fellowship.

References

Abdallah, V. D., J. L. Luque, D. R. Alves & A. R. Paraguassú, 2002. Aspectos quantitativos das infrapopulações de metazoários parasitos da cavalinha, *Scomber japonicus* (Osteichtyes: Scombridae) do litoral do Estado do Rio de Janeiro, Brasil. Revista Universidade Rural, Série Ciências da Vida 22: 103–107.

Acha, E. M., H. W. Mianzan, R. A. Guerrero, M. Favero & J. Bava, 2004. Marine fronts at the continental shelves of austral South America. Physical and ecological processes. Journal of Marine Systems 44: 83–105.

Alarcos, A. J., J. A. Etchegoin & F. Cremonte, 2008. Redescription of *Bacciger microacetabularis* (Martorelli & Suriano, 1983) nov. comb. parasitizing *Paralichthys orbignyanus* (Pisces, Paralichthyidae) from Argentina. Acta Parasitologica 53: 365–368.

Alvarez Colombo, G., H. W. Mianzan & A. Madirolas, 2003. Acoustic characterization of gelatinous plankton aggregations: four study cases from the argentine continental shelf. ICES Journal of Marine Science 60: 650–657.

Amato, J. F. R., 1982. Digenetic trematodes of percoid fishes of Florianópolis, southern Brazil – Fellodistomidae, Monascidae, Diplangidae, with description of two new species. Revista Brasileira de Biologia 42: 681–699.

Amato, J. F. R., 1983. Digenetic trematodes of percoid fishes of Florianópolis, southern Brazil – Homalometridae, Lepocreadiidae and Opecoelidae, with the description of seven new species. Revista Brasilera de Biologia 43: 73–98.

Arai, M. N., 1988. Interactions of fish and pelagic coelenterates. Canadian Journal of Zoology 66: 1913–1927.

Arai, M. N., 2005. Predation on pelagic coelenterates: a review. Journal of the Marine Biological Association of the United Kingdom 85: 523–536.

Arai, M. N., D. W. Welch, A. L. Dumsmuir, M. C. Jacobs & A. R. Ladouceur, 2003. Digestion of pelagic Ctenophora and Cnidaria by fish. Canadian Journal of Fisheries and Aquatic Sciences 60: 825–829.

Ates, R. M. L., 1988. Medusivorous fishes, a review. Zoologische Mededelingen 62: 29–42.

Ates, R. M. L., 1991. Predation on Cnidaria by vertebrates other than fishes. Hydrobiologia 216–217: 305–307.

Averbuj, A. & F. Cremonte, 2010. Parasitic castration of *Buccinanops cochlidium* (Gastropoda: Nassariidae) caused by a lepocreadiid digenean in San José Gulf, Argentina. Journal of Helminthology 84: 381–389.

Bush, A. O., K. D. Lafferty, J. M. Lotz & A. W. Shostak, 1997. Parasitology meets ecology on its own terms: Margolis et al., revisited. Journal of Parasitology 83: 575–583.

Cartwright, P., N. M. Evans, C. W. Dunn, A. C. Marques, M. P. Miglietta, M. P. Schuchert & A. G. Collins, 2008. Phylogenetics of Hydroidolina (Hydrozoa: Cnidaria). Journal of the Marine Biological Association of the United Kingdom 88: 1663–1672.

Chapman, M. G. & A. J. Underwood, 1999. Ecological patterns in multivariate assemblages: information and interpretation of negative values in ANOSIM tests. Marine Ecology Progress Series 180: 257–265.

Clarke, K. R. & R. M. Warwick, 2001. Change in Marine Communities: An Approach to Statistical Analysis and Interpretation, 2nd ed. PRIMER-E, Plymouth.

Collins, A. G., B. Bentlage, A. Lindner, D. Lindsay, S. H. D. Haddock, G. Jarms, J. L. Norenburg, T. Jankowski & P. Cartwright, 2008. Phylogenetics of Trachylina (Cnidaria: Hydrozoa) with new insights on the evolution of some problematical taxa. Journal of the Marine Biological Association of the United Kingdom 88: 1673–1685.

Costello, J. H. & H. W. Mianzan, 2003. Sampling field distributions of *Mnemiopsis leidyi* (Ctenophora, Lobata): planktonic or benthic methods? Journal of Plankton Research 25: 455–459.

Cremonte, F. & N. H. Sardella, 1997. The parasito fauna of *Scomber japoniscus* Houttuyn, 1782 (Pisces: Scombridae) in two zones of the Argentine Sea. Fisheries Research 31: 1–9.

Daponte, M. C., A. A. Gil de Pertierra, M. A. Palmieri & M. Ostrowski de Nuñez, 2006. Parasite infections (Trematoda, Digenea) of *Sagitta friderici* (chaetognatha) from the southwestern Atlantic Ocean: prevalence and distribution. Diseases of Aquatic Organisms 71: 225–232.

Daponte, M. C., A. A. Gil de Pertierra, M. A. Palmieri & M. Ostrowski de Nuñez, 2008. Monthly occurrence of parasites of the chaetognath *Sagitta friderici* of Mar del Plata, Argentina. Journal of Plankton Research 30: 567–576.

Fernandes, B. M. M., A. Kohn & R. M. Pinto, 1985. Aspidogastrid and digenetic trematodes parasites of marine fishes of the coast of Rio de Janeiro State, Brazil. Revista Brasileira de Biologia 45: 109–116.

Genzano, G. N., H. W. Mianzan & J. Bouillon, 2008. Hydromedusae (Cnidaria: Hydrozoa) from the temperate southwestern Atlantic Ocean: a review. Zootaxa 1750: 1–18.

Girola, C. V., S. R. Martorelli & N. H. Sardella, 1992. Presencia de Metacercarias de *Monascus filiformis* (Digenea, Fellodistomidae) en hidromedusas del Océano Atlántico Sur. Revista Chilena de Historia Natural 65: 409–415.

Guagliardo, S. E., C. B. Schwerdt, S. R. Martorelli, N. A. Galeano & R. D. Tanzola, 2010. Digenean trematodes of *Seriolella porosa* (Pisces, Centrolophidae) in San Matías Gulf, Argentina. Acta Parasitologica 55: 29–38.

Kohn, A., B. M. M. Fernandes & S. C. Cohen, 2007. South American Trematodes Parasites of Fishes. Imprinta Express Ltda, Rio de Janeiro.

Køie, M., 1975. On the morphology and life history of *Opechona bacillaris* (Molin, 1859) Loos, 1907 (Trematoda Lepocreadiidae). Ophelia 13: 63–86.

Lauckner, G., 1980. Diseases of Ctenophora. In Kinne, O. (ed.), Diseases of Marine Animals. Vol. 1. General Aspects, Protozoa to Gastropoda. John Wiley and Sons, New York: 167–253.

Lebour, M. V., 1916. Medusae as hosts for larval trematodes. Journal of the Marine Biological Association of the United Kingdom 11: 57–59.

Marcogliese, D. J., 1995. The role of zooplankton in the transmission of helminth parasites to fish. Reviews in Fish Biology and Fisheries 5: 336–371.

Marcogliese, D. J., 2002. Food webs and the transmission of parasites to marine fish. Parasitology 124: 83–99.

Marcogliese, D. J., 2004. Parasites: small players with crucial roles in the ecological theater. EcoHealth Journal Consortium 1: 151–164.

Marcogliese, D. J., 2005. Ecology. In Rohde, K. (ed.), Marine Parasitology. CSIRO Publishing, Collingwood: 279–321.

Marques, A. C. & A. G. Collins, 2004. Cladistic analysis of Medusozoa and cnidarian evolution. Invertebrate Biology 123: 23–42.

Martorelli, S. R., 1991. Primera cita de una Cercaria Tricocerca parásita de *Dorsanum Moniliferum* (Mollusca: Buccinidae) para el Atlántico sud-Occidental. Aportes al conocimiento de su ciclo de vida. Neotropica 37: 57–65.

Martorelli, S. R., 1996. First record of encysted metacercariae in hydrozoan jellyfishes and ctenophores of the Southern Atlantic. Journal of Parasitology 82: 352–353.

Martorelli, S. R., 2001. Digenea parasites of jellyfish and ctenophores of the Southern Atlantic. Hydrobiologia 451: 305–310.

Martorelli, S. R. & F. Cremonte, 1998. A proposed three-host life history of *Monascus filiformis* (Rudolphi, 1819) (Digenea: Fellodistomidae) in the southwest Atlantic Ocean. Canadian Journal of Zoology 76: 1198–1203.

Mianzan, H. W. & R. A. Guerrero, 2000. Environmental patterns and biomass distribution of gelatinous macrozooplankton. Three study cases in the South-western Atlantic Ocean. Scientia Marina 64: 215–224.

Mianzan, H. W., N. Marí, B. Prenski & F. Sanchez, 1996. Fish predation on neritic ctenophores from the Argentina continental shelf: a neglected food resource? Fisheries Research 27: 69–79.

Mianzan, H. W., M. Pájaro, G. Alvarez Colombo & A. Madirolas, 2001. Feeding on survival-food: gelatinous plankton as a source of food for anchovies. Hydrobiologia 451: 45–53.

Mianzan, H. W., P. Martos, J. H. Costello & R. A. Guerrero, 2010. Avoidance of hydrodynamically mixed environments by *Mnemiopsis leidyi* (Ctenophora: Lobata) in open-sea populations from Patagonia, Argentina. Hydrobiologia 645: 113–124.

Morandini, A. C., S. R. Martorelli, A. C. Marques & F. Lang da Silveira, 2005. Digenean metacercaria (Trematoda, Digenea, Lepocreadiidae) parasitizing "coelenterates" (Cnidaria, Scyphozoa and Ctenophora) from southeastern Brazil. Brazilian Journal of Oceanography 53: 39–45.

Pereira, J. Jr., B. M. M. Fernades & R. B. Robaldo, 2000. Digenea (Trematoda) of *Micropogonias furnieri* (Desmarest) (Perciformes, Sciaenidae) from Rio Grande do Sul, Brazil. Revista Brasileira de Zoologia 17: 681–686.

Rocka, A., 2006. Helminths of Antarctic fishes: life cycle biology, specificity and geographical distribution. Acta Parasitologica 51: 26–35.

Rohde, K., 1993. Ecology of Marine Parasites: An Introduction to Marine Parasitology, 2nd ed. CAB International, Wallingford.

Stunkard, H. W., 1967. The life cycle and developmental stages of a digenetic trematode whose unencysted metacercarial stages occur in medusae. Biological Bulletin 133: 488.

Stunkard, H. W., 1969. The morphology and life-history of *Neopechona pyriforme* (Linton, 1900) N. Gen., N. Comb. (Trematoda: Lepocreadiidae). Biological Bulletin 136: 96–113.

Travassos, L., J. F. T. Freitas & P. F. Bührnheim, 1965. Tramatódeos de peixes do litoral capixaba: *Monascus netoi* sp. n., parasita de vento leste. Atas da Sociedade de Biologia do Rio de Janeiro 9: 46–48.

Travassos, L., J. F. T. Freitas & P. F. Bührnheim, 1967. Relatório da excursão do Instituto Oswaldo Cruz ao estado do Espíritu Santo em novembro de 1964. Boletim do Museu de Biologia Mello Leitão 31: 1–5.

Wallet, M. & A. Kohn, 1987. Trematodes parasites de poissons marins du littoral de Rio de Janeiro, Brasil. Memórias do Instituto Oswaldo Cruz 82: 21–27.

Wiebe, P. H. & M. C. Benfield, 2003. From the Hensen net toward four-dimensional biological oceanography. Progress in Oceanography 56: 7–136.

Hydrobiologia (2012) 690:227–245
DOI 10.1007/s10750-012-1046-8

JELLYFISH BLOOMS

Review Paper

Jelly-falls historic and recent observations: a review to drive future research directions

Mario Lebrato · Kylie A. Pitt · Andrew K. Sweetman · Daniel O. B. Jones · Joan E. Cartes · Andreas Oschlies · Robert H. Condon · Juan Carlos Molinero · Laetitia Adler · Christian Gaillard · Domingo Lloris · David S. M. Billett

Published online: 21 March 2012
© Springer Science+Business Media B.V. 2012

Abstract The biological pump describes the transport of particulate matter from the sea surface to the ocean's interior including the seabed. The contribution by gelatinous zooplankton bodies as particulate organic matter (POM) vectors ("jelly-falls") has been neglected owing to technical and spatiotemporal sampling limitations. Here, we assess the existing evidence on jelly-falls from early ocean observations to present times. The seasonality of jelly-falls indicates that they mostly occur after periods of strong upwelling and/or spring blooms in temperate/subpolar zones and during late spring/early summer. A conceptual model helps to define a jelly-fall based on empirical and field observations of biogeochemical and ecological processes. We then compile and discuss existing strategic and observational oceanographic techniques that could be implemented to further jelly-falls research. Seabed video- and photography-based studies deliver the best results, and the correct use of fishing techniques, such as trawling, could provide comprehensive regional datasets. We conclude by considering the possibility of increased gelatinous biomasses in the future ocean induced by

Guest editors: J. E. Purcell, H. Mianzan & J. R. Frost / Jellyfish Blooms: Interactions with Humans and Fisheries

M. Lebrato (✉) · A. Oschlies · J. C. Molinero
GEOMAR, Helmholtz Centre for Ocean Research Kiel, Düsternbrooker Weg 20, 24105 Kiel, Germany
e-mail: mlebrato@ifm-geomar.de

K. A. Pitt
Australian Rivers Institute, Coast and Estuaries, Griffith University, Brisbane, QLD 4222, Australia

A. K. Sweetman
Norwegian Institute for Water Research, Thormøhlensgate 53D, 5006 Bergen, Norway

A. K. Sweetman
Centre for Geobiology, University of Bergen, Bergen, Norway

D. O. B. Jones · D. S. M. Billett
National Oceanography Centre, European Way, Southampton SO14 3ZH, UK

J. E. Cartes · D. Lloris
Institut de Ciències Del Mar de Barcelona, CSIC, Passeig Marítim de la Barceloneta 37-49, 08003 Barcelona, Spain

R. H. Condon
Dauphin Island Sea Lab, Dauphin Island, AL 36528, USA

L. Adler
Biocenter Grindel and Zoological Museum, Martin-Luther-King-Platz 3, 20146 Hamburg, Germany

Present Address:
L. Adler
School of Geological Sciences, University College Dublin, Belfield, Dublin 4, Ireland

C. Gaillard
Université de Lyon 1, UMR CNRS 5125, 2 rue Raphaël Dubois, 69622 Villeurbanne cedex, France

upper ocean processes favouring their populations, thus increasing jelly-POM downward transport. We suggest that this could provide a "natural compensation" for predicted losses in pelagic POM with respect to fuelling benthic ecosystems.

Keywords Biological pump · Gelatinous zooplankton · Jelly-fall · Organic matter

Introduction: particulate organic matter (POM) and jelly-falls

The input of POM drives secondary production and most benthic ecosystem processes in the deep-sea (Ruhl et al., 2008; Smith et al., 2008). POM inputs mainly include autochthonous particles from the euphotic zone, ranging in increasing size from phytodetritus (organic-rich material derived from phytoplankton blooms) (Beaulieu, 2002; Smith et al., 2008), marine snow (Caron et al., 1986; Alldredge & Silver, 1988), mucilaginous aggregates (Cartes et al., 2007; Martin & Miquel, 2010), mucous sheets from zooplankton (Robison et al., 2005; Lombard & Kiorbe 2010), faecal pellets (reviewed by Turner, 2002), wood particles (Turner, 1973), and macrophyte detritus (Vetter & Dayton, 1998, 1999; Cartes et al., 2010) to fish and whale carcasses (Soltwedel et al., 2003; Smith & Baco, 2003; Gooday et al., 2010).

Jelly-falls can be defined as point source organic matter inputs (as corpses/carcasses) that sink through the water column (remineralizing as dissolved organic/inorganic components), eventually causing an accumulation of jelly-POM (J-POM) at the seabed (Fig. 1). Numerous gelatinous zooplankton groups have been shown to accumulate at the ocean floor including the Cnidaria (Scyphozoa) and Thaliacea (Pyrosomida, Doliolida, and Salpida) (Table 1). The significance and magnitude of sinking material in the biological pump are primarily assessed by a variety of indirect techniques (Buesseler et al., 1992; Jahnke, 1996; Marchant et al., 1999) that cannot target the J-POM associated with jelly-falls. They include remote sensing algorithms (Behrenfeld & Falkowski, 1997; Balch et al., 2007), surface-tethered and neutrally buoyant sediment traps (Lampitt et al., 2001; Buesseler et al., 2007), and acoustic backscatter profiling sensors (ABS and ADCP) to study particles and biomass in the water column (Merckelbach & Ridderinkhof, 2006; Jiang et al., 2007). Sediment traps are the most used device, but they often underestimate the contribution of large particles and detritus (e.g. Rowe & Staresinic, 1979; but see Conte et al., 2003; Buesseler et al., 2007). Jelly-falls can only be sampled directly using techniques such as video (Wiebe et al., 1979; Lebrato & Jones, 2009), towed/still photography (Roe et al., 1990; Billett et al., 2006; Sweetman & Chapman, 2011), or benthic trawling (Sartor et al., 2003) (see Table 2 for other techniques/strategies). Therefore, although many sources of organic material have been widely studied and POM/DOM remineralization dynamics considered in biogeochemical models as a result (e.g. Burd et al., 2010), jelly-falls are relatively unexplored sources of POM, despite a significant fraction of the pelagic biomass being sequestered in the bodies of gelatinous zooplankton.

The study of jelly-falls represents a major challenge in the understanding of the biological pump mainly due to technical/sampling hurdles, and although there is no consensus that the oceans will turn into a "jelly-slime" ecosystem (e.g. Jackson, 2008), gelatinous zooplankton biomass appears to be increasing in certain areas of the world's oceans (Mills, 2001; Richardson et al., 2009; Purcell, 2012). As such, increased gelatinous biomass may translate into increased transfer of this material to the ocean floor and thus enhancing the magnitude and importance of the biogeochemical and ecological processes associated with jelly-falls. Thus, there is a pressing need for research on gelatinous zooplankton post-bloom processes.

Our primary objective is to provide a qualitative overview of historical and present records of jelly-falls, as well as the environmental context in which they were studied. Secondly, we define and conceptually model a general jelly-fall within the biological pump, including a synthesis of the factors triggering these events. We also assess the seasonality of jelly-falls from the available data and the benthic organisms that were observed feeding on the material. Our third objective is to discuss the possible consequences of increased gelatinous biomasses in the future ocean and provide a summary of the observational techniques and platforms that are, or could be used to study jelly-falls and their biogeochemical feedbacks.

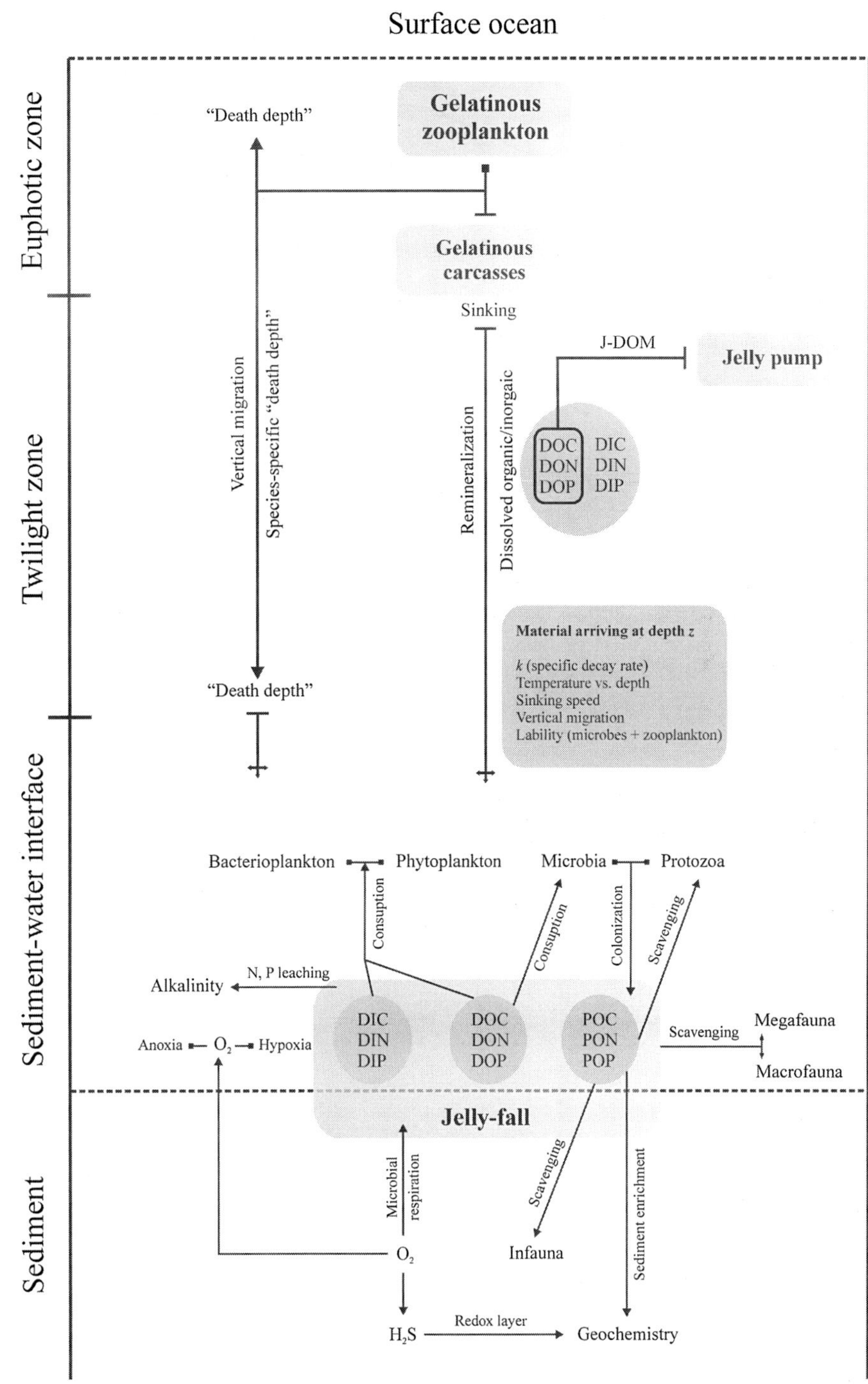

Fig. 1 Conceptual model of common processes beginning with sinking and the start of remineralization in the euphotic and twilight zones to deposition at the seabed followed by decomposition and scavenging. Under "material arriving at depth *z*", we have identified five critical factors that determine the amount of material reaching the seabed. The links to "bacterioplankton" and "phytoplankton" only proceed in the euphotic/twilight zone. Jelly-falls are linked to the "jelly-pump" concept (Condon & Steinberg, 2008; Condon et al., 2010) through the production of J-DOM in the water column and at the seabed. J-DOM is organic matter that fuels other trophic levels, which can occur while the organisms are still alive (e.g. Condon et al., 2011) or when dead (Hansson & Norrman, 1995)

Jelly-fall observations in the field

Thaliaceans

During the 1872–1876 H. M. S. Challenger expeditions, Moseley (1880) realized the potential importance of jellyfish in the biological pump by experimentally assessing the time it took a dead salp to sink 20 cm in a cylinder (~20 s). He then left the carcass in the cylinder for 1 month and noticed that it did not decompose completely. He subsequently wrote: "the deep-sea has to derive food for its inhabitants entirely

Table 1 A compilation of naturally occurring jelly-falls

Location	Origin	Species	Material state[a]	Latitude (range)[b]	Longitude (range)[b]	Depth (m)[c]	Survey device	Duration[d]	Reference
Norwegian Sea (Atlantic Ocean)	Likely Scyphozoa	Pending DNA analysis[e]	Det.	66.14°N	3.94°E	1,380 (8.3/−1)	ROV (video)	7 days (S)	Jones et al. (2010)
Norwegian Sea (Atlantic Ocean)	Scyphozoa	*P. periphylla*	F/Dec.	60.40°N–60.41°N	5.09°E–5.10°E	396–443 (–/7)	Yo–Yo (towed camera)	1 day (Spr.)	Sweetman & Chapman (2011)
Japan Sea (Pacific Ocean)	Scyphozoa	*A. limbata*	F	42.58°N	143.96°E	320 (17/2.2)	ROV (video)	Unknown (S)	Miyake et al. (2002)
Chesapeake Bay (Atlantic Ocean)	Scyphozoa	*C. quinquecirrha*	F	38.59°N	76.12°W	1.5–3 (15/14)	Visual (observers)	90 days (S)	Sexton et al. (2010)[f]
Japan Sea (Pacific Ocean)	Scyphozoa	*N. nomurai*	F	35.8°N–36.3°N	136°E–135.5°E	146–354 (22/10)	VTR system (towed camera)	35 days (S–A)	Yamamoto et al. (2008)
Japan Sea (Pacific Ocean)	Scyphozoa	*P. polylobata*	F	34.91°N	138.65°E	453 (16/8)	ROV (video)	Unknown (S)	Miyake et al. (2005)
Santa Catalina Basin (Pacific Ocean)	Scyphozoa	*Pelagia* sp.	Unk.	32.46°N	117.49°W	>1,000 (17/4)	Photographs	Unknown (N)	Jumars (1976)
Bermuda (Atlantic Ocean)	Scyphozoa	*Cassiopeia xamachana*	F/Dec./Det.	32.34°N	64.70°W	3 (25–25)	Photographs (quadrats)	Unknown (S)	M. Lebrato (unpublished)
Gulf of Aqaba (Red Sea)	Scyphozoa	*A. aurita*	F	29.50°N	34.91°E	20 (25–23)	Photographs (scuba diver)	Unknown (Sp. #)	Alamaru et al. (2009)
Arabian Sea (Indian Ocean)	Scyphozoa	Probably *C. orsini*	F	22.95°N	66.61°E	900 (25/9.5)	WASP (towed camera)	1 day (S #)	Murty et al. (2009)
Arabian Sea (Indian Ocean)	Scyphozoa	*C. orsini*	F/Dec./Det.	22.58°N–23.50°N	60.65°E–59.04°E	304 – 3,299 (25/2)	SHRIMP (towed camera)	17 days (W #)	Billett et al. (2006)
Japan Sea (Pacific Ocean)	Thaliacean (Doliolidae)	Not identified	F	34.40°N	150°E	150 (16/12)	Sediment trap	5 days (Sp.)	Takahashi et al. (2010)[g]
Tyrrhenian Sea (Mediterranean Sea)	Thaliacean (Pyrosomatidae)	*P. atlanticum*	F/Dec.	42.30°N	10.60°E	300–650 (25/12)	Bottom trawling	1995–1999 (Sp. S)	Sartor et al. (2003)[h]
Alboran Sea-Gulf of Lions (Mediterranean Sea)	Thaliacean (Pyrosomatidae)	*P. atlanticum*	F/Dec.	36.24°N–42.39°N	5.20°W–3.63°W	43–791 (18/13)	Bottom trawling (GOC 73)	1994–2005 (Sp. #)	Bertrand et al. (2002), MEDITS-ES[i]
Madeira Abyssal Plain (Atlantic Ocean)	Thaliacean (Pyrosomatidae)	*P. atlanticum*	F/Dec.	31.28°N	25.40°W	5,433 (20/2.2)	BATHYSNAP (fixed camera)	16 days (S)	Roe et al. (1990)
Cape Verde (Atlantic Ocean)	Thaliacean (Pyrosomatidae)	*P. atlanticum*	F	15.80°N	23.50°W	Unknown (26/–)	Unknown	Unknown (N)	Monniot & Monniot (1966)
Ivory Coast (Atlantic Ocean)	Thaliacean (Pyrosomatidae)	*P. atlanticum*	F/Dec.	5.15°N–4.94°N	4.51°W–4.49°W	26–1,275 (25/4)	ROV (video)	60 days (W #)	Lebrato & Jones (2009)

Table 1 continued

Location	Origin	Species	Material state[a]	Latitude (range)[b]	Longitude (range)[b]	Depth (m)[c]	Survey device	Duration[d]	Reference
Cook Strait (Pacific Ocean)	Thaliacean (Pyrosomatidae)	*P. atlanticum*	Dec.	41.73°S	174.3°E	100 (15/9)	Bottom trawling	Unknown (Sp.)	Hurley & McKnight (1959)
Tasman Sea (Pacific Ocean)	Thaliacean (Pyrosomatidae)	*P. atlanticum*	Unk.	42°S	148°E	330–640 (14/7)	Stomach content	Unknown (W)	Cowper (1960)
Gulf of Alaska (Pacific Ocean)	Thaliacean (Salpidae)	*S. fusiformis*	F	58.33°N	136.83°W	1–10 (7/4)	Visual (scuba diver)	120 days (Sp.)	Duggins (1981)
Sargasso Sea (Atlantic Ocean)	Thaliacean (Salpidae)	*S. aspera*	F/Dec.	38.60°N–39°N	71.4°E–71.1°E	2,500–3,000 (19/3)	ROV (video)	4 days (S)	Cacchione et al. (1978)
Sargasso Sea (Atlantic Ocean)	Thaliacean (Salpidae)	*S. aspera*	F/Dec.	38°N–40°N	72.5°E–70°E	2,000–3,000 (19/3)	ROV (video)	30 days (S)	Wiebe et al. (1979)
Gulf of Aqaba (Red Sea)	Thaliacean (Salpidae)	Not identified	F	34.90°N	29.50°E	20 (25/23)	Photographs (scuba diver)	Unknown (Sp. #)	Alamaru et al. (unpublished)[j]

[a] Material state refers to the condition in which the material was found: *Dec.* decomposing, *Det.* detritus, *F* fresh, *unk.* unknown (if not stated)

[b] Range for latitude, longitude and depth indicates that in some cases the material was retrieved along a gradient of depths and not in isolation (see reference paper for additional information)

[c] In situ surface and BT (°C) are included in parentheses when available in the original study or otherwise compiled from the GLODAP database (Key et al., 2004) and the World Ocean Atlas (http://odv.awi.de/en/data/ocean) in the nearest place available at the same depth

[d] Duration only indicates the time that the material was observed or surveyed at the seabed and does not indicate annual events; otherwise the time-series is given for annual depositions. The season is indicated as: *N* not available; *Sp.* spring, *S* summer, *A* autumn, *W* winter. # indicated when the event happened after seasonal upwelling and/or monsoon winds (e.g. tropical latitudes or specific cases like the Mediterranean Sea)

[e] Jelly material was unidentifiable to species level. Bar-coding with mtDNA and 18S rDNA ITS regions in progress to determine the affiliation

[f] The authors do not show seabed evidence. The potential POC flux to the sediments was relatively small (12.5–72.5 mg C m^{-2} $year^{-1}$) in comparison with the total annual flux to the sediments in the area (61.2 g C m^{-2} $year^{-1}$ Kemp et al., 1997)

[g] Carcasses recorded in sediment traps (export flux = 1.05 mg C m^{-2} day^{-1}, sinking speed = 4,000 m day^{-1}, small degradation observed)

[h] The data used for *P. atlanticum* correspond to the trawling catch from the seabed. The carcasses were dead at the seabed and decomposing

[i] The data were in the MEDITS-ES project (International bottom trawl survey in the Mediterranean) (http://www.sibm.it/SITO%20MEDITS/). The data for *P. atlanticum* correspond to the trawling catch from the seabed

[j] A. Alamaru also reports on the presence of salps at the seabed in the same area as *A. aurita*

Table 2 Sampling techniques and initiatives that may be available to monitor jelly-falls

Study method	Description	Advantages	Disadvantages
Large scale (regional)			
(1) Deep ocean observatories network (e.g. EUR-OCEANS, OceanSITES, ESONET) and offshore scientific platforms (e.g. PLOCAN)	Long-term reference network stations could be used to monitor seabed processes associated with jelly-falls. They could be used to study the organisms and carcasses in the water column and their arrival at the seabed by use of camera arrays. Global distribution	The moorings and cruises are in place so it is a matter of adapting the strategy. Could have camera systems throughout the year with periodic recoveries. In situ and real time monitoring	Challenging to study the jellies in the water column with camera devices. Possibility that jelly-falls do not ŏccur near the stations and/or the seabed may too deep for jelly-falls to be observed)
(2) Collaborations with industry (e.g. SERPENT project and similar)	The offshore oil and gas industry has regular access to expensive equipment (e.g. ROVs, camera systems) used in monitoring their own infrastructures. This equipment is not routinely used in scientific studies, but through collaboration it could be used to study jelly-falls at specific times of the year	Access to state-of-the-art expensive equipment to study the seabed in deep waters. ROVs used in industry operations follow paths, allowing transect study. They cover a larger seabed area than normal in a scientific study	Obtaining agreements with key industry personnel with access to the facilities. Confidentially and data release may take time to arrange. Surveys are confined to where the infrastructure exists. Cannot deviate greatly from established survey lines
Medium scale (local)			
(3) Scientific ROV surveys	Used at known sites of jelly-falls to monitor the depositions in transects	Real-time monitoring and quantitative or qualitative data available at the seabed	Time available to conduct the survey. Total area covered. Exclude water column processes. Expensive
(4) Towed and drop cameras from a vessel	Either used at known sites of jelly-falls or use to search for depositions in transects/specific locations	Real-time monitoring and quantitative or qualitative data available at the seabed. Can cover a relatively large area	Time available to conduct the survey. Camera angle of view much less than ROV camera. Exclude water column processes
(5) Time-lapse cameras (e.g. BATHYSNAP and benthic landers)	Used at known sites of jelly-falls to study the evolution of the material over time. A network of time-lapse cameras also feasible at specific locations	Real-time monitoring and qualitative data available at the seabed. Time component of the jelly-falls	Limited area covered. No biomass data. Jelly-fall may not occur where cameras are installed. Camera angle of view restricted. Exclude water column processes
(6) Trawling (from fisheries)	The fishery industry and other commercial species surveys (e.g. MEDITS) have records of bycatch organisms trawled at the seabed, including jelly-falls (carcasses)	Quantitative or qualitative data available at the seabed. Large areas covered over bathymetric gradients. Time component often available	Obtaining agreements with industry personnel that have access to the facility. Surveys confined to the industry study of commercial species. Excludes water column processes. Environmentally destructive
(7) Acoustic/electronic tagging studies	Living individuals in large blooms could be acoustically/electronically tagged to follow their fate	Real-time study of individuals in a jelly-fall. Time component of sinking and deposition	Difficulty of tag attachment to gelatinous body. Premature release of the tag. Limited information

Table 2 continued

Study method	Description	Advantages	Disadvantages
Small scale (local)			
(8) Large sediment traps (+5 m)	If a neutrally buoyant sediment trap is developed to follow blooms it may deliver data on the associated sinking material	Quantitative data in the water column. Possible to combine with a method at the seabed	Probably unable to catch much of the sinking jelly-fall. Problems with organisms that vertically migrate and are mistakenly trapped alive. Cost-effective problems
(9) Moored and free-drifting profilers	Some in development to measure water column properties over time (McLane labs, SeaCycler). If installed with a camera, study could cover the entire water column to the seabed	Possible to monitor entire water problem over time. Quantitative data. Possible to relate camera data and water column properties	Camera installation problems. Jelly-fall may not occur where the profilers are installed. Sinking speed of carcasses not tracked by profiler
(10) Genetic tools in sediments	Sediment proxy on freshly deposited gelatinous material can be tested using mtDNA and nuclear DNA	Possible to obtain a identification (general or specific) from jelly-falls in the sediment depending on the decomposition time. Possible to combine with camera studies if material is visible. Time component may be available	Limited to very fresh depositions. DNA contamination problems. Limited area covered

from debris of animals and plants falling to the bottom from the water above them. The dead pelagic animals must fall as a constant rain of food. It might be supposed that the animal carcasses would consume so long a time in dropping to the seabed that their soft tissues would be decomposed" (Mosely 1892). This is, to the best of our knowledge, the first mention of jelly-falls in the literature. A number of both quantitative and qualitative studies have followed since then (Table 1), but they still remain scarce when compared with studies that have assessed the importance of other POM vectors (Turner, 2002).

Hurley & McKnight (1959) were the first to report on a natural jelly-fall when they found the thaliacean *Pyrosoma atlanticum* Peron 1804 on the seabed off New Zealand. The organisms were sampled with a bottom trawl between 160- and 170-m depth (bottom temperature (BT) = 9°C) during spring and were described as "resting" on the seabed. Their observations are further supported by reports from the same area of seabed being covered in *P. atlanticum* carcasses in 1952 (H. B. Fell pers. obs. reported to Hurley & McKnight, 1959) and reports that local fishermen frequently trapped large quantities of moribund carcasses at certain times of the year. Similar fishermen's reports occur in the Mediterranean Sea (e.g. Sartor et al., 2003). Later, in the Tasman Sea, Cowper (1960) found that the stomachs of freshly caught carangid fish were full of *P. atlanticum* carcasses during winter. All fish were caught close to the bottom (BT = 7°C); therefore, the authors concluded that they were feeding either on recently settled carcasses or on moribund individuals on or near the seabed. A further analysis of stomach contents from the same fish species in the Tasman Sea from January to October revealed that the carcasses were most abundant in stomachs in January and March (Cowper, 1960). There are other observations in New South Wales, Australia of the giant pyrosomid *Pyrosoma spinosum* Herdman 1888 near to or deposited on rocky bottoms, and also portions of salps being recovered from stomachs of carangid fish feeding at the seabed (Griffin & Yaldwyn, 1970).

A considerable number of more recent studies document pyrosome falls. In the tropical Atlantic (off Cape Verde), Monniot & Monniot (1966) recorded moribund *P. atlanticum* at the seabed. In the deep Atlantic Madeira Abyssal Plain, high densities of pyrosomids were observed in the first 800 m of the water column (Roe et al., 1990). A survey using a fixed camera photographed a single carcass in an advanced state of decomposition at 5,433-m depth (BT = 2.2°C).

A starfish and a crustacean scavenged the carcass, which took >16 days to decompose completely. Recently, Lebrato & Jones (2009) reported a vast jelly-fall of *P. atlanticum* off the Ivory Coast, West Africa during ROV (remotely operated vehicle) surveys. Decomposing carcasses formed large patches (~1–20 m^2) and accumulated in troughs and channels (to a thickness of at least 0.5 m) from the shelf (<200 m) to the deep slope (>1,200 m) (BT = 4°C). The organic carbon contribution was estimated to be more than 20 g C m^{-1} in some areas, which is almost ten times the annual fluxes in the area, as measured by sediment traps (Wefer & Fischer, 1993). Carcasses were very abundant (707 individuals 100 m^{-2}) at the maximum depth surveyed (1,275 m) and the maximum depth of the deposit could not be determined. Megafauna (including echinoderms and crustaceans) were observed 63 times directly feeding on the material (Lebrato & Jones, 2009). In the Mediterranean Sea (Alboran Sea to the Catalan Sea), jelly-falls of *P. atlanticum* were identified and sampled from 1994 to 2005 (spring and summer) during bottom trawling down to 800 m (average BT = 13°C) in the MEDITS-ES surveys (Bertrand et al., 2002) (Fig. 2B). The catch often exceeded 300 carcasses per haul. This dataset provided the first evidence of jelly-falls encompassing entire continental margins during a period of 12 years (Fig. 2B). Living *P. atlanticum* were recovered from benthic trawls in canyon heads and walls (Cartes et al., 2009) near the wind-driven upwelling region of the Gulf of Lions (Johns et al., 1992). Sartor et al. (2003) reported catches in benthic trawls from 1995 to 1999 in the Mediterranean Sea (Tyrrhenian Sea) with numerous *P. atlanticum* occurring at the seafloor (100–500 g h^{-1} during >500 h over several km^2) at 300 and 650 m (BT = 12°C) (Fig. 2B). In the Mediterranean Sea, benthic deposits of *P. atlanticum* seem to be a common feature that are generally unnoticed.

Thaliaceans other than *P. atlanticum* have also been recorded at the seabed. Cacchione et al. (1978) described sinking living/moribound *Salpa aspera* Chamisso 1819 in the water column from a series of ROV observations below 2,500 m (BT = 3°C) in the Hudson Canyon, northwest Atlantic, over 30 days during summer. Salp bodies were observed rolling down the canyon. In the same area, Wiebe et al. (1979) observed a jelly-fall of *S. aspera* at >2,000 m (BT = 3°C). The carcasses accumulated in channels and furrows and formed string-like aggregations at the seabed (see Grassle & Morse-Porteus, 1987; Grassle & Grassle, 1994). In the Pacific Ocean, Duggins (1981) reported thousands of *Salpa fusiformis* Cuvier 1804 in the intertidal/subtidal environment (BT = 4°C) of the Alaska Gulf over several months. Echinoderms fed preferentially on the gelatinous resource as soon as it was available. Recently, in the Red Sea, salps have formed jelly-falls during spring and after upwelling (20 m, BT = 23°C) although these observations were not quantified (A. Alamaru pers. comm). In the Sea of Japan, a doliolid jelly-fall was studied in the water column by means of a sediment trap below 150 m (temperature = 12°C) (Takahashi et al., 2010).

Cnidarians

For Cnidaria, the first natural jelly-fall recorded was in a photographic survey (Jumars, 1976) below 1,000 m (BT = 4°C), where ophiuroids congregated around a *Pelagia* sp. carcass in the Santa Catalina basin (northeast Pacific). More recently, jelly-falls of *Aurelia limbata* Brandt 1835, *Parumbrosa polylobata* Kishinouye 1910, and *Nemopilema nomurai* Kishinouye 1922 were reported on the seafloor down to 400-m depth (BT = 2.2–10°C) in the Sea of Japan during summer and autumn (Miyake et al., 2002, 2005; Yamamoto et al., 2008, respectively). Thousands of *Crambionella orsini* Vanhöffen 1888 carcasses were photographed using a towed camera at the seabed during winter and after seasonal upwelling in the Arabian Sea (Billett et al., 2006). Carcasses were recorded as freshly deposited on the shelf, while 'jelly-lakes' of decomposing detritus were observed on the continental rise deeper than 3,000 m (BT = 2°C). White mats, assumed to be bacteria decomposing and remineralizing the organic material, covered the detritus. A scyphozoan jelly-fall (probably *C. orsini*) also was reported near the Pakistan Margin at 900 m (BT = 9.5°C) during summer and after seasonal upwelling (Murty et al., 2009). A large gelatinous mat covering the seabed, presumably scyphozoans in a very advanced state of decomposition, was surveyed for 7 days with a ROV in the Norwegian Sea at 1,380 m (BT = −1°C) during summer (Jones et al., 2010). A jelly-fall of *Periphylla periphylla* (Peron & Lesueur, 1810) was studied in spring 2011 in the Lurefjorden, Norway between 396 and 443 m (BT = 7°C) (Sweetman & Chapman, 2011). Carcasses were documented with a camera in two

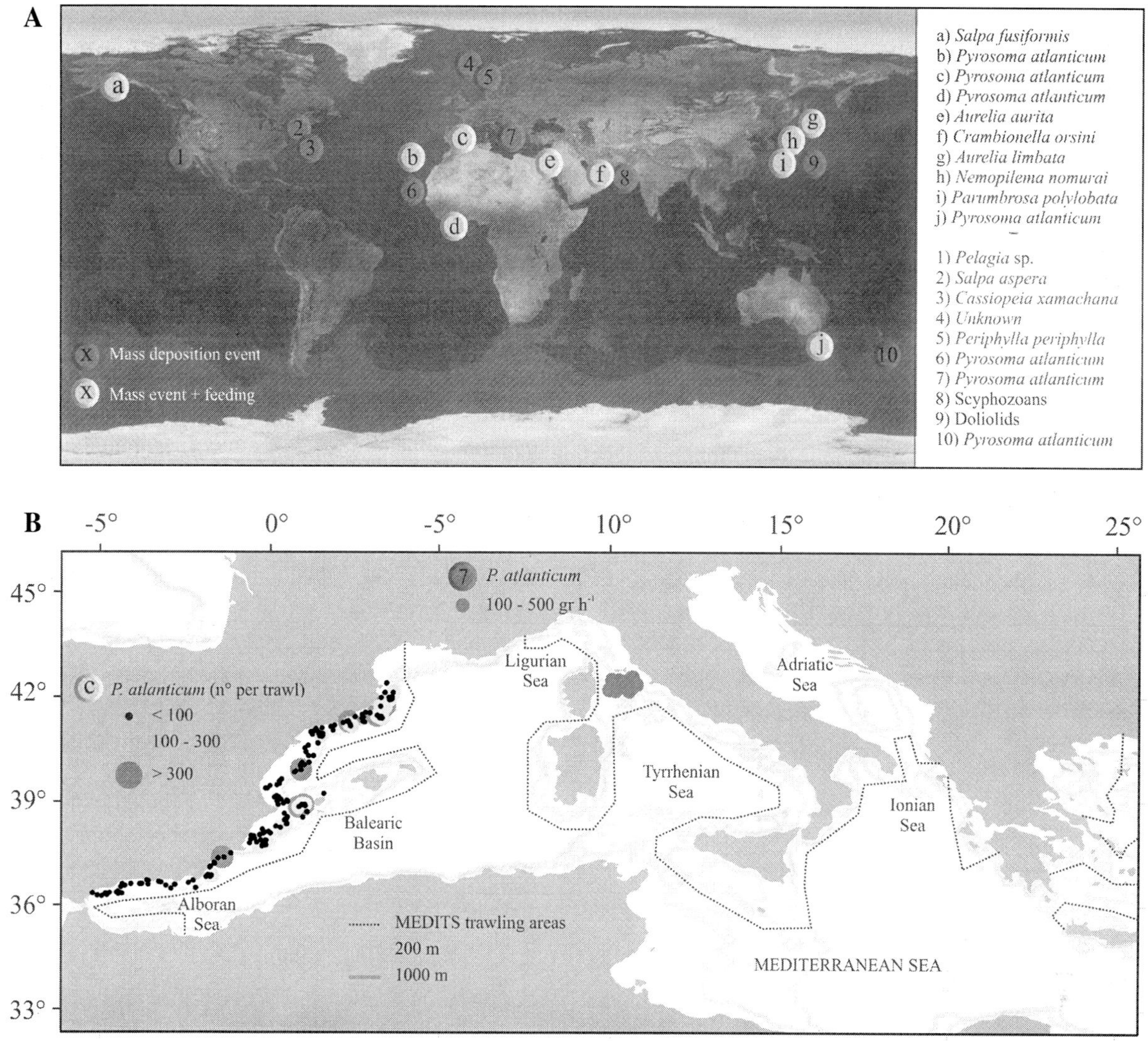

Fig. 2 **A** Global distribution of reported jelly-falls. Also included are the species that were recorded in each individual event (see Table 1 for detailed information). **B** Observations of *P. atlanticum* jelly-falls at the seabed in the Mediterranean Sea (from the MEDITS-ES project) (Bertrand et al., 2002). Numerous jelly-falls occur along the whole western Iberian Margin. The legend shows the average number of carcasses observed at each station from 1994 to 2005. Also included are observations of *P. atlanticum* in the Tyrrhenian Sea (northwest Mediterranean Sea) (Sartor et al., 2003). The bathymetric line (200 m) are from the general bathymetric chart of the oceans (GEBCO) Digital Atlas (IOC et al., 2003). The dotted line indicates the zones trawled in the MEDITS project that can be used to study jelly-falls from trawling data, as proposed in "Operational oceanography and exploration techniques" section

different areas in seven transects at very low densities (0.01 carcass m^{-2}), estimated to contribute <1% to the annual organic matter flux in the area. Numerous jelly-falls of *Aurelia aurita* Linnaeus 1758 occurred during spring and after upwelling events at 20-m depth in the Red Sea (BT = 23°C) (Alamaru et al., 2009). Sexton et al. (2010) reported a jelly-fall of *Chrysaora quinquecirrha* Desor 1848 medusae in a shallow sub-estuary of Chesapeake Bay during autumn.

Jelly-falls conceptualization

Processes from the euphotic zone to the seabed

A jelly-fall (Fig. 1) starts when gelatinous organisms die and sink from the so-called death depth subject to the organisms' vertical migration and displacement. Because gelatinous detritus is denser than the surrounding seawater, the corpses sink through the water column at a rate determined by the material's size and excess density (Stokes' Law) (e.g. Yamamoto et al., 2008). The organisms can settle at the seabed while still alive (Wiebe et al., 1979; Gili et al., 2006) and then die, thus remineralization can start on the seabed. As it sinks, the material can be consumed by scavengers and return to the faunal food web or be remineralized by bacteria (bacterioplankton) and enter the microbial loop (e.g. Hansson & Norrman, 1995). Dissolved organic matter (DOM) leaching from living or dead organisms provides a link to the "jelly-pump" concept (J-DOM) of microbial communities being fuelled by DOM excretion (Condon & Steinberg, 2008; Niggl et al., 2010; Condon et al., 2011) (Fig. 1). Microzooplankton and small zooplankton may also consume J-DOM (e.g. Iguchi et al., 2006; Titelman et al., 2006; West et al., 2009a). Laboratory incubations of scyphozoan material using deep (334 m) and shallow water (<10 m) differed in the remineralization time (Iguchi et al., 2006), which was attributed to the microbial community as well as temperature in situ. Differences in the lability of gelatinous tissues (C:N ratios; Larson, 1986), the various rates at which the different materials sink (Apstein, 1910; Mills, 1981), and rates of scavenging and bacterial mineralization (which may vary with temperature and depth) greatly influence the extent to which the jelly-fall is recycled within the water column versus at the seabed (Fig. 1). Jelly-falls that reach the seafloor may be transported elsewhere (e.g. along geomorphological features) (Billett et al., 2006), or retained in situ and consumed by the local faunal and microbial community (Lebrato & Jones, 2009). Leaching of dissolved compounds fuels production in higher trophic levels (West et al., 2009a) and biogeochemical processes such as oxygen consumption in the water and in the sediment proceed during the organic enrichment (West et al., 2009b; Sexton et al., 2010). Associated total alkalinity changes from excess DOM (Hansson & Norrman, 1995; Hoppe et al., 2010) and the non-Redfield stoichiometry of nitrogen and phosphorus leaching from the corpses (Pitt et al., 2009; Condon et al., 2010; Tinta et al., 2010) should also be considered. The decomposition dynamics has been the focus of several papers targeting a variety of species at different temperatures, thus decay rates (k) are available (e.g. Titelman et al., 2006). The turnover of J-POM is rapid during the first few days (Sempere et al., 2000) and then slows down, but it is highly dependent on temperature (Iguchi et al., 2006). These quantitative data on decomposition dynamics have enabled remineralization of sinking carcasses to be modelled in open ocean conditions (Lebrato et al., 2011). They provided a new metrics based on decay rate, temperature fields, 'death depth', and sinking speed that helps to understand why different gelatinous zooplankton groups transfer organic carbon to the seabed (e.g. scyphozoans and thaliaceans), while others may be completely remineralized in the water column.

The transport to the seafloor of J-POM is an important source of labile material to the whole size-spectrum of benthic communities in continental margins and the deep-sea (Table 1; Fig. 1). Evidence of organisms consuming J-POM at the seabed has accumulated slowly from photographs and videos (Table 1; Fig. 2A). Gelatinous material has a low energy content (0.5–6 gross energy kJ g dry mass^{-1}) compared to other types of carrion such as fish (5–22 gross energy kJ g dry mass^{-1}) or algae (>10 gross energy kJ g dry mass^{-1}) (Doyle et al., 2007). Among gelatinous species, the energy content is highest in salps and pyrosomids (4–6 gross energy kJ g dry mass^{-1}) (Davenport & Balazs, 1991; Clarke et al., 1992), which are important parts of the diets of numerous benthic organisms (Table 3). Although high energy resources are readily available on continental margins, food is a limiting factor in the deep-sea (Gage & Tyler, 1991). Thus, jelly-falls may represent a valuable nutritional input at certain times of the year (Table 3). Unlike other large food falls, which are usually sparsely scattered over the sea floor, gelatinous corpses accumulate in large patches (Billett et al., 2006; Lebrato & Jones, 2009) making it easier for scavengers to locate; however, scavengers traditionally observed around fish falls (such as isopods or fish) have not been observed around jelly-falls (Sweetman & Chapman, 2011). The reduced energy spent searching for food, and the lability of the gelatinous carrion relative to other sources of detritus, may compensate for the reduced energy density of the jelly-falls at least for some scavenger species (Doyle et al., 2007). Additionally, jelly-falls may provide an environment for macrofauna/microbial communities to proliferate, which, in turn, may be preyed upon by other taxa (Sweetman & Chapman, 2011). Sessile organisms (anthozoans, including hexacorallians, octocorallians, and scleractians) also consume J-POM

Table 3 Occurrences of jelly-falls and the megafaunal taxa feeding on them

Year	Depth (m)	Taxon	Timing[a]	Duration	Units	Feeding taxa	Reference
2011	396–443	*P. periphylla*	Mar—Sp.	1	days	Crustaceans[b]	Sweetman & Chapman (2011)
2010	150	Doliolids	May—Sp.	5	–	None	Takahashi et al. (2010)
2009	20	Salps	Post-upwelling	–	–	Anthozoans	Alamaru et al. (unpublished)
2009	20	*A. aurita*	Post-upwelling	–	–	Anthozoans	Alamaru et al. (2009)
2009	900	*C. orsini*	Post-upwelling	–	–	None	Murty et al. (2009)
2009	1,380	Scyphozoans	Jun—S	7	days	None	Jones et al. (2010)
2007	3	*C. xamachana*	Sep—S	–	–	None	M. Lebrato (unpublished)
2006	146–354	*N. nomurai*	Sep/Oct—S/A	30	days	Crustaceans, echinoderms	Yamamoto et al. (2008)
2006	26–1,275	*P. atlanticum*	Post-upwelling	60	days	Several[c]	Lebrato & Jones (2009)
2005	1.5–3	*C. quinquecirrha*	Jun/Sep—S	90	days	None	Sexton et al. (2010)
2003	304–3,299	*C. orsini*	Post-upwelling	17	–	Crustaceans, echinoderms	Billett et al. (2006)
2002	453	*P. polylobata*	Sep—S	–	–	Echinoderms	Miyake et al. (2005)
2001	320	*A. limbata*	Aug—S	–	–	Echinoderms	Miyake et al. (2002)
1999	300–650	*P. atlanticum*	Sp./S	3	months	None	Sartor et al. (2003)
1998	300–650	*P. atlanticum*	Sp./S	3	months	None	Sartor et al. (2003)
1997	300–650	*P. atlanticum*	Sp./S	3	months	None	Sartor et al. (2003)
1996	300–650	*P. atlanticum*	Sp./S	3	months	None	Sartor et al. (2003)
1995	300–650	*P. atlanticum*	Sp./Sum	3	months	None	Sartor et al. (2003)
1985	5,433	*P. atlanticum*	Jun/Jul—S	17	days	Crustaceans, echinoderms	Roe et al. (1990)
1978	1–10	*S. fusiformis*	Mar/Jun—Sp./S	3	months	Echinoderms	Duggins (1981)
1975	2,500–3,000	*S. aspera*	Aug—S	–	–	None	Cacchione et al. (1978)
1975	2,000–3,000	*S. aspera*	Aug—S	4	days	None	Wiebe et al. (1979)
1955	330–640	*P. atlanticum*	Jun/Jul—S	–	–	Fish	Cowper (1960)
1952	100	*P. atlanticum*	Oct–Sp.	–	–	Fish	Hurley & McKnight (1959)

[a] The month is abbreviated (when available), and the season is indicated as: *Sp.* spring, *S* summer, *A* autumn, *W* Winter

[b] Caridean shrimps grazed on carcasses, and density was higher around jelly-falls compared to non jelly-falls settings. Galatheid crabs were observed near carcasses, but no grazing was observed

[c] Anthozoans, crustaceans, echinoderms, fish, arthropods, polychaetes

(Gili et al., 2006; Alamaru et al., 2009; Lebrato & Jones, 2009). Echinoderms dominate scavenging observations at any depth, followed by crustaceans and fish (Table 3). Remains of J-POM (e.g. *Cymbulia peroni* De Blainville 1810) are commonly found in guts of numerous benthic decapods, such as the Norway lobster *Nephrops norvegicus* Linnaeus 1758, the crab *Geryon longipes* Milne-Edwards 1882 (in Cartes, 1993a), and the squat lobster *Munida tenuimana* Sars 1872 (in Cartes, 1993b). J-POM (*Iasis zonaria* Pallas 1774, *P. atlanticum*, *P. periphylla*) is also found in the guts of deep shrimps, such as *Plesionika martia* Milne-Edwards 1883 (in Fanelli & Cartes, 2008) or fish (Carrasson & Cartes, 2002; Drazen et al., 2008; Goldman & Sedberry, 2010). Any remaining material that is not channelled through macro/megafaunal scavenging will eventually be respired by microbial communities (Fig. 1). The build-up of impermeable gelatinous material (as in Billett et al., 2006) on the seafloor leads to reductions in O_2 flux into sediments (West et al., 2009b). This would favour microbial over metazoan biomass and remineralization processes, although low seawater O_2 combined with no light slows microbial decomposition of settling organic matter (Gooday et al., 2010), and toxic remineralization products (e.g. ammonium and free sulphides) could accumulate and seriously impact sediment biota as well as pelagic ecosystems (Titelman et al., 2006; Pitt et al., 2009) (Fig. 1). Ultimately, jelly-falls could induce spatial heterogeneity in the biodiversity of benthic communities (Gooday et al., 2010) as a consequence of the mass accumulation of undegraded labile material.

Causes and seasonality of jelly-falls

Factors driving the onset of jelly-falls are mostly linked to the ageing and end of a bloom (Purcell et al., 2001) and a long-term cumulative effect of negative factors, such as parasitism, starvation, infection, and predation (Mills, 1993), with subsequent deposition at the seabed if the material is not completely remineralized while sinking. In other cases, the material floats and it is washed ashore (e.g. Pakhomov et al., 2003; Houghton et al., 2007). The life history of individual species dictates their fate, although some generalities apply to all groups, such as seasonal disappearance from the waters (Mills, 1993). Life cycles are often completed within a year or a few months, with subsequent death (see Franqueville, 1971; Mills, 1993). For thaliaceans, there is evidence that high concentrations of particles and suspended organic matter [e.g. chlorophyll *a* >1 mg m^{-3}; Perissinotto & Pakhomov (1998)] clog their feeding apparatus causing death (Acuña, 2001) despite food being abundant (Harbison et al., 1986; Zeldis et al., 1995). This explains the salp jelly-fall studied by Duggins (1981) in the subtidal zone in Alaska and the beaching of salps reported by Pakhomov et al. (2003) in the Southern Ocean. Thaliacean jelly-falls tend to appear at the seabed after strong periods of upwelling (Lebrato & Jones, 2009) or after the spring bloom months when chlorophyll *a* levels are high (Wiebe et al., 1979; Duggins, 1981; Roe et al., 1990) (Table 3). Re-assessment of the season ($n = 24$) when carcasses of all groups arrive at the seabed indicates that >75% of jelly-falls occur after the spring bloom in temperate/subpolar areas and >25% in post-upwelling periods in the tropics. This happens irrespectively of the depth at which they are deposited. It remains unclear for thaliaceans if the concentration or particles per se causes clogging and subsequent death, or if the biological composition of the particles and autotroph community play a role. Potential connections between climate and gelatinous zooplankton populations in the water column and the jelly-falls at the seabed have not yet been investigated. In tropical areas, monsoon patterns trigger upwelling events that alter water column properties (e.g. lower temperature, high nutrients and chlorophyll *a*, high DOM levels, higher POM export) (Coble et al., 1998; Honjo et al., 1999), thus forcing in these zones is different than in temperate/subpolar latitudes.

In the Cnidaria, several variables may trigger the onset of jelly-falls, including sudden or sustained changes in temperature exceeding physiological performance (Gatz et al., 1973) (relevant in upwelling systems where organisms can experience rapid changes in the water mass properties due to physical forcing), ageing of the bloom followed by food depletion (causing starvation and poor nutrition) (Mills, 1993; Purcell et al., 2001; Sexton et al., 2010). The latter cause may have relevance for the *C. orsini* carcasses studied by Billett et al. (2006) and the depositions of *N. nomurai* observed by Yamamoto et al. (2008). The food exhaustion hypothesis would explain why we often observe scyphozoan jelly-falls after the spring bloom but predominantly in the late spring/early summer months (Table 3). Other factors include grazing damage (Arai, 2005), parasitism/injury/viral infections (Mills, 1993), senescence (Sexton et al., 2010), extreme weather events triggering large changes in physical properties of water (Cargo, 1976), and sinking driven by low temperatures and inducing deposition and later death owing to temperature changes (Sexton et al., 2010).

Operational oceanography and exploration techniques

The jelly-fall concept originates from a handful of studies undertaken in the field that either described accidental encounters or, in few cases, targeted known gelatinous depositions. In >80% of the cases ($n = 22$), ROV video and/or towed/still cameras were used as the sampling technique (Table 1). Unless a large area was covered and transects used to count individual carcasses (Billett et al., 2006; Lebrato & Jones, 2009), these techniques remain qualitative (Roe et al., 1990; Miyake et al., 2002, 2005; Yamamoto et al., 2008). Other techniques, including scuba diving and sediment traps account for <5% of the observations.

Trawling is the only other technique that allows large quantitative studies (MEDITS-ES dataset; Sartor et al., 2003). Field work has been accompanied by a series of laboratory or mesocosm studies that target associated biogeochemical processes (e.g. Sempere et al., 2000; Pitt et al., 2009; Tinta et al., 2010). Although we now have important information about the occurrence of jelly-falls and their potential influence on elemental cycling, we still lack combined effort and large-scale projects on this topic. Temporal monitoring can be addressed by 'ocean observatories' (Table 2; Claustre et al., 2010; Send et al., 2010). From these ocean observatory initiatives (e.g. EUR-OCEANS, OceanSITES, ESONET) and scientific projects that collaborate with offshore industries (e.g. SERPENT (Jones, 2009), DELOS (http://www.delos-project.org/), and HAUSGARTEN (Soltwedel et al., 2005), regular access to the deep-sea will increase our chances of making informative observations. We need to move beyond the present semi-empirical state of understanding to local or regional monitoring and quantification of jelly-falls. ROVs, AUVs (autonomous underwater vehicles), benthic landers, and towed, drop, and time-lapse cameras should be used (Table 2). In particular, the use of repeated AUV surveys or a network of time-lapse cameras strategically placed at the seabed in areas where jelly-falls have been observed could provide insights into seasonality and decomposition at the seabed. Benthic crawlers (Karpen et al., 2007) can survey inaccessible areas where jelly-falls have been observed via a optical cable from a shore-based station for long periods of time.

For large-scale quantification of jelly-falls, logbooks of bottom-trawling surveys from historical to present times are a unique tool that have not fully utilized. They mainly target commercial demersal fish and crustaceans species, but non-commercial or 'discarded' (bycatch) species, including gelatinous zooplankton, are sometimes consistently recorded (e.g. Sartor et al., 2003; Sanchez et al., 2003; Bastian et al., 2011). Data from jelly-falls have been collected in this way (Fig. 2B) and also data on living biomass (Bastian et al., 2011). Many benthic trawling programmes exist worldwide [e.g. MEDITS (International bottom trawl survey in the Mediterranean Sea) (Bertrand et al., 2002); Relini (2000) (Italian Seas); Sartor et al. (2003) (Tyrrhenian Sea); International Bottom Trawl Survey (ITBS); NOAA Gulf of Alaska bottom trawl survey; NEFSC bottom trawl survey (Gulf of Maine Area); Wilkins et al. (1998) (The 1995 Pacific West Coast bottom trawl survey); Bastian et al. (2011) (North Atlantic Ocean)]. Information could also be retrieved from fisheries information networks [e.g. PacFIN (http://pacfin.psmfc.org/index.php); AKFIN (http://www.akfin.org)]; and from state and wildlife agencies and fishery management councils (e.g. http://pacfin.psmfc.org/pacfin_pub/links.php)]. Trawling surveys normally cover specific depth ranges in the so-called trawlable areas in the shelves and slopes. The surveys do not normally work beyond the continental slope [(e.g. 0–800 m in the MEDITS-ES, 250–800 m in Sartor et al. (2003)] (Fig. 2B), but effectively sample the shelves consistently and repeatedly. The problem often is that to reduce cost and effort and increase efficiency, the size, weights, and numbers only of commercial species are recorded in logbooks, and the living and dead gelatinous component, if present, is overlooked. This issue was discovered in the MEDITS-ES project, where certain partners recorded the same data for commercial and for non-commercial species (jelly-fall data used in Fig. 2B), while the majority did not. Only through effective science-industry communication and collaboration can we make use of their potential to quantify jelly-falls and living biomass (Bastian et al., 2011).

At local scales, we suggest use of acoustic/electronic tagging (e.g. Seymour et al., 2004; Gordon & Seymour, 2008; Hays et al., 2008) on individuals found in blooms to discover their fate (Table 2). Tags can be mechanically secured in cnidarians in the bell area and peduncle, or using setting glue. Large neutrally buoyant sediment traps also could be used (Lampitt et al., 2008) that could drift under blooms, as well as free-drifting profilers with mounted cameras to investigate the water column. Genetic tools (e.g. Reusch et al., 2010) could also be used to characterize a jelly-fall signature in the sediment. Further research should quantify and study the diversity of the scavenging communities attracted to an 'artificial' jelly-fall (e.g. Yamamoto et al., 2008). This has traditionally been done with a bait in the field of view of a still camera (reviewed by Bailey et al., 2007). This can be combined with labelling studies to assess the fate of jelly-derived organic material, as for phytodetritus (Middelburg et al., 2000; Witte et al., 2003; Franco et al., 2008).

Can jelly-falls provide ecosystem services in the future?

The future ocean is expected to be a warmer, more-stratified, acidic, and oxygen-poor system characterized by reduced upwelling (Cox et al., 2000; Gregg et al., 2003). As a result, production exported to depth is expected to be reduced as phytoplankton communities shift from large diatom-based assemblages to picoplankton with lower export efficiency (Buesseler et al., 2007; Smith et al., 2008). Reduced export production and changes in community structure are expected to result in reduced delivery to, and an overall change in the composition of organic material reaching the abyssal ocean floor (Laws, 2004; Smith et al., 2008). This is expected to reduce food availability to the already food-limited deep-sea floor, causing a decline in deep-sea biomass and ecosystem changes (e.g. in faunal behaviour (Kaufmann & Smith, 1997; Wigham et al., 2003), bioturbation (Smith et al., 2008; Vardaro et al., 2009), faunal densities (Ruhl & Smith, 2004; Ruhl, 2007, 2008; Smith et al., 2008), reproductive traits (Tyler, 1988; Young, 2003; Ramirez-Llodra et al., 2005), faunal diversity (Levin et al., 2001), body size (McClain et al., 2005), taxonomic composition (Ruhl & Smith, 2004), sediment infaunal response (Sweetman & Witte, 2008a, b), and dominance (Cosson et al., 1997; Sweetman & Witte, 2008b). Reduced carbon export may also inhibit the ocean's ability to sequester carbon (Smith et al., 2008). The potential consequences of the combination of altered food inputs to the benthos and increased CO_2 content of seawater on ecosystem functioning and services (e.g. nutrient regeneration, energy transfer to higher trophic levels) could have large implications because recent studies suggest that an organism's ability to cope with acidification and elevated water temperatures may be regulated by food supply (Wood et al., 2008; Gooding et al., 2009).

Gelatinous zooplankton populations, on the other hand, may benefit from anthropogenic impacts on the marine environment (Purcell et al., 2007; Purcell, 2012). There is evidence of some populations increasing during the last decades, such as thaliaceans in the Southern Ocean (Loeb et al., 1997; Atkinson et al., 2004) and jellyfish in the Mediterranean Sea (Molinero et al., 2008). It has been suggested that jelly-biomass will become an increasingly important component in the future ocean (Purcell et al., 2007; Jackson, 2008; Richardson et al., 2009; Purcell, 2012). Therefore, if classic POM vectors (e.g. phytodetritus) become less important in the future ocean, an increased amount of J-POM sinking to the seabed could mitigate some of the losses of carbon from phytoplanktonic carbon sources, although it is likely to be much more heterogeneous at the seafloor (Gooday et al., 2010). Because the majority of jelly-falls deposits are located in deep, cold (<10°C) marine environments (Table 1), we hypothesize that J-POM:phytodetrital-POM flux ratios are likely to be higher in deep-sea and polar settings (Lebrato et al., 2011). This may maintain certain ecosystem functions in some areas (dependent on the threshold POM flux) by ensuring a continued minimum POM flux from the surface to the seafloor. It is also likely to have drastic implications for benthic community composition just as changes in surface phytoplankton community composition can substantially modify abyssal community composition.

Acknowledgments We are grateful to the scientific and environmental ROV partnership using existing industrial technology project (SERPENT) for enabling access to data off west Africa and in the deep Norwegian Sea. We thank the following contributions from individuals: L. Gil de Sola, C. García, J. Pérez Gil, and P. Abelló from the I.E.O (Fuengirola Oceanographic Center, Spain) for the facilities providing MEDITS-ES data, Brian J. Bett and R. S. Lampitt from the National Oceanography Center Southampton, UK provided unpublished data of *C. orsini* from Billett et al. (2006) and from Roe et al. (1990). This work was also supported by the "European Project on Ocean Acidification" (EPOCA), which is funded from the European Community's Seventh Framework Programme (FP7/2007–2013) under grant agreement no 211384. EPOCA is endorsed by the International Programmes IMBER, LOICZ and SOLAS. This work was funded by the grant Becas mineras. exp. 210001 to M. Lebrato and by the Kiel Cluster of Excellence "The Future Ocean" (D1067/87).

References

Acuña, J. L., 2001. Pelagic tunicates: why gelatinous? The American Naturalist 158: 100–107.

Alamaru, A., O. Bronstein, G. Dishon & Y. Loya, 2009. Opportunistic feeding by the fungid coral *Fungia scruposa* on the moon jellyfish *Aurelia aurita*. Coral Reefs 28: 865.

Alldredge, A. L. & M. W. Silver, 1988. Characteristics, dynamics and significance of marine snow. Progress in Oceanography 20: 41–82.

Apstein, C., 1910. Hat ein Organismus in der Tiefe Gelebt, in der er gefischt ist? Internationale Revue der gesamten Hydrobiologie und Hydrographie 3: 17–33.

Arai, M. N., 2005. Predation on pelagic coelenterates: a review. Journal of the Marine Biological Association of the United Kingdom 85: 523–536.

Atkinson, A., V. Siegel, E. Pakhomov & P. Rothery, 2004. Long-term decline in krill stock and increase in salps within the Southern Ocean. Nature 432: 100–103.

Bailey, D. M., N. J. King & I. G. Priede, 2007. Cameras and carcasses: historical and current methods for using artificial food falls to study deep-water animals. Marine Ecology Progress Series 350: 179–191.

Balch, W., D. Drapeau, B. Bowler & E. Booth, 2007. Prediction of pelagic calcification rates using satellite measurements. Deep Sea Research II 54: 478–495.

Bastian, T., D. Stokes, J. E. Kelleher, G. C. Hays, J. Davenport & T. K. Doyle, 2011. Fisheries bycatch data provide insights into the distribution of the mauve stinger (*Pelagia noctiluca*) around Ireland. ICES Journal of Marine Science 68: 436–443.

Beaulieu, S. E., 2002. Accumulation and fate of phytodetritus on the sea floor. Oceanography and Marine Biology Annual Review 40: 171–232.

Behrenfeld, M. J. & P. G. Falkowski, 1997. Photosynthetic rates derived from satellite-based chlorophyll concentration. Limnology and Oceanography 42: 1–20.

Bertrand, J. A., L. Gil de Sola, C. Papaconstantinou, G. Relini & A. Souplet, 2002. The general specifications of the Medits surveys. In Abello, P., J. A. Bertrand, L. Gil de Sola, C. Papaconstantinou, G. Relini & A. Souplet (eds), Mediterranean Marine Demersal Resources, The MEDITS International Trawl Survey (1994–1999). Scientia Marina 66: 9–17.

Billett, D. S. M., B. J. Bett, C. L. Jacobs, I. P. Rouse & B. D. Wigham, 2006. Mass deposition of jellyfish in the deep Arabian Sea. Limnology and Oceanography 51: 2077–2083.

Buesseler, K. O., M. P. Bacon, K. Cochran & H. D. Livingston, 1992. Carbon and nitrogen export during the JGOFS North Atlantic Bloom Experiment estimated from ^{234}Th, ^{238}U disequilibria. Deep Sea Research I 39: 1115–1137.

Buesseler, K. O., C. H. Lamborg, P. W. Boyd, P. J. Lam, T. W. Trull, R. R. Bidigare, J. K. B. Bishop, K. L. Casciotti, F. Dehairs, M. Elskens, M. Honda, D. M. Karl, D. A. Siegel, M. A. Silver, D. K. Steinberg, J. Valdes, B. Van Mooy & S. Wilson, 2007. Revisiting carbon flux through the ocean's twilight zone. Science 316: 567–570.

Burd, A. B., D. A. Hansell, D. K. Steinberg, T. R. Anderson, J. Aristegui, F. Baltar, S. R. Beaupre, K. O. Buesseler, F. DeHairs, G. A. Jackson, D. C. Kadko, R. Koppelmann, R. S. Lampitt, T. Nagata, T. Reinthaler, C. Robinson, B. H. Robison, C. Tamburini & T. Tanaka, 2010. Assessing the apparent imbalance between geochemical and biochemical indicators of meso- and bathypelagic biological activity, What the @$#! is wrong with present calculations and carbon budgets? Deep Sea Research II 57: 1557–1571.

Cacchione, D. A., G. T. Rowe & A. Malahoff, 1978. Submersible investigation of outer Hudson submarine canyon. In Stanley, D. J. & G. Kelling (eds), Sedimentation in Canyons, Fans and Trenches. Dowden, Hutchinson and Ross, Stroudsburg: 42–50.

Cargo, D. G., 1976. Some effects of tropical storm Agnes on the sea nettle population in the Chesapeake Bay. In Davis, J. & B. Laird (eds), The Effects of Tropical Storm Agnes on the Chesapeake Bay Estuarine System. Chesapeake Bay Consortium 54. Johns Hopkins University Press, Baltimore: 417–424.

Caron, D. A., P. G. Davis, L. P. Madin & J. M. Sieburth, 1986. Enrichment of microbial populations in macroaggregates (marine snow) from surface waters of the North Atlantic. Journal of Marine Research 44: 543–565.

Carrasson, M. & J. E. Cartes, 2002. Trophic relationships in a Mediterranean deep-sea fish community partition of food resources, dietary overlap and connections within the benthic boundary layer. Marine Ecology Progress Series 241: 41–55.

Cartes, J. E., 1993a. Diets of deep-water pandalid shrimps on the Western Mediterranean slope. Marine Ecology Progress Series 96: 49–61.

Cartes, J. E., 1993b. Diets of two deep-sea decapods *Nematocarcinus exilis* (Caridea: Nematocarcinidac) and *Munida tenuimana* (Anomura: Galatheidae) on the Western Mediterranean slope. Ophelia 37: 213–229.

Cartes, J. E., V. Papiol, A. Palanques, J. Guillen & M. Demestre, 2007. Dynamics of suprabenthos off the Ebro Delta (Catalan Sea, western Mediterranean), spatial and temporal patterns and relationships with environmental factors. Estuarine, Coastal and Shelf Science 75: 501–515.

Cartes, J. E., F. Maynou, E. Fanelli, C. Romano, V. Mamouridis & V. Papiol, 2009. The distribution of megabenthic, invertebrate epifauna in the Balearic Basin (Western Mediterranean) between 400 and 2300 m: environmental gradients influencing assemblages composition and biomass trends. Journal of Sea Research 66: 244–257.

Cartes, J. E., E. Fanelli, V. Papiol & F. Maynou, 2010. Trophic relationships at interannual spatial and temporal scales of macro and megafauna around a submarine canyon off the Catalonian coast (western Mediterranean). Journal of Sea Research 63: 180–190.

Clarke, A., L. J. Holmes & D. J. Gore, 1992. Proximate and composition of gelatinous zooplankton from the Southern Ocean. Journal of Experimental Marine Biology and Ecology 155: 55–68.

Claustre, H., J. Bishop, E. Boss, S. Bernard, J-F. Berthon, C. Coatanoan, K. Johnson, A. Lotiker, O. Ulloa, M. J. Perry, F. D'Ortenzio, O. H. F. D'andon & J. Uitz, 2010. Bio-optical profiling floats as new observational tools for biogeochemical and ecosystem studies: potential synergies with ocean colour remote sensing. OceanObs'09, Venice, September 21–25.

Coble, P. G., C. E. del Castillo & B. Avril, 1998. Distribution and optical properties of CDOM in the Arabian Sea during the 1995 Southwest Monsoon. Deep Sea Research II 45: 2195–2223.

Condon, R. H. & D. K. Steinberg, 2008. Development, biological regulation, and fate of ctenophore blooms in the York River estuary, Chesapeake Bay. Marine Ecology Progress Series 369: 153–168.

Condon, R. H., D. K. Steinberg & D. A. Bronk, 2010. Production of dissolved organic matter and inorganic nutrients by gelatinous zooplankton in the York River estuary, Chesapeake Bay. Journal of Plankton Research 32: 153–170.

Condon, R. H., D. K. Steinberg, P. A. del Giorgio, T. C. Bouvier, D. A. Bronk, W. A. Graham & H. W. Ducklow, 2011. Jellyfish blooms result in a major microbial respiratory sink of carbon in marine systems. Proceedings of the National Academy of Sciences of the United States of America 108: 10225–10230.

Conte, M. H., T. D. Dickey, J. C. Weber, R. J. Johnson & A. H. Knap, 2003. Transient physical forcing of pulsed export of bioreactive material to the deep Sargasso Sea. Deep Sea Research I 50: 1157–1187.

Cosson, N., M. Sibuet & J. Galeron, 1997. Community structure and spatial heterogeneity of the deep-sea macrofauna at three contrasting stations in the tropical northeast Atlantic. Deep Sea Research I 44: 247–269.

Cowper, T. R., 1960. Occurrence of *Pyrosoma* on the continental Slope. Nature 187: 878–879.

Cox, P. M., R. A. Bets, C. D. Jones, S. A. Spall & I. J. Totterdell, 2000. Acceleration of global warming due to carbon-cycle feedbacks in a coupled climate model. Nature 408: 184–187.

Davenport, J. & G. H. Balazs, 1991. "Fiery pyrosomas": are pyrosomas an important items in the diet of leatherback turtles? British Herpetological Society Bulletin 37: 33–38.

Doyle, T. K., D. R. Houghton, R. McDevitt, J. Davenport & G. C. Hays, 2007. The energy density of jellyfish: estimates from bomb-calorimetry and proximate-composition. Journal of Experimental Marine Biology and Ecology 343: 239–252.

Drazen, J. C., B. N. Popp, C. A. Choy, T. Clemente, L. de Forest & K. L. Smith, 2008. Bypassing the abyssal benthic food web: macrourid diet in the eastern North Pacific inferred from stomach content and stable isotopes analyses. Limnology and Oceanography 53: 2644–2654.

Duggins, D. O., 1981. Sea urchins and kelp, the effects of short term changes in urchin diet. Limnology and Oceanography 26: 391–394.

Fanelli, E. & J. E. Cartes, 2008. Spatio-temporal variability in the diet of two pandalid shrimps in the western Mediterranean evidenced from gut contents and stable isotope analysis: in-fluence on the reproductive cycle. Marine Ecology Progress Series 355: 219–233.

Franco, M. A., K. Soetaert, M. J. Costa, M. Vincx & J. Vanaverbeke, 2008. Uptake of phytodetritus by meiobenthos using ^{13}C labelled diatoms and Phaeocystis in two contrasting sediments from the North Sea. Journal of Experimental Marine Biology and Ecology 362: 1–8.

Franqueville, C., 1971. Macroplancton profond (invertébrés) de la Méditerranée nord-occidentale. Tethys 3: 11–56.

Gage, J. D. & P. A. Tyler, 1991. Deep-Sea Biology: A Natural History of Organisms at the Deep-sea Floor. Cambridge University Press, Cambridge.

Gatz, A. J., V. S. Kennedy & J. A. Mihursky, 1973. Effects of temperature on activity and mortality of the scyphozoan medusa *Chrysaora quinquecirrha*. Chesapeake Science 14: 171–180.

Gili, J.-M., S. Rossi, F. Pages, C. Orejas, N. Teixido, P. J. Lopez-Gonzalez & W. E. Arntz, 2006. A new trophic link between the pelagic and benthic systems on the Antarctic shelf. Marine Ecology Progress Series 322: 43–49.

Goldman, S. F. & G. R. Sedberry, 2010. Feeding habits of some demersal fish on the Charleston Bump off the southeastern United States. ICES Journal of Marine Science 68: 390–398.

Gooday, A. J., B. J. Bett, E. Escobar, B. Ingole, L. A. Levin, C. Neira, A. V. Raman & J. Sellanes, 2010. Habitat heterogeneity and its influence on benthic biodiversity in oxygen minimum zones. Marine Ecology 31: 125–147.

Gooding, R., C. Harley & E. Tang, 2009. Elevated water temperature and carbon dioxide concentration increase the growth of a keystone echinoderm. Proceedings of the National Academy of Sciences of the United States of America 106: 9316–9321.

Gordon, M. R. & J. E. Seymour, 2008. Quantifying movement of the tropical Australian cubozoan *Chironex fleckeri* using acoustic telemetry. Hydrobiologia 616: 87–97.

Grassle, J. F. & J. P. Grassle, 1994. Notes from the abyss, the effects of a patchy supply of organic material and larvae on soft-sediment benthic communities. In Giller, P. S., A. G. Hildrew & D. G. Raffaelli (eds), Aquatic Ecology, Scale, Pattern and Process, 34th Symposium of the British Ecological Society. Blackwell Scientific, Oxford: 499–515.

Grassle, J. F. & L. S. Morse-Porteus, 1987. Macrofaunal colonization of disturbed deep-sea environments and the structure of deep-sea benthic communities. Deep Sea Research I 34: 1911–1950.

Gregg, W., P. Ginoux, P. Schopf & N. Casey, 2003. Phytoplankton and iron, validation of a global three-dimensional ocean biogeochemical model. Deep Sea Research II 50: 3143–3169.

Griffin, D. J. G. & J. C. Yaldwyn, 1970. Giant colonies of pelagic tunicate (*Pyrosoma spinosum*) from SE Australia and New Zealand. Nature 226: 464.

Hansson, L. J. & B. Norrman, 1995. Release of dissolved organic carbon (DOC) by the scyphozoan jellyfish *Aurelia aurita* and its potential influence on the production of planktic bacteria. Marine Biology 121: 527–532.

Harbison, G. R., V. L. McAlister & R. W. Gilmer, 1986. The response of the salp, *Pegea confoederata,* to high levels of particulate material: starvation in the midst of plenty. Limnology and Oceanography 31: 371–382.

Hays, G. C., T. K. Doyle, J. D. R. Houghton, M. K. S. Lilley, J. D. Metcalfe & D. Righton, 2008. Diving behaviour of jellyfish equipped with electronic tags. Journal of Plankton Research 30: 325–331.

Honjo, S., J. Dymond, W. Prell & V. Ittekkot, 1999. Monsoon-controlled export fluxes to the interior of the Arabian Sea. Deep Sea Research II 46: 1859–1902.

Hoppe, C. J. M., G. Langer, S. D. Rokitta, D. A. Wolf-Gladrow & B. Rost, 2010. On CO_2 perturbation experiments: over-determination of carbonate chemistry reveals inconsisten cies. Biogeosciences Discussions 7: 1707–1726.

Houghton, J. D. R., T. K. Doyle, J. Davenport, M. K. S. Lilley, R. P. Wilson & G. C. Hays, 2007. Stranding events provide indirect insights into the seasonality and persistence of jellyfish medusae (Cnidaria: Scyphozoa). Hydrobiologia 589: 1–13.

Hurley, D. E. & D. G. McKnight, 1959. Occurrence of *Pyrosoma* on the sea-floor 160 metres deep. Nature 183: 554–555.

Iguchi, N., R. Ishikawa, O. Sato, T. Onishi, & T. Maeda, 2006. Decomposition rate of the giant jellyfish *Nemopilema nomurai* in Sado Island. FRA, Japan Sea National Fisheries Research Institute report. http://jsnfri.fra.affrc.go.jp/Kurage/kurage_hp18/Sado_bunkai.pdf.

IOC, IHO, and BODC, 2003. Centenary Edition of the GEBCO Digital Atlas, published on CD-ROM on behalf of the Intergovernmental Oceanographic Commission and the

International Hydrographic Organization as part of the General Bathymetric Chart of the Oceans. British Oceanographic Data Centre, Liverpool.
Jackson, J. B. C., 2008. Ecological extinction and evolution in the brave new ocean. Proceedings of the National Academy of Sciences of the United States of America 105: 11458–11465.
Jahnke, R. A., 1996. The global ocean flux of particulate organic carbon, areal distribution and magnitude. Global Biogeochemical Cycles 10: 71–88.
Jiang, S., T. D. Dickey, D. K. Steinberg & L. P. Madin, 2007. Temporal variability of zooplankton biomass from ADCP backscatter time series data at the Bermuda Testbed Mooring site. Deep Sea Research I 54: 608–636.
Johns, B., P. Marsaleix, C. Estournel & R. Vehil, 1992. On the wind-driven coastal upwelling in the Gulf of Lions. Journal of Marine Systems 3: 309–320.
Jones, D. O. B., 2009. Using existing industrial remotely operated vehicles for deep-sea science. Zoologica Scripta 38: 41–47.
Jones, D. O. B., C. N. Roterman & A. R. Gates, 2010. Shell Gro SERPENT final report. National Oceanography Centre Southampton, Southampton. National Oceanography Centre Southampton Research and Consultancy Report: 79.
Jumars, P. A., 1976. Deep-sea species diversity. Does it have a characteristic scale? Journal of Marine Research 34: 217–246.
Karpen, V., T. Viergutz & L. Thomsen, 2007. Baltic Observatory for Oceanographic Monitoring. Sea Technology, July 2007, 18–20.
Kaufmann, R. S. & K. L. Smith, 1997. Activity patterns of mobile epibenthic megafauna at an abyssal site in the eastern North Pacific: results from a 17-month time-lapse photographic study. Deep Sea Research I 44: 559–579.
Kemp, W. M., E. M. Smith, M. Marvin-DiPasquale & W. R. Boynton, 1997. Organic carbon balance and net ecosystem metabolism in Chesapeake Bay. Marine Ecology Progress Series 150: 229–248.
Key, R. M., A. Kozyr, C. L. Sabine, K. Lee, R. Wanninkhof, J. Bullister, R. A. Feely, F. Millero, C. Mordy & T. H. Peng, 2004. A global ocean carbon climatology, results from GLODAP. Global Biogeochemical Cycles 18: GB4031.
Lampitt, R. S., B. J. Bett, K. Kiriakoulakis, E. E. Popova, O. Ragueneau, A. Vangriesheim & G. A. Wolff, 2001. Material supply to the deep-seafloor in the Northeast Atlantic. Progress in Oceanography 50: 27–63.
Lampitt, R. S., B. Boorman, L. Brown, M. Lucas, I. Salter, R. Sanders, K. Saw, S. Seeyave, S. J. Thomalla & R. Turnewitsch, 2008. Particle export from the euphotic zone: estimates using a novel drifting sediment trap, ^{234}Th and new production. Deep Sea Research I 55: 150–1484.
Larson, R. J., 1986. Water content, organic content, and carbon and nitrogen composition of medusae from the northeast Pacific. Journal of Experimental Marine Biology and Ecology 99: 107–120.
Laws, E. A., 2004. Export flux and stability as regulators of community composition in pelagic marine biological communities: implications for regime shifts. Progress in Oceanography 60: 343–354.
Lebrato, M. & D. O. B. Jones, 2009. Mass deposition event of *Pyrosoma atlanticum* carcasses off Ivory Coast (West Africa). Limnology and Oceanography 54: 1197–1209.
Lebrato, M., M. Pahlow, A. Oschlies, K. A. Pitt, D. O. B. Jones, J. C. Molinero & R. H. Condon, 2011. Depth attenuation of organic matter export associated with jelly falls. Limnology & Oceanography 56:1917–1928.
Levin, L. A., R. J. Etter, M. A. Rex, A. J. Gooday, C. R. Smith, J. Pineda, C. T. Stuart, R. R. Hessler & D. Pawson, 2001. Environmental influences on regional deep-sea species diversity. Annual Review of Ecology Evolution and Systematics 32: 51–93.
Loeb, V., V. Siegel, O. Holm-Hansen, R. Hewitt, W. Fraser, W. Trivelpiece & S. Trivelpiece, 1997. Effects of sea-ice extent and krill or salp dominance on the Antarctic food web. Nature 387: 897–900.
Lombard, F. & T. P. Kiorbe, 2010. Marine snow originating from appendicularian houses: age-dependent settling characteristics. Deep Sea Research I 57: 1304–1313.
Marchant, M., D. Hebbeln & G. Wefer, 1999. High resolution planktic foraminiferal record of the last 13:300 years from the upwelling area off Chile. Marine Geology 61: 115–128.
Martin, J. & J.-C. Miquel, 2010. High downward flux of mucilaginous aggregates in the Ligurian Sea during summer 2002. Similarities with the mucilage phenomenon in the Adriatic Sea. Marine Ecology 31: 393–406.
McClain, C. R., M. A. Rex & R. Jabbour, 2005. Deconstructing bathymetric body size patterns in deep-sea gastropods. Marine Ecology Progress Series 297: 181–187.
Merckelbach, L. M. & H. Ridderinkhof, 2006. Estimating suspended sediment concentration from ADCP backscatterance at a site with strong tidal currents. Ocean Dynamics 56: 153–168.
Middelburg, J. J., C. Barranguet, H. T. S. Boschker, P. M. J. Herman, T. Moens & C. H. R. Heip, 2000. The fate of intertidal microphytobenthos carbon: an in situ ^{13}C-labeling study. Limnology Oceanography 45: 1224–1234.
Mills, C. E., 1981. Diversity of swimming behaviours in hydromedusae as related to feeding and utilization of space. Marine Biology 64: 185–189.
Mills, C. E., 1993. Natural mortality in NE Pacific coastal hydromedusae: grazing predation, wound healing and senescence. Bulletin of Marine Science 53: 194–203.
Mills, C. E., 2001. Jellyfish blooms: are populations increasing globally in response to changing ocean conditions? Hydrobiologia 451: 55–68.
Miyake, H., D. L. Lindsay, J. C. Hunt & T. Hamatsu, 2002. Scyphomedusa *Aurelia limbata* (Brandt, 1838) found in deep waters off Kushiro, Hokkaido, Northern Japan. Journal of Plankton Biology and Ecology 49: 44–46.
Miyake, H., D. L. Lindsay, M. Kitamura & S. Nishida, 2005. Occurrence of the scyphomedusa *Parumbrosa polylobata* Kishinouye, 1910 in Suruga Bay, Japan. Journal of Plankton Biology and Ecology 52: 58–66.
Molinero, J. C., M. Casini & E. Buecher, 2008. The influence of the Atlantic and regional climate variability on the long-term changes in gelatinous carnivore populations in the northwestern Mediterranean. Limnology and Oceanography 53: 1456–1467.
Monniot, C. & F. Monniot, 1966. A benthic pyrosoma, *Pyrosoma benthica* n. sp. French Academy of Science 263: 368–370.
Moseley, H. N., 1880. Deep-sea dredging and life in te deep sea III. Nature 21: 591–593.

Murty, S. J., B. J. Bett & A. J. Gooday, 2009. Megafaunal responses to strong oxygen gradients on the Pakistan margin of the Arabian Sea. Deep Sea Research I 56: 472–487.

Niggl, W., M. S. Naumann, U. Struck, R. Manasrah & C. Wild, 2010. Organic matter release by the benthic upside-down jellyfish *Cassiopea* sp. fuels pelagic food webs in coral reefs. Journal of Experimental Marine Biology and Ecology 384: 99–106.

Pakhomov, E. A., V. Fuentes, I. Schloss, A. Atencio & G. B. Snal, 2003. Beaching of the tunicate *Salpa thompsoni* at high levels of suspended particulate matter in the Southern Ocean. Polar Biology 26: 427–431.

Perissinotto, R. & E. A. Pakhomov, 1998. Contribution of salps to carbon flux of marginal ice zone of the Lazarev Sea, Southern Ocean. Marine Biology 131: 25–32.

Pitt, K. A., D. T. Welsh & R. H. Condon, 2009. Influence of jellyfish blooms on carbon, nitrogen, and phosphorus cycling and plankton production. Hydrobiologia 616: 133–149.

Purcell, J. E., 2012. Jellyfish and ctenophore blooms coincide with human proliferations and environmental perturbations. Annual Review of Marine Science 4: 209–235.

Purcell, J. E., W. M. Graham & H. Dumont, 2001. Jellyfish Blooms: Ecological and Societal Importance. Developments in Hydrobiology 155 (Hydrobiologia 451). Kluwer Academic, Dordrecht: 1–333.

Purcell, J. E., S.-I. Uye & W.-T. Lo, 2007. Anthropogenic causes if jellyfish blooms and their direct consequences for humans: a review. Marine Ecology Progress Series 350: 153–174.

Ramirez-Llodra, E., W. D. K. Reid & D. S. M. Billett, 2005. Long-term changes in reproductive patterns of the holothurian *Oneirophanta mutabilis* from the Porcupine Abyssal Plain. Marine Biology 146: 683–693.

Relini, G., 2000. Demersal trawl surveys in Italian seas: a short review. In Bertrand, J. A. & G. Relini (eds), Demersal Resources in the Mediterranean. Proceedings of the Symposium held in Pisa, 18–21 March 1998. Actes de Colloques 26, 76–93.

Reusch, T. B. H., S. Bolte, M. Sparwel, A. G. Moss & J. Javidpour, 2010. Microsatellites reveal origin and genetic diversity of Eurasian invasions by one of the world's most notorious marine invader, *Mnemiopsis leidyi* (Ctenophora). Molecular Ecology 19: 2690–2699.

Richardson, A. J., A. Bakun, G. C. Hays & M. J. Gibbons, 2009. The jellyfish joyride, causes, consequences and management responses to a more gelatinous future. Trends in Ecology & Evolution 24: 312–322.

Robison, B. H., K. R. Reisenbichler & R. E. Sherlock, 2005. Giant larvacean houses, rapid carbon transport to the deep-sea floor. Science 308: 1609–1611.

Roe, H. S. J., D. S. M. Billett & R. S. Lampitt, 1990. Benthic/midwater interactions on the Madeira Abyssal Plain; evidence for biological transport pathways. Progress in Oceanography 24: 127–140.

Rowe, G. T. & N. Staresinic, 1979. Source of organic matter to the deep-sea benthos. Ambio Special Report 6: 19–23.

Ruhl, H. A., 2007. Abundance and size distribution dynamics of abyssal epibenthic megafauna in the northeast Pacific. Ecology 88: 1250–1262.

Ruhl, H. A., 2008. Community change in the variable resource habitat of the abyssal northeast Pacific. Ecology 89: 991–1000.

Ruhl, H. A. & K. L. Smith, 2004. Shifts in deep-sea community structure linked to climate and food supply. Science 305: 513–515.

Ruhl, H. A., J. A. Ellena & K. L. Smith, 2008. Connections between climate, food limitation, and carbon cycling in abyssal sediment communities. Proceedings of the National Academy of Sciences of the United States of America 105: 17006–17011.

Sanchez, P., M. Demestre & P. Martin, 2003. Characterisation of the discards generated by bottom trawling in the northwestern Mediterranean. Fisheries Research 67: 71–80.

Sartor, P., M. Sbrana & B. Reale, 2003. Impact of the deep sea trawl fishery on demersal communities of the northern Tyrrhenian Sea (Western Mediterranean). Journal of the Northwest Atlantic Fishery Science 31: 275–284.

Sempere, R., S. C. Yoro, F. van Wambeke & B. Charriere, 2000. Microbial decomposition of large organic particles in the northwestern Mediterranean Sea: an experimental approach. Marine Ecology Progress Series 198: 61–72.

Send, U., R. Weller, D. Wallace, F. Chavez, R. S. Lampitt, T. Dickey, M. Honda, K. Nittis, R. Lukas, M. McPhaden & R. A. Feely, 2010. OceanSITES. OceanObs'09, Venice, September 21–25.

Sexton, M. A., R. R. Hood, J. Sarkodee-adoo & A. M. Liss, 2010. Response of *Chrysaora quinquecirrha* medusae to low temperature. Hydrobiologia 645: 125–133.

Seymour, J. E., T. J. Carrette & P. A. Sutherland, 2004. Do box jellyfish sleep at night? Medical Journal of Australia 181: 706.

Smith, C. R. & A. M. Baco, 2003. Ecology of whale falls at the deep-sea floor. Oceanography and Marine Biology Annual Review 41: 311–354.

Smith, C. R., F. C. De Leo, A. F. Bernardino, A. K. Sweetman & P. Martinez-Arbizu, 2008. Abyssal food limitation, ecosystem structure and climate change. Trends in Ecology & Evolution 23: 518–528.

Soltwedel, T., K. von Juterzenka, K. Premke & M. Klages, 2003. What a lucky shot! Photographic evidence for a medium-sized natural food-fall at the deep seafloor. Oceanologica Acta 26: 5–6.

Soltwedel, T., E. Bauerfeind, M. Bergmann, N. Budaeva, E. Hoste, N. Jaeckisch, K. von Juterzenka, J. Matthiessen, V. Mokievsky, E.-V. Nothig, N.-V. Queric, B. Sablotny, E. Sauter, M. Schewe, B. Urban-Malinga, J. Wegner, M. Wlodarska-Kowalczuk & M. Klages, 2005. HAUSGARTEN: multidisciplinary investigations at a deep-sea, long-term observatory in the Arctic Ocean. Oceanography 18: 46–61.

Sweetman, A. K. & A. Chapman, 2011. First observations of jelly-falls at the seafloor in a deep-sea fjord. Deep Sea Research I 58: 1206–1211.

Sweetman, A. K. & U. Witte, 2008a. Response of an abyssal macrofaunal community to a phytodetrital pulse. Marine Ecology Progress Series 355: 73–84.

Sweetman, A. K. & U. Witte, 2008b. Macrofaunal response to phytodetritus in a bathyal Norwegian fjord. Deep Sea Research I 55: 1503–1514.

Takahashi, K., H. Saito, S. Kakehi & K. Hidaka, 2010. Sapphirinid copepods as a predator of doliolids, their contribution to doliolids mortality and sinking flux. EOS Transactions AGU, 91, 26, Ocean Science Meeting. Abstract BO13A-08.

Tinta, T., A. Malej, M. Kos & V. Turk, 2010. Degradation of the Adriatic medusa Aurelia sp. by ambient bacteria. Hydrobiologia 1: 179–191.

Titelman, J., L. Riemann, T. A. Sornes, T. Nilsen, P. Griekspoor & U. Bamstedt, 2006. Turnover of dead jellyfish, stimulation and retardation of microbial activity. Marine Ecology Progress Series 325: 43–58.

Turner, R. D., 1973. Wood-boring bivalves, opportunistic species in the deep sea. Science 180: 1377–1379.

Turner, J. T., 2002. Zooplankton faecal pellets, marine snow and sinking phytoplankton blooms. Aquatic Microbiology and Ecology 27: 57–102.

Tyler, P. A., 1988. Seasonality in the deep sea. Oceanography and Marine Biology Annual. Review 18: 125–153.

Vardaro, M. F., H. A. Ruhl & K. L. Smith, 2009. Climate variation, carbon flux, and bioturbation in the abyssal North Pacific. Limnology and Oceanography 54: 2081–2088.

Vetter, E. W. & P. K. Dayton, 1998. Macrofaunal communities within and adjacent to a detritus-rich submarine canyon system. Deep Sea Research II 45: 25–54.

Vetter, E. W. & P. K. Dayton, 1999. Organic enrichment by macrophyte detritus, and abundance patterns of megafaunal populations in submarine canyons. Marine Ecology Progress Series 186: 137–148.

Wefer, G. & G. Fischer, 1993. Seasonal patterns of vertical particle flux in equatorial and coastal upwelling areas of the eastern Atlantic. Deep Sea Research I 40: 1613–1645.

West, E. J., K. A. Pitt, D. T. Welsh, K. Koop & D. Rissik, 2009a. Top-down and bottom-up influences of jellyfish on primary productivity and planktonic assemblages. Limnology and Oceanography 54: 2058–2071.

West, E. J., D. T. Welsh & K. A. Pitt, 2009b. Influence of decomposing jellyfish on sediment oxygen demand and nutrient dynamics. Hydrobiologia 616: 151–160.

Wiebe, P. H., L. P. Madin, L. R. Haury, G. R. Harbison & L. M. Philbin, 1979. Diel vertical migration by *Salpa aspera* and its potential for large-scale particulate organic matter transport to the deep-sea. Marine Biology 53: 249–255.

Wigham, B. D., P. A. Tyler & D. S. M. Billett, 2003. Reproductive biology of the abyssal holothurian *Amperima rosea*: an opportunistic response to variable flux of surface derived organic matter? Journal of the Marine Biology Association of the United Kingdom 83: 175–188.

Wilkins, M. E., M. Zimmernmann & K. L. Weinberg, 1998. The 1995 Pacific West Coast bottom trawl survey of groundfish resources: estimates of distribution, abundance, and length and age composition. NOAA technical Memorandum NMFS-AFSC-89.

Witte, U., N. Aberte, M. Sand & F. Wenzhofer, 2003. Rapid response of a deep-sea benthic community to POM enrichment, an in situ experimental study. Marine Ecology Progress Series 251: 27–36.

Wood, H. L., J. I. Spicer & S. Widdicombe, 2008. Ocean acidification may increase calcification rates, but at a cost. Proceedings of the Royal Society B 275: 1767–1773.

Yamamoto, J., M. Hirose, T. Ohtani, K. Sugimoto, K. Hirase, N. Shimamoto, T. Shimura, N. Honda, Y. Fujimori & T. Mukai, 2008. Transportation of organic matter to the sea floor by carrion falls of the giant jellyfish *Nemopilema nomurai* in the Sea of Japan. Marine Biology 153: 311–317.

Young, C. M., 2003. Reproduction, development and life-history traits. In Tyler, P. A. (ed.), Ecosystems of the World. Elsevier Science, Amsterdam: 381–426.

Zeldis, J., C. S. Davis, M. R. James, S. L. Ballara, W. E. Booth & F. H. Chang, 1995. Salp grazing: effects on phytoplankton abundance, vertical distribution and taxonomic composition in a coastal habitat. Marine Ecology Progress Series 126: 267–283.

Hydrobiologia (2012) 690:247–256
DOI 10.1007/s10750-012-1042-z

Pulse perturbations from bacterial decomposition of *Chrysaora quinquecirrha* (Scyphozoa: Pelagiidae)

Jessica R. Frost · Charles A. Jacoby · Thomas K. Frazer · Andrew R. Zimmerman

Published online: 6 March 2012
© Springer Science+Business Media B.V. 2012

Abstract Bacteria decomposed damaged and moribund *Chrysaora quinquecirrha* Desor, 1848 releasing a pulse of carbon and nutrients. Tissue decomposed in 5–8 days, with 14 g of wet biomass exhibiting a half-life of 3 days at 22°C, which is 3× longer than previous reports. Decomposition raised mean concentrations of organic carbon and nutrients above controls by 1–2 orders of magnitude. An increase in nitrogen (16,117 μg l^{-1}) occurred 24 h after increases in phosphorus (1,365 μg l^{-1}) and organic carbon (25 mg l^{-1}). Cocci dominated control incubations, with no significant increase in numbers. In incubations of tissue, bacilli increased exponentially after 6 h to become dominant, and cocci reproduced at a rate that was 30% slower. These results, and those from previous studies, suggested that natural assemblages may include bacteria that decompose medusae, as well as bacteria that benefit from the subsequent release of carbon and nutrients. This experiment also indicated that proteins and other nitrogenous compounds are less labile in damaged medusae than in dead or homogenized individuals. Overall, dense patches of decomposing medusae represent an important, but poorly documented, component of the trophic shunt that diverts carbon and nutrients incorporated by gelatinous zooplankton into microbial trophic webs.

Keywords Jellyfish · Scyphomedusae · Bacterial decomposition · Carbon · Nitrogen · Phosphorus

Guest editors: J. E. Purcell, H. Mianzan & J. R. Frost / Jellyfish Blooms: Interactions with Humans and Fisheries

J. R. Frost (✉)
Institute for Hydrobiology and Fisheries Science, University of Hamburg, Olbersweg 24, 22767 Hamburg, Germany
e-mail: frost.jessica.r@gmail.com

Present Address:
J. R. Frost
Fisheries and Aquatic Science Program, School of Forest Resources and Conservation, University of Florida, 7922 NW 71st Street, Gainesville, FL 32653, USA

C. A. Jacoby
Department of Soil and Water Science, University of Florida, 7922 NW 71st Street, Gainesville, FL 32653, USA
e-mail: cajacoby@ufl.edu

T. K. Frazer
Fisheries and Aquatic Science Program, School of Forest Resources and Conservation, University of Florida, 7922 NW 71st Street, Gainesville, FL 32653, USA
e-mail: frazer@ufl.edu

A. R. Zimmerman
Department of Geological Sciences, University of Florida, 241 Williamson Hall, P.O. Box 112120, Gainesville, FL 32611, USA
e-mail: azimmer@ufl.edu

Introduction

Medusae raise interesting questions for ecologists seeking to understand carbon and nutrient cycles in marine systems. Individual medusae do not sequester large quantities of carbon and macronutrients because their bodies typically comprise 97% water and 3% organic matter consisting of 72 ± 14% protein (mean ± standard deviation), 22 ± 12% lipids, and 7 ± 5% carbohydrates (Larson, 1986; Schneider, 1988; Arai et al., 1989; Clarke et al., 1992; Lucas, 1994, 2009; Doyle et al., 2007; Pitt et al., 2009). Nevertheless, dense aggregations and blooms of medusae have been recorded (Graham et al., 2001; Mills, 2001; Richardson et al., 2009), and in such numbers, medusae irrefutably perturb the flow of energy and cycles of elements in an ecosystem.

Mass occurrences of medusae generally last for weeks to months (Mills, 2001; Sexton et al., 2010), and during this time, they assimilate and release carbon and nutrients creating a relatively protracted, press perturbation (sensu Glasby & Underwood, 1996). For example, medusae can consume significant numbers of zooplankton and larval fish and assimilate up to 88% of the carbon in their prey (Mills, 1995; Arai, 1997; Purcell, 1997; Purcell & Arai, 2001; Pitt et al., 2009). In turn, medusae excrete and secrete carbon, nitrogen, and phosphorus, primarily as mucus, ammonium, and phosphate (Pitt et al., 2009; Condon et al., 2011). Excretion rates vary among species, as well as with temperature and feeding history, but releases of 1,114 μmol g dry weight (DW)$^{-1}$ d^{-1} of dissolved organic carbon, 187 μmol g DW^{-1} d^{-1} of ammonium, 137 μmol g DW^{-1} d^{-1} of dissolved organic nitrogen, 36 μmol g DW^{-1} d^{-1} of phosphate, and 12 μmol g DW^{-1} d^{-1} of dissolved organic phosphorus have been recorded (Morand et al., 1987; Malej, 1989, 1991; Schneider, 1989; Nemazie et al., 1993; Hansson & Norrman, 1995; Shimauchi & Uye, 2007; Pitt et al., 2009; Condon et al., 2010, 2011). Rapidly accumulating evidence suggests that when medusae reach sufficient densities, these releases create a press response (sensu Glasby & Underwood, 1996) by supporting phytoplanktonic and bacterial production, although the magnitude of their influence depends heavily on the availability of carbon and nutrients from other sources and rates of flushing (Pitt et al., 2005; Malej et al., 2007; Condon et al., 2010, 2011). Yet, what happens when the medusae in an aggregation or bloom are damaged or die?

Damaged, moribund, or dead medusae should begin to decompose as they sink or drift, in part because the continuous release of organic matter while they were alive insures they are surrounded by a thriving bacterial assemblage (Doores & Cook, 1976; Heeger et al., 1992; Hansson & Norrman, 1995; Riemann et al., 2006). In some cases, decomposition is not completed in the water column, and carcasses of medusae carry hundreds of grams of carbon to the seafloor (Miyake et al., 2002, 2005; Billett et al., 2006; Koppelmann & Frost, 2008; Yamamoto et al., 2008; Murty et al., 2009). Observations of a carrion fall of *Pyrosoma atlanticum* Peron, 1804 (Lebrato & Jones, 2009) suggest that benthic scavengers will feed on accumulations of dead and moribund medusae. In addition, one set of field observations in deep water and one set of mesocosm experiments in shallow water indicate that medusae will decompose (Billett et al., 2006; West et al., 2009). Over a period of days, bacterial decomposition and respiration created pulse perturbations (sensu Glasby & Underwood, 1996) consisting of increased nutrients and reduced oxygen concentrations as evidenced by the production of hydrogen sulfide (Billett et al., 2006; West et al., 2009). Depending on the system's response, these two perturbations could yield discrete or protracted responses (sensu Glasby & Underwood, 1996). For example, nutrients could stimulate a discrete increase in primary productivity if there is sufficient light, and hypoxia/anoxia could create a protracted decrease in the abundance of infauna that persists through multiple cycles of recruitment. Thus, an understanding of impacts from carrion falls of medusae requires an understanding of their decomposition.

The three previous studies that examined decomposition of Scyphomedusae used suffocated, frozen or homogenized medusae, without controls for the effects of these treatments (Titelman et al., 2006; West et al., 2009; Tinta et al., 2010). Nevertheless, all three studies indicate that the carcasses of scyphomedusae contain few refractory, structural compounds, so they decay readily as bacteria digest their tissue, thereby releasing carbon, nitrogen, and phosphorus (Titelman et al., 2006; West et al., 2009; Tinta et al., 2010). Bacteria typically increased in abundance by one or more orders of magnitude over a period of days in incubations of suffocated or homogenized medusae (Titelman et al., 2006; Tinta et al., 2010). Measurements of half-lives for carcasses depended on ambient temperatures and ranged from 0.6 to 1.0 d, but these measurements potentially included unquantified loss of tissue and consumption by scavengers (Titelman et al., 2006). In all cases, concentrations of carbon, nitrogen, and phosphorus in incubations of suffocated, frozen, or homogenized Scyphomedusae were significantly higher than those measured in controls (Titelman et al., 2006; West et al., 2009; Tinta et al., 2010).

This study tests hypotheses arising from the results of the three previous investigations. It employs a laboratory experiment to test the overall hypothesis that a pulse perturbation will be created by decomposing *Chrysaora quinquecirrha* Desor, 1848, a scyphomedusa with a circumglobal distribution and tendency to form aggregations and blooms (Graham, 2001; Graham et al., 2001; Mills, 2001; Purcell, 2005; Purcell & Decker, 2005; Hamner & Dawson, 2009; Sexton et al., 2010). In particular, the experiment addresses three related sub-hypotheses: (i) the half-life of tissue from damaged medusae will be longer than 24 h, (ii) the abundance and composition of bacterial assemblages will differ between incubations with and without tissue from damaged medusae, and (iii) concentrations of total organic carbon, total nitrogen, and total phosphorus will increase in incubations with tissue.

Materials and methods

Collection and handling

Individual Scyphomedusae, *Chrysaora quinquecirrha*, and ambient seawater were collected in 2–3 m of water near the mouth of the Steinhatchee River along Florida's west coast (N 29.6°; W 83.4°). Snorkelers allowed each medusa to swim into a stationary, hand-held, 3-l polypropylene beaker (Kartell) that previously had been washed with a 10% hydrochloric acid solution to minimize contamination from beyond the sampling site. Onboard a boat, bell diameters and any visible signs of damage were recorded before each individual was placed in a new, labeled, 12-l plastic bag containing ambient seawater (Table 1). Each bag with its single medusa, was placed in a covered, 190-l bin containing ambient seawater. To alleviate heat stress, the water in the bin was replaced every 2–3 h. A total of 13 medusae was collected. At the same time, ambient seawater for use in experimental and control incubations was collected in an acid-washed 75-l carboy.

Table 1 Results of analysis of variance for $\log_{10}$-transformed concentrations of various elements and ratios of total organic carbon to total nitrogen during decomposition of quarters of *Chrysaora quinquecirrha* medusae

Parameter	Anderson–Darling *P*	Cochran's *P*	Source	*df*	SS	MS	*F*	*P*
Log_{10}(TOC)	<0.01	<0.01	Tr	1	7.538	7.538	126.82	<0.001
			Dur	8	1.698	0.212	3.57	0.002
			Tr * Dur	8	1.052	0.131	2.21	0.041
			Error	54	3.209	0.059		
Log_{10}(TN)	>0.01	>0.05	Tr	1	37.375	37.375	1989.21	<0.001
			Dur	8	2.120	0.265	14.10	<0.001
			Tr * Dur	8	2.671	0.334	17.77	<0.001
			Error	54	1.015	0.019		
Log_{10}(TP)	>0.05	>0.05	Tr	1	70.307	70.307	1451.52	<0.001
			Dur	8	2.574	0.322	6.64	<0.001
			Tr * Dur	8	1.968	0.246	5.08	<0.001
			Error	54	2.616	0.048		
Log_{10}(TOC:TN)	<0.01	<0.01	Tr	1	11.344	11.344	317.60	<0.001
			Dur	8	0.663	0.083	2.32	0.032
			Tr * Dur	8	0.854	0.107	2.99	0.008
			Error	54	1.929	0.036		

TOC total organic carbon (mg l^{-1}); *TN* total nitrogen (μg l^{-1}); *TP* total phosphorus (μg l^{-1}); *TOC:TN* ratio of total organic carbon to total nitrogen; *Tr* treatment, either with or without medusae; *Dur* duration of incubation

Experimental design

In the laboratory, medusae and water were held for less than 10 h in a climate-controlled room at 22°C (similar to conditions at the collection site) under a 12:12 light:dark cycle. Before processing, individual *Chrysaora quinquecirrha* were examined to ensure they were pulsing, swimming, and free from trapped air bubbles or visible damage to their bells in an effort to establish an unbiased starting point for decomposition. Three of the 13 medusae were classified as unhealthy, and they were held as whole specimens during the experiment. The remaining 10 healthy medusae were treated to simulate damage.

Individual medusae were removed from their plastic bags with acid-washed forceps, placed on a tared foil pan, blotted gently with a paper towel to remove excess water, and weighed to the nearest 0.1 g using an analytical balance. The balance and tared foil pan were cleaned prior to each measurement. With gloved hands, a scalpel was used to section each healthy individual into four, nearly equal pieces that were weighed separately. The 40 quarters created from these 10 medusae varied in weight from 9.6 to 19.6 g because all medusae had oral arms that differed in length and bell diameters that differed by 1–3 cm. According to previously determined random numbers, four replicate quarters were allocated to each of 10 durations, i.e., 0, 2, 6, 10, 24, 48, 72, 96, 120, and 200 h. In addition, previously selected random numbers were used to allocate four replicate controls to each duration.

Individual quarters and each of the three whole *Chrysaora quinquecirrha* medusae were incubated in an acid-washed, 1-l Nalgene plastic bottle that was loosely capped and contained 200 ml of ambient seawater drawn from the well-mixed carboy and 800 ml of air-filled headspace. The controls consisted of 200 ml of ambient seawater in similar, acid-washed, 1-l Nalgene plastic bottles. The water in experimental replicates and controls was devoid of visible zooplankters. Bottles were incubated at 22°C under a 12:12 light:dark cycle.

Sampling and analysis

The progress of decomposition was tracked by recording the state of tissue samples and associated water at least once each day. At each of the selected durations, any remaining tissue in each experimental replicate was removed and weighed as described above.

w?>Water for counts of bacteria was removed from the appropriate control and experimental bottles, preserved with buffered formalin (final concentrations 2%), and stored at −4°C until analysis. Bacteria in 1-ml aliquots were stained with acridine orange and filtered onto 0.22-mm, black, polycarbonate, membrane filters (Osmonics). Subsequently, filters were mounted onto microscope slides and bacteria enumerated at 1,000× magnification using immersion oil and a Nikon Labophot epifluorescence microscope. Numbers of bacteria per milliliter were estimated from counts of morphotypes in haphazardly chosen sets of five grids, until at least 100 specimens of a single morphotype were counted. Raw counts were scaled to account for dilution, number of grids examined, and the area of each grid.

Samples for carbon and nutrient analyses were preserved with 2 N sulfuric acid to pH <1 and stored at −4°C. Concentrations of total organic carbon (TOC), total nitrogen (TN), and total phosphorus (TP) were measured in aliquots of the appropriate water samples; if the initial value was beyond the relevant detection range, a second aliquot was analyzed after being diluted with pre-filtered, distilled water. Samples for analysis of TOC concentrations (mg l^{-1}) were sparged with carbon dioxide free air for 2 min to remove inorganic carbon prior to high temperature catalytic oxidation using a Shimadzu TOC-5000 analyzer with infra-red carbon dioxide detection. Each sample was analyzed twice, and each analytical run comprised 3–5, 60 μl injections, with injections ceasing when the coefficients of variation among replicates were <5%. Potassium hydrogen phthalate was used as a standard. Samples yielding total nitrogen concentrations (μg l^{-1}) were oxidized with persulfate, and the resulting nitrate was measured with second derivative spectroscopy (Bachmann & Canfield, 1996). Concentrations of TP (μg l^{-1}) were determined using an acidified solution of ammonium molybdate and antimony following a persulfate digestion (Murphy & Riley, 1962; Menzel & Corwin, 1965).

Statistical analyses

An exponential decay model was fitted to mean changes in wet weights of tissue over time. A half-life was calculated using the resulting decay coefficient. The relationship between degradation rates and initial wet weights of quarters was evaluated by correlating proportional losses of wet weight with initial wet weights.

Linear regressions were used to compare temporal changes in counts of each bacterial morphotype between control and experimental treatments. Residuals were tested for normality with Anderson–Darling tests and equality of variance with Cochran's tests. Data were $\log_{10}$-transformed to improve normality and homoscedasticity. For each bacterial morphotype, an exponential growth curve was fitted to back-transformed mean counts and a doubling time calculated.

Differences in concentrations of TOC, TN, and TP, as well as ratios of TOC to TN, were tested with ANOVAs. Data were $\log_{10}$-transformed to improve normality and homoscedasticity, which were evaluated as described above. In ANOVAs, treatment (either control or experimental) and duration of incubation were treated as fixed factors. Data are presented as mean ± standard error.

Results

Evidence of decomposition began with a smell of rotting tissue that was noted in some incubations at 48 h, but one quarter was still pulsing at 72 h, which highlighted variation in responses to sectioning. The release of hydrogen sulfide was noted at 120 h in some incubations, which indicated the onset of anaerobic decomposition. Through time, wet weights decreased, with no tissue remaining in any sample at 200 h. An exponential decay curve fit to mean weight losses for 0–120 h yielded a half-life of 3 days or 72 h (wet weight = $13.3 \times e^{-0.229 \times \text{days}}$; $r^2 = 0.883$). Proportional rates of tissue loss were not correlated with initial wet weights ($r = 0.272$, $P = 0.089$). In addition, three whole animals weighing 39.6–47.0 g were 92–100% decomposed by 120–200 h.

Samples from four controls spanning 120 h of incubation contained coccoid bacteria ($2.7 \times 10^5 \pm 4.8 \times 10^4$ cells ml^{-1}). Numbers of cocci did not vary significantly throughout the incubation period according to a linear regression (Fig. 1A; $F_{1,2} = 1.67$, $P = 0.326$, $r^2 = 0.45$).

The bacterial assemblage in experimental replicates containing quarters of *Chrysaora quinquecirrha* medusae diverged from that in controls. In analyses of abundance, one anomalously low value from 6 h was excluded from the regressions to improve normality and homoscedasticity; the degrees of freedom were reduced accordingly. Cocci were the only bacteria observed through the second hour ($5.5 \times 10^5 \pm 5.0 \times 10^4$ cells ml^{-1}), and their numbers increased significantly throughout the incubation period (Fig. 1A; $F_{1,31} = 67.1$, $P < 0.001$, $r^2 = 0.68$). From the sixth hour, bacilli were visible, and their numbers also increased significantly (Fig. 1B; $F_{1,31} = 21.6$, $P < 0.001$, $r^2 = 0.41$). In fact, the slopes of regression lines for cocci (0.02) and for bacilli (0.03) were similar if the initial samples that did not contain bacilli were excluded (Fig. 1E; $F_{1,24} = 58.6$, $P < 0.001$, $r^2 = 0.70$). Near the end of the experiment, bacterial films formed in some replicates with quarters of *C. quinquecirrha*.

Furthermore, both forms of bacteria exhibited exponential growth through 200 h (Fig. 1C, D; r^2 for cocci = 0.83; r^2 for bacilli = 0.93), with bacilli predominating by an order of magnitude (raw counts at 200 h = $1.2 \times 10^8 \pm 8.8 \times 10^7$ bacilli ml^{-1} and $1.5 \times 10^7 \pm 1.2 \times 10^7$ cocci ml^{-1}). As expected given their predominance, bacilli had a shorter doubling time (24.8 h) than cocci (33.6 h).

As indicated by significant interactions in ANOVAs and comparisons of back-transformed means (Table 1; Fig. 2), the introduction and incubation of tissue from *Chrysaora quinquecirrha* medusae led to increased concentrations of TOC, TN, and TP in the seawater. Concentrations of all elements were elevated immediately, and increases in concentrations of TOC, TN, and TP followed different time courses to eventually become 1–2 orders of magnitude higher than the relatively stable concentrations measured in controls (Fig. 2). As tissue decomposed, TOC and TP concentrations exhibited a three-fold increase between 24 and 48 h, whereas, a similar increase in TN occurred 24 h later (Fig. 2).

The time courses followed by ratios of TOC to TN also differed significantly between control and experimental replicates (Table 1), with ratios in controls always being higher (Fig. 2; 16.9 ± 1.1 for control incubations and 3.1 ± 0.5 for experimental incubations). In experimental replicates, TN concentrations lagged TOC concentrations by 24 h, which led to a maximum ratio of 5.3 at 48 h (Fig. 2B–D). In control replicates, the maximum ratio of 31.1 at 72 h was due to a relatively low mean concentration of TN (193 $\mu g\ l^{-1}$ versus mean of all other values = 325 $\mu g\ l^{-1}$) and a

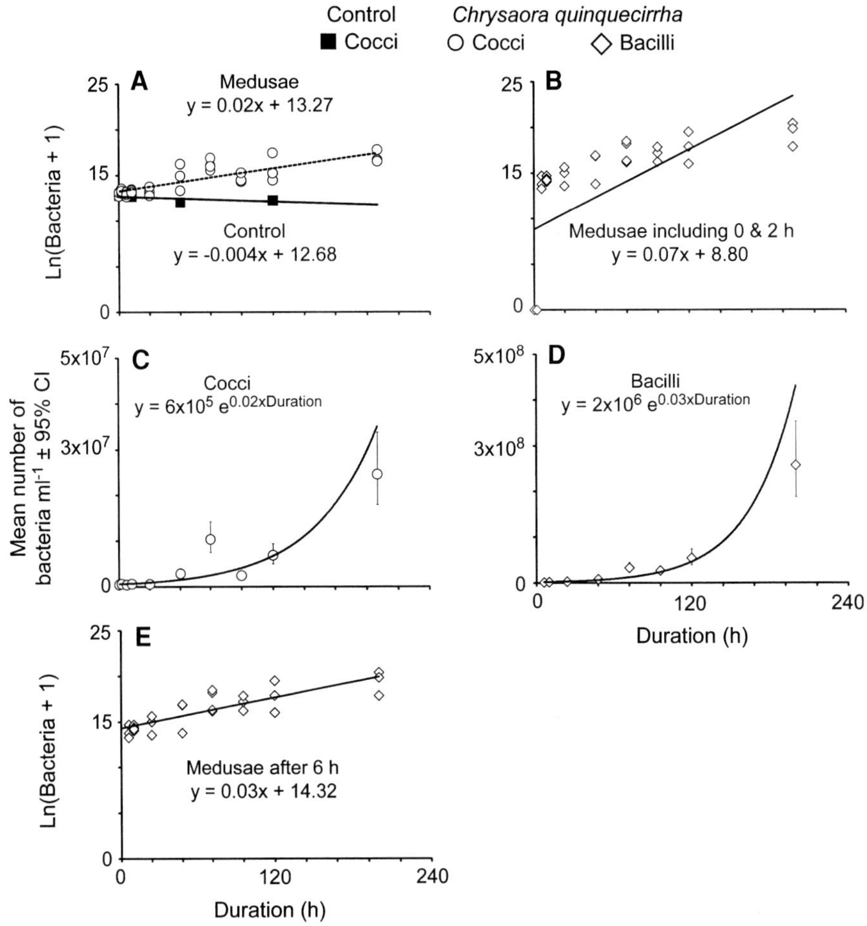

Fig. 1 Linear regressions based on natural log transformed counts of bacteria versus duration of the experiment and exponential growth curves fitted to back-transformed mean counts of bacilli and cocci in incubations of quarters of *Chrysaora quinquecirrha* medusae. **A** regressions for counts of cocci in seawater controls and experimental incubations with medusae; **B** regression for counts of bacilli in experimental incubations; **C** exponential growth curve fit to counts of cocci in experimental incubations; **D** exponential growth curve fit to counts of bacilli in experimental incubations; **E** regression for counts of bacilli in experimental incubations after they become established at 6 h. *CI* confidence interval

slightly higher mean concentration of TOC (6 mg l^{-1} versus mean of all other values = 5 mg l^{-1}).

Mean TOC, TN, and TP concentrations in experimental replicates yielded an elemental ratio of 23:8:1 (TOC:TN:TP by weight) immediately after the introduction of *C. quinquecirrha* tissue. At 96–120 h, a ratio of 14:8:1 was calculated from the relatively stable mean concentrations of TOC, TN, and TP (Fig. 2). In comparison to the Redfield ratio of 40:7:1, nitrogen and phosphorus concentrations remained near the theoretical balance, whereas carbon was limiting throughout the incubations.

Discussion

Quarters of *Chrysaora quinquecirrha* medusae lost weight exponentially and at rates that were statistically similar for pieces of tissue weighing 9.6–19.6 g. As hypothesized, these rates of tissue loss were slower than those reported for whole specimens of *Periphylla periphylla* Péron & Lesueur, 1810 (Titelman et al., 2006). In fact, the rate of tissue loss for quarters of *C. quinquecirrha* was 6.7–11.2× slower after using a Q_{10} of 2.3 to adjust the rates for *P. periphylla* from 10.1 or 12.5 to 22°C (Bidle et al., 2002; Titelman et al., 2006). Further evidence that tissue was lost over 8 days comes from observations that whole, unhealthy *C. quinquecirrha* decomposed in the same period of time as quarters of healthy individuals. In addition, previously frozen, whole *Catostylus mosaicus* Quoy & Gaimard, 1824 decomposed completely in approximately 9 days when resting on the bottom at 30°C (West et al., 2009). Although oxygen concentrations probably remained higher during the in situ incubations of *P. periphylla* in mesh bags, the most likely explanations for the more rapid change in wet weights would be the observed, but unquantified, loss of tissue during sample recovery and consumption by zooplankton that colonized the samples (Titelman et al., 2006).

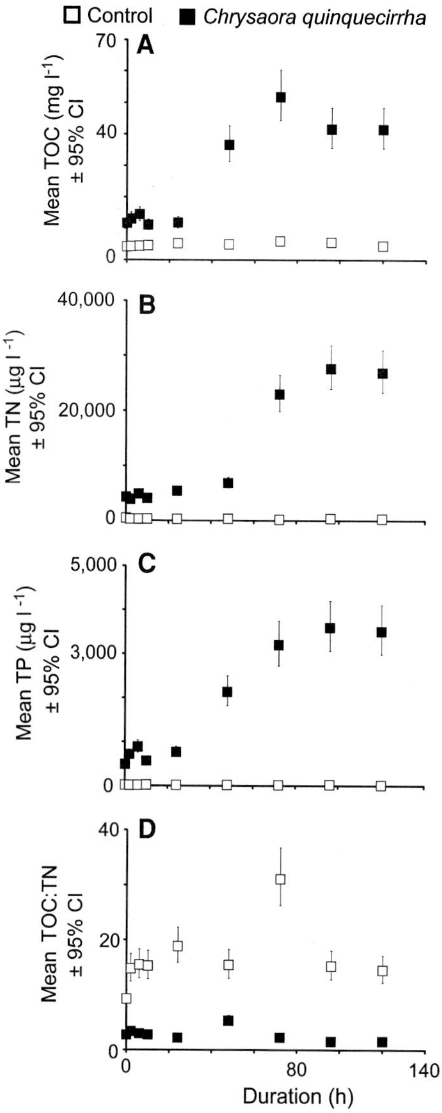

Fig. 2 Back-transformed mean concentrations of **A** total organic carbon (TOC in mg l^{-1}), **B** total nitrogen (TN in µg l^{-1}), and **C** total phosphorus (TP in µg l^{-1}), as well as **D** ratios of total organic carbon to total nitrogen in incubations containing quarters of *Chrysaora quinquecirrha* medusae and seawater controls. *CI* confidence interval

Tissue was not lost during *C. quinquecirrha* incubations, which did not contain visible zooplanktonic scavengers. Identifying the degree of consistency in decomposition rates and separating the consequences of scavenging and decomposition will be important for modeling of carbon and nutrient cycles.

Incubations with tissue from *Chrysaora quinquecirrha* exhibited changes in the abundance of bacteria as hypothesized. As the wet weight of tissue declined, numbers of two bacterial morphotypes increased exponentially, a bacterial film formed in some experimental replicates, and hydrogen sulfide was produced. The presence of a bacterial film resembled observations of "slime" on a massive carrion fall of *Crambionella orsini* Vanhöffen, 1888 in the Arabian Sea (Billett et al., 2006). Evidence of decreased oxygen concentrations in the experiment with *C. quinquecirrha* matched observations of anoxic sediments associated with the *C. orsini* fall (Billett et al., 2006) and other experiments on decomposition (West et al., 2009; Tinta et al., 2010). Overall, the experiment with *C. quinquecirrha* appeared to simulate conditions associated with carrion falls of Scyphomedusae, which makes the data on the release of carbon and nutrients valuable for predicting the magnitude and duration of these pulse perturbations. In fact, the results will have direct application for the Gulf of Mexico and Chesapeake Bay where *C. quinquecirrha* and several congeners are known to form blooms (Graham, 2001; Graham et al., 2001; Mills, 2001; Purcell, 2005; Purcell & Decker, 2005; Hamner & Dawson, 2009; Sexton et al., 2010).

The hypothesis that incubations with tissue will contain a different bacterial assemblage also was supported. Initially, control and experimental samples contained cocci, with bacilli observed in experimental bottles from 6 h onward. Growth coefficients and doubling times indicated that bacilli grew 1.4-times faster than cocci, and bacilli abundances became an order of magnitude greater. Naturally occurring bacteria from two locations in the Adriatic Sea, with temperatures from 10 to 19°C, grew rapidly on homogenates of *Aurelia* sp., with bacterial abundances typically increasing 100-fold in 1–3 days and only one experiment exhibiting a 6-day lag to maximum densities (Tinta et al., 2010). Bacilli also predominated when homogenized *Periphylla periphylla* medusae were incubated with natural bacterial assemblages (Titelman et al., 2006) and on moribund *Chrysaora quinquecirrha* medusae in Chesapeake Bay (Doores & Cook, 1976). In addition, analyses of bacterial DNA indicated that assemblages differed between seawater controls and incubations with homogenized tissue (Titelman et al., 2006; Tinta et al., 2010). Incubations of medusa tissue with naturally occurring microbial assemblages and specific bacterial isolates showed that numbers of some bacteria decreased or remained static, with the strongest inhibitory effect on bacteria attributed to the umbrella of *P. periphylla* (Titelman et al., 2006). Nevertheless, numerous bacteria decompose medusae because nine bacterial isolates increased in numbers over 10 h in other experiments (Titelman et al., 2006). In summary, available data indicate that certain bacteria may be primarily responsible for decomposition of medusae and grow more rapidly than other forms that may benefit from the ensuing release of carbon and nutrients. This interpretation receives further support from field observations of higher abundances of certain bacteria in depth zones where *P. periphylla* were abundant (Riemann et al., 2006) and a laboratory experiment demonstrating that *Brevibacterium* sp. JCM 6894, but not *Escherichia coli* ATCC 9637, decomposed tissue of an unspecified medusa (Mimura & Nagata, 2001). In our experiments with *C. quinquecirrha*, coccoid bacteria may have utilized carbon and nutrients released during decomposition driven by bacilli.

Incubations with tissue of damaged medusae yielded the hypothesized increases in carbon and macronutrients. In fact, the introduction of *Chrysaora quinquecirrha* tissue raised concentrations of TOC, TN, and TP by 3-, 9-, and 35-fold, respectively, within ~1 h, suggesting that damaged medusae leak carbon and nutrients. Concentrations of carbon and nutrients were elevated by approximately 1–2 orders of magnitude after 24–48 h in incubations containing tissue being decomposed by bacteria. A 16,117 $\mu g\ l^{-1}$ increase in TN lagged a 1,365 $\mu g\ l^{-1}$ increase in TP and a 25 $mg\ l^{-1}$ increase in TOC by 24 h, which resulted in a maximum TOC:TN ratio at 72 h. Thus, it appeared that nitrogen-rich proteins were not degraded faster than polysaccharides and other carbon-rich compounds, as previously hypothesized (Titelman et al., 2006). Perhaps, the proteins in pieces of tissue were less labile than those in homogenized tissue (Titelman et al., 2006), and the observation that one quarter continued to pulse for 72 h suggests that resistance to decomposition varied among individual *C. quinquecirrha*. Ultimately, the overall TOC:TN ratio in the water from experimental replicates was 2.7 ± 0.4, which was slightly lower than ratios of 3.4–4.1 previously reported for tissue of other medusae (Larson, 1986; Schneider, 1988; Clarke et al., 1992; Doyle et al., 2007). This discrepancy may have been due to mineralization of carbon during production of bacterial biomass, which would yield inorganic carbon that was not measured in our analyses. In fact, carbon appeared to be the limiting element throughout the incubations, and changes in TOC:TN:TP ratios between 0 h and 96–120 h indicated that it became increasingly limiting. In combination with increases in bacteria, this relatively large decrease compared to changes in concentrations of nitrogen and phosphorus signified a conversion of "pelagic" TOC to "benthic" bacterial biomass.

In combination with previous reports, our data demonstrate that damaged medusae decompose readily, with carbon, nitrogen, and phosphorus released from relatively labile compounds. Certain bacteria appear to drive decomposition, and the resulting dissolved carbon and nutrients support the growth of other bacteria and phytoplankton as shown here and in other studies (Pitt et al., 2005; Riemann et al., 2006; Malej et al., 2007; Shimauchi & Uye, 2007; Pitt et al., 2009; Condon et al., 2010, 2011). Although they decompose rapidly, carcasses of medusae may reach the sea floor as carrion falls that should provide food for benthic scavengers, will create hot spots with elevated concentrations of key elements in the sediment and adjacent water column, and will contribute to low oxygen concentrations in bottom waters and sediments (Billett et al., 2006; Yamamoto et al., 2008; West et al., 2009; this study). In addition, the results of this and previous studies suggest that pulse perturbations generated by carrion falls of Scyphomedusae should be integrated into the conceptual model of a trophic shunt that diverts carbon, nitrogen, and phosphorus through living gelatinous zooplankton and into microbial trophic webs (Condon et al., 2011). Further work determining rates of decomposition in diverse and realistic situations will yield an improved understanding of differences among species; the influence of temperature, water movement, and other environmental conditions; and differences between dead and living yet damaged medusae.

Acknowledgments Special thanks go to E. Phlips for providing access to his laboratory and microscope for bacterial counts. Further appreciation goes to C. Brown and personnel in the water chemistry laboratory of the Fisheries and Aquatic Sciences Program for analyzing nutrient concentrations. J. Jin at the Department of Geological Sciences analyzed carbon concentrations. S. Barry, M. Edwards, A. Krzystan, J. Lockwood, and D. Saindon provided valuable support during this study, and the comments of three anonymous reviewers led to significant improvements in the manuscript. This article is a contribution to EUR-OCEANS Network of Excellence (WP4-SYSMS-1101).

References

Arai, M. N., 1997. A functional biology of Scyphozoa. Chapman and Hall, London: 316 pp.

Arai, M. N., J. A. Ford & J. N. C. Whyte, 1989. Biochemical composition of fed and starved *Aequorea victoria* (Murbach et Shearer, 1902) (Hydromedusa). Journal of Experimental Marine Biology and Ecology 127: 289–299.

Bachmann, R. W. & D. E. Canfield Jr., 1996. Use of an alternative method for monitoring total nitrogen concentrations in Florida lakes. Hydrobiologia 323: 1–8.

Bidle, K. D., M. Manganelli & F. Azam, 2002. Regulation of oceanic silicon and carbon preservation by temperature control on bacteria. Science 298: 1980–1984.

Billett, D. S. M., B. J. Bett, C. L. Jacobs, I. P. Rouse & B. D. Wigham, 2006. Mass deposition of jellyfish in the deep Arabian Sea. Limnology and Oceanography 51: 2077–2083.

Clarke, A., L. J. Holmes & D. J. Gore, 1992. Proximate and elemental composition of gelatinous zooplankton from the Southern Ocean. Journal of Experimental Marine Biology and Ecology 155: 55–68.

Condon, R. H., D. K. Steinberg & D. A. Bronk, 2010. Production of dissolved organic matter and inorganic nutrients by gelatinous zooplankton in the York River estuary, Chesapeake Bay. Journal of Plankton Research 32: 153–170.

Condon, R. H., D. K. Steinberg, P. A. del Giorgio, T. C. Bouvier, D. A. Bronk, W. M. Graham & H. W. Ducklow, 2011. Jellyfish blooms result in a major microbial respiratory sink of carbon in marine systems. Proceedings of the National Academy of Sciences 108: 10225–10230.

Doores, S. & T. M. Cook, 1976. Occurrence of *Vibrio* and other bacteria on the sea nettle, *Chrysaora quinquecirrha*. Microbial Ecology 3: 31–40.

Doyle, T. K., J. D. R. Houghton, R. McDevitt, J. Davenport & G. C. Hays, 2007. The energy density of jellyfish: estimates from bomb-calorimetry and proximate-composition. Journal of Experimental Marine Biology and Ecology 242: 239–252.

Glasby, T. M. & A. J. Underwood, 1996. Sampling to differentiate between pulse and press perturbations. Environmental Monitoring and Assessment 42: 241–252.

Graham, W. M., 2001. Numerical increases and distributional shifts of *Chrysaora quinquecirrha* (Desor) and *Aurelia aurita* (Linné) (Cnidaria: Scyphozoa) in the northern Gulf of Mexico. Hydrobiologia 451: 97–111.

Graham, W. M., F. Pagès & W. M. Hamner, 2001. A physical context for gelatinous zooplankton aggregations: a review. Hydrobiologia 451: 199–212.

Hamner, W. M. & M. N. Dawson, 2009. A review and synthesis on the systematic and evolution of jellyfish blooms: advantageous aggregations and adaptive assemblages. Hydrobiologia 616: 161–191.

Hansson, L. J. & B. Norrman, 1995. Release of dissolved organic carbon (DOC) by the scyphozoan jellyfish *Aurelia aurita* and its potential influence on the production of planktonic bacteria. Marine Biology 121: 527–532.

Heeger, T., U. Piatkowski & H. Möller, 1992. Predation on jellyfish by the cephalopod *Argonauta argo*. Marine Ecology Progress Series 88: 293–296.

Koppelmann, R. & J. Frost, 2008. The ecological role of zooplankton in the twilight and dark zones of the ocean. In Mertens, L. P. (ed.), Biological Oceanography Research Trends. Nova Science Publishers, Inc., New York: 67–130.

Larson, R. J., 1986. Water content, organic content, and carbon and nitrogen composition of medusae from the northeast Pacific. Journal of Experimental Marine Biology and Ecology 99: 107–120.

Lebrato, M. & D. O. B. Jones, 2009. Mass deposition event of *Pyrosoma atlanticum* carcasses off Ivory Coast (West Africa). Limnology and Oceanography 54: 1197–1209.

Lucas, C. H., 1994. Biochemical composition of *Aurelia aurita* in relation to age and sexual maturity. Journal of Experimental Marine Biology and Ecology 183: 179–192.

Lucas, C. H., 2009. Biochemical composition of the mesopelagic coronate jellyfish *Periphylla periphylla* from the Gulf of Mexico. Journal of the Marine Biological Association of the United Kingdom 89: 77–81.

Malej, A., 1989. Respiration and excretion rates of *Pelagia noctiluca* (Semaeostomeae, Scyphozoa). In Polish Academy of Sciences (eds), Proceedings of the 21st EMBS. Institute of Oceanology, Gdansk: 107–113.

Malej, A., 1991. Rates of metabolism of jellyfish as related to body weight, chemical composition and temperature. In Proceedings of the II Workshop on Jellyfish in the Mediterranean Sea. UNEP, Athens: 253–259.

Malej, A., V. Turk, D. Lučić & A. Benović, 2007. Direct and indirect trophic interactions of *Aurelia* sp. (Scyphozoa) in a stratified marine environment (Mljet Lakes, Adriatic Sea). Marine Biology 151: 827–841.

Menzel, D. W. & N. Corwin, 1965. The measurement of total phosphorus in seawater based on the liberation of organically bound fractions by persulfate oxidation. Limnology and Oceanography 10: 280–282.

Mills, C. E., 1995. Medusae, siphonophores, and ctenophores as planktivorous predators in changing global ecosystems. ICES Journal of Marine Science 52: 575–581.

Mills, C. E., 2001. Jellyfish blooms: are populations increasing globally in response to changing ocean conditions? Hydrobiologia 451: 55–68.

Mimura, H. & S. Nagata, 2001. Degradation of water-soluble fraction of jellyfish by a marine bacterium, *Brevibacterium* sp. JCM 6894. Microbes and Environments 16: 121–123.

Miyake, H., D. J. Lindsay, J. C. Hunt & T. Hamatsu, 2002. Scyphomedusa *Aurelia limbata* (Brandt, 1838) found in deep waters off Kushiro, Hokkaido, Northern Japan. Plankton Biology and Ecology 49: 44–46.

Miyake, H., D. J. Lindsay, M. Kitamura & S. Nishida, 2005. Occurrence of the scyphomedusa *Parumbrosa polylobata* Kishinouye, 1910 in Suruga Bay, Japan. Plankton Biology and Ecology 52: 58–66.

Morand, P., C. Carré & D. C. Biggs, 1987. Feeding and metabolism of the jellyfish *Pelagia noctiluca* (scyphomedusae, semaeostomeae). Journal of Plankton Research 9: 651–665.

Murphy, J. & J. P. Riley, 1962. A modified single solution method for the determination of phosphate in natural waters. Analytica Chimica Acta 27: 31–36.

Murty, S. J., B. J. Bett & A. J. Gooday, 2009. Megafaunal responses to strong oxygen gradients on the Pakistan margin of the Arabian Sea. Deep-Sea Research II 56: 472–487.

Nemazie, D. A., J. E. Purcell & P. M. Gilbert, 1993. Ammonium excretion by gelatinous zooplankton and their contribution to the ammonium requirements of microplankton in Chesapeake Bay. Marine Biology 116: 451–458.

Pitt, K. A., K. Koop & D. Rissik, 2005. Contrasting contributions to inorganic nutrient recycling by the co-occurring jellyfishes, *Catostylus mosaicus* and *Phyllorhiza punctata* (Scyphozoa, Rhizostomeae). Journal of Experimental Marine Biology and Ecology 315: 71–86.

Pitt, K. A., D. T. Welsh & R. H. Condon, 2009. Influence of jellyfish blooms on carbon, nitrogen and phosphorus cycling and plankton production. Hydrobiologia 616: 133–149.

Purcell, J. E., 1997. Pelagic cnidarians and ctenophores as predators: selective predation, feeding rates, and effects on prey populations. Annales de l'Institut Oceanographique, Paris 73: 125–137.

Purcell, J. E., 2005. Climate effects on formation of jellyfish and ctenophore blooms: a review. Journal of the Marine Biological Association of the United Kingdom 85: 461–476.

Purcell, J. E. & M. N. Arai, 2001. Interactions of pelagic cnidarians and ctenophores with fish: a review. Hydrobiologia 451: 27–44.

Purcell, J. E. & M. B. Decker, 2005. Effects of climate on relative predation by scyphomedusae and ctenophores on copepods in Chesapeake Bay during 1987–2000. Limnology and Oceanography 50: 376–387.

Richardson, A. J., A. Bakun, G. C. Hays & M. J. Gibbons, 2009. The jellyfish joyride: causes, consequences and management responses to a more gelatinous future. Trends in Ecology and Evolution 24: 312–322.

Riemann, L., J. Titelman & U. Båmstedt, 2006. Links between jellyfish and microbes in a jellyfish dominated fjord. Marine Ecology Progress Series 325: 29–42.

Schneider, G., 1988. Chemische zusammensetzung und biomasseparameter der ohrenqualle *Aurelia aurita*. Helgoländer Meeresuntersuchungen 42: 319–327.

Schneider, G., 1989. The common jellyfish *Aurelia aurita*: standing stock, excretion and nutrient regeneration in the Kiel Bight, Western Baltic. Marine Biology 100: 507–514.

Sexton, M. A., R. R. Hood, J. Sarkodee-adoo & A. M. Liss, 2010. Response of *Chrysaora quinquecirrha* medusae to low temperature. Hydrobiologia 645: 125–133.

Shimauchi, H. & S. Uye, 2007. Excretion and respiration rates of the scyphomedusa *Aurelia aurita* from the Inland Sea of Japan. Journal of Oceanography 63: 27–34.

Tinta, T., A. Malej, M. Kos & V. Turk, 2010. Degradation of the Adriatic medusa *Aurelia* sp. by ambient bacteria. Hydrobiologia 645: 179–191.

Titelman, J., L. Riemann, T. A. Sørnes, T. Nilsen, P. Griekspoor & U. Båmstedt, 2006. Turnover of dead jellyfish: stimulation and retardation of microbial activity. Marine Ecology Progress Series 325: 43–58.

West, E. J., D. T. Welsh & K. A. Pitt, 2009. Influence of decomposing jellyfish on the sediment oxygen demand and nutrient dynamics. Hydrobiologia 616: 151–160.

Yamamoto, J., M. Hirose, T. Ohtani, K. Sugimoto, K. Hirase, N. Shimamotoa, T. Shimura, N. Honda, Y. Fujimori & T. Mukai, 2008. Transportation of organic matter to the sea floor by carrion falls of the giant jellyfish *Nemopilema nomurai* in the Sea of Japan. Marine Biology 153: 311–317.

Hydrobiologia (2012) 690:257–268
DOI 10.1007/s10750-012-1041-0

JELLYFISH BLOOMS

Abundance patterns of cubozoans on and near the Great Barrier Reef

M. J. Kingsford · J. E. Seymour · M. D. O'Callaghan

Published online: 15 March 2012
© Springer Science+Business Media B.V. 2012

Abstract The ecology of cubozoans is poorly understood and there are few quantitative studies on their distribution patterns. Sampling was designed to test first for variation in abundance with distance across the continental shelf in waters of the Great Barrier Reef, Australia. Second, we tested for the possible influence of islands versus submerged reefs on the abundances of cubozoan jellyfishes. Jellyfishes were collected after attraction to tethered night lights. Additional sampling focused on turbid near-shore waters. Carybdeid jellyfishes were found at mainland, inner, and mid-shelf reefs during summers between 2007 and 2010. No cubozoan medusae were found at outer reef sites. *Copula sivickisi* and *Carukia barnesi* were more abundant near reefs with islands than at fully submerged reefs. There was no evidence of lunar periodicity in abundance for these cubozoan taxa. *Chironex fleckeri* medusae were only found close to shore near the mainland, but they were rarely observed when riverine runoff was high. All taxa were characterized by high spatial and temporal variation and there was some evidence for small populations at spatial scales of less than one kilometer. "Blooms" and related intensity of predation and risk to humans are most likely at small spatial scales.

Keywords *Chironex* · Irukandji · *Carukia* · *Alatina* · Abundance · Runoff

Guest editors: J. E. Purcell, H. Mianzan & J. R. Frost / Jellyfish Blooms: Interactions with Humans and Fisheries

M. J. Kingsford (✉) · M. D. O'Callaghan
School of Marine and Tropical Biology, James Cook University, Townsville, QLD 4811, Australia
e-mail: Michael.Kingsford@jcu.edu.au

M. J. Kingsford · M. D. O'Callaghan
ARC Centre of Excellence in Coral Reef Studies, James Cook University, Townsville, QLD 4811, Australia

J. E. Seymour
School of Marine and Tropical Biology, James Cook University, Cairns, QLD 4870, Australia

J. E. Seymour
Queensland Emergency Medical Research Foundation, James Cook University, Cairns, QLD 4870, Australia

Introduction

Jellyfishes of the Class Cubozoa (box jellyfish) are of great biological interest (Bentlage et al., 2010) and are a great risk to users of tropical waters (Barnes, 1966; Gershwin et al., 2010). Despite their low species diversity (40–50 species, Bentlage et al., 2010), they are morphologically diverse and have fast growth rates (Gordon et al., 2004), interesting life histories (Hartwick, 1991a; Straehler-Pohl & Jarms, 2005), strong swimming abilities (Gordon & Seymour, 2009), complex eyes that are used to hunt (Coates & Theobald, 2003; Nilsson et al., 2005), and powerful venom (e.g., Kintner et al., 2005). The nematocysts of "Stingers" (*Chironex fleckeri* Southcott) cause life threatening stings and have been responsible for many deaths in Australia alone (Gershwin et al., 2010). Other taxa are also a threat. For example, "Irukandji Syndrome" is an envenoming reaction in humans that results from stings of several species of box jellyfish (Little et al., 2006), which on rare occasions results in death (Pereira et al., 2010). Cubozoans that pose threats to humans occur in tropical waters of many parts of the world (Fenner & Lippmann, 2009). The threat of cubozoans has given great focus to the nature of venoms (Nagai et al., 2000; Underwood & Seymour, 2007), geographic variation in venoms (Winter et al., 2009), affects on patients (Winter et al., 2008; Tiong, 2009), and the development of antivenoms. Although there is a diversity of dangerous cubozoan medusae in tropical waters, knowledge of their ecology is poor.

Scyphozoan and cubozoan jellyfishes are notoriously patchy in distribution at spatial scales ranging from meters to tens of kilometers (Pitt & Kingsford, 2000; Gordon et al., 2004). Other species are simply very difficult to find, which has been a major problem for cubozoan research. For example, Hartwick (1991b) completed 47 cruises across the continental shelf on the Great Barrier Reef (GBR) near Townsville. On each cruise, multiple Tucker Trawls and neuston tows were done, but only eight *C. fleckeri* Southcott and about 82 other cubozoans were caught in total. Similarly, in 700 tows he collected only 100 *C. fleckeri*, with maximum densities <3 medusae per 100 m^3 within four estuaries. Other approaches have included casual observations (Matsumoto, 1995) and the sampling of beach wrack for jellyfishes, which have yielded few medusae (Yamada et al., 2010). There are few data on temporal variation in abundance, but new cohorts of one species, *Chiropsella bronzie*, appear after rain events (Gordon et al., 2004). Furthermore, in Hawaii regular occurrences of *Alatina moseri* Mayer appear 9–13 days after the full moon and this is thought to relate to spawning (Thomas et al., 2001).

Physical forcing often has a significant role in the population dynamics of jellyfishes. Variation in factors such as salinity, temperature, and abundance of food can directly affect abundance of jellyfishes. These factors often correlate with variation in nutrient levels, riverine runoff, and upwelling, which may affect the release of medusae from polyps and the survival of medusae (Kingsford et al., 2000). For example, medusae of the semaestostome *Phyllorhiza punctata* von Lendenfeld die if the salinity drops below 12 (Rippingale & Kelly, 1995). Greater knowledge of the environmental conditions required by jellyfishes is especially important because there is growing speculation about the affects of climate change on populations of jellyfish (e.g., Lynam et al., 2005, 2010)—specifically are blooms more likely?

Many cubozoans are photopositive and anecdotal accounts suggested that they could be attracted to lights, potentially allowing quantitative measures as for pre-settlement reef fishes collected with light traps (Doherty, 1987). Our objective was to focus on shallow waters near the reefs at different distances from the coast across the GBR. Shallow waters are important biologically and also are the areas of highest risk for swimmers.

The specific aims of this project were as follows:

1. To use a mensurative experimental design to test the null hypotheses that abundances of cubozoan medusae do not vary with distance across the GBR and that patterns would be consistent in multiple cross-shelf transects;
2. To test that broad-scale patterns of abundance of cubozoan medusae do not vary with lunar phase;
3. To use mensurative experimental designs to test the null hypothesis that cubozoan medusa abundances are not different between islands and submerged reefs;
4. To use opportunistic sampling and data from Surf Life Saving Australia (SLSA) to obtain data on rarer species;
5. Test if patterns of abundance of *C. fleckeri* and riverine runoff are correlated.

Mensurative experimental designs are used to test hypotheses about patterns, where the sites are not selected by random (Hurlbert, 1984).

Methods

Abundances cross-shelf

The hypothesis that abundances of cubozoans do not vary cross-shelf was tested within the framework of a mensurative multi-factorial experimental design. Three cross-shelf cruises (transects) were completed annually during the summers of: (1) 2007–2008, (2) 2008–2009, and (3) 2009–2010 between December

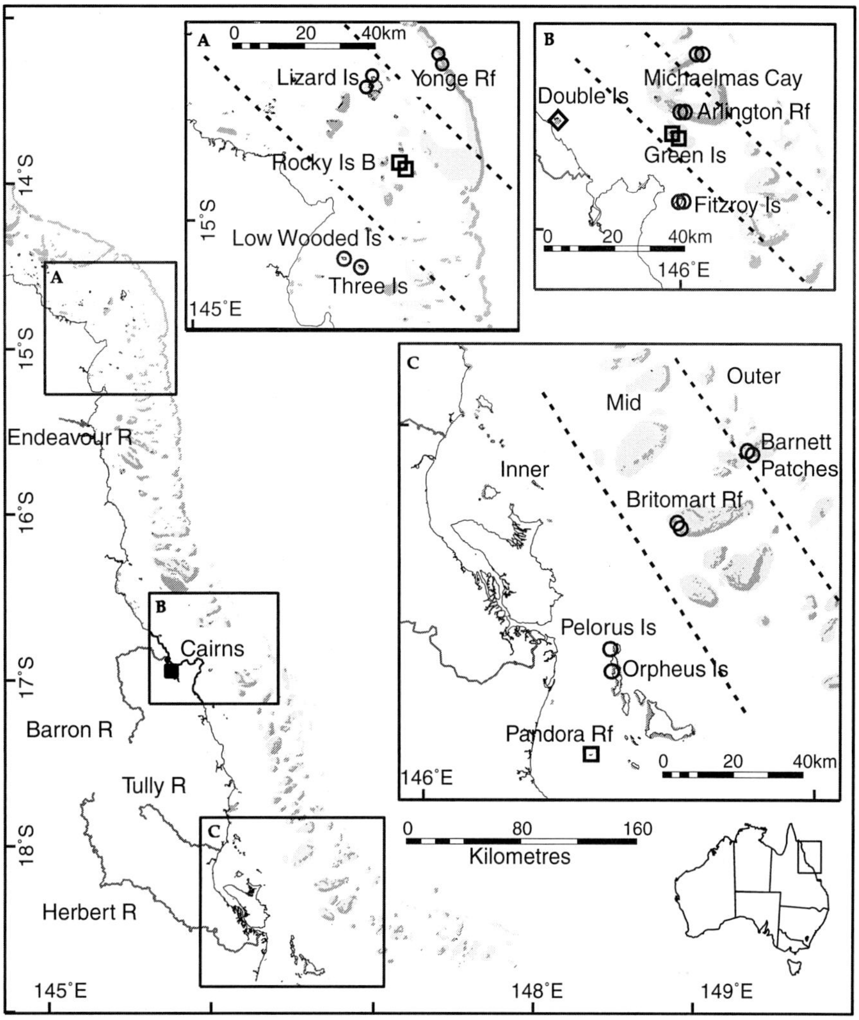

Fig. 1 Map of areas in northeastern Australia for study of distribution patterns of cubozoan medusae. *Circles* indicate sites of night lighting in the cross-shelf-sampling design; *squares* indicate sites in the island verses reef design; the *diamond* is an additional site used in the temporal study. Lizard Transect (*A*), Cairns Transect (*B*), and Palms Transect (*C*)

and February (see Fig. 1). Transects were Lizard Island, Cairns, and the Palm Island Group, and extended from 14°35′ E to 18°33′ E, about 450 km North–South (Table 1). Three categories were defined according to distance strata across the shelf (Distance strata: inner, mid, and outer). For each transect at the three distances, sampling was completed at two sites separated by 0.7–3 km. Cubozoan medusae were sampled by light attraction (1× 1,000 W bulbs positioned within the top 1 m of the water column). At each site, two replicate 1-h samples were taken for abundance data; replicates were taken from two anchored vessels that were separated by 50–200 m so that pools of light did not overlap. Additional sampling time was to collect more jellyfishes for size frequency determinations. The physical characteristics of the water column were also measured at most sites using a CTD (*Seabird,* SBE 19 Plus).

Temporal variation in medusa abundances was determined from Mermaid Bay (over four summers) and Double Island (near Palm Cove, 16°43′32 S; 145°41′00 E; Fig. 1). Sampling was completed on multiple nights within each season.

The influence of geology on abundances of cubozoan medusae

Patterns found in 2007–2008 suggested that carybdeid jellyfishes would be most abundant around islands. In two summers we chose distances from shore where carybdeids were found in year one. We sampled two reefs mid-shelf on the Lizard Island transect (Lizard

Table 1 Locations of cubozoan jellyfish collection sites in northeastern Australia and frequency of sampling

Location	Transect	Shelf	Seasons sampled	Latitude	Longitude	Distance from mainland (km)	Geological description
Inner Islands	Lizard	Inner	1, 2, 3	15°06.147 S	145°24.030 E	14	Wooded coral cay
Rocky Islets B	Lizard	Mid	2, 3	14°52.715 S	145°31.459 E	22	Reef
Lizard Island	Lizard	Mid	1, 2, 3	14°38.865 S	145°27.235 E	30	Granite island
Yonge Reef	Lizard	Outer	1, 2, 3	14°35.685 S	145°37.153 E	50	Reef
Fitzroy Island	Cairns	Inner	1, 2, 3	16°53.926 S	145°57.415 E	6	Granite island
Green Island	Cairns	Mid	2, 3	16°45.312 S	145°57.966 E	12	Wooded coral cay
Arlington Reef	Cairns	Mid	1, 2, 3	16°42.103 S	145°58.122 E	19	Reef
Michaelmas Reef	Cairns	Outer	1, 2, 3	16°36.480 S	145°58.022 E	35	Reef
Orpheus Island	Orpheus	Inner	1, 2, 3	18°36.060 S	146°29.011 E	16	Granite island
Pelorus Island	Orpheus	Inner	1, 2, 3	18°33.266 S	146°29.170 E	16	Granite island
Britomart Reef	Orpheus	Mid	1, 2, 3	18°14.590 S	146°38.159 E	48	Reef
Barnett Patches	Orpheus	Outer	1, 2, 3	18°04.748 S	146°51.023 E	64	Reef
Pandora Reef	Orpheus	Inner	3	18°48.691 S	146°26.030 E	16	Coral cay

Jellyfish season is from November to March each year; sampling was done December to February: 1, 2007–2008; 2, 2008–2009; 3, 2009–2010

Island, a granite island and, the Rocky Islets) and two sites per reef. Rocky Islets are a group of reefs at a similar distance from the mainland as is Lizard Island, but the reefs rarely emerge at any state of tide and they are made up of a coral matrix. Similarly, on the Cairns transect, Green Island (a coral cay) was sampled mid-shelf to compare with Michaelmas, a largely submerged reef. In the summer of 2009–2010, a comparison was also made between Pandora Reef (Coral cay and mid to low tide) and the granitic islands of Pelorus and Orpheus (Table 1).

Sampling at mainland beaches and estuaries

Sampling by night light was not done near the mainland because the waters were very turbid. Trawling, visual counts, and opportunistic sampling were used in these waters at all cross-shelf transects. The beam trawl (1.5 m × 0.5 m mouth, 8-mm mesh) was towed in very shallow water (<2 m deep) and deeper waters (3–5 m deep) adjacent to the mainland, at the entrance of rivers, and 1 km from the rivers ($n = 2$ five-minute tows at each depth and location). The two depths of sampling allowed for variation in movement of jellyfishes with the tide. We measured the distance of the trawl with a General Oceanics 2030 flowmeter (200–250 m^3 filtered per tow). Visual estimates and samples are also taken during the trawls (3-m-wide visual swath × distance of the trawl). Sampling was completed between December and February in the summers of: (1) 2007–2008, (2) 2008–2009, and (3) 2009–2010. The Cairns transect was not sampled near shore in 2007–2008 due to high freshwater flows and an abundance of floating logs.

Data on temporal variation in medusa abundance were collected using trawls and visual sampling from Townsville in 2008–2009 and 2009–2010 at three sites separated by 1–2 km at Rowes Bay (19°14.170 S, 146°47.379 E). An additional site was added in 2009–2010, on the Strand (19°14.762 S, 146°48.802 E); all trawl and visual count procedures were the same as above. Opportunistic sampling was done on each occasion that we boated along the edge of beaches and around the marina and harbor at the entrance to Ross Creek.

Additional sampling was undertaken at Port Douglas, Double Island near Palm Cove, Mission Beach, Balgal Beach, Townsville beaches, and Magnetic Island; some of these samples were provided by SLSA, which samples popular beaches daily between 16°28.394 S, 145°27.468 E and 19°15.638 S, 146°50.934 E. Samples were provided from Keeper Reef by the MV Kalinda. Verbal records of *C. fleckeri* were obtained from Palm Cove in Cairns, Mission Beach to Cardwell, the Strand in Townsville, and Magnetic Island. Data were categorized as: early (October–December), mid (January–February), and late (March–May) in the summer period.

Taxonomy

Identification of cubozoans has been problematic and uncertainty about taxonomic names still exists (Bentlage et al., 2010). We based identifications on Gershwin, (2005a, b) and Gershwin & Kingsford (2008). The majority of specimens were the carybdeids *C. barnesi* Southcott and *Copula sivickisi* Stiasny (recently changed from *Carybdea sivickisi* Stiasny; Bentlage et al., 2010). Some *Carukia* spp. were identified where we were not certain they were *C. barnesi;* it is possible they are undescribed taxa. *Carybdea xaymacana* Conant was identified according to Gershwin (2005b), but this identification has been questioned as being based on coincidental evidence (Bentlage et al., 2010); therefore, in this paper it should be considered a type. The size of cubozoans was measured as inter-pedalial distances (IPD).

Rainfall

The relationship between the abundances of jellyfishes collected near shore and riverine runoff was tested with a Pearson's correlation. Data on freshwater outflow was obtained for major watersheds that drain onto the shelf adjacent to the areas where the abundance of *C. fleckeri* medusae was measured. River gauging station data were obtained from the Barron River (Cairns area) and the Tully River (between Mission Beach and the Herbert River). Data were expressed in megalitres (Ml); web source from Department of Environment and Resources Management (www.derm.qld.gov.au/watershed).

Treatment of the data

Data were analysed using a mixed model ANOVA; Distance (treatments: inner, mid, and outer shelf) was a fixed factor, and latitude (treatments: Lizard, Cairns, and Palms) and sites (nested in distance) were random factors. Data were sometimes $\ln(x + 1)$ transformed, but if they were still heterogeneous, according to Cochran's tests, we proceeded with analyses on $\ln(x + 1)$ data as ANOVA is robust to heterogeneity (Underwood, 1997). Variance components were calculated only for raw data and with fully-nested designs and random factors (Kingsford, 1998). Because there are accounts of lunar periodicity in the occurrence of *A. moseri* Mayer medusae in Hawaii, we used a pattern-seeking approach with all cross-shelf data by plotting the abundance of *Carukia* spp. by phase of the lunar month (i.e., days 1–30; full moon on day 15). Catches of jellyfish from Double Island (part of the temporal study) were also compared with lunar phase.

Results

Abundance cross-shelf

A total of 208 cubozoans were collected during the first hour of sampling at sites and cross-shelf sampling programs over the three summer seasons December 2007 to February 2010; an additional 55 specimens were collected within the second hour (Table 2). The species breakdown was as follows (first hour sampling only): *C. barnesi* (76), *Carukia* other (6), and *C. sivickisi* (125); also see Table 2.

Catches of *Carukia* spp. were low in 2008–2009 and 2009–2010 compared to 2007–2008. In 2007–2008 we collected 69 *C. barnesi* (three cruises combined) in 1 h counts and an additional 43 specimens were collected. In 2008–2009, we collected four *C. barnesi* (three cruises combined) in 1 h counts and no additional specimens. Similarly in 2009–2010 we collected two *C. barnesi* (all transects combined); an additional four specimens were obtained that were not collected in 1 h counts (Fig. 2) and six *C. barnesi* were all caught at additional sites in the islands versus reefs design.

Table 2 Total numbers of cubozoan medusae collected in three cross-shelf transects in northeastern Australia during three summers from 2007 to 2010, distance strata (mainland, inner, mid, and outer shelf) are progressively further offshore (Table 1)

Taxa\distance	Mainland	Inner	Mid	Outer
C. xaymacana	2	0	0	0
C. barnesi	0	15	108	0
Other *carybdeids*	0	9	0	0
Alatina sp.	0		(2)	0
C. sivickisi	0	(45)	129	0
C. fleckeri	(255)	0	0	0

Data from sampling with night lights (1 h of sampling and additional sampling after the first hour) and trawls (only near the mainland). Numbers in brackets refer to incidental counts (e.g., Surf Life Saving and other sources)

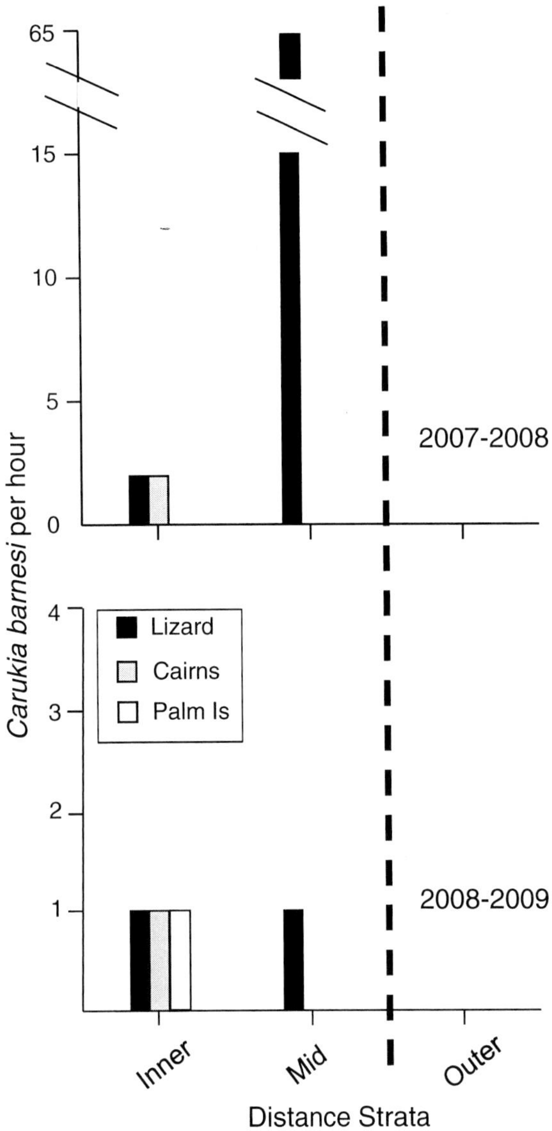

Fig. 2 Total abundance of *Carukia* spp. medusae (most were *C. barnesi*) collected in cross-shelf transects in northeastern Australia during three summers from 2007 to 2010. All counts were done with night lighting. Data were pooled by position on shelf within transects

Although no irukandji jellyfishes were found offshore at any latitude or in any summer, cross-shelf patterns varied by latitude. When *C. barnesi* medusae were abundant in 2007–2008, variation across the shelf resulted in a significant latitude × distance interaction (Table 3). This was largely because no jellyfish were found in the Palm Island transect and high abundance only occurred mid-shelf on the Lizard Island transect.

Great differences were found among sites within distance strata (Fig. 3). For example, at Lizard Island (mid-shelf), means of 14.5 *C. barnesi* occurred at one site and 3.5 at the other. Forty-one percent of variation in abundance was explained by variation at the level of site. This suggested that *C. barnesi* populations may be very localised at small spatial scales such as within bays; the greatest variation was found among replicates (51%).

Carukia spp. collected in the cross-shelf study ranged from 3 to 18 mm in IPDs (mean 8.5 mm). The majority of these jellyfish were collected at Lizard Island (95.3%) and the great variation in size suggested that the medusae had been released from polyps over many nights, rather than in a distinct pulse.

Other cubozoans were collected in the three summers and all were collected from the mainland to mid-shelf reefs (Table 2). Irukandji jellyfishes other than *C. barnesi* were as follows: Two *C. xaymacana* were caught in trawls on beaches near Cooktown. Two *Alatina* sp. were collected from a charter boat at a mid-shelf reef (Keeper) near the Palms transect, November 2009.

Copula sivickisi is a carybdeid cubozoan that has a mild sting that does not result in "Irukandji syndrome". A total of 125 *C. sivickisi* were collected, 34 in 2007–2008 and 90 in 2008–2009, and 1 in 2009–2010. All were collected at mid-shelf reefs, Lizard Island, and Green Island. Casual counts with lights at Magnetic Island (Inner) also detected *C. sivickisi* in shallow water (Table 3). Although no *C. fleckeri* were collected around lights, many were found near the mainland in shallow water during the 3-year study.

Table 3 ANOVA, *C. barnesi*, $\ln(x + 1)$ transformed

Factor	Source of variation	df	MS	F	Denom	*P*
R	Latitude	2	2.556	0.55	Site (L × D)	NS
F	Distance	2	1.833	0.86	L × D	NS
	L × D	4	2.144	4.15	Site (L × D)	*
R	Site (L × D)	9	0.517	4.64	Residual	**
	Residual	18	0.112			

F fixed, *R* Random, *Denom* denominator mean square

** $P < 0.01$; * $P < 0.05$; *NS* not significant

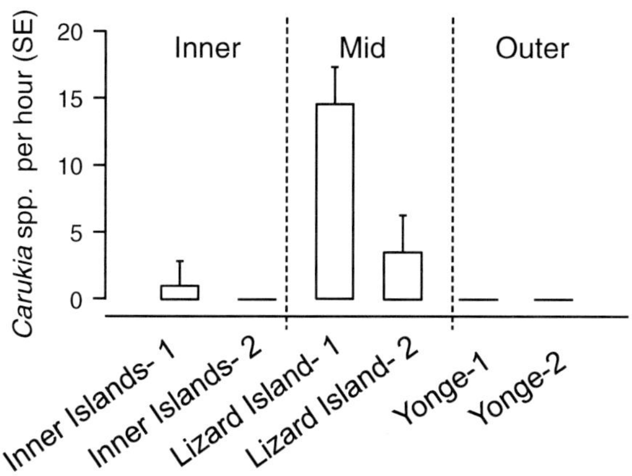

Fig. 3 Mean abundance of *Carukia* spp. medusae (most were *C. barnesi*) collected in the Lizard Island cross-shelf transect at *inner*, *mid*, and *outer* shelf locations in northeastern Australia in 2007. All counts were done with night lighting. Variance components from nested ANOVA: distance 7.8%, df = 2, MS = 341.58; site (distance) 41.1%, df = 3, MS = 252.41; residual 51.1, df = 6, MS = 96.75. *df* degrees of freedom, *MS* mean square

There was no evidence for lunar periodicity in the abundances of *Carukia* spp. and *C. sivickisi* (Fig. 4). Relatively high abundances of *Carukia* spp. and *C. sivickisi* were found at more than one phase of the lunar cycle.

Where physical data were available and *Carukia* spp. were collected, they were found in waters with salinity ranging from 31.6 to 35.1 and temperatures of 28.1–30.0°C. With the exception of Fitzroy Island in 2009, where the water column was relatively fresh at the surface, the water columns were generally well mixed between top and bottom.

Does the presence of islands influence the abundance of cubozoans?

All carybdeid jellyfishes were collected near islands (Low Wooded Isle—Inner 1, Three Islands—Inner 2, Fitzroy Island—Inner, Lizard Island-mid) in the cross-shelf transects during the study, with the greatest numbers collected near granite islands (e.g., Lizard Island). Few *C. barnesi* were collected in 2008–2009 (one at Lizard Island) and 2009–2010 (one at Lizard Island, four at Pandora Reef (Inner, Palm Island Group). Pandora is not a granite-based reef, but it emerges at low tide and probably should be considered to be geologically similar to Three Islands (Inner, Lizard Transect), where we also collected carybdeids.

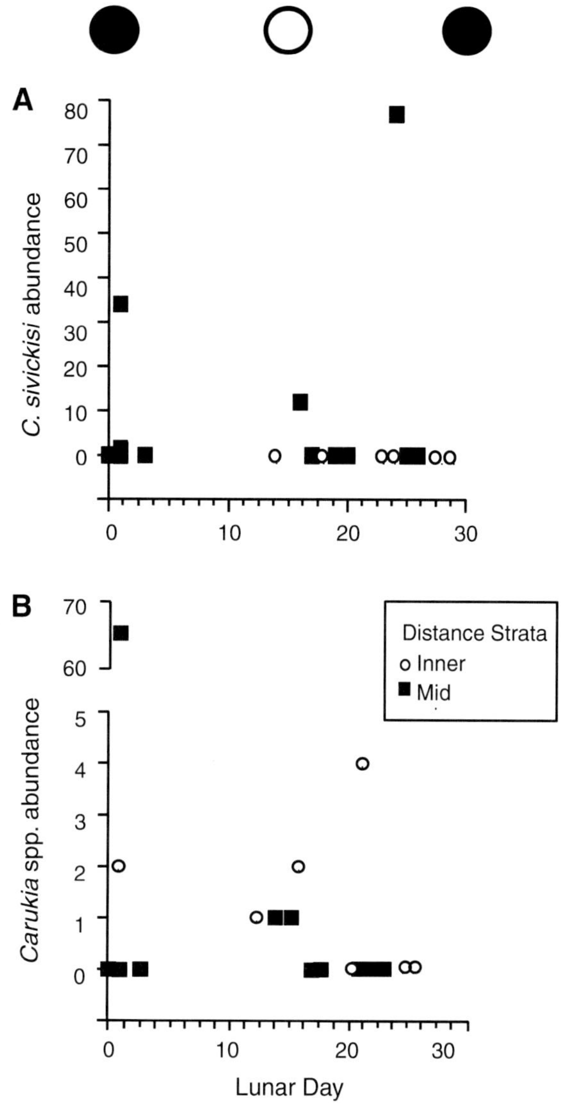

Fig. 4 Lunar patterns for *Carukia* spp. and *C. sivickisi* medusae totaled for all three summers in 2007–2010 for *inner* and *mid* shelf reefs in northeastern Australia (1 h fishing at each sampling)

Multiple *C. sivickisi* medusae were collected for the paired comparisons. The relationship with the presence of islands was a clear; 90 were collected at islands and only one at reefs without islands (at Rocky Islets B; Fig. 5). There was great variation between replicates.

Temporal variation of *C. barnesi*

There was great temporal variation in abundance of *Carukia* spp. caught at Mermaid Bay, Lizard Island

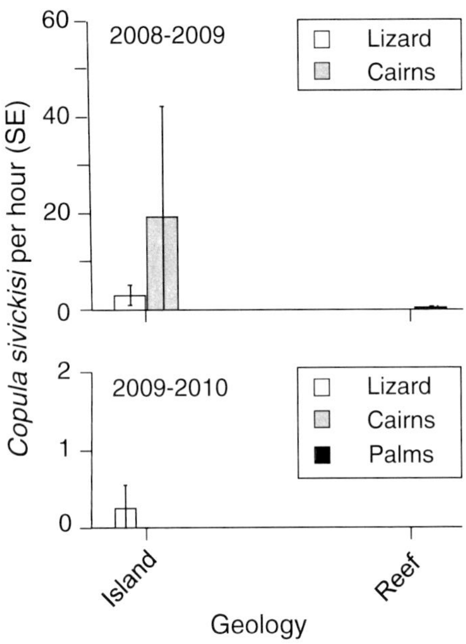

Fig. 5 Abundance of *C. sivickisi* medusae near islands and at mostly-immersed coral reefs in northeastern Australia

over four seasons; however, there was always a high probability of collecting *Carukia* spp. there. Even with this variation, differences among years, as described earlier (i.e., highest abundance in (2006–2008)), were robust (Table 4). Temporal variation in abundance also was great at Double Island (<1 km from the mainland); nevertheless, nine sampling days showed that the probability of detecting *Carukia* spp. was high regardless of year. We compared catches at Double Island (by day, $n = 19$) by lunar day (i.e., 1–30) and found no patterns, which concurred with the broad-scale study.

Sampling near the mainland

Few cubozoans were found in trawls and transects near the mainland. We collected only two *C. xaymacana* near Cooktown (Lizard Island Transect) and one *C. fleckeri* in transects over three summers; however, 255 *C. fleckeri* were observed or collected near the mainland (Table 2). Pooled data from near-shore surveys, our casual observations, some trained observers, and information from SLSA showed that the most *C. fleckeri* were collected between October and December in the summers of 2007–2008 and 2008–2009. Between 10 and just over 100 individuals were found in locations including Trinity Beach, Cairns, Hinchinbrook Channel near Cardwell, Townsville, and Magnetic Island then.

Chironex fleckeri was rare in January to February in the first two summers and absent from March to May. In contrast, *C. fleckeri* were sparse from October 2009 to May 2010, but a few medusae were seen early, mid, and late in the season. In all years, medusae were found in shallow water, usually 0.5–5 m deep and within 100 m of shore. An exception to this was near Townsville, *C. fleckeri* were found near the mainland

Table 4 Temporal variation in abundance of *C. barnesi* medusae for 1 h of fishing with a 1,000 W light (*n* samples) at two island locations

Season	Mermaid Bay		Double Island	
	Date	Mean (range) *n*	Date	Mean (range) *n*
2006–2007	18 Dec 2006	3 (–) 1		
	19 Dec 2006	14 (13–15) 2		
	20 Dec 2006	18 (9–27) 2		
2007–2008	6 Dec 2007	0.3 (0–1) 2	20 Oct 2007	1 (–) 1
	11 Dec 2007	30 (13–47) 4	12–17 Dec 2007	2.5 (2–3) 2
			3 Jan 2008	5 (–) 1
2008–2009	4 Dec 2008	0 (–) 2	12–28 Nov 2008	0 (–) 2
	13 Dec 2008	0 (–) 2	7–19 Dec 2008	1.7 (0–4) 3
			2–6 Jan 2009	0 (–) 2
2009–2010	16 Dec 2009	0 (–) 2	26–30 Dec 2009	9 (0–24) 5
	21 Apr 2010	0 (–) 2	2–6 Jan 2010	7.7 (1–20) 3
	25 Apr 2010	0 (–) 2	6 Feb 2010	14 (–) 1

Mermaid Bay data are by day, Double Island data are individual replicates collected over 3–4 days within a month in shallow waters, but they were also found at Magnetic Island (about 10 km from the mainland) where they appear to have colonized near-shore waters (19°06.921, 146°51.698). However, waters separating the island from the mainland are less than 5 m deep. Jellyfish were found in estuaries and marinas (e.g., Port Douglas, Hinchinbrook Island, and Townsville) and on beaches that were exposed to the sea (e.g., Townsville). At all locations temporal persistence of *C. fleckeri* was low.

The influence of riverine flow on *C. fleckeri* abundance

Chironex fleckeri was not observed when freshwater outflows were high in mid- to late-summer, in years one and two (Fig. 6). River flow varied greatly among rivers and years. Flow was low in all years near the Cairns transect (Tully River) and was lowest early in the season in all rivers. The correlation between riverine runoff and abundance of *C. fleckeri* was not significant (Fig. 6), probably because jellyfish mostly disappeared after the first period of sustained heavy rain. Year three had the lowest flows in all rivers and some *C. fleckeri* medusae were found in early-, mid-, and late-summer.

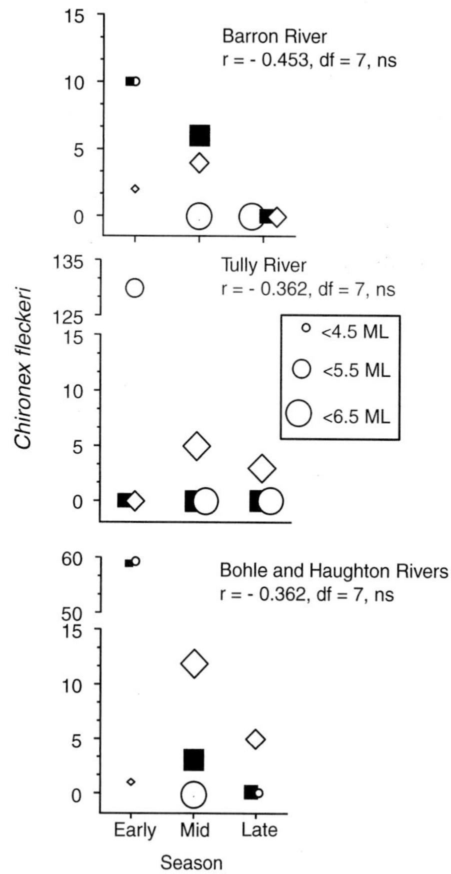

Fig. 6 Abundance of *C. fleckeri* medusae in northeastern Australia during three summers (*open circles* 2007–2008; *solid squares* 2008–2009; *diamonds* 2009–2010). Data were pooled by area for early, mid, and last within summers. Riverine flow in ML is provided for the rivers adjacent to areas were jellyfish were observed (locations and latitudinal range for the area affected by each river); Barron River (Cairns Regions; 16°13′522 E, 145°28.309 S to 16°57.664 E, 145°50.544 S); Tully River (Mission Beach and Hinchinbrook channel; 17°51.133 S, 146°08.071 E to 18°17.041 E, 146°03.051 S); Bole and Haughton rivers (Balgal Beach to Townsville/ Magnetic Island; 19°01.354 S, 146°24.938 E to 19°15.638 S, 146°50.934 E)

Discussion

There was great variation in abundance patterns of cubozoan medusae cross-shelf. *C. fleckeri* medusae were restricted to near-shore waters, estuarine areas, and mainland beaches. All other cubozoans, the irukandji species (*C. barnesi*, *C. xaymacana*, *Alatina* sp.) and the relatively innocuous *C. sivickisi* were found near the mainland and/or at inner and mid-shelf reefs. Variation in cross-shelf patterns have been found for some scyphomedusan jellyfish species (Lynam et al., 2005), but there were no previous data for cubozoans.

It was possible that we failed to detect some jellyfish because of the sampling design. The abundance of *A. moseri* medusae are most abundant on beaches of Hawaii 9–13 days after the full moon (Thomas et al., 2001). *Alatina* sp. medusae were found during the study, but none were collected in lights at outer reefs despite multiple samples being collected after the full moon. There are anecdotal accounts of lunar pulses of *Alatina* near reefs of the Coral Sea. However, it is likely that their spatial distribution is very patchy, even given possible lunar periodicity. We found no evidence of lunar periodicity in *C. barnesi* or *C. sivickisi*.

The greatest numbers of *C. barnesi* were found near granite islands. Although orthogonal comparisons near and away from granite islands were inconclusive, the probability that *Carukia* spp. would be collected was much greater at granite islands. We also received photographs, each with as many as eight carybdeids, from a site (by Macona Inlet, 20°148.21 S and 148°55.47 E) near the granitic Hayman Island, the Whitsundays (Inner); 2 Feb 2010. There was also strong evidence that *C. sivickisi* were most abundant around islands. Possible explanations for an island effect include: (1) there is more suitable habitat for polyps, (2) oceanographic and wind effects around islands facilitate retention (Wolanski et al., 1984), and (3) for *C. sivickisi, Sargassum* spp., which is the preferred substratum for the jellyfish polyps (Hartwick, 1991a) is abundant. Even near islands, aggregations of *Carukia* spp. were rare. The highest concentrations, and a broad size range of individuals, were found only at a few sites, such as Mermaid Bay. This suggested that populations are highly localized due to local retention and supply of medusae. High variance between sites was common and great differences were found among replicates, which is typical for jellyfishes (Pitt & Kingsford, 2000).

There was strong evidence that freshwater flow influenced the abundance of *C. fleckeri* medusae, primarily during the wet season on north Queensland (December–April). In the first two summers, most *C. fleckeri* were collected early (November–December) and few were found as the runoff of freshwater increased from January to April. Relatively high abundance in January at the Barron River was found just before the heavy rain fall starting about 10 January 2009. Although river runoff and time within a season are confounded, more *C. fleckeri* were found mid- and late-season when the lowest flows occurred in the final season (2009–2010). To clarify this issue, experiments are required to test the affects of salinity on different life history stages.

It well known that changes in salinity can trigger the production of jellyfish from polyps in and influence the survival of scyphozoan medusa (reviews: Kingsford et al., 2000; Purcell et al., 2007). Although the paradigm for the cubozoan *C. fleckeri* is that the release of medusae is triggered by an input of freshwater (Hartwick, 1991b), and it has been assumed that the source of medusae is in estuaries, our data also suggest that there is a lower limit for salinity. This concurs with occasional observations of dead *C. fleckeri* on beaches after periods of strong river runoff. We suggest, therefore, that seasons of high rainfall may be a high risk to *C. fleckeri* populations. Due to global climate change, the frequency and intensity of cyclones is predicted to increase in north Queensland (Lough, 2008). Although experimental testing of critical salinities is required, we suggest that increased rainfall may have a negative affect on *C. fleckeri* populations. It is also likely that the affects of climate change will vary by species and region; both positive and negative effects on population sizes probably will be found (Lynam et al., 2005).

In conclusion, our data on cubozoan distributions and abundances tested hypotheses about the possible effects of distance from shore and the influence of islands. There were clear cross-shelf patterns in the abundance of cubozoans. The risk of envenomation to humans was greatest from the mainland to mid-shelf reefs, and especially around granite islands. There was no evidence for lunar-related variation in abundance, but physical forcing by freshwater input apparently had a strong influence on the abundance of *C. fleckeri*. This, combined with its near-shore distribution, suggests strong possibilities for biophysical modeling. The greatest challenge for reliable long-term data on cubozoan medusae is the extreme variation in their spatial and temporal distributions.

Acknowledgments Counting cubozoans under lights is often arduous and we thank our many volunteers for assisting, especially Shelley Templeman and Christopher Mooney. The crews of MV *Piscean*, MV *Kalinda*, and MV *Phoenix* provided critical support on the cross-shelf cruises. For the Double Island samples we thank Teresa Carrette, Avril Underwood, Glenda Seymour, and Richard Fitzpatrick for their assistance. We also thank SLSA for providing data and specimens of cubozoans collected on beaches in North Queensland. We also appreciate photographs of cubozoans provided by John Sinclair. Funding was provided by a Marine Science Tropical Science Research Facility (MTSRF) and a LIONS Foundation grant.

References

Barnes, J. H., 1966. Studies on three venomous Cubomedusae. In Rees, W. J. (ed.), The Cnidaria and Their Evolution. Academic Press, London: 307–332.

Bentlage, B., P. Cartwright, A. A. Yanagihara, C. Lewis, G. S. Richards & A. G. Collins, 2010. Evolution of box jellyfish (Cnidaria: Cubozoa), a group of highly toxic invertebrates. Proceedings of the Royal Society B-Biological Sciences 277: 493–501.

Coates, M. C. & J. C. Theobald, 2003. Optimal visual parameters for a cubozoan jellyfish in the mangrove environment. Integrative and Comparative Biology 43: 1016.

Doherty, P. J., 1987. Light-traps: selective but useful devices for quantifying the distributions and abundance of larval fishes. Bulletin of Marine Science 41: 423–431.

Fenner, P. J. & J. Lippmann, 2009. Severe Irukandji-like jellyfish stings in Thai waters. Diving and Hyperbaric Medicine 39: 175–177.

Gershwin, L. A., 2005a. Taxonomy and phylogeny of Australian Cubozoa. PhD Thesis, James Cook University, Townsville.

Gershwin, L. A., 2005b. Two new species of jellyfishes (Cnidaria: Cubozoa: Carybdeida) from tropical Western Australia, presumed to cause Irukandji syndrome. Zootaxa 1084: 1–30.

Gershwin, L. A. & M. J. Kingsford, 2008. Pelagic Cnidaria and Ctenophora. In Hutchings, P., M. Kingsford & O. Hoegh-Guldberg (eds), The Great Barrier Reef: Biology, Environment and Management. CSIRO Publishing, Collingwood: 188–198.

Gershwin, L. A., M. De Nardi, K. D. Winkel & P. J. Fenner, 2010. Marine stingers: review of an under-recognized global coastal management issue. Coastal Management 38: 22–41.

Gordon, M. R., C. Hatcher & J. E. Seymour, 2004. Growth and age determination of the tropical Australian cubozoan *Chiropsalmus* sp. Hydrobiologia 530: 339–345.

Gordon, M. R. & J. E. Seymour, 2009. Quantifying movement of the tropical Australian cubozoan *Chironex fleckeri* using acoustic telemetry. Hydrobiologia 616: 87–97.

Hartwick, R. F., 1991a. Observations on the anatomy, behavior, reproduction and life-cycle of the cubozoan *Carybdea sivickisi*. Hydrobiologia 216: 171–179.

Hartwick, R. F., 1991b. Distributional ecology and behavior of the early life stages of the box-jellyfish *Chironex fleckeri*. Hydrobiologia 216: 181–188.

Hurlbert, S. H., 1984. Pseudoreplication and the design of ecological field experiments. Ecological Monographs 54: 187–211.

Kingsford, M. J., 1998. Analytical aspects of sampling design. In Kingsford, M. J. & C. N. Battershill (eds), Studying Temperate Marine Environments: A Handbook for Ecologists. University of Canterbury Press, Christchurch: 49–83.

Kingsford, M. J., K. A. Pitt & B. M. Gillanders, 2000. Management of jellyfish fisheries, with special reference to the order Rhizostomeae. Oceanography and Marine Biology: An Annual Review 38: 85–156.

Kintner, A., S. Edwards & J. E. Seymour, 2005. Variation in lethality and effects of two Australian chirodropid jellyfish venoms, *Chironex fleckeri* and *Chiropsalmus* sp., in fish. Toxicon 46: 699–708.

Little, M., P. Pereira, T. Carrette & J. Seymour, 2006. Jellyfish responsible for Irukandji syndrome. Quarterly Journal of Medicine 99: 425–427.

Lough, J. M., 2008. A changing climate for coral reefs. Journal of Environmental Monitoring 10: 21–29.

Lynam, C. P., S. J. Hay & A. S. Brierley, 2005. Jellyfish abundance and climatic variation: contrasting responses in oceanographically distinct regions of the North Sea, and possible implications for fisheries. Journal of the Marine Biological Association of the United Kingdom 85: 435–450.

Lynam, C. P., M. J. Attrill & M. D. Skogen, 2010. Climatic and oceanic influences on the abundance of gelatinous zooplankton in the North Sea. Journal of the Marine Biological Association of the United Kingdom 90(special issue): 1153–1159.

Matsumoto, G. I., 1995. Observations on the anatomy and behaviour of the cubozoan *Carybdea rastonii* Haacke. Marine and Freshwater Behaviour and Physiology 26: 139–148.

Nagai, H., K. Takuwa, M. Nakao, B. Sakamoto, G. L. Crow & T. Nakajima, 2000. Isolation and characterization of a novel protein toxin from the Hawaiian box jellyfish (sea wasp) *Carybdea alata*. Biochemical and Biophysical Research Communications 275: 589–594.

Nilsson, D. E., L. Gislen, M. M. Coates, C. Skogh & A. Garm, 2005. Advanced optics in a jellyfish eye. Nature 435: 201–205.

Pereira, P., J. Barry, M. Corkeron, P. Keir, M. Little & J. E. Seymour, 2010. Intracerebral hemorrhage and death after envenoming by the jellyfish *Carukia barnesi* death due to Irukandji syndrome. Clinical Toxicology 48: 390–392.

Pitt, K. A. & M. J. Kingsford, 2000. Geographic separation of stocks of the edible jellyfish, *Catostylus mosaicus* (Rhizostomeae) in New South Wales, Australia. Marine Ecology Progress Series 196: 143–155.

Purcell, J. E., S. Uye & W.-T. Lo, 2007. Anthropogenic causes of jellyfish blooms and their direct consequences for humans: a review. Marine Ecology Progress Series 350: 153–174.

Rippingale, R. J. & S. J. Kelly, 1995. Reproduction and survival of *Phyllorhiza punctata* (Cnidaria: Rhizotomeae) in a seasonally fluctuating salinity regime in Western Australia. Marine & Freshwater Research 46: 1145–1151.

Straehler-Pohl, I. & G. Jarms, 2005. Life cycle of *Carybdea marsupialis* Linnaeus, 1758 (Cubozoa, Carybdeidae) reveals metamorphosis to be a modified strobilation. Marine Biology 147: 1271–1277.

Thomas, C. S., S. A. Scott, D. J. Galanis & R. S. Goto, 2001. Box jellyfish *(Carybdea alata)* in Waikiki: their influx cycle plus the analgesic effect of hot and cold packs on their stings to swimmers at the beach: a randomized, placebo-controlled, clinical trial. Hawaii Medical Journal 60: 100–107.

Tiong, K., 2009. Irukandji syndrome, catecholamines, and mid-ventricular stress cardiomyopathy. European Journal of Echocardiography 10: 334–336.

Underwood, A. J., 1997. Experiments in Ecology: Their Logical Design and Interpretation Using Analysis of Variance. Cambridge University Press, Cambridge: 504 pp.

Underwood, A. H. & J. E. Seymour, 2007. Venom ontogeny, diet and morphology in *Carukia barnesi*, a species of Australian box jellyfish that causes Irukandji syndrome. Toxicon 49: 1073–1082.

Winter, K. L., G. K. Isbister, J. J. Schneider, N. Konstantakopoulos, J. E. Seymour & W. C. Hodgson, 2008. An examination of the cardiovascular effects of an 'Irukandji' jellyfish, *Alatina* nr *mordens*. Toxicology Letters 179: 118–123.

Winter, K. L., G. K. Isbister, S. McGowan, J. J. Schneider, N. Konstantakopoulos, J. E. Seymour & W. C. Hodgson,

2009. A pharmacological and biochemical examination of the geographical variation of *Chironex fleckeri* venom. Toxicology Letters 192: 419–424.
Wolanski, E., J. Imberger & M. L. Heron, 1984. Island wakes in shallow coastal waters. Journal of Geophysical Research 89: 553–569.
Yamada, T., T. Takeda & S. Kubota, 2010. Temporal patterns of jellyfish species occurrence at the Suma Coast, Kobe City, Hyogo Prefecture (Years 2003–2009). Kuroshio Biosphere 6: 27–30.

Hydrobiologia (2012) 690:269–277
DOI 10.1007/s10750-012-1049-5

JELLYFISH BLOOMS

Sources and movements of *Chironex fleckeri* medusae using statolith elemental chemistry

C. J. Mooney · M. J. Kingsford

Published online: 8 March 2012
© Springer Science+Business Media B.V. 2012

Abstract *Chironex fleckeri* medusae metamorphose from sessile polyps, possibly in estuarine environments, and migrate into coastal waters. The objective of this study was to critically test the anecdotal paradigm that the medusae originate in lower salinity waters. Laser-ablation inductively coupled plasma-mass spectrometry was used on *C. fleckeri* statoliths to test the hypothesis that *C. fleckeri* medusae only originate from low salinity tidal creeks. Statoliths were extracted from *C. fleckeri* medusae collected from multiple locations around tropical Australia. Strontium:Calcium (Sr:Ca) ratios were used as a proxy for salinity; where salinity remained consistent in the field, the ratio was compared with the elemental chemistry in statoliths. Sr:Ca ratios of the statolith core and edge zones showed some evidence that medusae originated in lower Sr:Ca levels and moved to higher levels as expected under the hypothesis. That pattern was not consistent, however, and sources from multiple oceanographic regimes were indicated. Core-to-edge elemental profiles of statoliths and concentric increments showed high variability in Sr:Ca ratios both within and between individuals. The ratios suggested that many jellyfish had been exposed to a wide range of oceanographic regimes, while others had spent their whole lives in high Sr:Ca ratio waters. Elemental chemistry and concentric increments in the $CaSO_4$ matrix of cubomedusan statoliths provide a tool to study cubozoan ecology.

Keywords Jellyfish · LA-ICPMS

Introduction

The potentially lethal cubozoan, the box jellyfish (*Chironex fleckeri* Southcott), poses a significant threat to users of tropical Australian coastal waters. Stings from *C. fleckeri* medusae were attributed to approximately 70 fatal envenomations in Australia during the twentieth century (Currie & Jacups, 2005; Coughlan et al., 2006). It is thought that the metamorphosis from polyp to medusa phases results in a shift in preferred habitat from tidal estuaries to adjacent beaches (Hartwick, 1991). *C. fleckeri* medusae are considered to be littoral (sensu Tibballs, 2006) and are found mostly in shallow water mangrove channels, creek mouths and along sandy beaches (Hamner et al., 1995; Coates, 2003). Knowledge of cubomedusan distribution and movements is limited (but see Gordon & Seymour, 2009; Kingsford et al., 2012).

Guest editors: J. E. Purcell, H. Mianzan & J. R. Frost / Jellyfish Blooms: Interactions with Humans and Fisheries

C. J. Mooney (✉) · M. J. Kingsford
School of Marine and Tropical Biology and ARC Centre of Excellence in Coral Reef Studies, James Cook University, Townsville, QLD 4811, Australia
e-mail: christopher.mooney@my.jcu.edu.au

Our objective was to test the hypothesis that young *C. fleckeri* medusae originate in low salinity tidal creeks and disperse to coastal waters. Natural geochemical signatures based on the elemental or isotopic composition of calcified structures of living organisms can allow the natal sources of individuals to be determined (Gillanders & Kingsford, 1996; Thorrold et al., 2002). Differences in the trace elemental composition of calcified structures are largely influenced by exposure to different water masses or different environments (Arkhipkin et al., 2004), such as estuaries. Elemental signatures have been used to distinguish environmentally-segregated groups of a variety of organisms with calcifying structures (Gillanders & Kingsford, 1996; Gillanders, 2005). Calcifying structures that previously have been used include statoliths of molluscs (Clarke, 1978; Jackson, 1990; Ikeda et al., 1998; Arkhipkin et al., 2004; Arkhipkin, 2005), coral skeletons (Fallon et al., 2002), bivalve shells (Richardson, 1988; Leng & Pearce, 1999), walrus teeth (Evans et al., 1995), and in fish, scales, vertebrae, and most often, otoliths (Campana, 1999; Gillanders, 2001).

The promise for the potential use of the elemental composition of carbonate structures as a reliable method to understand animal movements among water masses and to distinguish among populations or stocks is founded largely on two key properties (Campana et al., 1999; Rooker et al., 2003): first, the structure is metabolically inert with no resorption (Campana & Neilson, 1985), and second, trace element uptake onto the growing structure reflects the physical and chemical environment of the organism during that period of time (Gillanders & Kingsford, 1996; Daverat et al., 2005; Swan et al., 2006). Otolith growth occurs within a semi-permeable membrane not subjected to resorption (Campana & Thorrold, 2001) and so is maintained even through periods of little or no somatic growth (Maillet & Checkley, 1990). Time calibration of any movements between water masses of different chemical composition is possible by comparison of changes in otolith elemental composition in daily (Pannella, 1971) or annual (Campana & Thorrold, 2001) growth increments; these key properties of otoliths are shared by cubozoan statoliths.

Statoliths are crucial to cubomedusa orientation (Sötje et al., 2011). The statolith forms within the statocyst, which consists of epidermal and gastrodermal cells and mesoglea, in the distal part of the rhopalium (Sötje et al., 2011). Daily growth increments have been found in statoliths of the cubomedusae *Carybdea rastoni* Haacke (Ueno et al., 1995), *Chironex yamaguchii* Lewis & Bentlage (as *Chiropsalmus quadrigatus* Haeckel in Kawamura et al., 2003), and *Chiropsella bronzie* Gershwin (as *Chiropsalmus* sp. Haeckel in Gordon et al., 2004), with the size of the statolith being positively correlated with medusa age. Daily growth rings and protection from the aqueous environment via the statocyst (Sötje et al., 2011) suggest that statolith elemental composition may prove a useful chronological measure of environmental exposure as for the otoliths of fishes. Cubomedusan statoliths have a $CaSO_4$ matrix (bassanite; Tiemann et al., 2006; Sötje et al., 2011) and not a $CaCO_3$ matrix as in fish otoliths and mollusc statoliths.

Hartwick (1991) suggested that *C. fleckeri* polyps replicate and young medusae form in tidal estuarine creeks. Then the medusae are thought to enter the sea at the beginning of each season, occurring during October/November in Queensland waters (Hartwick, 1991) and coinciding with the onset of the tropical wet season. We used elemental chemistry, as a proxy for salinity, and concentric growth increments to test the anecdotal paradigm that has evolved from Hartwick (1991) suggesting that medusae originate from lower salinity waters. The objectives of our study were to (1) test laser ablation-inductively coupled plasma mass spectrometry (LA-ICPMS) as a technique to elucidate movements of *C. fleckeri* medusae, (2) compare the elemental chemistry of core and edge zones of the statolith, and (3) measure elemental profiles from the beginning of statolith formation to medusa capture. The concentric increments allowed a within-statolith chronology in *C. fleckeri* to test the hypothesis that *C. fleckeri* medusae originate from low salinity estuarine creeks.

Materials and methods

Sampling sites

The generality of low salinity origins was tested by collecting medusae from ten locations among five main regions (Fig. 1) around Australia's northern coastline that represented a wide range of salinity regimes, ranging from marine (waters with no fresh water influence) to highly estuarine (water partially

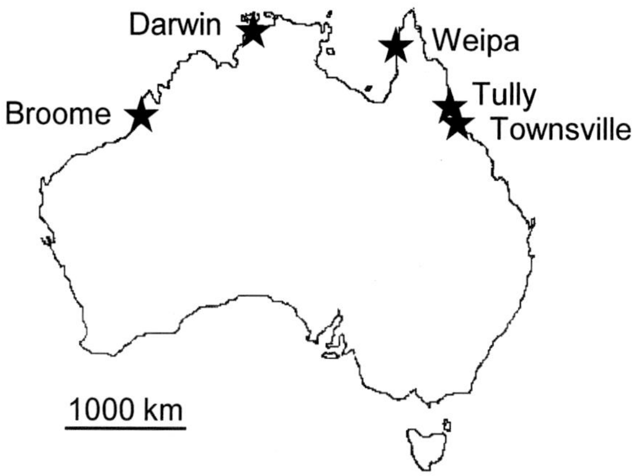

Fig. 1 Sampling regions (*stars*) for *Chironex fleckeri* medusae around Australia's northern coastline

surrounded by land where fresh river and ocean water mixed; Table 1). Medusae were caught in a seine net or individually from boats. All the rhopalial niches were extracted from medusae and preserved in 100% ethanol.

Statolith preparation

Each statolith was extracted from rhopalial tissue and placed on a glass slide. Crystal Bond adhesive was applied to a heated slide, which then was lowered onto the statolith so that it adhered. Once mounted and the Crystal Bond had cooled, each statolith was polished using 0.3 μm lapping film until a transverse section showing a smooth surface with concentric increments could be seen under magnification of an OLYMPUS CX31 compound microscope (Fig. 2). Statoliths were measured using Leica IM50 software coupled with a Leica DC300 camera fitted to a Leica DMLB microscope.

LA-ICPMS operation and data analysis

Statoliths were analysed using a Coherent GeolasPro excimer laser system ($\lambda = 193$ nm) coupled with a Varian 820-MS ICPMS. Statoliths were placed in a sealed sample cell mounted on an automated X–Y sample stage and ablation occurred in a helium (He) gas environment.

A range of parameters first were tested to determine the most accurate and precise measures of element concentration and to provide high temporal resolution by daily increments. The He carrier gas flow rate was kept constant at 240 ml min^{-1}, while other parameters were manipulated. Laser energies greater than 100 mJ caused major fracturing of the statolith material, even

Table 1 Sampling locations and dates for *Chironex fleckeri* medusae (*n*)

Location	Date	Site description	Latitude	Longitude	*n*
Roebuck Bay (Broome)	Feb 2007	Tidal beachfront ~5 km from Dampier Creek; marine	17°58′20S	122°14′19E	5
Fannie Bay (Darwin)	Apr 2007	Tidal flats of shallow bay >5 km from the nearest creek; marine	12°25′00S	130°49′45E	5
Wooldrum Point (Weipa)	Dec 2010	Uninterrupted stretch gently sloping sand/mud beachfront, extending ~9 km south of Embley Estuary mouth; strong marine	12°41′56S	141°47′33E	5
Duyfken Point (Weipa)	Jan 2006	Headland at northern end of Albatross Bay ~10 km from Pine River mouth; strong marine	12°35′42S	141°38′19E	5
Andermon Point (Weipa)	Dec 2009	Mangrove-lined mud flats near mouth of Mission River inside Albatross Bay; estuarine	12°36′01S	141°49′11E	5
Hey Point (Weipa)	Dec 2009	Intersection of Hey River & Embley Estuary, ~13 km inside mouth of Embley Estuary; strong estuarine	12°44′05S	141°53′31E	5
South Mission Beach (Tully)	Mar 2007	Gently sloping sand/mud beachfront ~7 km north of Hull River; little estuarine	17°56′13S	146°05′48E	3
Lugger Bay (Tully)	Jan 2007	Tidal flats ~5 km north of Hull River; little estuarine	17°57′39S	146°05′52E	5
Cardwell (Tully)	Feb 2010	Mangrove-lined mud tidal flats, >5 km from Hinchinbrook Channel; little estuarine	18°15′42S	146°01′39E	3
Townsville (Townsville)	Apr 2010	Sand/mud tidal flats & man-made marina in Cleveland Bay—catchment for numerous estuaries; estuarine	19°15′02S	146°49′20E	4

Site descriptions include water mass influences, which were categorized as estuarine or marine

Fig. 2 Polished *Chironex fleckeri* medusa statolith with concentric growth rings radiating (*dashed lines*) from statolith core

at a low number of pulses/spot and a mask aperture of 23 μm. An energy of 80 mJ and mask aperture of 16 μm resulted in little fracturing.

In the resulting protocol, all samples first were cleaned to remove any possible etching of the surface during preparation by a step-repeat path of 1 Hz at one pulse/spot with a 16 μm aperture. Then analysis was performed with a step-repeat path of 5 Hz at 10 pulses/spot with 16 μm aperture along the same core to edge transect. The mean (±SE) width of concentric growth increments from *C. fleckeri* statoliths equalled 6.69 ± 0.21 μm ($n = 38$ statoliths; three increments per statolith), indicating each measure of elemental composition was ~2.4 increments per spot. A continuous transect was not used as it was considered likely to result in fracturing and not maintain high temporal resolution by concentric increments.

Elements above detection limits included: sulphur (S), calcium (Ca), strontium (Sr) and barium (Ba), although Ba was generally very close to the detection limit. Sulphur was found at very high levels and was not included in further analysis because it is the major component of the cubozoan statolith matrix (bassanite, $CaSO_4{\cdot}0.5H_2O$; Tiemann et al., 2006; Sötje et al., 2011) and remained relatively consistent across statoliths.

For LA-ICPMS analyses, the ICPMS was started 30 s before laser ablation. Calibration of the ICPMS was achieved using the certified reference material NIST 610 (National Institute of Standards and Technology, Maryland), a synthetic glass containing known levels of elements. The same step-repeat cleaning and analysis procedure was performed twice on NIST 610 before and after every eight statolith samples to correct for elemental fractionation and instrument drift in the ICPMS. Iolite 2.11 (Hellstrom et al., 2008) was used for the subtraction of baselines to express raw counts as counts per second (CPS), internal standardization of elements to Ca, and the calculation of ppm values relative to NIST 610 glass data. As Iolite 2.11 used ^{44}Ca as the internal standard during calibration, ^{44}Ca now needed to be converted from CPS to ppm. The mean CPS of ^{44}Ca within the statolith sample was divided by the known concentration of ^{44}Ca in the $CaSO_4{\cdot}0.5H_2O$ matrix (determined by mass percent of ^{44}Ca in chemical formula expressed as ppm) to determine the sensitivity of ^{44}Ca in the statolith sample. ^{44}Ca CPS values were then divided by the sensitivity to determine ^{44}Ca ppm values (after Longerich et al., 1996). ^{44}Ca and ^{88}Sr concentrations (ppm) were then divided by their respective Molar mass to produce concentrations in μmol g^{-1}. Ratios of ^{88}Sr:^{44}Ca were presented as mmol mol^{-1}.

Laser ablation of 10 pulses per spot at 5 Hz equated to 2 s per spot during step-repeat transects. ICPMS scanning of plasma occurred ~8 times in a 2 s period. The mean of the middle six scans of 10 scans per spot was taken as the mean value per spot, accounting for ICPMS scans during time lag of laser mask movement between spots.

Elemental composition of statoliths

The source of *C. fleckeri* medusae was hypothesized to be from low salinity estuarine waters. Sr was used as a proxy for salinity as Sr is positively correlated with water salinities; this is particularly evident along estuarine/marine gradients (Daverat et al., 2005; McCulloch et al., 2005).

Salinity versus statolith elemental chemistry was calibrated at two locations separated by 1.5 km at the Wooldrum Point beachfront on the ocean side of the

Weipa estuary complex. Over three consecutive days, readings from a conductivity temperature and depth device (CTD; *Seabird*, SBE 19 Plus) were taken near shore throughout the water column of 1.5 m depth to determine the stability of salinity levels at time of medusa capture. Because the water column was well mixed, salinity readings were averaged for the water column at this depth. Five *C. fleckeri* medusae were collected from the same 1.5 km stretch at Wooldrum Point for this calibration and all were collected greater than 5 km from estuarine influence and in clear water; i.e. it was unlikely they moved 10 km to and from the river during the experiment. Medusa sizes were measured as inter-pedalial distance (IPD), the distance between the bases of pedalia on one side of bell (mean ± SE = 51.4 ± 4.4 mm). LA-ICPMS sampling for calibration was done only at the edge of the statolith over the last ~2.4 growth increments.

Two approaches were used: (1) Elemental comparisons were made between the core and edge of statoliths from jellyfish collected at multiple sites. The statolith core forms early in metamorphosis (Straehler-Pohl & Jarms, 2005), suggesting that the core would be a good indicator of medusa origin (Fig. 2). Mean core values were calculated from the average of two spots at the core. Those were compared with the mean of nine spots taken around the edge of the statolith (<4.4% variation between spots). (2) Sources and exposure of medusae to different inferred salinities were investigated by producing elemental ratio profiles along core to edge transects of each statolith.

Statistical analyses

The model that there would be consistent differences between the centre and edge of statoliths was tested with two-factor analysis of variance (ANOVA). The factor zone was treated as fixed (treatments: core and edge), site was treated as random (treatments: 10 sites, Table 1), and the analysis resulted in a zone*site interaction. Data met the assumptions for a parametric test. When investigating relationships between statolith core and edge zones, ANOVA assumes independent samples. There was no reason that the chemical composition at the core and edge should be dependent, because the bassanite is inert and layers deposited concentrically.

Results

Over three consecutive days of CTD measurements at two locations at Wooldrum Point, salinity remained consistent and variation was low (34.23 ± 0.07). This corresponded to a mean edge Sr:Ca ratio of 3.38 ± 0.19 mmol mol^{-1} for statoliths from this location and time (Fig. 3a).

The Sr:Ca ratios at the core of statoliths compared to the edge differed among sites (Fig. 3); this resulted in significantly different Sr:Ca ratios among sites ($F_{9,70} = 6.632$, $P < 0.001$) and a significant interaction between statolith zone and site ($F_{9,70} = 1.591$, $P < 0.001$), yet no significant difference was found between statolith core and edge ($F_{1,9} = 1.558$, $P > 0.05$). For example, ratios in statoliths from sites at Weipa and Tully (excluding Cardwell) had relatively low ratios at the core, all corresponding to salinities below 34 (Fig. 3). In contrast, statoliths from Roebuck Bay, Fannie Bay and, to a lesser extent,

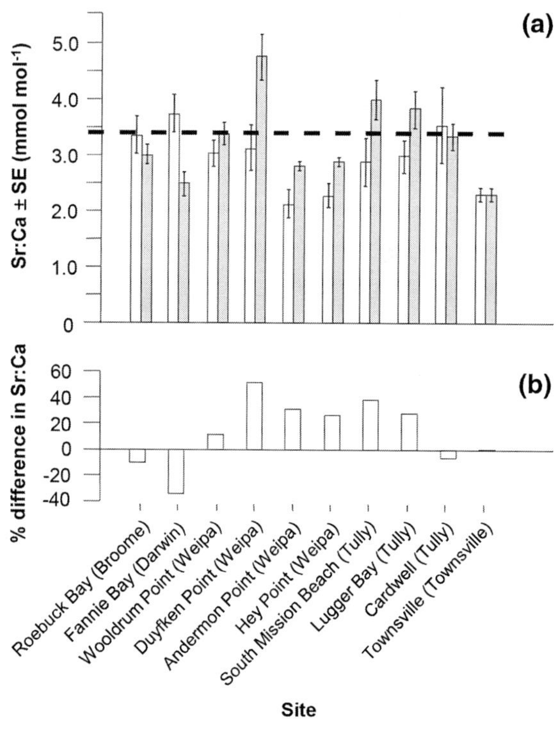

Fig. 3 **a** Mean Sr:Ca ratios ± SE in *Chironex fleckeri* medusa statoliths among sites between core (*white bars*) and edge (*grey bars*) zones relative to a salinity of 34 (*dashed line*), *n* varied: $3 \le n \le 5$; **b** percentage difference in Sr:Ca ratios from core to edge zones of statolith, positive indicates change from lower to higher ratios; negative indicates from higher to lower ratios

Cardwell all showed decreased mean Sr:Ca ratios from core to edge, suggesting medusa origination where Sr:Ca ratios corresponded to salinities at or above 34. In statoliths from Townsville, ratios were similar at edge and centre.

There was great variation in statolith elemental profiles among *C. fleckeri* medusae during the period between their production from polyps and their capture. Although a wide range of Sr:Ca ratios were found, elemental profiles were rarely below 2 mmol mol^{-1} or above 5 mmol mol^{-1} (Fig. 4). Statolith Sr:Ca ratios in some individuals showed little variation, either at relatively high or low ratios (e.g. Fig. 4a, e). Some medusae showed a great range of ratios during their life (e.g. Fig. 4b). Other medusa statoliths started at lower Sr:Ca ratios in early life and ratios increased with age (e.g. Fig. 4c, d).

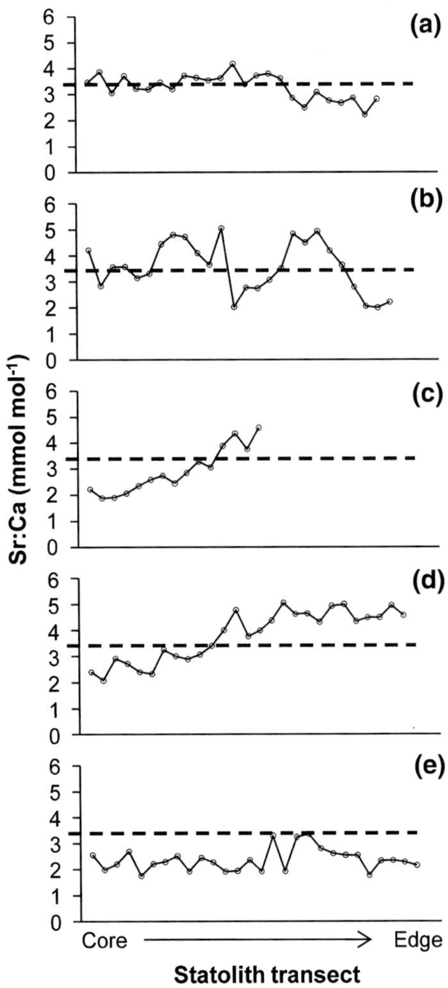

Fig. 4 Statolith elemental profiles of Sr:Ca ratios from *Chironex fleckeri* medusae relative to a salinity of 34 (*dashed line*); examples from individual medusa caught at **a** Roebuck Bay (Broome), **b** Fannie Bay (Darwin), **c** Duyfken Point (Weipa), **d** South Mission Beach (Tully), and **e** Townsville (Townsville)

Discussion

It was demonstrated that LA-ICPMS can be used to measure chronological records of elemental chemistry with high resolution in the statoliths of *C. fleckeri*. Furthermore, the techniques show the potential to address important ecological questions relating to cubozoan jellyfish. Previous use of LA-ICPMS studies for application in population identification of marine organisms has mainly focused on calcified structures consisting of a $CaCO_3$ crystal matrix (aragonite) in fish otoliths (Campana, 1999; Thorrold et al., 2001; Fowler et al., 2005) and cephalopod statoliths (Arkhipkin et al., 2004; Arkhipkin, 2005). Cubozoan statoliths consist of a $CaSO_4$ crystal matrix and not $CaCO_3$. Cubozoan statoliths were thought to be constructed of a $CaSO_4$ crystal matrix in the dihydrate form (gypsum; $CaSO_4{\cdot}2H_2O$), reported in the statoliths of *C. rastoni* by Ueno et al. (1995) and *C. bronzie* (as *Chiropsalmus* sp. by Chapman, 1985). Recent use of X-ray diffraction of *Carybdea* sp. Péron & Lesuer statoliths showed that the material is actually the water deficient (hemihydrate) form of $CaSO_4$ (bassanite; $CaSO_4{\cdot}0.5H_2O$; Tiemann et al., 2006). Single crystal X-ray diffraction also showed this to be true for *C. fleckeri* statoliths (Sötje et al., 2011). Statoliths of both cubozoans and scyphozoans have been shown to be constructed of the rare biomineral bassanite (Sötje et al., 2011).

The basic structural feature of all $CaSO_4$ subhydrates involves chains of alternating Ca^{2+} ions and SO_4^{2-} polyhedra that build up structural channels capable of accommodating different quantities of H_2O molecules (Abriel & Nesper, 1993) and trace elements as guest phases. Hydrogen bonding of H_2O to SO_4^{2-} groups is responsible for the relative stability of bassanite (Abriel & Nesper, 1993), meaning that elemental exchange of Ca^{2+} ions and elements from the surrounding water column may occur. This is thought to be the mechanism in fish otoliths that leads to the incorporation of contaminants and their utility for stock discrimination (Campana,

1999). Experiments are required to calibrate elemental ratios in the cubozoan bassanite statolith matrix with salinity and temperature, as done for fish otoliths (de Vries et al., 2005; Elsdon & Gillanders, 2005).

Field calibrations could be done only at locations with stable salinity where medusae were caught. Salinity varied only 0.2% from the mean over 3 days at Wooldrum Point where our calibration was done. The Sr:Ca ratios of statolith edge zones of medusae caught then varied only 5.7% from the mean, suggesting a relatively consistent exposure to water among individuals. Acoustic telemetry tracking of *C. fleckeri* medusae showed that six out of seven medusae tagged in coastal habitats remained in coastal habitats (Gordon & Seymour, 2009); the two medusae they tagged in Wooldrum Point remained there for the duration of the tracking period (up to 38 h). Although it is possible that the medusae caught in our study travelled out of the 1.5 km sampling area along Wooldrum Point during 3 days, the tracking experiment results plus the fact that exposure of medusae to estuarine influence would require a round trip of more than 10 km, gives us confidence that the Sr:Ca ratios in the outer ~2.4 growth increments of statoliths from this time reflected the salinity measured at Wooldrum Point. Because Sr:Ca ratios corresponding to a salinity ~34 measured at this marine site were among the highest ratios recorded, it is reasonable to conclude that lower ratios in statoliths was due to exposure to lower salinities.

Elemental profiles of *C. fleckeri* statoliths were highly variable among individuals and some showed great variation in the water masses they would have encountered. Adult medusae tracked with acoustic telemetry within coastal and estuarine habitats showed variable movements over short periods of time (Gordon & Seymour, 2009). Of twelve large medusae tagged, eleven remained in their respective coastal or estuarine habitats, while only one individual (IPD = 180 mm) travelled ~10 km over 26 h up and down a beachfront and in and out of an estuary (Gordon & Seymour, 2009). Given the movement of medusae in and out of estuarine areas and river plumes (Kingsford & Suthers, 1994; Grimes & Kingsford, 1996) over short periods of time, the great variation we found in the elemental chemistry of statoliths would be expected. The patterns of medusa movement over short periods (up to 38 h) shown by Gordon & Seymour (2009) resemble trends shown in statoliths that represented much longer time periods. The 16 μm laser spot we used covered ~2.4 growth rings. If the growth increments in *C. fleckeri* prove to be daily as in other cubomedusae (Ueno et al., 1995; Kawamura et al., 2003; Gordon et al., 2004), each reading would equal ~2.4 days. Extensive travel by medusae over short periods, as seen in one large medusa by Gordon & Seymour (2009), or medusae staying in one location during ebb and flood tides, would thus have less affect on these readings than multi-day movements by medusae between different water masses. Thus, trends we saw in Sr:Ca ratio elemental profiles would likely show mesoscale movements, or lack thereof, of medusae between different water masses, because microscale movements would not be shown by averages of ~2.4 days.

Not all sites showed an increase in Sr:Ca ratios from statolith core to edge. Hartwick (1991) suggested that *C. fleckeri* metamorphose from sessile polyps in tidal estuarine creeks to move as medusae into coastal waters. Given this anecdotal paradigm, one would expect to see increasing Sr:Ca ratios from statolith core to edge indicating movement of medusae from relatively lower to higher salinity waters. Medusae from most sites in the Weipa and Tully regions did show that trend; however, others did not. The differences between edge and core ratios varied greatly, suggesting that a consistent story of low salinity origins is unlikely. Thus, we rejected the hypothesis that *C. fleckeri* medusae only originate from low salinity estuarine creeks.

The use of Sr:Ca ratios in otoliths to establish fish movements in or out of estuaries is well established. Sr:Ca ratios are positively correlated with salinity in otoliths of euryhaline species such as barramundi, *Lates calcarifer* Bloch (McCulloch et al., 2005; Milton et al., 2008) and the European eel, *Anguilla anguilla* Linnaeus (Daverat et al., 2005), with the Sr:Ca ratios specific to site regardless of the history of previous salinity exposure (Daverat et al., 2005). Milton et al. (2008) found Sr:Ca to be the only trace metal that showed a consistent pattern of concentrations in fish from habitats of known salinity. They found that Sr:Ca levels within the otolith < 1.5 mmol mol^{-1} were characteristic of freshwater (salinity 0) and Sr:Ca > 2.25 mmol mol^{-1} characteristic of estuarine conditions (salinities 26–35). Our study found Sr:Ca > 2 mmol mol^{-1} in all *C. fleckeri* statoliths, suggesting that if the statolith bassanite matrix behaves like the otolith aragonite matrix, then *C. fleckeri* may never venture below mid-level salinities. Reliable low salinity values to correspond to statolith edge Sr:Ca ratios of wild-caught medusae, as we did for a salinity of 34, would be difficult because of salinity fluctuations in tidal areas. This further highlights the need for calibration experiments to establish if the bassanite matrix behaves like the otolith aragonite matrix.

In conclusion, this study showed that LA-ICPMS is a useful technique to elucidate elemental chronologies within *C. fleckeri* statoliths. High variability in Sr:Ca ratios was seen both within and between individual elemental profiles. The hypothesis that *C. fleckeri* medusae originate from low salinity estuarine creeks was rejected. Although Sr:Ca ratios of the statolith showed that some medusae had lower ratios at the core and higher ratios at the edge, many did not. Our results suggest that *C. fleckeri* medusae may originate from estuaries with low salinity and also from coastal environments with high salinity, and possibly hypersaline estuaries. Suitable habitat for polyps, therefore, is likely to be broader than previously considered. Experiments are required to determine relationships between elemental ratios in the bassanite statolith matrix of cubozoans and known salinities and temperatures.

Acknowledgments The authors thank Jamie Seymour and Matt Gordon for providing some archived statolith samples. The authors also thank Yi Hu at Advanced Analytical Centre, James Cook University for assistance with LA-ICPMS analyses, and the reviewers and editors for their help. Financial supports from the Marine and Tropical Science Research Facility and James Cook University are duly acknowledged.

References

Abriel, W. & R. Nesper, 1993. Determination of crystal structure of $CaSO_4(H_2O)0.5$ by X-ray diffraction and potential profile calculations. Zeitschrift für Kristallographie 205: 99–113.

Arkhipkin, A. I., 2005. Statoliths as 'black boxes' (life recorders) in squid. Marine & Freshwater Research 56: 573–583.

Arkhipkin, A. I., S. E. Campana, J. Fitzgerald & S. R. Thorrold, 2004. Spatial and temporal variation in elemental signatures of statoliths from the Patagonian longfin squid (*Loligo gahi*). Canadian Journal of Fisheries and Aquatic Sciences 61: 1212–1224.

Campana, S. E., 1999. Chemistry and composition of fish otoliths: pathways, mechanisms and applications. Marine Ecology Progress Series 188: 263–297.

Campana, S. E. & J. D. Neilson, 1985. Microstructure of fish otoliths. Canadian Journal of Fisheries and Aquatic Sciences 42: 1014–1032.

Campana, S. E. & S. R. Thorrold, 2001. Otoliths, increments and elements: keys to a comprehensive understanding of fish populations? Canadian Journal of Fisheries and Aquatic Sciences 58: 30–38.

Campana, S. E., G. A. Chouinard, J. M. Hanson & A. Frechet, 1999. Mixing and migration of overwintering cod stocks near the mouth of the Gulf of St. Lawrence. Canadian Journal of Fisheries and Aquatic Sciences 56: 1873–1881.

Chapman, D. M., 1985. X-ray microanalysis of selected coelenterate statoliths. Journal Marine Biology Association UK 65: 617–627.

Clarke, M. R., 1978. The cephalopod statolith – an introduction to its form. Journal Marine Biology Association UK 58: 701–712.

Coates, M. M., 2003. Visual ecology and functional morphology of Cubozoa (Cnidaria). Integrative and Comparative Biology 43: 542–548.

Coughlan, J. P., J. Seymour & T. F. Cross, 2006. Isolation and characterisation of seven polymorphic microsatellite loci in the box jellyfish (*Chironex fleckeri*, Cubozoa, Cnidaria). Molecular Ecology Notes 6: 41–43.

Currie, B. J. & S. P. Jacups, 2005. Prospective study of *Chironex fleckeri* and other box jellyfish stings in the "Top End" of Australia's Northern Territory. Medical Journal Australia 183: 631–636.

Daverat, F., J. Tomas, M. Lahaye, M. Palmer & P. Elie, 2005. Tracking continental habitat shifts of eels using otolith Sr/Ca ratios: validation and application to the coastal, estuarine and riverine eels of the Gironde–Garonne–Dordogne watershed. Marine & Freshwater Research 56: 619–627.

de Vries, M. C., B. M. Gillanders & T. S. Elsdon, 2005. Facilitation of barium uptake into fish otoliths: influence of strontium concentration and salinity. Geochimica et Cosmochimica Acta 69: 4061–4072.

Elsdon, T. S. & B. M. Gillanders, 2005. Consistency of patterns between laboratory experiments and field collected fish in otolith chemistry: an example and applications for salinity reconstructions. Marine & Freshwater Research 56: 609–617.

Evans, R. D., P. Richner & P. M. Outridge, 1995. Micro-spatial variations in heavy metals in the teeth of walrus as determined by laser ablation ICP-MS: the potential for reconstructing a history of metal exposure. Archives of Environmental Contamination and Toxicology 28: 55–60.

Fallon, S. J., J. C. White & M. T. McCulloch, 2002. Porites corals as recorders of mining and environmental impacts: Misima Island, Papua New Guinea. Geochemica et Cosmochimica Acta 66: 45–62.

Fowler, A. J., B. M. Gillanders & K. C. Hall, 2005. Relationship between elemental concentration and age from otoliths of adult snapper (*Pagrus auratus*, Sparidae): implications for movement and stock structure. Marine & Freshwater Research 56: 661–676.

Gillanders, B. M., 2001. Trace metals in four structures of fish and their use for estimates of stock structure. Fishery Bulletin 99: 410–419.

Gillanders, B. M., 2005. Using elemental chemistry of fish otoliths to determine connectivity between estuarine and coastal habitats. Estuarine, Coastal and Shelf Science 64: 47–57.

Gillanders, B. M. & M. J. Kingsford, 1996. Elements in otoliths may elucidate the contribution of estuarine recruitment to sustaining coastal reef populations of a temperate reef fish. Marine Ecology Progress Series 141: 13–20.

Gordon, M. R. & J. E. Seymour, 2009. Quantifying movement of the tropical Australian cubozoan *Chironex fleckeri* using acoustic telemetry. Hydrobiologia 616: 87–97.

Gordon, M., C. Hatcher & J. Seymour, 2004. Growth and age determination of the tropical Australian cubozoan *Chiropsalmus* sp. Hydrobiologia 530: 339–345.

Grimes, C. B. & M. J. Kingsford, 1996. How do riverine plumes of different sizes influence fish larvae: do they enhance recruitment? Marine & Freshwater Research 47: 191–208.

Hamner, W. M., M. S. Jones & P. P. Hamner, 1995. Swimming, feeding, circulation, and vision in the Australian box jellyfish, *Chironex fleckeri* (Cnidaria: Cubozoa). Marine & Freshwater Research 46: 985–990.

Hartwick, R. F., 1991. Distribution ecology and behaviour of the early life stages of the box-jellyfish *Chironex fleckeri*. Hydrobiologia 216: 181–188.

Hellstrom, J., C. Paton, J. D. Woodhead & J. M. Hergt, 2008. Iolite: software for spatially resolved LA-(quad and MC) ICPMS analysis. In Sylvester, P. (ed.), Laser Ablation ICP-MS in the Earth Sciences: Current Practices and Outstanding Issues. Mineralogical Association of Canada Short Course Series, Vol. 40: 343–348.

Ikeda, Y., N. Arai, W. Sakamoto, H. Kidokoro & K. Yoshida, 1998. Microchemistry of the statoliths of the Japanese common squid *Todarodes pacificus* with special reference to its relation to the vertical temperature profiles of squid habitat. Fisheries Science 64: 179–184.

Jackson, G. D., 1990. Age and growth of the tropical nearshore loliginid squid *Sepioteuthis lessoniana* determined from statolith growth-ring analysis. Fishery Bulletin US 88: 113–118.

Kawamura, M., S. Ueno, S. Iwanaga, N. Oshiro & S. Kubota, 2003. The relationship between fine rings in the statolith and growth of the cubomedusa *Chiropsalmus quadrigatus* (Cnidaria: Cubozoa) from Okinawa Island, Japan. Plankton Biology & Ecology 50: 37–42.

Kingsford, M. J. & I. M. Suthers, 1994. Dynamic estuarine plumes and fronts: importance to small fish and plankton in coastal waters of NSW, Australia. Continental Shelf Research 14: 655–672.

Kingsford, M. J., J. E. Seymour & M. D. O'Callaghan, 2012. Abundance patterns of cubozoans on and near the Great Barrier Reef. Hydrobiologia. doi:10.1007/s10750-012-1041-0

Leng, M. J. & N. J. G. Pearce, 1999. Seasonal variation of trace element and isotopic composition in the shell of a coastal mollusk, *Mactra isabelleana*. Journal of Shellfish Research 18: 569–574.

Longerich, H. P., S. E. Jackson & D. Günther, 1996. Laser ablation inductively coupled plasma mass spectrometric transient signal data acquisition and analyte concentration calculation. Journal of Analytical Atomic Spectrometry 11: 899–904.

Maillet, G. L. & J. D. M. Checkley, 1990. Effects of starvation on the frequency of formation and width of growth increments in sagittae of laboratory-reared Atlantic menhaden *Brevoortia tyrannus* larvae. Fishery Bulletin US 88: 155–165.

McCulloch, M., M. Cappo, J. Aumend & W. Muller, 2005. Tracing the life history of individual barramundi using laser ablation MC-ICP-MS Sr-isotopic and Sr/Ba ratios in otoliths. Marine & Freshwater Research 56: 637–644.

Milton, D. A., I. Halliday, M. Sellin, R. Marsh, J. Staunton-Smith & J. Woodhead, 2008. The effect of habitat and environmental history on otolith chemistry of barramundi *Lates calcarifer* in estuarine population of a regulated tropical river. Estuarine, Coastal and Shelf Science 78: 301–315.

Pannella, G., 1971. Fish otoliths: daily growth layers and periodical patterns. Science 173: 1124–1127.

Richardson, C. A., 1988. Exogenous and endogenous rhythms of band formation in the shell of the clam *Tapes philippinarum*. Journal Experimental Marine Biology and Ecology 122: 105–126.

Rooker, J. R., D. H. Secor, V. S. Zdanowicz, G. De Metrio & L. O. Relini, 2003. Identification of Atlantic bluefin tuna (*Thunnus thynnus*) stocks from putative nurseries using otolith chemistry. Fishery Oceanography 12: 75–84.

Sötje, I., F. Neues, M. Epple, W. Ludwig, A. Rack, M. Gordon, R. Boese & H. Tiemann, 2011. Comparison of statolith structures of *Chironex fleckeri* (Cnidaria, Cubozoa) and *Periphylla periphylla* (Cnidaria, Scyphozoa): a phylogenetic approach. Marine Biology 158: 1149–1161.

Straehler-Pohl, I. & G. Jarms, 2005. Life cycle of *Carybdea marsupialis* Linnaeus, 1758 (Cubozoa, Carybdeidae) reveals metamorphosis to be a modified strobilation. Marine Biology 147: 1271–1277.

Swan, S. C., A. J. Geffen, B. Morales-Nin, J. D. M. Gordon, T. Shimmield, T. Sawyer & E. Massuti, 2006. Otolith chemistry: an aid to stock separation of *Helicolenus dactylopterus* (bluemouth) and *Merluccius merluccius* (European hake) in the Northeast Atlantic and Mediterranean. ICES Journal of Marine Science 63: 504–513.

Thorrold, S. R., C. Latkoczy, P. K. Swart & C. M. Jones, 2001. Natal homing in a marine fish metapopulation. Science 291: 297–299.

Thorrold, S. R., G. P. Jones, M. E. Hellberg, R. S. Burton, S. E. Swearer, J. E. Niegel, S. G. Morgan & R. R. Warner, 2002. Quantifying larval retention and connectivity in marine populations with artificial and natural markers. Bulletin Marine Science 70: 291–308.

Tibballs, J., 2006. Australian venomous jellyfish, envenomation syndromes, toxins and therapy. Toxicon 48: 830–859.

Tiemann, H., I. Sötje, A. Becker, G. Jarms & M. Epple, 2006. Calcium sulfate hemihydrate (bassanite) statoliths in the cubozoan *Carybdea* sp. Zoologischer Anzeiger 245: 13–17.

Ueno, S., C. Imai & A. Mitsutani, 1995. Fine growth rings found in statolith of a cubomedusa *Carybdea rastoni*. Journal of Plankton Research 17: 1381–1384.

Hydrobiologia (2012) 690:279–290
DOI 10.1007/s10750-012-1051-y

JELLYFISH BLOOMS

Variation in soft tissue chemistry among scyphozoan and cubozoan jellyfishes from the Great Barrier Reef, Australia

Michelle A. Templeman · Michael J. Kingsford

Published online: 9 March 2012
© Springer Science+Business Media B.V. 2012

Abstract Bioaccumulation of trace elements in jellyfish has so far received little attention, despite their being prey for many animals from multiple trophic levels and targeted by commercial jellyfish fisheries. Scyphozoan and cubozoan jellyfish were collected over a three year period from across-shelf and along the northern and central Great Barrier Reef, Australia. To test the hypotheses that jellyfishes were able to accumulate elements above ambient background levels, and if there were spatial or temporal variations among species, soft tissue concentrations of 14 trace elements were compared with ambient seawater concentrations. Most elements, including aluminium, arsenic, barium, cadmium, chromium, copper, iron, manganese and zinc were measured at concentrations above ambient seawater levels indicating bioaccumulative capacity. Results showed some regulation of lithium in *Cassiopea* sp., *Cyanea* sp. and *Mastigias* sp., while calcium, magnesium and strontium reflected ambient conditions for all species. Accumulation varied significantly among species and sampling locations. For *Mastigias* sp. and *Netrostoma* sp., tissue concentrations of Al, As, Cu, Fe and Zn decreased with distance from the mainland. The hypothesis that jellyfishes are capable of accumulating trace elements was accepted, and their use as biomonitors should be investigated further.

Keywords Jellyfish · Scyphozoa · Cubozoa · Trace elements · Heavy metals · Great Barrier Reef

Introduction

Knowledge of environmental levels of dissolved metals in marine waters is important for monitoring ecosystem health. A large body of research exists on the levels, fluxes and cycling of chemicals in marine environments (e.g. Sadiq, 1992; Luoma & Rainbow, 2008); however, concentrations of metals do not necessarily reflect their ecological significance to biota. The use of organisms as marine biomonitors is an important tool for understanding how time-integrated measures of changes in water quality can affect the diversity and abundance of local biota (Bresler et al., 2003; Luoma & Rainbow, 2008; Creighton & Twining, 2010). The term, biomonitor, in this context, is defined as the ability to accumulate metals from the surrounding environment in an organism's tissues (Luoma & Rainbow, 2008).

The ability to absorb, store and detoxify metals is important for animals exposed to dissolved metals in

Guest editors: J. E. Purcell, H. Mianzan & J. R. Frost / Jellyfish Blooms: Interactions with Humans and Fisheries

M. A. Templeman (✉) · M. J. Kingsford
School Marine & Tropical Biology and ARC Centre of Excellence for Coral Reef Studies, James Cook University, Townsville, QLD 4811, Australia
e-mail: Shelley.Templeman@jcu.edu.au

the marine environment, as exposure and accumulation above a threshold can be damaging or deleterious to the animal (e.g. Chapman, 1995). A wide variety of invertebrates and vertebrates have been investigated to determine their ability to regulate or accumulate dissolved metals (e.g. Benson & Summons, 1981; Rainbow & Phillips, 1993; Ruus et al., 2005). The ability of some species (e.g. barnacles) to readily accumulate metals makes them very useful as marine biomonitors (e.g. Rainbow & Phillips, 1993; Bresler et al., 2003). Unfortunately, there are few published data on accumulation of dissolved metals in scyphozoan jellyfishes (Templeman & Kingsford, 2010) and no reported elemental tissue concentrations for cubozoan jellyfishes. Jellyfishes seem to have the capacity to absorb an elemental load that can significantly influence metal concentrations in the surrounding seawater (Heymans & Baird, 2000; Kingsford et al., 2000; Fukuda & Naganuma, 2001; Hay, 2006; Pitt et al., 2009; Jantzen et al., 2010). Based on the limited information available, it is hypothesised that jellyfishes are capable of accumulating metals and providing an elemental signal that may reflect ambient exposures. In addition, due to their ability to consume plankton and turn over large volumes of water, low ambient concentrations of dissolved metals may still result in high body loads in jellyfish. However, it is not known how extensive this accumulation is among species, and whether there is measurable accumulation on a spatial or temporal basis in jellyfishes.

Metals in tissues can be classified as essential and non-essential elements. Essential elements include arsenic (As), copper (Cu), cobalt (Co), iron (Fe), manganese (Mn) and zinc (Zn) (Luoma & Rainbow, 2008). Trace amounts of these elements are essential to metabolic activity in organisms, but levels below a threshold can result in sub-optimal health. For many of these elements, organisms are able to regulate them to meet their metabolic requirements, store them in a detoxified form or excrete them if thresholds are exceeded (Rainbow, 2007). Non-essential elements including aluminium (Al), cadmium (Cd), lead (Pb) and mercury (Hg) that have no identified role in metabolic activity can also be accumulated in tissues above ambient concentrations. Accumulation of non-essential elements can be harmful unless detoxification mechanisms are available.

Accumulation of dissolved metals in jellyfish has the potential to indirectly influence the health of higher-ordered predators through trophic transfer (Kingsford et al., 2000). Despite historical arguments of jellyfish as 'trophic dead ends', evidence (Purcell & Arai, 2001; Arai, 2005; Pauly et al., 2009) suggests that jellyfish form a significant proportion of the diet of many marine animals, including other gelatinous zooplankton (Purcell, 1991), cephalopods (Heeger et al., 1992), *Fungia* sp. (Alamaru et al., 2009), nudibranchs (reviewed in Arai, 2005), turtles (Caurant et al., 1999), seabirds (Harrison, 1984) and fish (Pauly et al., 2009). Furthermore, jellyfish fisheries are important industries providing dried tissue for human consumption, particularly in Asian cuisine (Kingsford et al., 2000; Kitamura & Omori, 2010).

To confirm whether jellyfish play a role in the transfer of metals up the food chain requires information on the elemental loads in jellyfish tissues. Thus, the objectives of this study were to test the null hypotheses that (1) elemental concentrations do not accumulate in scyphozoan and cubozoan jellyfishes collected from multiple locations along the Great Barrier Reef (GBR), (2) the extent of elemental accumulation above ambient seawater did not differ among jellyfish species, and (3) there were no differences in tissue elements among species, times or location.

Materials and methods

Specimen and water collection and handling

Jellyfish and water samples were collected between December 2007 and March 2010 from multiple locations along the GBR, Australia (Fig. 1; Table 1). In order to obtain a range of jellyfish species, sampling was conducted at four latitudes and multiple distances from the mainland (coast, inner, mid- and outer-shelves). Medusae of five species of Scyphozoa and two species of Cubozoa were collected at or near the surface using either dip or seine nets, with targeted collections during the day or under lights (1,000 W) at night. *Cassiopea* sp. (Péron & Lesueur 1810) were collected in plastic bags by SCUBA divers at depths between 7 and 12 m. At least three medusae were collected at each sampling location except one (Table 1). Surface water samples were collected at the same time as the medusae, except for a single collection of *Chironex fleckeri* (Southcott 1956), when

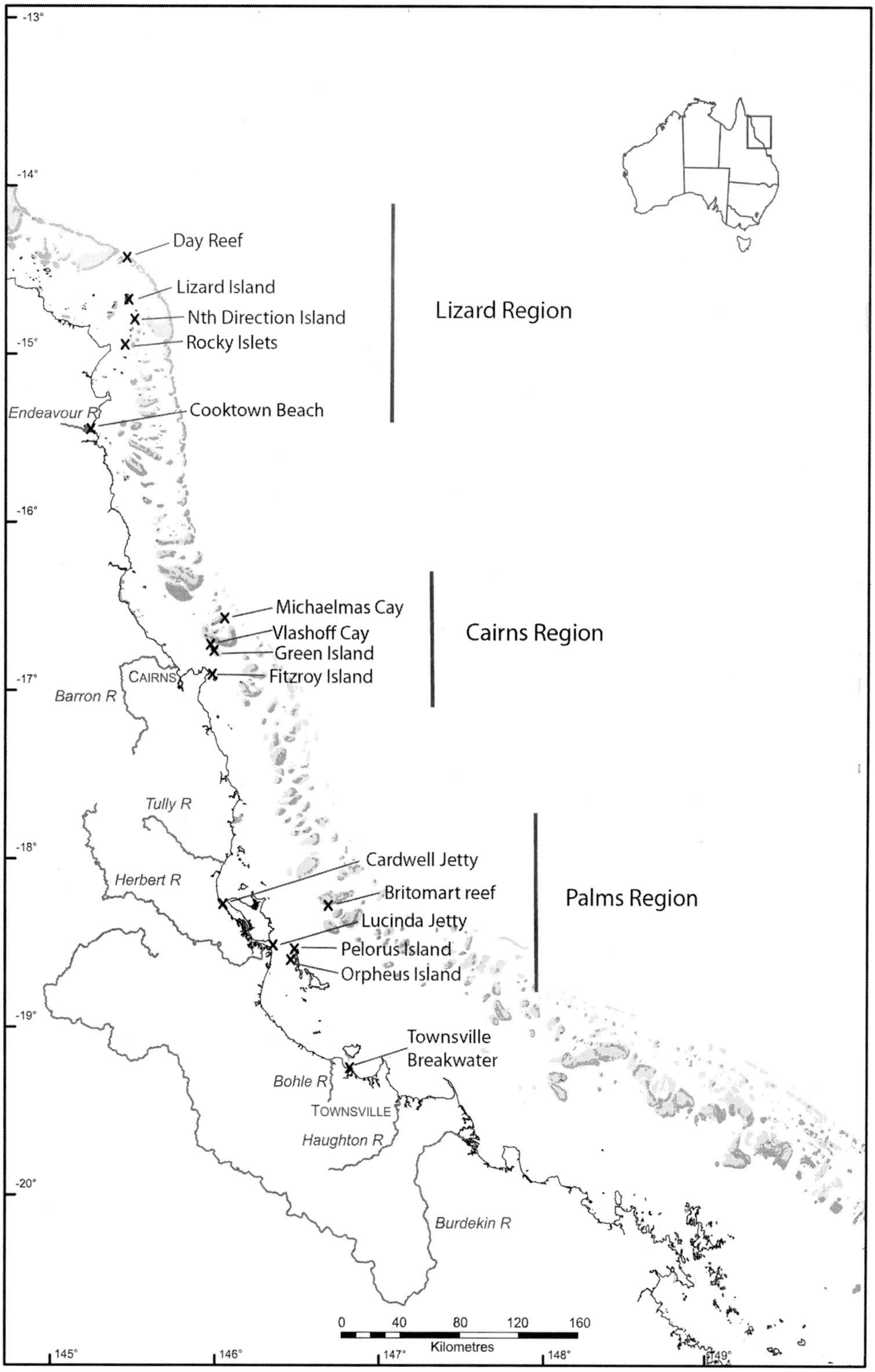

Fig. 1 Field sampling locations between 2007 and 2010 along the GBR, Australia. Locations are separated by region

Table 1 Sampling locations for elemental analysis of medusae collected between 2007 and 2010

Collection date	Sampling site	Region	Cross-shelf location	Species	No. medusae	Sample coding
10 Dec 2007	North Direction Island	Lizard	Mid	*Aurelia* sp.	5	A1
11 Dec 2007	Day Reef	Lizard	Outer	*Aurelia* sp.	5	A2
07 Jan 2008	Michaelmas Reef	Cairns	Outer	*Aurelia* sp.	5	A3
09 Feb 2008	Britomart Reef	Palms	Mid	*Aurelia* sp.	5	A4
10 Feb 2010	Britomart Reef	Palms	Mid	*Aurelia* sp.	3	A5
11 Dec 2007	Mermaid Bay	Lizard	Mid	*C. sivickisi*	3	B1
13 Dec 2008	Mermaid Bay	Lizard	Mid	*C. sivickisi*	4	B2
20 Jan 2009	Green Island	Cairns	Mid	*C. sivickisi*	5	B3
11 Dec 2007	Mermaid Bay	Lizard	Mid	*Mastigias* sp.	5	C1
14 Dec 2008	Mermaid Bay	Lizard	Mid	*Mastigias* sp.	2	C2
17 Dec 2009	Rocky Islets	Lizard	Inner	*Mastigias* sp.	1	C3
18 Dec 2009	Cooktown Beach	Lizard	Coastal	*Mastigias* sp.	11	C4
29 Dec 2009	Townsville Breakwater	Townsville	Coastal	*Mastigias* sp.	3	C5
09 Jan 2008	Fitzroy Island	Cairns	Coastal	*Netrostoma* sp.	5	D1
22 Jan 2009	Michaelmas Reef	Cairns	Outer	*Netrostoma* sp.	5	D2
09 Feb 2010	Pelorus Island	Palms	Inner	*Netrostoma* sp.	3	D3
09 Feb 2010	Orpheus island	Palms	Inner	*Netrostoma* sp.	4	D4
07 Feb 2008	Pelorus Island	Palms	Inner	*Cyanea* sp.	5	E1
15 Mar 2008	Townsville Breakwater	Townsville	Coastal	*Cyanea* sp.	5	E2
09 Feb 2010	Lucinda Jetty	Palms	Coastal	*Cyanea* sp.	10	E3
14 Dec 2008	Lizard Island Lagoon 1	Lizard	Mid	*Cassiopea* sp.	5	F1
14 Dec 2008	Lizard island Lagoon 2	Lizard	Mid	*Cassiopea* sp.	5	F2
21 Jan 2009	Vlashoff Cay	Cairns	Mid	*Cassiopea* sp.	6	F3
17 Dec 2009	Lizard Island Lagoon 1	Lizard	Mid	*Cassiopea* sp.	5	F4
17 Dec 2009	Lizard island Lagoon 2	Lizard	Mid	*Cassiopea* sp.	11	F5
05 Jan 2010	Vlashoff Cay	Cairns	Mid	*Cassiopea* sp.	9	F6
18 Feb 2010	Cardwell Jetty	Palms	Coastal	*C. fleckeri*	3	G1

Shelf locations represent position across the GBR. Sample coding used in the legend of Fig. 4

no water samples could be taken. All equipment and containers used in jellyfish collections and processing were cleaned using 10% nitric acid (HNO_3), triple rinsed in deionised water and allowed to air dry in a Class 100 laminar flow unit before use. Sampling containers were stored in clean plastic bags until needed to avoid metal contamination.

After collection, jellyfish were rinsed with ambient water to remove any visible sediment or other material adhering to the animals. Animals < 40 mm in diameter were placed in clean, acid-washed vials. Animals > 40 mm in diameter were sub-sampled using a corer consisting of an acid-washed 30-ml plastic vial. Five random cores were taken from the swimming bell, stomach, gonads (if present), oral arms and tentacles of each medusa. Owing to their small sizes, 4–5 *Copula sivickisi* (Stiasny, 1926) medusae were pooled per replicate sample. Tissue samples were frozen as soon as possible and kept at −18°C until they were processed.

Duplicate 30-ml water samples were collected at the water surface, immediately filtered through a 0.45-μm-pore size syringe filter, and stored in acid-washed vials. Water samples were acidified on-site with 20% Suprapur grade HNO_3 and stored at 4°C until analysis.

Tissue processing and analysis

Tissue samples for subsequent chemical analysis were digested with a heated acid solution. Samples ranging from 0.3 to 3.0 g wet weight were digested in 5 ml concentrated (69%) Suprapur grade HNO_3 on a hot plate for approximately 2 h. Then, 3–5 ml of AR grade hydrogen peroxide was added to remove any residual organic carbon and colour. Once all samples were clear with no residual colour, they were evaporated again to a final volume of approximately 2 ml. Samples were cooled to room temperature and then brought to a final volume of 25 ml with Milli-Q water. This digestion method is similar to that used previously for jellyfish tissue digestion (Templeman & Kingsford, 2010).

Both digested tissue and water samples were analysed using a Varian 820-MS Inductively Coupled Plasma Mass Spectrometer (ICP-MS) and Varian Liberty Series II Inductively Coupled Plasma Atomic Emission Spectrometer (ICP-AES). ICP-MS was used to determine aluminium (Al), arsenic (As), barium (Ba), copper (Cu), cadmium (Cd), chromium (Cr), lithium (Li), manganese (Mn), lead (Pb), strontium (Sr) and Zn, while ICP-AES was used to measure calcium (Ca), magnesium (Mg) and Fe. The detection limits were 0.1 $\mu g\ l^{-1}$ for Ba, Cr, Cu, Li, Mg, Mn and Sr; 0.05 $\mu g\ l^{-1}$ for Pb and Cd; 0.5 $\mu g\ l^{-1}$ for Al; 1.0 $\mu g\ l^{-1}$ for As and Fe; 2.0 $\mu g\ l^{-1}$ for Zn; and 10.0 $\mu g\ l^{-1}$ for Ca. Elements chosen for analysis were based on either their importance as essential metabolic elements or their consideration as anthropogenic or priority pollutants. Owing to issues with signal suppression, it was necessary to dilute seawater samples 1:10 (seawater:diluent) before analysis.

Indium, gallium and yttrium were used as internal standards to correct for potential instrument drift and matrix effects. Subsets of samples were spiked with known concentrations of all elements for quality control and to determine recoveries (72–116%). Analytical data were checked to ensure that signal strength of results exceeded three standard deviations for all analyses. Digestion blanks were included to ensure integrity of the digestion process. Digestion blanks had low levels of contamination; therefore, tissue data were corrected for blank results before statistical analyses.

Data analysis

Bioconcentrations were calculated by dividing the metal concentration in tissue by the metal concentration in seawater for each species (Sadiq, 1992; Parametrix, 1995). Univariate data for distance from the mainland were analysed with a one-way ANOVA (Statistica Version 9.0); data were transformed where necessary to meet assumptions of normality and homogeneity of variance (Bartlett's Test). Principal components analysis (PCA) was performed using SYSTAT Version 10 (Crane Software) after ln + 1 transformation to describe spatial and temporal variation in multi-element signatures, following the recommendations of Legendre & Legendre (2003).

Results

Elemental concentrations in jellyfish

Concentrations of elements in jellyfish tissues varied among species, with a range of values found among both years and locations (Table 2). The cubozoan *C. sivickisi* had the highest mean concentration of most elements except for the osmoconforming elements Ca, Mg and Sr (Table 2). The other cubozoan species, *C. fleckeri,* had much lower concentrations of all elements except for Mg, Ca, Sr and Fe than did *C. sivickisi.*

Among the scyphozoans, the rhizostome jellyfishes with symbiotic zooxanthellae generally had higher mean tissue element concentrations than asymbiotic species. Among the symbiotic species, *Cassiopea* sp. and *Mastigias* sp. had higher concentrations of most elements than did *Netrostoma* sp. (Table 2). *Cassiopea* sp. also had higher concentrations of Al, Ba, Cd, Cu, Fe, Mn and Sr than all other scyphozoan species. Among the Semaestomeae, *Cyanea* sp. typically had higher concentrations of most elements than *Aurelia* sp. The major osmoconforming elements of Ca, Li, Mg and Sr were similar across all scyphozoan species (Table 2).

The concentration range for individual elements differed among species in a similar way in terms of the mean concentrations. *Cassiopea* sp. and *C. sivickisi* had high variation in elemental concentrations for many elements. For *Cassiopea* sp., there were large variations in Al (4–2,840 $\mu g\ kg^{-1}$), Cr (1.6–276 $\mu g\ kg^{-1}$) and Cu (48–261 $\mu g\ kg^{-1}$). In *C. sivickisi*, the greatest variation in elements occurred in Al (65–5,883 $\mu g\ kg^{-1}$), Cr (12–349 $\mu g\ kg^{-1}$) and Fe (751–5,216 $\mu g\ kg^{-1}$) (Table 2). *Cyanea* sp. also showed high variation among locations and years for As (53–2,100 $\mu g\ kg^{-1}$) and Fe (74–3,440 $\mu g\ kg^{-1}$), while the concentration ranges tended to be lower for *Aurelia* sp. and *Netrostoma* sp. for most elements (Table 2).

Table 2 Mean and range of elemental tissue concentrations by species and in the water

Species	Al	As	Ba	Ca	Cd	Cr	Cu	Fe	Li	Mg	Mn	Pb	Sr	Zn
Aurelia sp. (23)														
Mean	11.4	63.0	32.9	348	5.72	6.58	15.0	82.3	154	1,118	12.8	1.51	6,380	93.7
Range	1.78–58.9	5.45–135	9.19–125	293–400	2.50–12.6	1.86–9.80	8.10–43.0	12.4–455	131–208	970–1,230	6.80–22.2	0.55–6.56	4,800–8,750	47.0–313
Water	<DL–1.41	<DL	<DL–4.88	359–395	0.45–7.16	<DL–7.66	<DL–3.03	<DL	127–145	1,130–1,235	<DL–2.48	<DL	6.40–8.70	<DL–8.21
Cassiopea sp. (41)														
Mean	519	186	173	438	33.2	44.9	86.0	2,756	120	1,116	193	9.49	28,267	572
Range	4.24–2,840	65.4–313	18.7–1,670	295–811	10.4–69.6	1.57–276	48.2–261	1,147–5,510	11.4–226	809–1,620	68.3–522	0.61–60.0	11,800–73,900	298–1,980
Water	<DL	<DL	5.22–5.94	363–444	<DL–3.77	<DL	<DL–5.23	<DL	144–182	1,037–1,280	<DL–3.71	<DL–0.14	7.72–9.11	<DL–2.35
C. fleckeri (3)														
Mean	157	236	17.9	260	0.52	3.21	65.3	681	81.2	765	57.5	3.05	5,023	553
Range	92.5–206	139–365	15.2–22.9	255–268	0.50–0.57	2.39–4.49	42.4–84	374–901	71.3–90.6	737–816	44.2–84.0	2.06–4.57	4,680–5,380	314–714
*C. sivickisi** (12)														
Mean	1,266	4,137	50.9	609	410	90.0	475	3,391	173	1,357	224	115	10,913	6,028
Range	64.6–5,883	2,380–5,806	13.4–144	372–1,040	220–691	11.8–349	234–751	751–5,216	123–282	900–1,980	102–382	12.1–207	6,564–19,400	4,280–9,292
Water	<DL–10.9	<DL	<DL–6.14	364–378	0.70–2.96	<DL	<DL	<DL	145–180	1,030–1,175	<DL–2.52	<DL–1.44	6.60–8.77	<DL
Cyanea sp. (20)														
Mean	75.3	473	37.3	263	38.7	11.1	63.3	908	114	865	22.2	2.41	5,437	537
Range	7.35–636	52.8–2,100	6.63–180	210–329	4.23–184	1.19–22.7	24.2–116	73.9–3,440	89.3–142	6,652–1,110	5.11–85.1	0.74–7.55	4,880–6,290	109–1,600
Water	<DL–1.17	<DL	<DL–8.11	317–380	<DL–1.48	<DL–8.6	<DL–1.97	<DL	107–182	974–1,120	2.09–7.52	<DL–0.74	5.80–7.23	<DL–11.62
Mastigias sp. (22)														
Mean	160	177	59.5	353	12.0	17.3	50.8	981	144	1,124	91.8	18.6	8,002	586
Range	3.99–468	61.5–542	7.88–142	326–385	5.68–23.1	0.80–35.2	19.6–74.3	130–1,940	123–170	1,030–1,210	12.8–161	2.27–35	6,690–9,460	203–1,100
Water	<DL–10.9	<DL	<DL–13.05	364–378	0.70–3.23	<DL	<DL–1.78	<DL	146–188	1,030–1,175	<DL–16.95	<DL–1.44	6.60–8.77	<DL–5.83
Netrostoma sp. (17)														
Mean	37.1	104	27.3	335	12.0	7.94	31.8	280	140	1,056	20.7	3.83	6,974	295
Range	12.4–115	9.84–200	5.57–79.3	282–419	3.01–33.6	1.20–21.0	13.3–80.3	26.2–1,490	103–171	872–1,320	5.76–41.1	0.56–29.8	5,750–9,270	65.7–954
Water	<DL–2.58	<DL	<DL–5.50	317–374	0.51–1.43	<DL–4.58	<DL–4.41	<DL	123–151	1,045–1,132	<DL–2.53	<DL	6.06–8.59	<DL–8.36

Results are combined data from all years and locations. All concentrations in jellyfish tissues are given in μg kg^{-1} wet weight except Ca and Mg, which were measured as mg kg^{-1} wet weight. Numbers in parentheses alongside species represents the number of animals collected. For water samples, all concentrations are given in μg l^{-1} except Ca, Mg and Sr, which were measured in mg l^{-1}

<DL less than reported detection limit

* The number of pooled samples

Bioaccumulation of elements by jellyfish

Bioaccumulation differed among elements and among species (Fig. 2). Except for a few individual samples, tissue concentrations of all elements were present above detectable levels at all locations in all years (Table 2). Water concentrations of As and Fe were below the detection limit for all water samples and were excluded from the analysis to avoid skewing the results (Table 2). In addition, other elements were below detection in water samples at some locations (Table 2). Samples where water concentrations were below detection, but measurable in the tissues, also were excluded from the analysis because bioaccumulation was unable to be calculated. Tissue concentrations of Cr, Cu and Zn were above detection in *C. sivickisi*, but these elements were below detection in the seawater samples where this species was collected (Table 2). Similarly, although concentrations of Pb were above detection in *Aurelia* sp. and *Netrostoma* sp. tissues and Al in *Cassiopea* sp. tissue, water concentrations were not. Water concentrations of Cr for *Cassiopea* sp., *C. sivickisi* and *Mastigias* sp. were below the measured detection level at all sites (Table 2).

C. sivickisi had the greatest accumulation among species for Al and Cd (Fig. 2). Accumulations of Ba, Cu, Mn, Pb and Zn were the highest in *Cassiopea* sp. (28, 151, 392, 67 and 221 times seawater concentrations, respectively). *Mastigias* sp. also had high accumulation

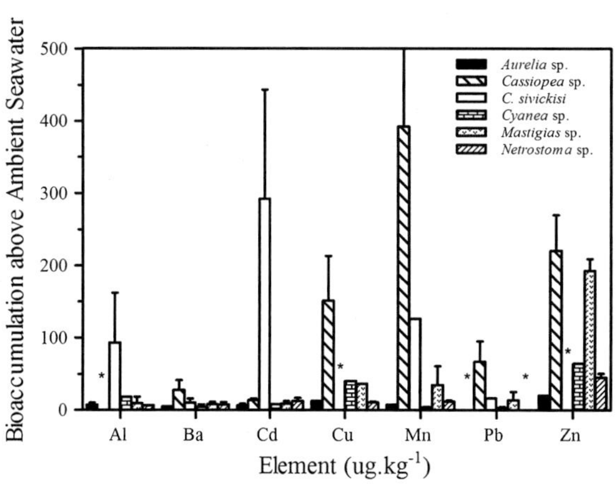

Fig. 2 Bioaccumulation of elements above ambient seawater concentration in jellyfish tissues. Data pooled from all locations and all years for each species. *Asterisk* indicates seawater concentration below detection level for individual species (data removed from plot). *Error bars* represent standard errors

of Zn (193 times seawater). Cu accumulation was similar for *Cyanea* sp. (41 times) and *Mastigias* sp. (37), but much lower for *Aurelia* sp. (13) and *Netrostoma* sp. (11). *Netrostoma* sp. and *Aurelia* sp. generally had the lowest levels of accumulation among species, while the extent of accumulation in *Cyanea* sp. was element dependent (Fig. 2). Li seemed to be actively regulated by *Cassiopea* sp., *Cyanea* sp. and *Mastigais* sp., with tissue concentrations of 0.71, 0.88 and 0.89 times the concentration of the ambient seawater, respectively. Ca and Mg were measurable in the tissues and present at concentrations similar to ambient water concentrations. Sr was 3.3 times seawater in *Cassiopea* sp. tissues as compared to 0.75–1.42 times seawater concentration for other species. Concentrations of Cr were twice that in water concentrations for *Aurelia* sp., *Cyanea* sp. and *Netrostoma* sp.

Variation in elemental concentrations relative to distance offshore

Only *Mastigias* sp. and *Netrostoma* sp. were found at more than two distances from the mainland over the 3-year sampling period. *Mastigias* sp. had an inverse relationship between tissue concentration and distance from the mainland for Al, As, Cu, Zn and Fe (Fig. 3a). The effect of distance from the mainland was significant for Al ($F_{2,18} = 101.8, P < 0.001$), As ($F_{2,18} = 10.712$, $P < 0.001$), Cu ($F_{2,18} = 107.1$, $P < 0.001$), Fe ($F_{2,18} = 185.4$, $P < 0.001$) and Zn ($F_{2,18} = 17.48$, $P < 0.001$). Other elements (Cd, Cr and Pb) showed no relationship with distance from the mainland.

In contrast to *Mastigias* sp., *Netrostoma* sp. tissue concentrations differed with distance from the mainland in that coastal and inner-shelf locations had similar concentrations while outer-shelf locations had lower concentrations (Fig. 3b). Significant differences among distances were found for As ($F_{2,14} = 84.736$, $P < 0.001$), Cu ($F_{2,14} = 46.458$, $P < 0.001$), Fe ($F_{2,14} = 34.503$, $P < 0.001$) and Zn ($F_{2,14} = 18.563$, $P < 0.001$); however, Al was not significantly different ($P > 0.05$). Results were only considered significant if $P < 0.01$ because the data were heterogeneous (Underwood, 1997).

Patterns in elemental fingerprints

There was greater variation in elemental fingerprints among species than among years and locations

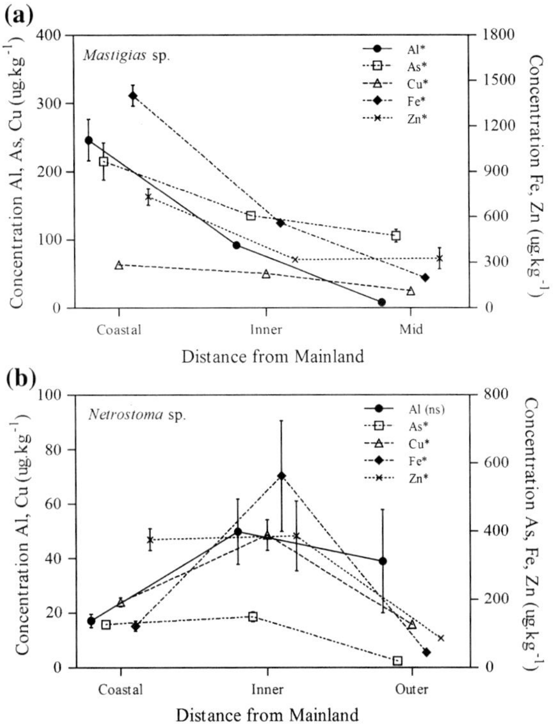

Fig. 3 Variation in tissue concentrations of Al, As, Cu, Fe and Zn in μg kg^{-1} wet weight with distance from shore for *Mastigias* sp. (**a**) and *Netrostoma* sp. medusae (**b**). Data pooled from all years. *Asterisk* represent significant differences ($P < 0.001$) among location for each element; *ns* no significant difference. Data $\log_n + 1$ transformed to meet ANOVA assumptions. *Error bars* represent standard errors

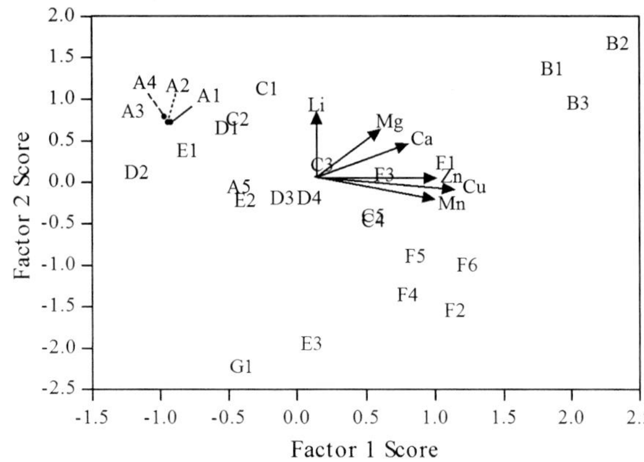

Fig. 4 Results of multivariate PCA for multi-element signatures in μg kg^{-1} (wet weight) in jellyfish among locations and years. Percent of variation explained by Factor 1 = 58% and Factor 2 = 16%. Data $\log_n + 1$ transformed before analysis to reduce contribution from elements with highest concentrations. Sample coding from Table 1

(Fig. 4). *C. sivickisi* (B1–B3) and *Cassiopea* sp. (F1–F6) had different elemental fingerprints, but within each species there were no differences by location or time (Fig. 4). In contrast, elemental fingerprints of *Cyanea* sp. (E1–E3) differed in terms of both location and time. The elemental fingerprints of both *Mastigias* sp. (C1–C5) and *Netrostoma* sp. (D1–D4) were more similar spatially, with distance from the mainland showing closer affinities than among locations in general (Fig. 4). Despite having limited spatial association, *Aurelia* sp. (A1–A5) had a distinct temporal fingerprint, with 2007/2008 samples (A1–A4) being more similar to each other than to other years (A5). Analyses of the ambient water collected with *Aurelia* sp. showed temporal variation between 2007/2008 and 2010 also.

Overall, total variance in the matrix was 74%, with 58% being explained by Factor 1 and 16% by Factor 2. Factor 1 was characterized by positive loadings for Cu, Mn and Zn (0.933, 0.915 and 0.894, respectively). Elements that were readily accumulated (i.e. Cu, Mn and Zn) influenced loadings on Factor 1. Factor 2 was characterized by positive loadings of 0.923, 0.732 and 0.516 for the osmoconforming elements Li, Mg and Ca, respectively. These results indicate that differences in salinity may be the driver for variations along the Factor 2 axis.

Discussion

Previous studies (e.g. Hanaoka et al., 2001; Fowler et al., 2004; Templeman & Kingsford, 2010) have demonstrated that jellyfishes and other gelatinous plankton are capable of absorbing trace elements from the environment in measurable concentrations, but most studies were limited in extent and species. This study demonstrated that both cubozoan and scyphozoan species accumulate elements above ambient water concentrations thus rejecting the null hypothesis. The extent of accumulation varied among species, depending on the element. There was limited evidence for temporal variation, suggesting that more study would be needed to reject the null hypothesis.

Accumulation of elements in tissues is dependent upon the species, speciation of the metal, uptake route and organism sensitivity. For many marine fish, the bioconcentration factors for metals generally are less than 100 (Parametrix, 1995); however, that is not typical for all marine organisms. Some invertebrates, including barnacles (Rainbow & Wang, 2001) and the mussel *Mytilus edulis* (Talbot, 1987) are very efficient bioconcentrators of metals, sometimes exceeding 1,000 times ambient metal concentrations. The accumulative capacity was very high in two of the three symbiotic species (*Cassiopea* sp. and *Mastigias* sp.) and greater for some metals than that measured in many fish species (Parametrix, 1995) (Fig. 2). *C. sivickisi* also readily accumulated Cd and Al above ambient water concentrations (Fig. 2).

Typically, ideal biomonitor species should meet a number of criteria including behaviour, abundance, robustness and bioaccumulative capacity (Luoma & Rainbow, 2008). Sedentary behaviour is one criterion, as it can provide a time-integrated measure of bioavailable metals from a defined location (Rainbow & Phillips, 1993). For this reason, biomonitoring research has often focused on sessile species like mussels and barnacles, although other species that are representative of the study location also can be used (Luoma & Rainbow, 2008). As a group, jellyfishes are not sedentary, except for the upside-down jellyfish *Cassiopea* sp. Despite the patchy nature of jellyfish distributions, their ability to form large conspicuous blooms means they can constitute a significant portion of the biomass in local areas and may be the most visible representative at a given location (e.g. Graham et al., 2003; Dybas, 2006). This aggregating behaviour, combined with the ability to accumulate metals, suggests that sedentary traits may not be a strong requirement for jellyfish to be considered useful biomonitors.

For some elements (e.g. Al, As, Fe and Zn), *Mastigias* sp. and *Netrostoma* sp. collected at coastal and inshore locations had greater variation in tissue concentrations than animals collected from mid- or off-shore locations (Fig. 3). These results supported the alternate hypothesis that there were spatial variations in elemental concentrations for these two species. This variation is typical of coastal locations because terrestrial inputs, both natural and anthropogenic, and riverine contributions can result in greater fluctuations in water quality and metal bioavailability (e.g. Lopes et al., 2007). Thus, the greater variability in tissue concentrations in animals from coastal locations implies they are probably reflecting local water quality variability and could potentially be useful in monitoring coastal water quality.

Water quality monitoring on the GBR has indicated that overall contaminant levels are low; however, areas adjacent to urban activity and intensive agriculture have elevated levels of contaminants (Haynes & Johnson, 2000). In addition, the distribution of dissolved metals can be strongly influenced by the presence of suspended particulate matter (Balls, 1988), which may be influencing metal concentrations in coastal regions (in particular) along the GBR. Water samples were filtered before analysis in this study and therefore the contribution of particulate matter was not assessed, although it may have influenced uptake in the medusae. This may have affected the bioavailability of elements to jellyfish collected from coastal locations. Due to the patchiness of sampling that is inherent with jellyfish collections, it was not possible to obtain samples at multiple distances from the mainland for most species. *Mastigias* sp. and *Netrostoma* sp. were the only species where spatial variation could be measured, but pooling of data from multiple years was still required (Fig. 3). With the exception of Cu and Zn in *Netrostoma* sp., the tissue elemental concentrations did not track the water concentrations. This in part was due to the low concentrations of the elements in the water but also indicated that elemental bioaccumulation was occurring. The change in tissue elemental concentration with distance from the mainland showed that animals may be either maintaining their position in given locations or drifting but maintaining exposure to a given water quality type. The higher concentrations of Al, As, Cu, Fe and Zn at coastal and inner locations for these species indicated the presence of a general nearshore metal signal from either anthropogenic inputs or riverine plumes (Haynes & Johnson, 2000; Haynes & Michalek-Wagner, 2000).

In the body, some trace elements have an essential role in maintaining organism health and a minimum tissue concentration is required to maintain health. Cu is an essential trace element required by all organisms including symbiotic jellyfish, which use Cu for inducing superoxide dismutase activity in defense of oxygen radicals (Harland & Nganro, 1990). Zn is a component of another enzyme, carbonic anhydrase, which is particularly abundant in organisms with symbiotic zooxanthellae (Furla et al., 2000). Zn was present in all species (Table 2), with the highest accumulation found in two (*Cassiopea* sp. and *Mastigias* sp.) of the three symbiotic jellyfish species, but not in *Netrostoma* sp. (Fig. 2); however, other elements with no identified essential role (e.g. Al, Cd and, Pb) also were accumulated by multiple jellyfish species (Fig. 2).

Jellyfish may use multiple routes to absorb essential trace elements to maintain health. Elements that were present in elevated concentrations in tissues but below detection in ambient water (e.g. As and Fe) may have alternate uptake routes, such as from surface-adsorbed particles, diet or, in the case of *Cassiopea* sp., from sediment. Jantzen et al. (2010) found that *Cassiopea* sp. demonstrated active bioturbation of sediment, which may potentially expose them to elevated metal concentrations found in pore waters or adsorbed to sedimentary particles. Dietary source as a metal uptake route also has been identified as a potentially significant exposure route (e.g. Depledge & Rainbow, 1990; Rainbow & Wang, 2001) and may have been a source of accumulated metals measured in the jellyfish tissues.

The multi-element signature in both *Cassiopea* sp. and *C. sivickisi* discriminated them from other species and each other (Fig. 4); however, there was minimal evidence of any temporal or spatial patterns in either of these two species. Spatially, this may be due to limited sampling locations because both species were only collected in two locations (Table 1). The multi-element signatures in *Cassiopea* sp. and *C. sivickisi* supported the alternate hypotheses that they accumulated metals above ambient concentrations and there was also variation among species, but did not reject the null hypotheses with respect to temporal or spatial variations.

Rejection of the null hypotheses for spatial or temporal variation was strongly species dependent. *Aurelia* sp. displayed similarity among locations, although there was temporal separation between 2007/2008 and 2010 collections (Fig. 4). In contrast, *Mastigias* sp. and *Netrostoma* sp. showed spatial variation in elemental fingerprints. Because they were collected at a greater number of locations, this implied that tissue concentrations may reflect local environmental exposure; however, spatial data are limited, and additional collections would be useful to elucidate this relationship further.

The extent of accumulation of elements in jellyfish should be considered important given the number of species that have been identified to prey on them (Pauly et al., 2009). Trophic transfer is an important route for accumulation of contaminants, although accumulation is dependent on how and where contaminants are stored in the prey. Caurant et al. (1999) demonstrated that jellyfish may be a major source of Cd accumulation in the diet of leatherback turtles. There has been minimal direct evidence of contaminant accumulation through jellyfish diets; however, given that nudibranchs selectively absorb and utilize the nematocysts from jellyfish (Arai, 2005) and incorporate and concentrate pigments (Bayer, 1963 cited in Arai, 2005), the potential exists for contaminant accumulation.

The behaviour of metals in the environment and the organismal response are affected by both spatial and temporal factors. Climatic conditions, residence time of elements in the water column, and fluxes between water, air, and sediment can all change elemental load and exposure. These changes can occur as short duration called ‘pulses’ or more sustained long-term ‘press’ events, and the degree and duration of exposure affect metabolic function in different ways (Ives & Carpenter, 2007). In marine systems, proximity to the coast can affect elemental loads with both biotic and abiotic factors cycling elements between the water, sediment, biota and atmosphere (Dauer et al., 2000). Depending on the element, terrestrial inputs can strongly influence both the presence and persistence of dissolved metals (Balls, 1988; Haynes & Michalek-Wagner, 2000).

The ability to accumulate trace elements also can be useful as it provides the opportunity to monitor pulse or press measures of contaminant loads in marine ecosystems. The extent and duration of accumulation is critical to establishing time-integrated measures of marine water quality. A previous study on *Cassiopea* sp. showed they were able to accumulate metals above ambient seawater concentrations (Templeman & Kingsford 2010). In addition, there was distinct spatial variation in elemental concentrations between populations at both small (<1 km) and large (1,000 km) distances (Templeman & Kingsford, 2010). In the present study, we did not determine the duration of retention of elements. Further experimental studies are required to elucidate and fully characterize the extent and duration of the accumulative capacity.

In conclusion, multiple elements were found in tissues of seven scyphozoan and cubozoan jellyfish species on the GBR. Except for the major osmoconforming elements (Ca, Mg and Li), tissue concentrations of elements in all species exceeded that of ambient seawater, which implied accumulation and the alternate hypothesis of accumulation was accepted. The alternate hypothesis on accumulation among species was accepted, as there was a demonstration of the variation in accumulative capacity among species, although this was element dependent to some extent. Rejection of the null hypotheses for temporal and spatial variation in tissue concentrations was species dependent, but overall the results indicated that jellyfish can accumulate and retain elements. With experimental validation, therefore, they show potential as marine biomonitors.

Acknowledgments Special thanks go to Dr. Yi Hu for performing the ICP-MS/AES analyses as well as the numerous volunteers who assisted in sampling. Funding for the project was provided by The Marine and Tropical Research Fund (MTSRF), with additional fundings from the Australian Research Council (ARC) to MJ Kingsford, and The Noel and Kate Monkman Postgraduate Award in Marine Biology to MA Templeman.

References

Alamaru, A., O. Bronstein, G. Dishon & Y. Loya, 2009. Opportunistic feeding by the fungiid coral *Fungia scruposa* on the moon jellyfish *Aurelia aurita*. Coral Reefs 28: 865.

Arai, M. N., 2005. Predation on pelagic coelentrerates: a review. Journal of the Marine Biological Association of the United Kingdom 85: 523–536.

Balls, P. W., 1988. The control of trace metals concentrations in coastal seawater through partition onto suspended particulate matter. Netherlands Journal of Sea Research 22: 213–218.

Benson, A. & R. Summons, 1981. Arsenic accumulation in Great Barrier Reef invertebrates. Science 211: 482–483.

Bresler, V., A. Abelson, L. Fishelson, T. Feldstein, M. Rosenfeld & O. Mokady, 2003. Marine molluscs in environmental monitoring I. Cellular and molecular responses. Helgolander Marine Research 57: 157–165.

Caurant, F., P. Bustamante, M. Bordes & P. Miramand, 1999. Bioaccumulation of cadmium, copper and zinc in some tissues of three species of marine turtles stranded along the French Atlantic Coasts. Marine Pollution Bulletin 38: 1085–1091.

Chapman, P. M., 1995. Ecotoxicology and pollution – Key issues. Marine Pollution Bulletin 31: 167–177.

Creighton, N. & J. Twining, 2010. Bioaccumulation from food and water of cadmium, selenium and zinc in an estuarine fish, *Ambassis jacksoniensis*. Marine Pollution Bulletin 60: 1815–1821.

Dauer, D. M., J. A. Ranasinghe & S. B. Weisberg, 2000. Relationships between benthic community condition, water quality, sediment quality, nutrient loads, and land use patterns in Chesapeake Bay. Estuaries and Coasts 23: 80–96.

Depledge, M. H. & P. S. Rainbow, 1990. Models of regulation and accumulation of trace metals in marine invertebrates. Comparative Biochemistry and Physiology 97C: 1–7.

Dybas, S. L., 2006. On a collision course: ocean plankton and climate change. BioScience 56: 642–646.

Fowler, S. W., J.-L. Teyssie, O. Cotret, B. Danis, C. Rouleau & M. Warnau, 2004. Applied radiotracer techniques for studying pollutant bioaccumulation in selected marine organisms (jellyfish, crabs and sea stars). Nucleonika 49: 97–100.

Fukuda, Y. & T. Naganuma, 2001. Potential dietary effects on the fatty acid composition of the common jellyfish *Aurelia aurita*. Marine Biology 138: 1029–1035.

Furla, P., D. Allemand & M.-N. Orsenigo, 2000. Involvement of H+-ATPase and carbonic anhydrase in inorganic carbon uptake for endosymbiotic photosynthesis. American Journal of Physiology – Regulatory, Integrative and Comparative Physiology 278: 870–881.

Graham, W. M., D. L. Martin, D. L. Felder, V. L. Asper & H. M. Perry, 2003. Ecological and economic implications of a tropical jellyfish invader in the Gulf of Mexico. Biological Invasions 5: 53–69.

Hanaoka, K., H. Ohno, N. Wada, S. Ueno, W. Goessler, D. Kuehnelt, C. Schlagenhaufen, T. Kaise & K. J. Irgolic, 2001. Occurrence of organo-arsenicals in jellyfishes and their mucus. Chemosphere 44: 743–749.

Harland, A. D. & N. R. Nganro, 1990. Copper uptake by the sea anemone *Anemonia viridis* and the role of zooxanthellae in metal regulation. Marine Biology 104: 297–301.

Harrison, N. M., 1984. Predation on jellyfish and their associates by seabirds. Limnology and Oceanography 29: 1335–1337.

Hay, S., 2006. Marine ecology: gelatinous bells may ring change in marine ecosystems. Current Biology 16: R679–R682.

Haynes, D. & J. E. Johnson, 2000. Organochlorine, heavy metal and polyaromatic hydrocarbon pollutant concentrations in the Great Barrier Reef (Australia) environment: a review. Marine Pollution Bulletin 41: 267–278.

Haynes, D. & K. Michalek-Wagner, 2000. Water quality in the Great Barrier Reef World Heritage Area: past perspectives, current issues and new research directions. Marine Pollution Bulletin 41: 428–434.

Heeger, T., U. Piatkowski & H. Möller, 1992. Predation on jellyfish by the cephalopod *Argonauta argo*. Marine Ecology Progress Series 88: 293–296.

Heymans, J. J. & D. Baird, 2000. A carbon flow model and network analysis of the northern Benguela upwelling system, Namibia. Ecological Modelling 126: 9–32.

Ives, A. R. & S. R. Carpenter, 2007. Stability and diversity of ecosystems. Science 317: 58–62.

Jantzen, C., C. Wild, M. Rasheed, M. El-Zibdah & C. Richter, 2010. Enhanced porewater nutrient fluxes by the upside-down jellyfish *Cassiopea* sp. in a Red Sea coral reef. Marine Ecology Progress Series 411: 117–125.

Kingsford, M. J., K. A. Pitt & B. M. Gillanders, 2000. Management of jellyfish fisheries, with special reference to the order Rhizostomeae. Oceanography and Marine Biology: An Annual Review 38: 85–156.

Kitamura, M. & M. Omori, 2010. Synopsis of edible jellyfishes collected from Southeast Asia, with notes on jellyfish fisheries. Plankton and Benthos Research 5: 106–118.

Legendre, P. & L. Legendre, 2003. Numerical Ecology. Elsevier Science, Amsterdam.

Lopes, C. B., M. E. Pereira, C. Vale, A. I. Lillebo, M. A. Pardal & A. C. Duarte, 2007. Assessment of spatial environmental quality status in Ria de Aveiro (Portugal). Scientia Marina 71: 293–304.

Luoma, S. N. & P. S. Rainbow, 2008. Metal Contamination in Aquatic Environments: Science and Lateral Management. Cambridge University Press, Cambridge.

Parametrix, 1995. Persistence, Bioaccumulation and Toxicity of Metals and Metal Compounds. ICME, Ottawa: 93 pp.

Pauly, D., W. Graham, S. Libralato, L. Morissette & M. L. D. Palomares, 2009. Jellyfish in ecosystems, online databases, and ecosystem models. Hydrobiologia 616: 67–85.

Pitt, K. A., D. T. Welsh & R. H. Condon, 2009. Influence of jellyfish blooms on carbon, nitrogen and phosphorus cycling and plankton production. Hydrobiologia 616: 133–149.

Purcell, J. E., 1991. A review of cnidarians and ctenophores feeding on competitors in the plankton. Hydrobiologia 216(217): 335–342.

Purcell, J. E. & M. N. Arai, 2001. Interactions of pelagic cnidarians and ctenophores with fishes: a review. Hydrobiologia 451: 27–44.

Rainbow, P. S., 2007. Trace metal bioaccumulation: models, metabolic availability and toxicity. Environment International 33: 576–582.

Rainbow, P. S. & D. J. H. Phillips, 1993. Cosmopolitan biomonitors of trace metals. Marine Pollution Bulletin 26: 593–601.

Rainbow, P. S. & W.-X. Wang, 2001. Comparative assimilation of Cd, Cr, Se, and Zn by the barnacle *Elminius modestus* from phytoplankton and zooplankton diets. Marine Ecology Progress Series 218: 239–248.

Ruus, A., M. Schaanning, S. Øxnevad & K. Hylland, 2005. Experimental results on bioaccumulation of metals and organic contaminants from marine sediments. Aquatic Toxicology 72: 273–292.

Sadiq, M., 1992. Toxic Metal Chemistry in Marine Environments. CRC Press, New York.

Talbot, V., 1987. Relationship between lead concentrations in seawater and in the mussel *Mytilus edulis*: a water quality criterion. Marine Biology 94: 557–560.

Templeman, M. A. & M. J. Kingsford, 2010. Trace element accumulation in *Cassiopea* sp. (Scyphozoa) from urban marine environments in Australia. Marine Environmental Research 69: 63–72.

Underwood, A. J., 1997. Experiments in Ecology: Their Logical Design and Interpretation using Analysis of Variance. Cambridge University Press, Cambridge.

Printed by Publishers Graphics Canada Inc
CASO20121024.19.28.38